Edited by
Henner Schmidt-Traub,
Michael Schulte, and
Andreas Seidel-Morgenstern

Preparative Chromatography

Related Titles

Seidel-Morgenstern, A. (ed.)

Membrane Reactors

Distributing Reactants to Improve Selectivity and Yield

2010

ISBN: 978-3-527-32039-4

Miller, J. M.

Chromatography

Concepts and Contrasts

2008

E-Book

ISBN: 978-0-470-35781-1

Sinaiski, E.G., Lapiga, E. J.

Separation of Multiphase, Multicomponent Systems

2007

ISBN: 978-3-527-40612-8

Afonso, C. A. M., Crespo, J. P. G. (eds.)

Green Separation Processes

Fundamentals and Applications

2005

ISBN: 978-3-527-30985-6

Sundmacher, K., Kienle, A., Seidel-Morgenstern, A. (eds.)

Integrated Chemical Processes

Synthesis, Operation, Analysis, and Control

2005

ISBN: 978-3-527-30831-6

Edited by Henner Schmidt-Traub, Michael Schulte, and Andreas Seidel-Morgenstern

Preparative Chromatography

Second, Completely Revised and Updated Edition

WILEY-VCH Verlag GmbH & Co. KGaA

The Editors

Prof. Dr.-Ing. Henner Schmidt-Traub
TU Dortmund
Fakultät für Bio- und
Chemieingenieurwesen
Lehrstuhl für Anlagen- und
Prozesstechnik
Emil-Figge-Str. 70
44227 Dortmund
Germany

Dr. Michael Schulte
Merck KGaA
R&D Performance & Life Science Chemicals
Frankfurter Str. 250
64293 Darmstadt
Germany

Prof. Dr.-Ing. Andreas Seidel-Morgenstern
Otto-von-Guericke-Universität
Institut für Verfahrenstechnik
Lehrstuhl für Chemische Verfahrenstechnik

and

Max-Planck-Institut für Dynamik komplexer
technischer Systeme
Sandtorstraße 1
Universitätsplatz 2
39106 Magdeburg
Germany

Cover
The cover figure has been kindly provided by
Novasep, France.

All books published by **Wiley-VCH** are carefully produced. Nevertheless, authors, editors, and publisher do not warrant the information contained in these books, including this book, to be free of errors. Readers are advised to keep in mind that statements, data, illustrations, procedural details or other items may inadvertently be inaccurate.

Library of Congress Card No.: applied for

British Library Cataloguing-in-Publication Data
A catalogue record for this book is available from the British Library.

Bibliographic information published by the Deutsche Nationalbibliothek
The Deutsche Nationalbibliothek lists this publication in the Deutsche Nationalbibliografie; detailed bibliographic data are available on the Internet at http://dnb.d-nb.de.

© 2012 Wiley-VCH Verlag & Co. KGaA, Boschstr. 12, 69469 Weinheim, Germany

All rights reserved (including those of translation into other languages). No part of this book may be reproduced in any form – by photoprinting, microfilm, or any other means – nor transmitted or translated into a machine language without written permission from the publishers. Registered names, trademarks, etc. used in this book, even when not specifically marked as such, are not to be considered unprotected by law.

Composition Thomson Digital, Noida, India

Printing and Binding Markono Print Media Pte Ltd, Singapore

Cover Design Formgeber, Eppelheim

Print ISBN: 978-3-527-32898-7
ePDF ISBN: 978-3-527-64931-0
ePub ISBN: 978-3-527-64930-3
mobi ISBN: 978-3-527-64929-7
oBook ISBN: 978-3-527-64928-0

Printed in Singapore
Printed on acid-free paper

Contents

Preface *XV*
About the Editors *XVII*
List of Contributors *XIX*
List of Abbreviations *XXI*
Notations *XXV*

1 **Introduction** *1*
Henner Schmidt-Traub and Reinhard Ditz
1.1 Development of Chromatography *1*
1.2 Focus of the Book *3*
1.3 Recommendation to Read this Book *4*
References *6*

2 **Fundamentals and General Terminology** *7*
Andreas Seidel-Morgenstern, Michael Schulte, and Achim Epping
2.1 Principles of Adsorption Chromatography *7*
2.1.1 Adsorption Process *9*
2.1.2 Chromatographic Process *10*
2.2 Basic Effects and Chromatographic Definitions *11*
2.2.1 Chromatograms and Parameters *11*
2.2.2 Voidage and Porosity *12*
2.2.3 Influence of Adsorption Isotherms on Chromatogram Shapes *15*
2.3 Fluid Dynamics *18*
2.3.1 Extra Column Effects *18*
2.3.2 Column Fluid Distribution *19*
2.3.3 Packing Nonidealities *19*
2.3.4 Sources for Nonideal Fluid Distribution *20*
2.3.5 Column Pressure Drop *21*
2.4 Mass Transfer Phenomena *22*
2.4.1 Principles of Mass Transfer *22*
2.4.2 Efficiency of Chromatographic Separations *24*
2.4.3 Resolution *27*
2.5 Equilibrium Thermodynamics *30*

2.5.1	Definition of Isotherms	30
2.5.2	Models of Isotherms	32
2.5.2.1	Single-Component Isotherms	32
2.5.2.2	Multicomponent Isotherms Based on the Langmuir Model	34
2.5.2.3	Competitive Isotherms Based on the Ideal Adsorbed Solution Theory	35
2.5.2.4	Steric Mass Action Isotherms for Ion Exchange Equilibria	38
2.6	Thermodynamic Effects on Mass Separation	40
2.6.1	Mass Load	40
2.6.2	Linear and Nonlinear Isotherms	41
2.6.3	Elution Modes	43
	References	45
3	**Stationary Phases and Chromatographic Systems**	47
	Michael Schulte, Matthias Jöhnck, Romas Skudas, Klaus K. Unger, Cedric du Fresne von Hohenesche, Wolfgang Wewers, Jules Dingenen, and Joachim Kinkel	
3.1	Column Packings	47
3.1.1	Survey of Packings and Stationary Phases	47
3.1.2	Generic, Designed, and Customized Adsorbents	48
3.1.2.1	Generic Adsorbents	48
3.1.2.2	Designed Adsorbents	54
3.1.2.3	Customized Adsorbents	62
3.1.3	Reversed Phase Silicas	66
3.1.3.1	Silanisation of the Silica Surface	67
3.1.3.2	Chromatographic Characterization of Reversed Phase Silicas	69
3.1.4	Cross-Linked Organic Polymers	72
3.1.4.1	General Aspects	73
3.1.4.2	Hydrophobic Polymer Stationary Phases	76
3.1.4.3	Hydrophilic Polymer Stationary Phases	76
3.1.4.4	Ion Exchange (IEX)	77
3.1.4.5	Mixed Mode	85
3.1.5	Chiral Stationary Phases	85
3.1.5.1	Antibiotic CSP	91
3.1.5.2	Synthetic Polymers	91
3.1.5.3	Targeted Selector Design	92
3.1.5.4	Further Developments	93
3.1.6	Properties of Packings and their Relevance to Chromatographic Performance	95
3.1.6.1	Chemical and Physical Bulk Properties	95
3.1.6.2	Mass Loadability	101
3.1.6.3	Comparative Rating of Columns	102
3.1.7	Sorbent Maintenance and Regeneration	103
3.1.7.1	Cleaning in Place (CIP)	103
3.1.7.2	Conditioning of Silica Surfaces	106

3.1.7.3	Sanitization in Place (SIP) *108*
3.1.7.4	Column and Adsorbent Storage *108*
3.2	Selection of Chromatographic Systems *109*
3.2.1	Definition of the Task *114*
3.2.2	Mobile Phases for Liquid Chromatography *118*
3.2.2.1	Stability *118*
3.2.2.2	Safety Concerns *118*
3.2.2.3	Operating Conditions *121*
3.2.2.4	Aqueous Buffer Systems *123*
3.2.3	Adsorbent and Phase Systems *125*
3.2.3.1	Choice of Phase System Dependent on Solubility *127*
3.2.3.2	Improving Loadability for Poor Solubilities *128*
3.2.3.3	Dependency of Solubility on Sample Purity *130*
3.2.3.4	Generic Gradients for Fast Separations *131*
3.2.4	Criteria for Choosing NP Systems *132*
3.2.4.1	Pilot Technique Thin-layer Chromatography *134*
3.2.4.2	Retention in NP Systems *134*
3.2.4.3	Solvent Strength in Liquid–Solid Chromatography *136*
3.2.4.4	Selectivity in NP Systems *138*
3.2.4.5	Mobile-Phase Optimization by TLC Following the PRISMA Model *139*
3.2.4.6	Strategy for an Industrial Preparative Chromatography Laboratory *148*
3.2.5	Criteria for Choosing RP Systems *153*
3.2.5.1	Retention and Selectivity in RP Systems *155*
3.2.5.2	Gradient Elution for Small amounts of Product on RP Columns *156*
3.2.5.3	Rigorous Optimization for Isocratic Runs *157*
3.2.5.4	Rigorous Optimization for Gradient Runs *161*
3.2.5.5	Practical Recommendations *164*
3.2.6	Criteria for Choosing CSP Systems *167*
3.2.6.1	Suitability of Preparative CSP *168*
3.2.6.2	Development of Enantioselectivity *169*
3.2.6.3	Optimization of Separation Conditions *171*
3.2.6.4	Practical Recommendations *172*
3.2.7	Downstream Processing of Mabs using Protein A and IEX *174*
3.2.8	Size Exclusion (SEC) *179*
3.2.9	Overall Chromatographic System Optimization *181*
3.2.9.1	Conflicts During Optimization of Chromatographic Systems *181*
3.2.9.2	Stationary Phase Gradients *184*
	References *189*
4	**Chromatography Equipment: Engineering and Operation** *199*
	Abdelaziz Toumi, Jules Dingenen, Joel Genolet, Olivier Ludemann-Hombourger, Andre Kiesewetter, Martin Krahe, Michele Morelli, Henner Schmidt-Traub, Andreas Stein, and Eric Valery
4.1	Introduction *199*
4.2	Engineering and Operational Challenges *201*

4.3	Chromatography Columns Market	207
4.3.1	Generalities – The Suppliers	207
4.3.2	General Design	208
4.3.3	High- and Low-Pressure Columns	210
4.3.3.1	Chemical Compatibility	211
4.3.3.2	Frits Design	211
4.3.3.3	Special Aspects of Bioseparation	215
4.4	Chromatography Systems Market	217
4.4.1	Generalities – The Suppliers	217
4.4.2	General Design Aspects – High Performance and Low-Pressure Systems	217
4.4.3	Material	219
4.4.4	Batch Low-Pressure Liquid Chromatography (LPLC) Systems	220
4.4.4.1	Inlets	220
4.4.4.2	Valves to Control Flow Direction	220
4.4.4.3	Pumps	221
4.4.4.4	Pump(s) Valves and Gradient Formation	222
4.4.5	Batch High-Pressure Liquid Chromatography (HPLC) Systems	224
4.4.5.1	General Layout	224
4.4.5.2	Inlets and Outlets	224
4.4.5.3	Pumps	226
4.4.5.4	Valves and Pipes	227
4.4.6	Batch SFC Systems	228
4.4.6.1	General Layout	228
4.4.6.2	Inlets	230
4.4.6.3	Pumps, Valves, and Pipes	231
4.4.7	Continuous Systems – Simulated Moving Bed	231
4.4.7.1	General Layout	231
4.4.7.2	A Key Choice: The Recycling Strategy	232
4.4.7.3	Pumps, Inlets, and Outlets	233
4.4.7.4	Valves and Piping	233
4.4.8	Auxiliary Systems	233
4.4.8.1	Slurry Preparation Tank	234
4.4.8.2	Slurry Pumps and Packing Stations	234
4.4.8.3	Cranes and Transport Units	235
4.4.8.4	Filter Integrity Test	235
4.5	Process Control	236
4.5.1	Standard Process Control	236
4.5.2	Advanced Process Control	237
4.5.3	Detectors	240
4.6	Packing Methods	243
4.6.1	Column and Packing Methodology Selection	243
4.6.2	Slurry Preparation	244
4.6.3	Column Preparation	246
4.6.4	Flow Packing	246

4.6.5	Dynamic Axial Compression (DAC) Packing	249
4.6.6	Stall Packing	250
4.6.7	Combined Method (Stall + DAC)	250
4.6.8	Vacuum Packing	252
4.6.9	Vibration Packing	253
4.6.10	Column Equilibration	254
4.6.11	Column Testing and Storage	254
4.6.11.1	Test Systems	254
4.6.11.2	Hydrodynamic Properties and Column Efficiency	256
4.6.11.3	Column and Adsorbent Storage	257
4.7	Process Troubleshooting	257
4.7.1	Technical Failures	258
4.7.2	Loss of Performance	259
4.7.2.1	Pressure Increase	259
4.7.2.2	Loss of Column Efficiency	262
4.7.2.3	Variation of Elution Profile	263
4.7.2.4	Loss of Purity/Yield	264
4.7.3	Column Stability	265
4.8	Disposable Technology for Bioseparations	265
4.8.1	Market Trend	265
4.8.2	Prepacked Columns	266
4.8.3	Membrane Chromatography	267
4.8.4	Membrane Technology	269
	References	270
5	**Process Concepts**	**273**
	Malte Kaspereit, Michael Schulte, Klaus Wekenborg, and Wolfgang Wewers	
5.1	Discontinuous Processes	273
5.1.1	Isocratic Operation	273
5.1.2	Flip-Flop Chromatography	275
5.1.3	Closed-Loop Recycling Chromatography	276
5.1.4	Steady-State Recycling Chromatography	278
5.1.5	Gradient Chromatography	279
5.1.6	Chromatographic Batch Reactors	281
5.2	Continuous Processes	283
5.2.1	Column Switching Chromatography	283
5.2.2	Annular Chromatography	283
5.2.3	Multiport Switching Valve Chromatography (ISEP/CSEP)	284
5.2.4	Isocratic Simulated Moving Bed (SMB) Chromatography	286
5.2.5	SMB Chromatography with Variable Process Conditions	290
5.2.5.1	VariCol	290
5.2.5.2	PowerFeed	291
5.2.5.3	Partial-Feed, Partial-Discard, and Fractionation-Feedback Concepts	292
5.2.5.4	Improved/Intermittent SMB (iSMB)	293

5.2.5.5	ModiCon	294
5.2.5.6	FF-SMB	294
5.2.6	SMB Chromatography with Variable Solvent Conditions	294
5.2.6.1	Gradient SMB Chromatography	295
5.2.6.2	Supercritical Fluid SMB Chromatography	296
5.2.7	Multicomponent Separations	296
5.2.8	Multicolumn Systems for Bioseparations	298
5.2.8.1	Sequential Multicolumn Chromatography (SMCC)	298
5.2.8.2	Multicolumn Countercurrent Solvent Gradient Purification (MCSGP)	299
5.2.9	Countercurrent Chromatographic Reactors	301
5.2.9.1	SMB Reactor	301
5.2.9.2	Processes with Distributed Functionalities	302
5.3	Choice of Process Concepts	304
5.3.1	Scale	305
5.3.2	Range of k'	306
5.3.3	Number of Fractions	306
5.3.4	Example 1: Lab Scale; Two Fractions	306
5.3.5	Example 2: Lab Scale; Three or More Fractions	308
5.3.6	Example 3: Production Scale – Wide Range of k'	309
5.3.7	Example 4: Production Scale; Two Main Fractions	310
5.3.8	Example 5: Production Scale; Three Fractions	311
5.3.9	Example 6: Production Scale; Multi-Stage Process	312
	References	315
6	**Modeling and Model Parameters**	**321**
	Andreas Seidel-Morgenstern, Henner Schmidt-Traub, Mirko Michel, Achim Epping, and Andreas Jupke	
6.1	Introduction	321
6.2	Models for Single Chromatographic Columns	322
6.2.1	Classes of Chromatographic Models	322
6.2.2	Derivation of the Mass Balance Equations	324
6.2.2.1	Mass Balance Equations	325
6.2.2.2	Convective Transport	327
6.2.2.3	Axial Dispersion	327
6.2.2.4	Intraparticle Diffusion	327
6.2.2.5	Mass Transfer	328
6.2.2.6	Adsorption Kinetics	329
6.2.2.7	Adsorption Equilibrium	329
6.2.3	Equilibrium ("Ideal") Model	330
6.2.4	Models with One Band Broadening Effect	334
6.2.4.1	Dispersive Model	334
6.2.4.2	Transport Model	336
6.2.4.3	Reaction Model	337
6.2.5	Lumped Rate Models	338

6.2.5.1	Transport-Dispersive Model	*338*
6.2.5.2	Reaction-Dispersive Model	*339*
6.2.6	General Rate Models	*340*
6.2.7	Initial and Boundary Conditions of the Column	*343*
6.2.8	Models of Chromatographic Reactors	*344*
6.2.9	Stage Models	*344*
6.2.10	Assessment of Different Model Approaches	*346*
6.2.11	Dimensionless Model Equations	*348*
6.3	Modeling HPLC Plants	*350*
6.3.1	Experimental Setup and Simulation Flow Sheet	*350*
6.3.2	Modeling Extra Column Equipment	*351*
6.3.2.1	Injection System	*351*
6.3.2.2	Piping	*352*
6.3.2.3	Detector	*352*
6.4	Calculation Methods	*353*
6.4.1	Analytical Solutions	*353*
6.4.2	Numerical Solution Methods	*353*
6.4.2.1	General Solution Procedure	*353*
6.4.2.2	Discretization	*354*
6.5	Parameter Determination	*357*
6.5.1	Parameter Classes for Chromatographic Separations	*357*
6.5.1.1	Design Parameters	*357*
6.5.1.2	Operating Parameters	*358*
6.5.1.3	Model Parameters	*358*
6.5.2	Determination of Model Parameters	*359*
6.5.3	Evaluation of Chromatograms	*361*
6.5.3.1	Moment Analysis and HETP Plots	*362*
6.5.3.2	Parameter Estimation	*369*
6.5.3.3	Peak Fitting Functions	*370*
6.5.4	Detector Calibration	*374*
6.5.5	Plant Parameters	*375*
6.5.6	Determination of Packing Parameters	*376*
6.5.6.1	Void Fraction and Porosity of the Packing	*376*
6.5.6.2	Axial Dispersion	*377*
6.5.6.3	Pressure Drop	*378*
6.5.7	Isotherms	*379*
6.5.7.1	Determination of Adsorption Isotherms	*379*
6.5.7.2	Determination of the Henry Coefficient	*382*
6.5.7.3	Static Isotherm Determination Methods	*382*
6.5.7.4	Dynamic Methods	*385*
6.5.7.5	Frontal Analysis	*385*
6.5.7.6	Analysis of Disperse Fronts (ECP/FACP)	*390*
6.5.7.7	Peak Maximum Method	*391*
6.5.7.8	Minor Disturbance/Perturbation Method	*392*
6.5.7.9	Curve Fitting of the Chromatogram	*394*

6.5.7.10	Prediction of Mixture Behavior from Single-Component Data 395
6.5.7.11	Data Analysis and Accuracy 396
6.5.8	Mass Transfer 398
6.5.9	Identification of Isotherms and Mass Transfer Resistance by Neural Networks 399
6.6	Experimental Validation of Column Models 401
6.6.1	Batch Chromatography 401
6.6.2	SMB Chromatography 404
6.6.2.1	Model Formulation and Parameters 404
6.6.2.2	Experimental Validation of SMB Models 410
	References 418

7	**Model-Based Design, Optimization, and Control** 425
	Henner Schmidt-Traub, Malte Kaspereit, Sebastian Engell, Arthur Susanto, Achim Epping, and Andreas Jupke
7.1	Basic Principles and Definitions 425
7.1.1	Performance, Costs, and Optimization 425
7.1.1.1	Performance Criteria 426
7.1.1.2	Economic Criteria 428
7.1.1.3	Objective Functions 429
7.1.2	Degrees of Freedom 430
7.1.2.1	Optimization Parameters 430
7.1.2.2	Dimensionless Operating and Design Parameters 430
7.1.3	Scaling by Dimensionless Parameters 435
7.1.3.1	Influence of Different HETP Coefficients for Every Component 436
7.1.3.2	Influence of Feed Concentration 437
7.1.3.3	Examples for a Single Batch Chromatographic Column 438
7.1.3.4	Examples for SMB Processes 440
7.2	Batch Chromatography 442
7.2.1	Fractionation Mode (Cut Strategy) 442
7.2.2	Design and Optimization of Batch Chromatographic Columns 444
7.2.2.1	Design and Optimization Strategy 444
7.2.2.2	Process Performance Depending on Number of Stages and Loading Factor 447
7.2.2.3	Other Strategies 452
7.3	Recycling Chromatography 453
7.3.1	Design of Steady-State Recycling Chromatography 454
7.3.2	Scale-Up of Closed Loop Recycling Chromatography 457
7.4	Conventional Isocratic SMB Chromatography 461
7.4.1	Optimization of Operating Parameters 462
7.4.1.1	Process Design Based on TMB Models (Shortcut Methods) 463
7.4.1.2	Process Design Based on Rigorous SMB Models 471
7.4.2	Optimization of Design Parameters 476

7.5	Isocratic SMB Chromatography under Variable Operating Conditions *481*	
7.6	Gradient SMB Chromatography *490*	
7.7	Multicolumn Systems for Bioseparations *495*	
7.8	Advanced Process Control *497*	
7.8.1	Online Optimization of Batch Chromatography *498*	
7.8.2	Advanced Control of SMB Chromatography *501*	
7.8.2.1	Purity Control for SMB Processes *502*	
7.8.2.2	Direct Optimizing Control of SMB Processes *503*	
7.8.3	Advanced Parameter and State Estimation for SMB Processes *509*	
	References *510*	

Appendix A: Data of Test Systems *519*

Index *527*

Preface

Over 7 years have passed since the 1st edition of this book was published, and practical application as well as theoretical research on preparative chromatography has since then progressed rapidly. This motivated us to revise the content of the 1st edition.

We decided to rearrange the structure in this 2nd edition. Our intention was to present the aspects of practical equipment design and operation together in a separate chapter, to merge the discussion on stationary phases and the selection of chromatographic systems in one chapter, and to reduce the content concerning chromatographic reactors because of their specific features and the still limited practical relevance. These changes provided room for important new sections on ion exchange, bioseparation, and new process concepts and calculation methods.

What else is new in this revised second edition? First of all, the team did change significantly. Besides the additional editors, there are several new authors from industry and academia. The former crew from Dortmund University went to industries and is now active in other fields of chemical engineering. Their names as well as the names of other authors of the first edition are marked by asterisk in the byline of the corresponding chapters.

We are grateful to Klaus Unger, Jules Dingenen, and Reinhard Ditz that they agreed to join us as senior authors. The most challenging task to tackle is presented in Chapter 4 that has been efficiently handled by Abdelaziz Toumi, Joel Genolet, Andre Kiesewetter, Martin Krahe, Michele Morelli, Olivier Ludemann-Hombourger, Andreas Stein, and Eric Valery. It is in the nature of practical design and plant operation that the experience and interests are sometimes different. Additionally, the limited volume further constrains the content. But we hope to meet most of the practical aspects related to design and operation of chromatographic plants.

In Chapter 3, Matthias Jöhnck and Romas Skudas with the team of Michael Schulte combined the formerly separated topics on stationary phases and chromatographic systems to a unique and completely revised chapter and also extended it to ion exchange. We are especially indebted to Malte Kaspereit for his valuable contributions to Chapters 5 and 7. Sebastian Engell provided in Chapter 7 an overview of the latest research results on advanced process control. We hope that this will motivate practitioners to have a closer look at these promising methods.

Finally, we want to acknowledge the assistance of Fabian Thygs, who produced the new drawings and was patient enough to handle all our revisions.

As in the 1st edition, we have summarized the recently published results. In addition, we have made efforts to address preparative and process chromatographic issues from both the chemist and the process engineer viewpoints in order to improve the mutual understanding and to transfer knowledge between both disciplines.

With this book we want to reach colleagues from industries as well as universities interested in chromatographic separation with preparative purpose. Students and other newcomers looking for detailed information about design and operation of preparative chromatography are hopefully other users. Our message to all of them is that chromatography is nowadays rather well understood and not that difficult and expensive as it is often said and perceived. On the other hand, it is of course not the solution for all separation problems.

We would like to thank all authors for their contributions. We apologize for sometimes getting on their nerves pressing them to meet time limits. Last but not least, we thank our families and friends for their patience and cooperation in bringing out this book.

August 2012

Henner Schmidt-Traub
Michael Schulte
Andreas Seidel-Morgenstern

About the Editors

Henner Schmidt-Traub was Professor of Plant and Process Design at the Department of Biochemical and Chemical Engineering, TU Dortmund University, Germany, until his retirement in 2005. He is still active in the research community and his main areas of research focus on preparative chromatography, downstream processing, integrated processes, and plant design. Prior to his academic appointment, Prof. Schmidt-Traub gained 15 years of industrial experience in plant engineering.

Michael Schulte is Senior Director, Emerging Businesses Energy, at Merck KGaA Performance Materials, Darmstadt, Germany. In his PhD thesis at the University of Münster, Germany, he developed new chiral stationary phases for chromatographic enantioseparations. In 1995 he joined Merck and since then he has been responsible for research and development in the area of preparative chromatography, including the development of new stationary phases, new separation processes, and the implementation of Simulated Moving Bed technology at Merck. In his current position, one of the areas of his research is the use of ionic liquids for separation processes.

Andreas Seidel-Morgenstern is Director at the Max Planck Institute for Dynamics of Complex Technical Systems, Magdeburg, Germany, and holds the Chair in Chemical Process Engineering at the Otto-von-Guericke-University, Magdeburg, Germany. He received his PhD in 1987 at the Institute of Physical Chemistry of the Academy of Sciences in Berlin. From there he went on to work as postdoctoral fellow at the University of Tennessee, Knoxville, TN. In 1994 he finished his habilitation at the Technical University in Berlin. His research is focused on new reactor concepts, chromatographic reactors, membrane reactors, selective crystallization, adsorption and preparative chromatography, and separation of enantiomers among others.

List of Contributors

Jules Dingenen
Horststraat 51
2370 Arendonk
Belgium

Reinhard Ditz
Merck KGaA
Technology Office Chemicals (TO-I)
Frankfurter Str. 250
64291 Darmstadt
Germany

Sebastian Engell
TU Dortmund
Fakultät Bio- und
Chemieingenieurwesen
Lehrstuhl für Systemdynamik und
Prozessführung
Emil-Figge-Str. 70
44227 Dortmund
Germany

Joel Genolet
Merck Serono S.A.
Corsier sur Vevey
Zone Industrielle B
1809 Fenil sur Corsier
Switzerland

Matthias Jöhnck
Merck KGaA
R&D Performance & Life Science
Chemicals
Frankfurter Str. 250
64291 Darmstadt
Germany

Malte Kaspereit
Friedrich-Alexander-Universität
Erlangen-Nürnberg
Lehrstuhl für Thermische
Verfahrenstechnik
Egerlandstr. 3
91058 Erlangen
Germany

Andre Kiesewetter
Merck KGaA
PC-SRG-Bioprocess
Chromatography
Frankfurter Str. 250
64293 Darmstadt
Germany

Martin Krahe
Bideco AG
Bankstr. 13
8610 Uster
Switzerland

Olivier Ludemann-Hombourger
Polypeptide laboratories France
7 rue de Boulogne
67100 Strasbourg
France

Michele Morelli
Merck-Millipore SAS
39 Route Industrielle de la
Hardt – Bldg E
67120 Molsheim
France

Henner Schmidt-Traub
TU Dortmund
Fakultät für Bio- und
Chemieingenieurwesen
Lehrstuhl für Anlagen- und
Prozesstechnik
Emil-Figge-Str. 70
44227 Dortmund
Germany

Michael Schulte
Merck KGaA
R&D Performance & Life Science
Chemicals
Frankfurter Str. 250
64291 Darmstadt
Germany

Andreas Seidel-Morgenstern
Otto-von-Guericke-Universität
Lehrstuhl für Chemische
Verfahrenstechnik
Universitätsplatz 2

and

Max-Planck-Institut für Dynamik
komplexer technischer Systeme
Sandtorstraße 1
39106 Magdeburg
Germany

Romas Skudas
Merck KGaA
R&D Performance & Life Science
Chemicals
Frankfurter Str. 250
64291 Darmstadt
Germany

Andreas Stein
Merck KGaA
Chromatography Global Applied
Technology
Frankfurter Str. 250
64291 Darmstadt
Germany

Abdelaziz Toumi
Merck Serono S.A.
Corsier sur Vevey
Zone Industrielle B
1809 Fenil sur Corsier
Switzerland

Klaus K. Unger
Am alten Berg 40
64342 Seeheim
Germany

Eric Valery
Novasep Process
Boulevard de la Moselle
BP 50
54340 Pompey
France

List of Abbreviations

ACD:	At-column dilution
AIEX:	Anion exchanger
ARX:	Autoregressive exogenous
ATEX:	Explosion proof (French: ATmospheres EXplosibles)
BET:	Brunauer–Emmet–Teller
BJH:	Barrett–Joyner–Halenda
BR:	Chromatographic batch reactor
BV:	Bed volume
CACR:	Continuous annular chromatographic reactor
CD:	Circular dichroism (detectors)
CEC:	Capillary electrochromatography
CFD:	Computational fluid dynamics
cGMP:	Current good manufacturing practice
CIEX:	Cation exchanger
CIP:	Cleaning in place
CLP:	Column liquid chromatography
CLRC:	Closed-loop recycling chromatography
COGS:	Cost of goods sold
CPG:	Controlled pore glass
CSEP®:	Chromatographic separation
CSF:	Circle suspension flow
CSP:	Chiral stationary phase
CTA:	Cellulose triacetate
CTB:	Cellulose tribenzoate
DAC:	Dynamic axial compression
DAD:	Diode array detector
DMF:	Dimethyl formamide
DMSO:	Dimethyl sulfoxide
DTA:	Differential thermal analysis
DVB:	Divinylbenzene
EC:	Elution consumption
ECP:	Elution by characteristic points
EDM:	Equilibrium dispersive model

EMG:	Exponential modified Gauss (function)
FACP:	Frontal analysis by characteristic points
FDM:	Finite difference methods
FFT:	Forward flow test
FT:	Flow through
GC:	Gas chromatography
GMP:	Good manufacturing practice
GRM:	General rate model
HCP:	Health care provider
HETP:	Height of an equivalent theoretical plate
HFCS:	High fructose corn syrup
HIC:	Hydrophobic interaction chromatography
H-NMR:	Hydrogen nuclear magnetic resonance (spectroscopy)
HPLC:	High-performance liquid chromatography
HPW:	Highly purified water
IAST:	Ideal adsorbed solution theory
ICH:	International Guidelines for Harmonization
IEX:	Ion exchange
IMAC:	Immobilized metal affinity chromatography
IR:	Infrared
ISEC:	Inverse size exclusion chromatography
ISEP®:	Ion exchange separation
ISMB:	Improved/intermittent simulated moving bed
LC:	Liquid chromatography
LGE:	Linear gradient elution
LHS:	Liquid-handling station
LOD:	Limit of detection
LOQ:	Limit of quantification
LPLC:	Low-pressure liquid chromatography
LSB:	Large Scale Biotech project
MaB:	Monoclonal antibody
mAbs:	monoclonal antibodies
MD:	Molecular dynamics
MPC:	Model predictive control
MS:	Mass spectroscopy
MW:	Molecular weight
NMPC:	Nonlinear model predictive control
NMR:	Nuclear magnetic resonance (spectroscopy)
NN:	Neural network
NP:	Normal phase
NPLC:	Normal-phase liquid chromatography
NSGA:	Non-dominating sorting generic algorithm
OC:	Orthogonal collocation
OCFE:	Orthogonal collocation on finite elements
ODE:	Ordinary differential equation

PAT:	Process analytical technology
PDE:	Partial differential equation
PDT:	Pressure decay test
PEEK:	Poly(ether ether ketone)
PES:	Poly(ethoxy)siloxane
PLC:	Programmable logic controller
PMP:	Polymethylpentene
PSD:	Particle size distribution
QC:	Quality control
R&D:	Research and Development
RI:	Refractive index
RMPC:	Repetitive model predictive control
RP:	Reversed phase
S/N:	Signal-to-noise ratio
SEC:	Size exclusion chromatography
SEM:	Scanning electron microscopy
SFC:	Supercritical fluid chromatography
SIP:	Sanitization in place
SIP:	Steaming in place
SMB:	Simulated moving bed
SMBR:	Simulated moving bed reactor
SOP:	Standard operation procedure
SQP:	Sequential quadratic programming
SSRC:	Steady-state recycling chromatography
St-DVB:	Styrene-divinylbenzene
TDM:	Transport dispersive model
TEM:	Transmission electron microscopy
TEOS:	Tetraethoxysilane
TFA:	Trifluoroacetic acid
TG/DTA:	Thermogravimetric/differential thermal analysis
THF:	Tetrahydrofuran
TLC:	Thin-layer chromatography
TMB:	True moving bed process
TMBR:	True moving bed reactor
TPX^{TM}:	Transparent polymethylpentene
UPLC:	Ultrahigh-performance liquid chromatography
USP:	United States pharmacopoeia
UV:	Ultraviolet
VSP:	Volume-specific productivity
WFI:	Water for injection
WIT:	Water intrusion test

Notation

Symbols

Symbol	Description	Units
a_i	Coefficient of the Langmuir isotherm	$cm^3 \, g^{-1}$
a_s	Specific surface area	$cm^2 \, g^{-1}$
A	Area	cm^2
A_c	Cross section of the column	cm^2
A_i	Coefficient in the Van Deemter equation	cm
A_s	Surface area of the adsorbent	cm^2
ASP	Cross section-specific productivity	$g \, cm^{-2} \, s^{-1}$
b_i	Coefficient of the Langmuir isotherm	$cm^3 \, g^{-1}$
B	Column permeability	m^2
B_i	Coefficient in the Van Deemter equation	$cm^2 \, s^{-1}$
c_i	Concentration in the mobile phase	$g \, cm^{-3}$
$c_{p,i}$	Concentration of the solute inside the particle pores	$g \, cm^{-3}$
C	Annual costs	€
C_i	Coefficient in the Van Deemter equation	s
$C_{DL,i}$	Dimensionless concentration in the liquid phase	—
$C_{p,DL,i}$	Dimensionless concentration of the solute inside the particle pores	—
C_{spec}	Specific costs	$€ \, g^{-1}$
d_c	Diameter of the column	cm
d_p	Average diameter of the particle	cm
d_{pore}	Average diameter of the pores	cm
D_{an}	Angular dispersion coefficient	$cm^2 \, s^{-1}$
$D_{app,i}$	Apparent dispersion coefficient	$cm^2 \, s^{-1}$
$D_{app,pore}$	Apparent dispersion coefficient inside the pores	$cm^2 \, s^{-1}$

Symbol	Description	Units
D_{ax}	Axial dispersion coefficient	cm² s⁻¹
D_m	Molecular diffusion coefficient	cm² s⁻¹
$D_{pore,i}$	Diffusion coefficient inside the pores	cm² s⁻¹
$D_{solid,i}$	Diffusion coefficient on the particle surface	cm² s⁻¹
Da	Damkoehler number	—
EC	Eluent consumption	cm³ g⁻¹
F	Prices	€ l⁻¹, € g⁻¹
f_i	Fugacity	—
h	Reduced plate height	—
hR_f	Retardation factor	—
Δh_{vap}	Heat of vaporization	kJ mol⁻¹
H_i	Henry coefficient	—
H_p	Prediction horizon	—
H_r	Control horizon	—
HETP	Height of an equivalent theoretical plate	cm
$k_{ads,i}$	Adsorption rate constant	cm³ g⁻¹ s⁻¹
$k_{des,i}$	Desorption rate constant	cm³ g⁻¹ s⁻¹
$k_{eff,i}$	Effective mass transfer coefficient	cm² s⁻¹
K_{eq}	Equilibrium constant	Miscellaneous
K_{EQ}	Dimensionless equilibrium coefficient	—
$k_{film,i}$	Boundary or film mass transfer coefficient	cm s⁻¹
k'_i	Retention factor	—
$\tilde{k}'_i$	Modified retention factor	—
k_0	Pressure drop coefficient	—
k_{reac}	Rate constant	Miscellaneous
LF	Loading factor	—
L_c	Length of the column	cm
$\dot{m}_i$	Mass flow	g s⁻¹
m_i	Mass	g
m_j	Dimensionless mass flow rate in section j	—
m_s	Total mass	g
n_i	Molar cross section of component i	—
n_T	Pore connectivity	—
N	Column efficiency, number of plates	—
N_{col}	Number of columns	—
N_{comp}	Number of components	—
N_p	Number of particles per volume element	—
Δp	Pressure drop	Pa
Pe	Péclet number	—
Pr_i	Productivity	g cm³ h⁻¹
P_s	Selectivity point	—
Pu_i	Purity	%

Symbol	Description	Units
q_i	Solid load	g cm^{-3}
q_i^*	Total load	g cm^{-3}
$\bar{q}_i^*$	Averaged particle load	g cm^{-3}
$q_{\text{sat},i}$	Saturation capacity of the stationary phase	g cm^{-3}
$Q_{\text{DL},i}$	Dimensionless concentration in the stationary phase	—
r	Radial coordinate	cm
r_i	Reaction rate	Miscellaneous
r_p	Particle radius	cm
R_f	Retardation factor	—
R_i	Regulation term	—
R_s	Resolution	—
Re	Reynolds number	—
S_{BET}	Specific surface area	$\text{m}^2 \text{g}^{-1}$
Sc	Schmidt number	—
Sh	Sherwood number	—
St	Stanton number	—
t	Time	s
t_0	Dead time of the column (for total liquid holdup)	s
$t_{0,\text{int}}$	Dead time of the column (for interstitial liquid holdup)	s
t_{cycle}	Cycle time	s
t_g	Gradient time	s
t_{inj}	Injection time	s
t_{life}	Lifetime of adsorbent	h
t_{plant}	Dead time of the plant without column	s
$t_{R,i}$	Retention time	s
$t_{R,i,\text{net}}$	Net retention time	s
t_{shift}	Switching time of the SMB plant	s
t_{total}	Total dead time	s
T	Temperature	K
T	Degree of peak asymmetry	—
u_0	Velocity in the empty column	cm s^{-1}
u_{int}	Interstitial velocity in the packed column	cm s^{-1}
u_m	Effective velocity (total mobile phase)	cm s^{-1}
v_{sp}	Specific pore volume	$\text{cm}^3 \text{g}^{-1}$
V	Volume	cm^3
$\dot{V}$	Volume flow	$\text{cm}^3 \text{s}^{-1}$
V_{ads}	Volume of the stationary phase within a column	cm^3
V_c	Total volume of a packed column	cm^3
V_i	Molar volume	$\text{cm}^3 \text{mol}^{-1}$
V_{int}	Interstitial volume	cm^3
V_m	Overall fluid volume	cm^3

Symbol	Description	Units
V_{pore}	Volume of the pore system	cm^3
V_{solid}	Volume of the solid material	cm^3
VSP	Volume-specific productivity	$g\,cm^3\,s^{-1}$
w_i	Velocity of propagation	$cm\,s^{-1}$
x	Coordinate	cm
x_i	State of the plant	—
X_i	Mole fraction	—
X	Conversion	%
X_{cat}	Fraction of catalyst of the fixed bed	—
Y_i	Yield	%
Z	Dimensionless distance	—

Greek Symbols

Symbol	Description	Units
α	Selectivity	—
α_{exp}	Ligand density	$\mu mol\,m^{-2}$
β	Modified dimensionless mass flow rate	—
γ	Obstruction factor for diffusion or external tortuosity	—
Γ	Objective function	—
ε	Void fraction	—
ε^0	Solvent strength parameter	—
ε_p	Porosity of the solid phase	—
ε_t	Total column porosity	—
η	Dynamic viscosity	mPa s
Θ	Angle of rotation	°
Λ	Total ion exchange capacity	mM
λ	Irregularity in the packing	—
μ_i	Chemical potential	$J\,mol^{-1}$
μ_t	First absolute moment	—
ν	Kinematic viscosity	$cm^2\,s$
ν_i	Stoichiometric coefficient	—
π	Spreading pressure	Pa
ϱ	Density	$g\,cm^{-3}$
σ_t	Standard deviation	—
σ_i	Steric shielding parameter	—
ι	Dimensionless time	—

Symbol	Description	Units
ϕ	Bed voidage	—
φ	Running variable	
ψ	Friction number	—
ψ_{reac}	Net adsorption rate	$g\,cm^{-3}\,s^{-1}$
ω_j	Coefficient in the triangle theory	—
ω	Rotation velocity	$°s^{-1}$

Subscripts

Symbol	Description
1, 2	Component 1/component 2
I, II, III, IV	Section of the SMB or TMB process
acc	Accumulation
ads	Adsorbent
c	Column
cat	Catalyst
conv	Convection
crude	Crude loss
des	Desorbent
diff	Diffusion
disp	Dispersion
DL	Dimensionless
eff	Effective
el	Eluent
exp	Experimental
ext	Extract
feed	Feed
het	Heterogeneous
hom	Homogeneous
i	Component i
in	Inlet
inj	Injection
j	Section j of the TMB or SMB process
l	Liquid
lin	Linear
max	Maximum
min	Minimum

Notation

Symbol	Description
mob	Mobile phase
mt	Mass transfer
opt	Optimal
out	Outlet
p	Particle
pore	Pore
pipe	Pipe within HPLC plant
plant	Plant without column
prod	Product
raf	Raffinate
reac	Reaction
rec	Recycle
sat	Saturation
sec	Section
shock	Shock front
SMB	Simulated moving bed process
solid	Solid adsorbent
spec	Specific
stat	Stationary phase
tank	Tank within HPLC plant
theo	Theoretical
TMB	True moving bed process

Definition of Dimensionless Parameters

Péclet number	$Pe = \dfrac{u_{int} L_c}{D_{ax}}$	Convection to dispersion (column)
Péclet number of the particle	$Pe_p = \dfrac{u_{int} d_p}{D_{ax}}$	Convection to dispersion (particle)
Péclet number of the plant	$Pe_p = \dfrac{u_{plant} L_{plant}}{D_{ax, plant}}$	Convection to dispersion (plant without column)
Reynolds number	$Re = \dfrac{u_{int} d_p \rho_l}{\eta_l}$	Inertial force to viscous force
Schmidt number	$Sc = \dfrac{\eta_l}{\rho_l D_m}$	Kinetic viscosity to diffusivity
Sherwood number	$Sh = \dfrac{k_{film} d_p}{D_m}$	Mass diffusivity to molecular diffusivity
Stanton number (modified)	$St_{eff,i} = k_{eff,i} \cdot \dfrac{6}{d_p} \cdot \dfrac{L_c}{u_{int}}$	Mass transfer to convection

1
Introduction
Henner Schmidt-Traub and Reinhard Ditz

1.1
Development of Chromatography

Adsorptive separations have been in use well before the twentieth century. Tswett (1905, 1906), however, was the first who coined the term "Chromatography" in 1903 for the isolation of chlorophyll constituents. Kuhn and Brockmann, in the course of their research recognized the need for more reproducible and also more selective adsorbents, specially tuned for specific separation problems. This recognized demand for reproducible stationary phases led to the development of first materials standardized for adsorption strength and describes the first attempt toward reproducible separations (Unger *et al.*, 2010).

Liquid Chromatography (LC) was first applied as a purification tool and has thereby been used as a preparative method. It is the only technique that enables to separate and identify both femtomoles of compounds out of complex matrices in life sciences, and also allows the purification and isolation of synthetic industrial products in the ton range. The development of modern LC methodology and the corresponding technologies are based on three main pillars, which have developed over different time scales (Figure 1.1).

In the field of preparative and process chromatography the "restart" after the dormant period between the 1930s and the 1960s was not induced by the parallel emergence of analytical HPLC, but from engineering in search of more effective purification technologies. High selectivity of HPLC in combination with the principle to enhance mass transfer by counter current flow significantly increased the performance of preparative chromatography in terms of productivity, eluent consumption, yield, and concentration. The first process of this kind was the Simulated Moving Bed (SMB) chromatography for large-scale separation in the petrochemical area and in food processing. The development of new processes was accompanied by theoretical modeling and process simulation which are a prerequisite for better understanding of transport phenomena and process optimization.

In the 1980s, highly selective adsorbents were developed for the resolution of racemates into enantiomers. These adsorbents were mainly employed in analytical

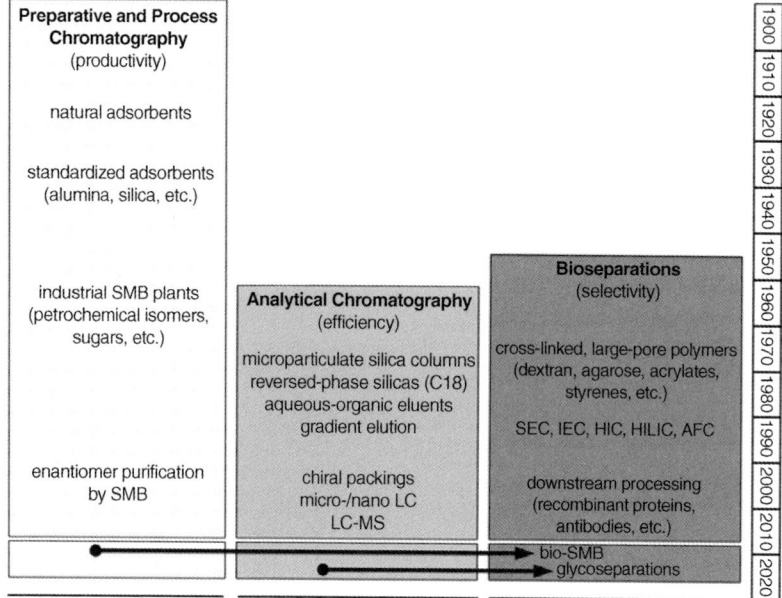

Figure 1.1 Development of chromatography (reproduced from Unger et al., 2010).

HPLC (Allenmark, 1992). However, the availability of enantioselective packings in bulk quantities also enabled the production of pure enantiomers by the SMB technology in the multi ton range. Productivities larger than 10 kg of pure product per kilogram of packing per day were achieved in the following years. In the 1990s the SMB concept was adapted and down-sized for the production of pharmaceuticals.

While preparative as well as analytical liquid chromatography were heavily relying on equipment and engineering and on the physical aspects of their tools for advancement in their fields, the domain Bioseparation was built around a different key aspect, namely, selective materials that allowed the processing of biopolymers, for example, recombinant proteins under nondegrading conditions, thus maintaining bioactivity. Much less focus in this area was on process engineering aspects, leading to the interesting phenomenon, that large-scale production concepts for proteins were designed around the mechanical instability of soft gels (Janson and Jönsson, 2010). The separation of proteins and other biopolymers has some distinctly different features compared with the separation of low molecular weight molecules from synthetic routes or from natural sources. Biopolymers have a molecular weight (MW) ranging from several thousand to several million. They are charged and characterized by their isoelectric point. More importantly, they have a dynamic tertiary structure that can undergo conformational changes. These changes can influence or even destroy the bio-activity in case of a protein denaturation. Biopolymers are separated in aqueous buffered eluents under conditions that maintain their bioactivity. Moreover, these large molecules exhibit approximately 100 times lower diffusion coefficients and consequently slower mass transfer than small molecules (Unger *et al.*, 2010). Due to these conditions, processes for biochromatography differ substantially from the separation of low-weight molecules. For instance, process pressure which is in many cases much lower for bio-processes than for HPLC requires a different plant design. Selectivity makes another difference; due to the very different retention times of bio-solutes an effective separation is only possible with solvent gradients.

Taking a peek into the future reveals a technology trend toward the use of continuous process operations and also downstream processing. Costs and production capacities will have to be addressed, asking for more integrated and efficient approaches. Adapting counter current solvent gradient concepts for the isolation of antibodies from complex fermentation broths will probably allow for more cost effective downstream processing of biopharmaceuticals within the next couple of years. A similar path might be useful to consider for dealing with the "glyco"-issue. Knowing that glycosylation plays a significant role in therapeutic drug efficacy, the analytical approaches developed around mixed-mode separation methods might be transferred to the process scale within a short time.

Validation of methods and assays will become a key issue. This fits directly with the Process Analytical Technology (PAT) initiative launched already years ago by the Food and Drug Administration (FDA), calling for a better process understanding. Among other things, this requires a much deeper insight into the underlying interactions using model-based approaches, which should finally allow "predictable" process design and monitoring strategies in the future to enhance process robustness and safety.

1.2
Focus of the Book

The general objective of preparative chromatography is to isolate and purify products independent of the amount of material to be separated. During this process, the products have to be recovered in the exact condition that they were in before undergoing the separation. In contrast to this analytical chromatography, which is beyond the content of this book, focuses on the qualitative and quantitative determination of a compound, that is, the sample can be processed, handled, and modified in any way suitable to generate the required information, including degradation, labeling, or otherwise changing the nature of the compounds.

The book describes and develops access to chromatographic purification concepts through the eyes of both engineers and chemists. This includes on one side the fundamentals of natural science and the design of matter and functionalities and on the other side mathematical modeling, simulation and plant design, as well as joined intersections in characterizing matter, process design, and plant operation. Such a joint view is necessary as the earliest possible interaction and cooperation between chemists and engineers is important to achieve time and cost-effective solutions and develop consistent methods that can be scaled up to a process environment.

With the second edition of this book the focus on fine chemicals and small pharmaceutical molecules is expanded to ion-exchange chromatography and the separation of biopolymers such as proteins. In accordance with the first edition these topics are restricted to those applications that can be modeled and simulated by current methods and procedures.

1.3
Recommendation to Read this Book

For most readers it is not necessary to read all chapters in sequence. For some readers the book may be a reference to answer specific questions depending on actual tasks, for others it may be a guide to acquire new fields of work in research or industrial applications. The different chapters are complementary to each other; therefore, it is recommended to be familiar with basic definitions explained in Chapter 2. The book may not provide answers to all questions. In which case, the reader can obtain further information from the cited literature.

Chapter 2 presents the basic principles of chromatography and defines the most important parameters such as retention, retention factor, selectivity, and resolution. It also explains the main model parameters as well as different kinds of isotherm equations including the IAS theory, and the determination of pressure drop. Other passages are devoted to plate numbers, HETP values as well as their determination based on first and second moments. The experienced reader may pass quickly through this chapter to become familiar with definitions used. For beginners this chapter is recommended in order to learn the general terminology and acquire a basic understanding. A further goal of this chapter is the harmonization of general chromatographic terms between engineers and chemists.

Chapter 3 focuses on stationary phases and the selection of chromatographic systems. The first part explains the structure and specifies the properties of stationary phases such as generic and designed phases, reversed-phase silicas, cross-linked organic polymers, and chiral phases, and gives instructions for their maintenance and regeneration. This part may be used as reference for special questions and will help those looking for an overview of attributes of different stationary phases. The second part deals with the selection of chromatographic systems, that is, the optimal combination of stationary phases and eluent or mobile phases for a given separation task. Criteria for choosing NP-, RP-, and CSP-systems are explained and are completed by practical recommendations. Other topics discussed are the processing of monoclonal antibodies and size exclusion. Finally, practical aspects of the overall optimization of chromatographic systems are discussed.

The selection of chromatographic systems is the most critical for process productivity and thus process economy. On one hand, the selection of the chromatographic system offers the biggest potential for optimization but, on the other hand, it is a potential source of severe errors in developing separation processes.

Chapter 4 focuses on practical aspects concerning equipment and operation of chromatographic plants for the production and purification of fine chemicals and small pharmaceutical molecules as well as proteins and comparable bio-molecules. It starts with the market of chromatographic columns followed by chromatography systems, that is, all equipment required for production. This includes high performance as well as low-pressure batch systems and SFC plants as well as continuous SMB systems, supplemented by remarks on auxiliary equipment. Further topics are standard process control and detailed procedures for different methods of column packing. The section on trouble shooting might be an interesting source for practitioners. Especially for the manufacturing of bio-therapeutics special disposable technologies such as prepacked columns and single-use membrane chromatography are exemplified.

Chapter 5 gives an overview of process concepts available for preparative chromatography. Depending on the operating mode, several features distinguish chromatographic process concepts: batch-wise or continuous feed introduction, operation in single- or multicolumn mode, integration of reaction and separation in one process step, elution under isocratic or gradient conditions, recycling of process streams, withdrawal of two or a multitude of fractions, and SMB processes under variable conditions. It finishes with guidelines for the choice of a process concept.

In Chapter 6, modeling and determination of model parameters are key aspects. "Virtual experiments" by numerical simulations can considerably reduce the time and amount of sample needed for process analysis and optimization. To reach this aim, accurate models and precise model parameters for chromatographic columns are needed. Validated models can be used predictively for optimal plant design. Other possible fields of application for process simulation include process understanding for research purposes as well as training of personnel. This includes the discussion of different models for the column and plant peripherals. Besides modeling, a major part of this chapter is devoted to the consistent determination of the model parameters, especially those for equilibrium isotherms. Methods of different complexity and experimental effort are presented which allow a variation of the desired accuracy, on the one hand, and the time needed on the other hand. Chapter 6 ends with a selection of different examples showing that an appropriate model combined with consistent parameters can simulate experimental data within high accuracy.

After general criteria and parameters for process optimization are defined, Chapter 7 focuses first on single-column processes. Design and scaling procedures for batch as well as recycle processes are described and a step-by-step optimization procedure is exemplified. In case of isocratic and gradient SMB processes, rigorous process simulations combined with short-cut calculations based on the TMB-model are useful tools for process optimization, which is illustrated by different example cases. Further sections discuss the improvements of SMB chromatography by variable operating conditions as given by Varicol, PowerFeed, or ModiCon processes. Finally, the latest scientific results on model-based advanced control of SMB processes are presented which are thought to be of increasing importance for practical applications.

References

Allenmark, S. (1992) *Chromatographic Enantioseparation: Methods and Applications*, Wiley, New York.

Janson, J. and Jönsson, J.-A. (2010) *Protein Purification* (ed. J.C. Janson), VCH, Weinheim.

Tswett, M.S. (1905) O novoy kategorii adsorbtsionnykh yavleny i o primenenii ikh k biokkhimicheskomu analizu (On a new category of adsorption phenomena and on its application to biochemical analysis), Trudy Varhavskago Obshchestva Estestvoispytatelei, Otdelenie Biologii (Proceedings of the Warsaw Society of Naturalists [i.e., natural scientists], Biology Section), 14, no. 6, 20–39.

Tswett, M.S. (1906) Physical chemical studies on chlorophyll adsorption. *Berichte der Deutschen botanischen Gesellschaft*, **24**, 316–323 (as translated and excerpted in H.M. Leicester, Source Book in Chemistry 1900–1950, Cambridge, MA: Harvard, 1968).

Unger, K., Ditz, R., Machtejevas, E., and Skudas, R. (2010) Liquid chromatography – its development and key role in life sciences applications. *Angew. Chemie*, **49**, 2300–2312.

2
Fundamentals and General Terminology

Andreas Seidel-Morgenstern, Michael Schulte*, and Achim Epping*

This chapter introduces fundamental aspects and basic equations for the characterization of chromatographic separations. Starting from the simple description of an analytical separation of different compounds, the influences of fluid dynamics, mass transfer, and thermodynamics are explained. The important separation characteristics for preparative and process chromatography, for example, the optimization of resolution and productivity as well as the differences compared with chromatography for analytical purposes, are described. The importance of understanding the behavior of substances in the nonlinear range of the adsorption isotherm is highlighted.

One further goal of this chapter is to contribute to the harmonization of general chromatographic terms between engineers and chemists.

2.1
Principles of Adsorption Chromatography

Chromatography belongs to the thermal separation processes used to separate homogeneous molecular dispersive mixtures. The separations can be divided in general into three steps (Sattler, 1995):

1) In addition to the initially homogeneous mixture phase, a second phase is generated by introduction of energy (e.g., during distillation) or an additive (e.g., during extraction or adsorption).
2) An exchange of mass and energy occurs between the single phases. The driving force for these transport processes is a deviation from thermodynamic equilibrium.
3) After completion of the exchange procedures, the two phases are characterized by different compositions and can be separated. Result of this phase separation is a partial separation of the initially homogeneous mixture.

*These authors have contributed to the first edition.

Preparative Chromatography, Second Edition. Edited by H. Schmidt-Traub, M. Schulte, and A. Seidel-Morgenstern.
© 2012 Wiley-VCH Verlag GmbH & Co. KGaA. Published 2012 by Wiley-VCH Verlag GmbH & Co. KGaA.

2 Fundamentals and General Terminology

Regarding chromatographic separation, the homogeneous mixture phase is a fluid. The additional second phase is a solid or a second immiscible fluid. The driving force for transport between the phases is the deviation of the compositions from the equilibrium state. The mixture introduced is separated by selective relative movements of the components between the two phases. Usually, the solid phase is fixed and designated accordingly as stationary phase. The fluid phase moves and is therefore called mobile phase. Chromatographic behavior is determined by the specific interactions of all single components present in the system with the mobile and stationary phases. The mixture of substances to be separated, the carrier (solvent), which is used for dissolution and transport, and the adsorbent (stationary phase) are summarized as the chromatographic system (Figure 2.1). In laboratory practice a chromatographic system suitable to solve a given separation problem is selected by a process typically called method development.

According to the state of aggregation of the fluid phase, chromatographic systems can be divided into several categories. If the fluid phase is gaseous, the process is called gas chromatography (GC). If the fluid phase is a liquid, the process is called liquid chromatography (LC). For a liquid kept at temperature and pressure conditions above its critical point the process is called supercritical fluid chromatography (SFC). Liquid chromatography can be further divided according to the geometrical orientation of the phases. A widely used process for analytical purposes as well as rapid method development and, in some cases, even a preparative separation process is thin-layer chromatography (TLC). The adsorbent is fixed onto a support (glass, plastic, or aluminum foil) in a thin layer. The solute is placed onto the adsorbent in small circles or lines. In a closed chamber one end of the thin-layer plate is dipped into the mobile phase, which then progresses along the plate due to capillary forces. Individual

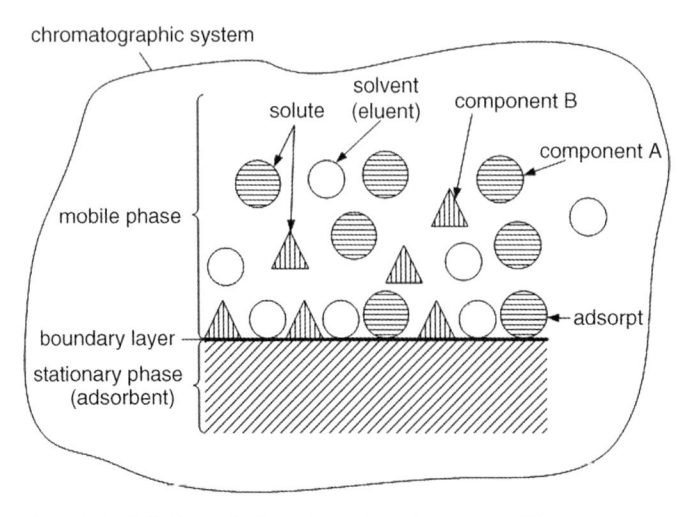

Figure 2.1 Definitions of adsorption and the chromatographic system.

substances can be visualized either by fluorescence quenching or after chemical reaction with detection reagents. The advantages of TLC are the visualization of all substances, even those sticking heavily to the adsorbent, as well as the easiness of parallel development.

In GC and LC the adsorbent is fixed into a cylinder (column) that is usually made of glass, polymer, or stainless steel (column). In this column the adsorbent is present as a porous or nonporous randomly arranged packing or as a monolithic block. If the columns are well packed with small particles and liquid mobile phases are used, a successful technique, designated as high-performance liquid chromatography (HPLC), is frequently applied.

2.1.1
Adsorption Process

Depending on the kind of interface between mobile and stationary phases, the following types of chromatographic systems can be distinguished:

- gaseous mobile phases/solid stationary phase;
- liquid mobile phases/solid stationary phase;
- gaseous mobile phases/liquid stationary phase;
- liquid mobile phases/liquid stationary phase.

Most chromatographic processes exploit the principle of adsorption and the separation process is based on the deposition and accumulation of molecules on solid surfaces.

Examples of industrial relevance for the first two phase combinations are the adsorption of pollutants from waste air or water onto activated carbon. Combinations three and four are relevant, for example, related to foam formation and stabilization in the presence of surfactants on water/air interfaces or at the interface of two immiscible liquids (e.g., oil and water). This book deals mainly with the case most typical for preparative chromatographic separations, that is, the exploitation of solid surfaces, liquid mobile phases, and dissolved feed mixtures. The following definitions are made: The solid onto which adsorption occurs is defined as the adsorbent. The adsorbed molecule is defined in its free state as the adsorptive and in its adsorbed state as the adsorpt. There are typically different solutes, which are often called components (for example, A and B, Figure 2.1).

On a molecular level the adsorption process is the formation of binding forces between the surface of the adsorbent and the molecules of the fluid phase. The binding forces can be different in nature and strength. Basically, two different types of binding can be distinguished (Atkins, 1990; IUPAC, 1972):

1) Physisorption or physical adsorption, which is a weak binding based essentially on van der Waals forces, for example, dipole, dispersion, or induction forces. These forces are weaker than the intramolecular binding forces of molecular species. Therefore, physisorbed molecules maintain their chemical identity.

Table 2.1 Adsorption enthalpies.

	Physisorption	Chemisorption	
Gas phase	1.5 Δh_{vap}	2–3 Δh_{vap}	Ruthven (1984)
Liquid phase	<50 kJ mol^{-1}	≥60–450 kJ mol^{-1}	Kümmel and Worch (1990)

2) The stronger binding type is chemisorption or chemical adsorption. It is caused by valence forces, equivalent to chemical, mainly covalent, bindings.

Table 2.1 gives estimated values for adsorption enthalpies. In gaseous systems these enthalpies are proportional to the heat of vaporization h_{vap}.

As periodic chromatographic processes require complete reversibility of the adsorption step, only adsorption processes based on physisorption can be exploited. The related energies are sufficient to increase the temperature of gases due to their low volumetric heat capacities (Ruthven, 1984). Fluids, however, are characterized by volumetric heat capacities 10^2–10^3 times higher. Thus, the energies connected to adsorption processes have not much influence on the local temperatures. All of the following processes can, therefore, be considered to be approximately isothermal.

2.1.2
Chromatographic Process

In liquid phase chromatography the mobile phase is forced through a column, packed with a multitude of adsorbent particles (the stationary phase). Figure 2.2 illustrates the injection of a homogeneous binary mixture (components A and B represented by triangles and circles) into the system at the column inlet. Hereby the triangles represent the component B with the higher affinity to the stationary phase. Therefore, the mean adsorption time of this component on the stationary phase surface is higher than that of component A with the lower affinity (circles).

The difference in affinities, and thus adsorption time, results in a reduced migration speed through the column of the more adsorbed component B. This delays its arrival at the column outlet compared with the less adsorbed component A. If the

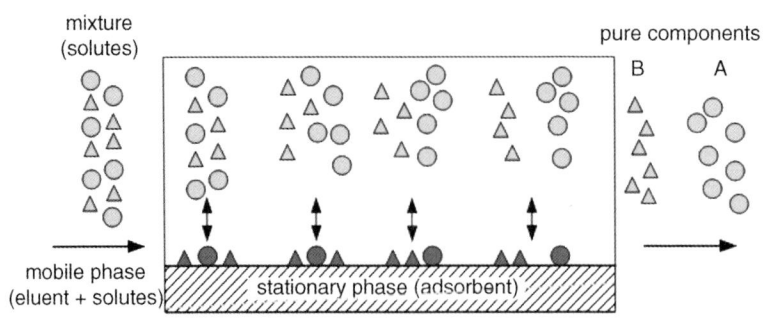

Figure 2.2 Principle of adsorption chromatography.

process conditions are chosen well, the two substances can be completely separated and collected as pure components at the column outlet.

2.2
Basic Effects and Chromatographic Definitions

2.2.1
Chromatograms and Parameters

Basic information regarding the development of a chromatographic separation process provides the chromatogram. A typical chromatogram resulting from the injection of three different components in analytical amounts is shown in Figure 2.3.

The interaction strength of each component with the stationary phase is proportional to its retention time $t_{R,i}$. Instead of the retention time, it is often useful to consider the retention volume, which is obtained by multiplying $t_{R,i}$ with the volumetric flow rate.

In the case of symmetrical peak shapes, the retention time can be determined from the peak maximum. For well-packed columns symmetrical peaks should be achieved as long as the amounts injected into the column are small, restricting the concentrations to the linear range of the adsorption isotherms. If increased amounts of substances are injected, that is, in the nonlinear range of the adsorption isotherms, the peaks are often distorted and asymmetrical. In that case the retention time has to be calculated from the point of gravity of the peaks applying the method of moments (Equation 2.38).

The total dead time t_{total} is the time a nonretained substance needs from the point of sample introduction to the point of detection. It depends on the holdup within the column itself as well as on further holdup in the system or plant, which is, for instance, influenced by pipe lengths and diameters, pump head volumes

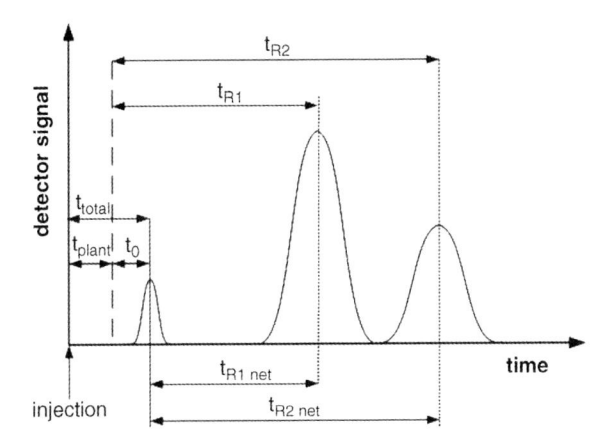

Figure 2.3 Chromatogram of two retained and one unretained component.

(if the sample is introduced via a pump), and detector volumes. Therefore, the dead time of the plant without the column t_{plant} has to be determined as well to get the correct dead time of the column t_0 by subtraction of t_{plant} from t_{total}. Typically, tracer molecules are used to determine the dead time. These molecules should not be retained by the solid phase and should have the similar molecular sizes as the components to be separated to penetrate the pore system in a comparable way.

The overall retention time $t_{R,i}$ of a retained component i is the sum of the dead time t_0 and the net retention time $t_{R,i,net}$. The net retention time represents the time during which molecules of a substance are adsorbed onto the surface of the adsorbent.

Because the retention time of a substance depends on the column geometry, the porosity and the mobile phase flow rate, a normalization is expedient. For this the capacity factor k'_i, which is also called the retention factor, is defined according to Equation 2.1:

$$k'_i = \frac{t_{R,i} - t_0}{t_0} \tag{2.1}$$

The capacity factor depends only on the distribution of the component of interest between the mobile and the stationary phases. It indicates the ratio of the time a component is adsorbed to the time it is in the fluid phase. It can also be expressed as a ratio of amounts of substance i in the stationary and mobile phases (Equation 2.2):

$$k'_i = \frac{n_{i,stat}}{n_{i,mob}} \tag{2.2}$$

Both the capacity factor and the net retention time might depend on the nature of the tracer substance, which is used to determine t_0. Therefore, only k'_i values that are based on experiments with the same tracer substance should be directly compared and used to calculate selectivities.

Every LC process aims to separate dissolved components. Therefore, the distance between the peak maxima of two or more components is of great importance. A selectivity α of a separation of two components i and j can be quantified by dividing their capacity factors (Equation 2.3):

$$\alpha = \frac{k'_j}{k'_i} = \frac{t_{R,j} - t_0}{t_{R,i} - t_0} \tag{2.3}$$

Hereby, by convention, the capacity factor of the more retained component is placed in the numerator. Thus, the selectivity of two separated components is always >1. The selectivity is also called separation factor.

2.2.2
Voidage and Porosity

Other important information that can and should be derived from standard chromatograms is the column porosities and the void fractions. These parameters have to be carefully determined as an important basis for quantification and

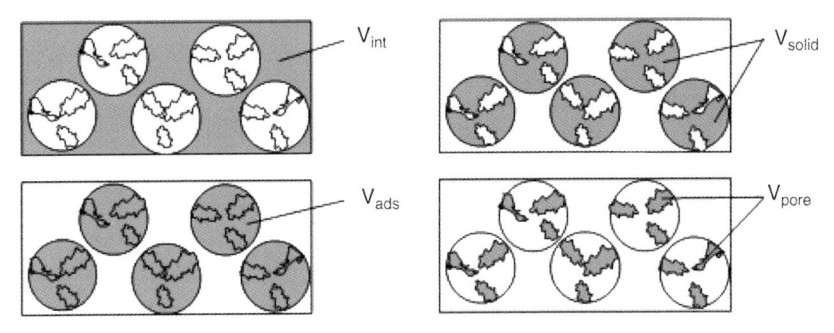

Figure 2.4 Structure of packed beds.

simulation of the purification process. As illustrated in Figure 2.4, the total volume of a packed column (V_c) is divided into two subvolumes: the interstitial volume of the fluid phase (V_{int}) and the volume of the stationary phase (V_{ads}) (Equation 2.4):

$$V_c = \pi \frac{d_c^2}{4} L_c = V_{ads} + V_{int} \tag{2.4}$$

Further, V_{ads} consists of the volume of the solid material (V_{solid}) and the volume of the pore system, V_{pore} (Equation 2.5):

$$V_{ads} = V_{solid} + V_{pore} \tag{2.5}$$

From these volumes different porosities can be calculated (Equations 2.6–2.8):

$$\text{Void fraction of the column}: \quad \varepsilon = \frac{V_{int}}{V_c} \tag{2.6}$$

$$\text{Porosity of the solid phase}: \quad \varepsilon_p = \frac{V_{pore}}{V_{ads}} \tag{2.7}$$

$$\text{Total porosity}: \quad \varepsilon_t = \frac{V_{int} + V_{pore}}{V_c} = \varepsilon + (1-\varepsilon)\varepsilon_p \tag{2.8}$$

Practical determination of the porosities often experiences difficulties. The most common method for determining the total porosity is the injection of nonretained, pore-penetrating tracer substances (gray and small black circles in Figure 2.5). In normal phase chromatography, toluene or 1,3,5-tri-*tert*-butylbenzene is often used, while in reversed phase chromatography uracil is typically selected as the component of choice.

The interstitial velocity of the mobile phase is given by Equation 2.9:

$$u_{int} = \frac{\dot{V}}{\varepsilon \cdot \pi \cdot (d_c^2/4)} \tag{2.9}$$

tracer components

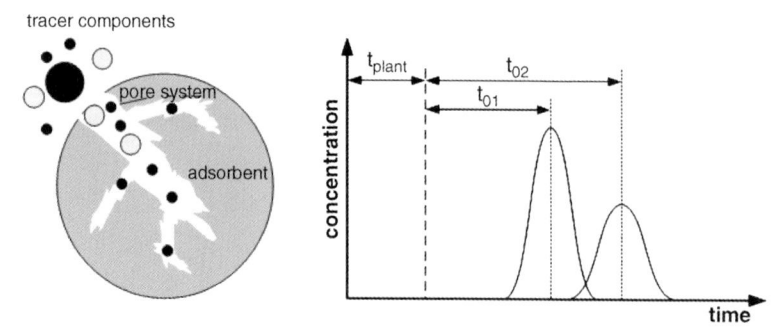

Figure 2.5 Chromatography of tracer components.

It is useful to measure the volumetric flow rate delivered by the pump during the determination of the column dead time. The total porosity is then calculated according to Equation 2.10:

$$\varepsilon_t = \frac{t_0 \cdot \dot{V}}{V_c} \tag{2.10}$$

To obtain the void fraction or interstitial porosity of a column the same methodology can be used based on injecting a high molecular weight substance that is unable to penetrate into the pores (large black circles in Figure 2.5). Figure 2.5 also shows an ideal chromatogram obtained from the injection of two tracer substances characterized by different molecular weights. For a nonretained and nonpenetrating molecule the dead time of the column and the interstitial velocity are connected by Equation 2.11:

$$t_{0,\text{int}} = \frac{L_c}{u_{\text{int}}} \tag{2.11}$$

Severe experimental problems often occur during the injection of high molecular weight tracers. Ideal high molecular weight substances should be globular and exhibit no adsorption or penetration into any pore but should, however, be highly mobile. Obviously, from these prerequisites, there is no ideal high molecular weight volume marker. Therefore, alternative methods for determining the void fraction are of interest. If the adsorbent parameters can be determined with high accuracy from physical measurements, the porosities can also be calculated from the mass of the adsorbent m_{solid}, its density ϱ_{solid}, and the specific pore volume v_{sp} (determined, for example, by nitrogen adsorption or mercury porosimetry). The volume of the solid material is then calculated by Equation 2.12:

$$V_{\text{solid}} = \frac{m_{\text{solid}}}{\varrho_{\text{solid}}} \tag{2.12}$$

The pore volume can be obtained from Equation 2.13:

$$V_{\text{pore}} = v_{\text{sp}} \cdot m_{\text{solid}} \tag{2.13}$$

The sum of solid volume and pore volume gives the volume of the stationary phase, which, subtracted from the column volume, leads to the interstitial volume (Equation 2.14):

$$V_{int} = V_c - (V_{solid} + V_{pore}) \tag{2.14}$$

The void fraction can then be determined using Equation 2.6. Subsequently, the other porosities can be obtained by Equations 2.7 and 2.8.

An alternative method for easy and exact determination of total porosities can be used for small-scale columns. As long as a column can be weighed exactly, the mass difference of the same column filled with two solvents of different densities can be used to determine the porosity. The column is first completely flushed with one solvent and then weighed; afterwards, the first solvent is completely displaced by a second solvent of different density. For normal phase systems methanol and dichloromethane can be used; for reversed phase systems water and methanol are quite commonly employed. The volume of the solvent, representing the sum of the interstitial volume and the pore volume, is determined by Equation 2.15:

$$V_{int} + V_{pore} = \frac{m_{s,1} - m_{s,2}}{\varrho_{s,1} - \varrho_{s,2}} \tag{2.15}$$

The total porosity is again calculated by Equation 2.8.

High-efficiency adsorbents are very often spherical and monodisperse in order to reduce pressure drops. In this case void fractions for spherical particles lie theoretically in the range $0.26 < \varepsilon < 0.48$ and mean values of ε of approximately 0.37 can be roughly applied (Brauer, 1971).

Particle porosities lie in the range $0.50 < \varepsilon_p < 0.90$, meaning that 50–10% of the solid is impermeable skeleton. In practice, the total porosity is often in the range $0.65 < \varepsilon_t < 0.80$. With monolithic columns total porosities lie in the range $0.80 < \varepsilon_t < 0.90$. At large porosities, sufficient stability of the bed has to be ensured.

2.2.3
Influence of Adsorption Isotherms on Chromatogram Shapes

A chromatogram is influenced by several factors, such as the fluid dynamics inside the packed bed, mass transfer phenomena, and, most importantly, the equilibrium of the adsorption processes taking place.

The adsorption equilibrium is typically described by isotherms, which give the correlations between the loadings of the solute on the adsorbent, q_i, as a function of the fluid phase concentrations, c_i, at a given temperature. Single-component and competitive adsorption isotherms are discussed in more detail in Section 2.5.

The elution profile of an ideal chromatogram depends only on the courses of the equilibrium functions characterizing the chromatographic system. To understand and predict real chromatograms additional mass transfer resistances and details regarding the fluid dynamics have to be taken into account.

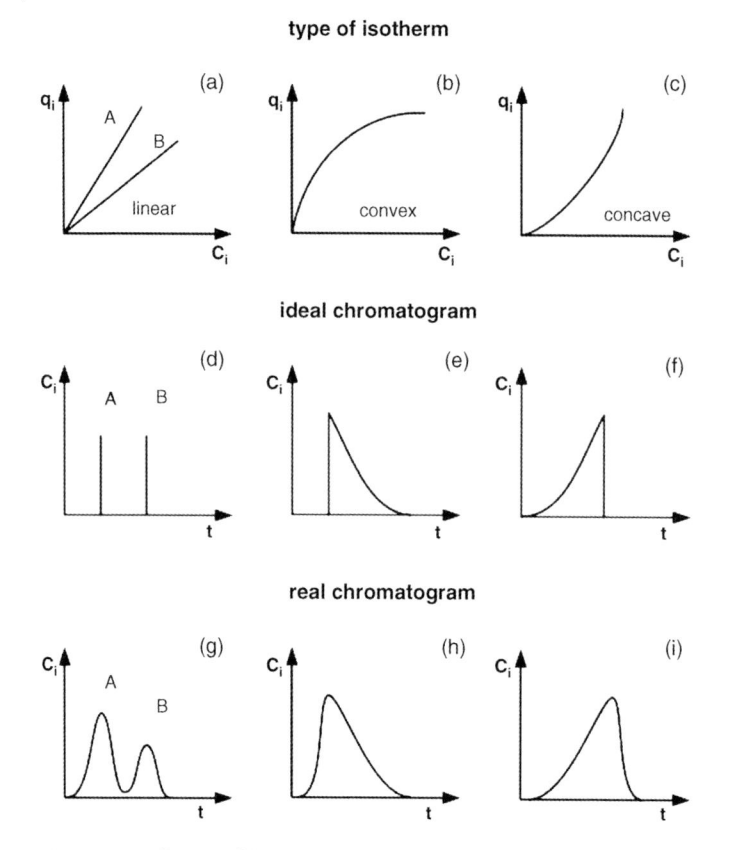

Figure 2.6 Influence of the type of equilibrium isotherm on the chromatogram.

The influence of the shape of the adsorption isotherm on the shape of a chromatogram is illustrated in Figure 2.6.

In an ideal analytical chromatogram, that is, in the linear range of the adsorption isotherms, the retention time is a function of the isotherm slope, quantified by the Henry constant H_i (Figure 2.6a), according to the basic equation of chromatography (Equation 2.16) (Guiochon et al. 2006):

$$t_{R,i} = t_0 \left(1 + \frac{1 - \varepsilon_t}{\varepsilon_t} H_i \right) \tag{2.16}$$

Thus, under linear conditions the retention time does not depend on the fluid phase concentration (Figure 2.6d).

In contrast, under nonlinear conditions peak deformations occur and the retention times become functions of concentration. This leads under thermodynamically controlled ideal conditions to the formation of disperse waves and shock fronts (Figure 2.6e and f).

2.2 Basic Effects and Chromatographic Definitions

The "retention times" of specific concentrations c_i^+ in waves are then related to the local slopes of the specific adsorption isotherm as follows:

$$t_{R,i}(c_i^+) = t_0 \left(1 + \frac{1 - \varepsilon_t}{\varepsilon_t} \frac{dq_i}{dc_i} \bigg|_{c_i^+} \right) \tag{2.17}$$

For convex isotherms (Figure 2.6b) the desorption branch of the peak (the "tail") forms a wave (Figure 2.6e), whereas for concave isotherms (Figure 2.6c) the adsorption branch forms a wave ("peak fronting," Figure 2.6f).

The second type of branches of overloaded ideal peaks is characterized by shock fronts having retention times described by Equation 2.18:

$$t_{R,i}^{shock} = t_0 \left(1 + \frac{1 - \varepsilon_t}{\varepsilon_t} \frac{q_{after} - q_{before}}{c_{after} - c_{before}} \bigg|_{shock} \right) \tag{2.18}$$

Real chromatograms illustrated in Figure 2.6g–i are shaped not only by thermodynamics but also by kinetics of mass transfer and deviations from plug flow conditions. A rectangular concentration profile of the solute injected at the entrance of the column soon changes into a bell-shaped Gaussian distribution, if the isotherm is linear. Figure 2.7a shows this distribution and some characteristic values, which will be referred to in subsequent chapters. The degree of peak asymmetry T can be evaluated from the difference between the two peak halves at 1/10th of the peak height (Figure 2.7b) (Equation 2.19):

$$T = \frac{b_{0.1}}{a_{0.1}} \tag{2.19}$$

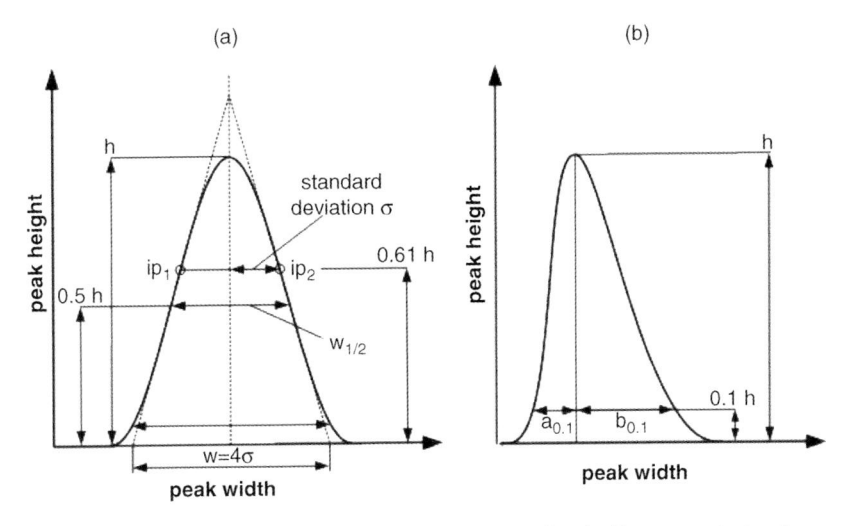

Figure 2.7 Determination of peak asymmetry: (a) symmetrical peak; (b) asymmetrical peak.

Another calculation method to evaluate peak asymmetry can be found in the different pharmacopoeias, for example, United States Pharmacopoeia (USP) or Pharmacopoeia Europea (PhEUR). These pharmacopoeias calculate peak tailing from an analysis at 5% of the peak height (Equation 2.20):

$$T_{USP} = \frac{a_{0.05} + b_{0.05}}{2 \cdot a_{0.05}} \tag{2.20}$$

2.3
Fluid Dynamics

All preparative and production-scale chromatographic separations aim to collect target components as highly concentrated as possible. The ideal case would be a rectangular signal with the same width as the pulse injected into the column. This behavior cannot be achieved in reality. In every chromatographic system nonidealities of fluid distribution occur, resulting in broadenings of the residence time distributions of the solutes. Hydrodynamic effects that contribute to the total band broadening are frequently captured in an axial dispersion term. Figure 2.8 illustrates the effect of the axial dispersion. The rest of the band broadening results from finite adsorption rates and mass transfer resistance.

Figure 2.8 shows that the rectangular pulse, which is introduced at the column inlet ($x = 0$), is symmetrically broadened as it travels along the column. As a consequence of the band broadening, the maximum concentration of the solute is decreased. This causes an unfavorable dilution of the target component fraction.

Main factors that influence axial dispersion are discussed below.

2.3.1
Extra Column Effects

Every centimeter of tubing, as well as any detector, between the point of solute injection and the point of fraction withdrawal contributes to the axial dispersion of

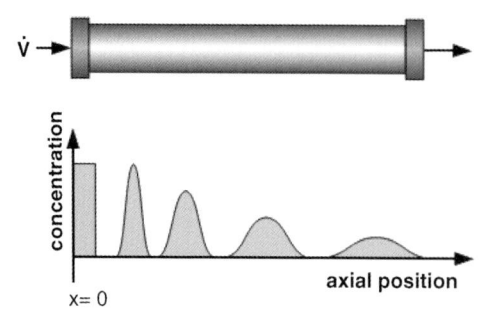

Figure 2.8 Band broadening in a column due to axial dispersion.

the samples and thus decreases the concentration and separation efficiency. Obviously, the length of connecting lines should be minimized. Critical are smooth connections of tubing and column, avoiding any dead space where fluid and, especially, sample can accumulate. The tubing diameter depends on the flow rate of the system and has to be chosen in accordance with the system pressure. The tubing should contribute as little as possible to the system pressure drop, without adding additional holdup volumes.

2.3.2
Column Fluid Distribution

Critical points for axial dispersion in column chromatography are the fluid distribution at the column head and the fluid collection at the column outlet (Section 4.3.3.2). Especially with large-diameter columns, these effects have to be carefully considered. Fluid distribution in the column head is widely driven by the pressure drop of the packed bed, which forces the sample to be radially distributed within the inlet frit. It is, therefore, important to use high-quality frits, which ensure an equal radial fluid distribution. In low-pressure chromatography with large dimensions, as well as with new types of adsorbents, such as monolithic packings, which exhibit much lower pressure drops, the fluid distribution is of even greater importance. Several approaches to optimize the distribution have been made by column manufacturers to overcome this problem of low-pressure flow distribution. The introduction of specially designed fluid distributors has greatly improved the situation (Section 4.3.3.2).

2.3.3
Packing Nonidealities

The packed bed of a chromatographic column will never attain optimum hexagonal dense packing due to the presence of imperfections. Those imperfections can be divided into effects due to the packing procedure and influences from the packing material. During the packing procedure several phenomena occur. The two most important are wall effects and particle bridges. Knox, Laird, and Raven (1976) determined a layer of about $30d_p$ thicknesses where wall effects influence the column efficiency. Therefore, attention has to be paid to this effect with small-diameter columns. The larger the column diameter, the less severe is this effect. The second reason why imperfect packing may occur in columns is inappropriate packing procedures. If the particles cannot arrange themselves optimally during the packing process, they will form bridges, which can only be destroyed by immense axial pressure, with the danger of damaging the particles. One way to reduce particle bridging is to pack the column by vacuum and, afterwards, by axial compression of the settled bed. For more details see Section 4.6.

Insufficient bed compression, which leads to void volumes at the column inlet, is another source of axial dispersion. Preparative columns should therefore preferably possess a possibility for adjusting the compression of the packed bed.

2.3.4
Sources for Nonideal Fluid Distribution

The reasons for nonidealities in fluid distribution can be divided into macroscopic, mesoscopic, and microscopic effects (Tsotsas, 1987). The different effects are illustrated in Figure 2.9.

Microscopic fluid distribution nonideality is caused by fluid dynamic adhesion between the fluid and the adsorbent particle inside the microscopic channels of the packed bed. Adhesion results in a higher fluid velocity in the middle of the channel than at the channel walls (Figure 2.9a). Solute molecules in the middle of the channel thus have a shorter retention time than those at the channel walls.

Local inhomogeneities of the voidage are a second source of broadening the mean residence time distribution. For small particles, the formation of particle agglomerates, which cannot be penetrated by the fluid, is another reason for axial dispersion. Figure 2.9b illustrates this mesoscopic fluid distribution nonideality. This effect results in local differences in fluid velocity and differences in path length of the solute molecules traveling straight through the particle agglomerates compared with those molecules moving around the particle aggregates. The second effect is also known as eddy diffusion (Figure 2.9d). Due to its similar origin, this statistical phenomenon is related to mesoscopic fluid dynamic nonidealities.

Macroscopic fluid distribution nonidealities are caused by local nonuniformities of the void space between the particles, which might occur especially in the wall region (Figure 2.9c).

All the above-mentioned effects cause an increase of the peak width and contribute, besides mass transfer resistances discussed below, to total band broadening.

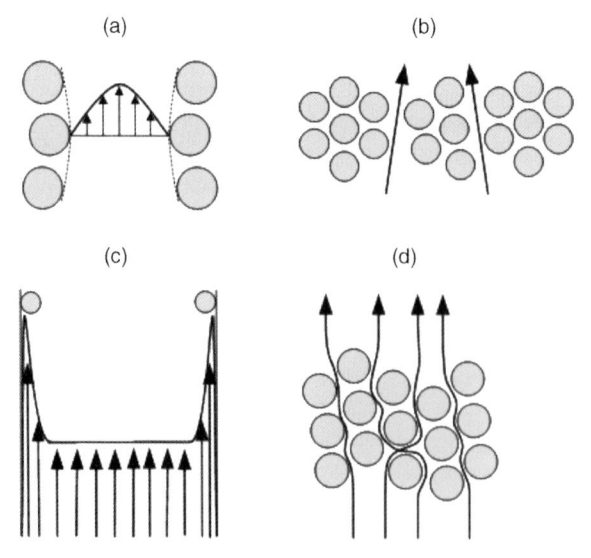

Figure 2.9 Fluid distribution nonidealities according to Tsotsas (1987): (a) microscopic; (b) mesoscopic; (c) macroscopic; (d) eddy diffusion.

2.3.5
Column Pressure Drop

In chemical engineering the Ergun equation (Equation 2.21) is well known for the estimation of friction numbers and corresponding pressure drops for fixed beds with granular particles:

$$\psi = \frac{150}{Re} + 1.75 \qquad (2.21)$$

It covers the broad span from fine particles to coarse materials (Brauer, 1971). For chromatographic columns small, more or less spherical particles are used. Therefore, the Reynolds numbers are very small and inertial forces can be neglected. Equation 2.21 then reduces to its first term, which represents the pressure drop because of viscous forces only.

The friction number is defined (Equation 2.22) as

$$\psi = \frac{\varepsilon^3}{(1-\varepsilon)^2} \frac{\Delta p \, d_p}{\varrho u_0^2 \, L_c} \qquad (2.22)$$

where u_0 is the velocity in the empty column (superficial velocity) (Equation 2.23):

$$u_0 = \frac{\dot{V}}{A_c} = \frac{\dot{V}}{\pi d_c^2/4} = \varepsilon \cdot u_{int} \qquad (2.23)$$

Introducing Equation 2.22 in the reduced Equation 2.21 leads to Equation 2.24:

$$\Delta p = 150 \frac{(1-\varepsilon)^2}{\varepsilon^3} \frac{\eta u_0 L_c}{d_p^2} \qquad (2.24)$$

This equation is identical to Darcy's law (Equation 2.25), which is well known among chromatographers (Guiochon, Golshan-Shirazi, and Katti, 1994):

$$\Delta p = \frac{1}{k_0} \frac{\eta u_0 L_c}{d_p^2} \qquad (2.25)$$

with

$$k_0 = \frac{\varepsilon^3}{150 \cdot (1-\varepsilon)^2} \qquad (2.26)$$

The coefficient k_0 typically lies between 0.5×10^{-3} and 2×10^{-3}. This agrees with Equation 2.26 where a void fraction of 0.4 results in a corresponding value of 1.2×10^{-3}. If more precise values are required, k_0 has to be measured for a given packing.

Another parameter derived also from Darcy's equation, which is often used to characterize a column, is the column permeability B (Equation 2.27):

$$B = \frac{\eta u_0 L_c}{\Delta p} \qquad (2.27)$$

Comparing Equations 2.25 and 2.27, the permeability is related to k_0 by

$$B = k_0 \cdot d_p^2 \tag{2.28}$$

Pressure drop measurements are used to check the stability of a column bed as a function of eluent flow rate. Deviations from linearity at ascending and descending flow rates serve as a strong indication of irreversible changes in the column bed. It is also important to notice the linear dependency between pressure drop and viscosity. The pressure drop for high-performance columns may exceed 100 bar. Therefore, the viscosity is also an important criterion in selecting the eluent. To pack column beds properly a liquid with low viscosity has to be chosen. If ethanol, for instance, is the eluent, methanol may be used for the packing procedure as its viscosity is 1.9 times lower. Equation 2.24 indicates the influence of the void fraction on the pressure drop. For particles with varying diameters the void fraction of the column can decrease tremendously. Therefore, the span of the particle size distribution for chromatographic adsorbents should be chosen in the narrow range between 1.7 and 2.5 for analytical columns (Section 3.1.6.1). For preparative columns, packed frequently with 10–20 µm particles, the distribution can be much broader or even bimodal.

2.4
Mass Transfer Phenomena

2.4.1
Principles of Mass Transfer

Most chromatographic production systems use particulate adsorbents with defined pore structures due to the higher loadability of mesoporous adsorbents compared with nonporous adsorbents. Adsorption of the target molecules on the inner surface of a particle has a significant influence on the efficiency of a preparative separation. Regarding mass transfer, several factors that contribute to band broadening can be distinguished (Figure 2.10), namely

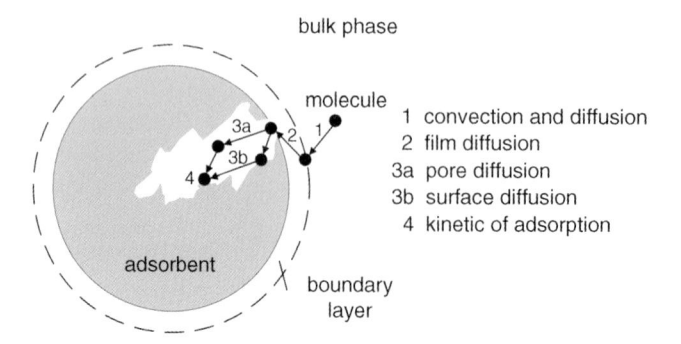

Figure 2.10 Mass transfer phenomena during the adsorption of a molecule.

- convective and diffusive transport toward the particles;
- film diffusion in the laminar boundary layer around the particles;
- pore (a) and surface (b) diffusion in the particles;
- kinetics of the actual adsorption and desorption steps.

Individual adsorbent particles within a packed bed are surrounded by a boundary layer, which is looked upon as a stagnant liquid film of the fluid phase. The thickness of the film depends on the fluid distribution in the bulk phase of the packed bed. Molecular transport toward the boundary layer of the particle by convection or diffusion is the first step (1) of the separation process.

The second relevant step is diffusive transport of the solute molecule through the film layer, which is called film diffusion (2). Transport of the solute molecules toward the adsorption centers inside the pore system of the adsorbent particles is the third step of the adsorption process. This step can follow two different transport mechanisms, which can occur separately or parallel to each other: pore diffusion (3a) and surface diffusion (3b).

If mass transport occurs by surface diffusion (3b), a solute molecule is adsorbed and transported deeper into the pore system by movement along the pore surface. During the whole transport process, molecules are within the attraction forces of the adsorbent surface. Notably, the attractive forces between the surface and adsorbed molecules are often so strong (Ruthven, 1984) that, for many common adsorpt–adsorbent systems encountered in preparative chromatography, surface diffusion is not very relevant.

Pore diffusion (3a) is driven by restricted Fickian diffusion of the solute molecules within the free pore volume. During the transport process, the solute molecules are outside the attraction forces of the adsorbent surface.

Once a solute molecule reaches a free adsorption site on the surface, actual adsorption takes place (4). If the adsorption processes (1–4) are slow compared with the convective fluid flow through the packed bed, the eluted peaks show a non-symmetrical band broadening. Severe tailing can be observed, due to the slower movement of those solute molecules, which penetrate deeper into the pore system.

External mass transfer (1) and kinetics of adsorption (4) are normally comparatively fast in the fluid phase (Guiochon, 2006; Ruthven, 1984). Rate-limiting processes are typically the film diffusion (2) and the transport inside the pore system (3a, 3b).

For particle diameters >5 µm, which are applied in most preparative chromatographic separations, transport resistances within the pore are the dominant contribution to the total mass transfer resistance (Ludemann-Hombourger, Bailly, and Nicoud, 2000a, 2000b).

The transport factors described above are influenced by the morphology of the adsorbent particles as well as by their surface chemistry. Pore sizes influence chromatographic properties in two main ways. The first effect is due to steric hindrance. Molecules can only penetrate into pores of sizes above a certain threshold. For bigger molecules the theoretically available surface area is only partly accessible. The second factor is related to pore size distributions.

Because transport within a mesoporous system is diffusive, band broadening increases dramatically when pores with large differences in sizes are present within an adsorbent backbone. This is due to the fact that molecules in the larger pores have a shorter residence time than molecules adsorbed in the smaller pores.

Within an ideal adsorbent all adsorption sites should be equal and accessible within short distances, thus keeping diffusion pathways short. This prerequisite favors that all pores should be of equal size, that no deep dead-end pores are present, and that the accessibility of the pores from the region of convective flow within a packed bed should be equal for all pores.

The above-mentioned factors indicate that ideal chromatographic adsorbents would exhibit high surface areas but moderate pore volumes and pore sizes (which are large enough to ensure good accessibility for the adsorptive molecules), as these three physical parameters are interlinked and cannot be independently adjusted. Their interdependence is summarized in the Wheeler equation, which assumes constant pore diameters within the particle. In Equation 2.29, d_{pore} is the pore diameter (nm), V_{pore} the pore volume (ml g^{-1}), and S_{BET} the specific surface area (m^2 g^{-1}):

$$d_{pore} = 4 \times 10^3 \frac{V_{pore}}{S_{BET}} \tag{2.29}$$

The design of an adsorbent for preparative chromatography is always a compromise between a high surface area (with good loadability and limited mechanical strength) and a high mechanical strength (with reduced surface area).

2.4.2
Efficiency of Chromatographic Separations

One of the key parameters that characterizes a preparative chromatographic system is the plate number N, which is also designated as column efficiency.

The plate number as well as the corresponding height of an equivalent theoretical plate (HETP) is well known in chromatography and chemical engineering as a common measure to quantify mass transfer effects and to capture deviations from ideal behavior within packed beds (Equation 2.30):

$$\text{HETP} = \frac{L_c}{N} \tag{2.30}$$

The plate height lumps together the contributions of fluid dynamic nonidealities (e.g., axial dispersion), mass transfer resistances, and finite adsorption and desorption rates, which all contribute to undesired band broadening. It can be defined as the rate of the local gradient of the width of a Gaussian peak (Equation 2.31):

$$\text{HETP} = \frac{\partial \sigma_z^2}{\partial z} \tag{2.31}$$

For uniformly packed columns and incompressible eluents this expression can be integrated, providing Equation 2.32:

$$\text{HETP} = \frac{\sigma_z^2}{L_c} \tag{2.32}$$

In practice, peak profiles are not measured over axial coordinates but detected at the column outlets over time. Therefore, a corresponding standard deviation of the time-dependent concentration profile σ_t can be used (Equation 2.33):

$$\frac{\sigma_t}{t_R} = \frac{\sigma_z}{L_c} \tag{2.33}$$

The plate height can then be calculated by Equation 2.34:

$$\text{HETP} = \left(\frac{\sigma_t}{t_R}\right)^2 L_c \tag{2.34}$$

The smaller the plate height of a column, that is, the higher the efficiency, the narrower is the peak width and the closer the peak shape approaches to the ideal rectangular injection profile shape. Owing to limited rates of mass transfer and nonidealities of fluid dynamics, real peaks are characterized by limited efficiencies. Narrower peaks result in a good peak resolution, small elution volumes, and thus higher outlet concentrations. All these facts provide favorable conditions for preparative applications of chromatography.

For the quantitative determination of the column efficiency typically small amounts of retained test components are injected into the column. From the resulting chromatograms the column efficiencies can be calculated.

For highly efficient columns symmetrical peaks are detected, which offer the possibility to estimate the plate number quickly if the width of the Gaussian distribution w (Figure 2.7) and Equation 2.30 are introduced in Equation 2.34:

$$N_i = \left(\frac{t_R}{\sigma_t}\right)^2 = 16\left(\frac{t_{R,i}}{w_i}\right)^2 \tag{2.35}$$

The index i indicates that the plate number can be different for each component. Another equation to determine the same plate number exploits the easy to measure peak width at half-height $w_{1/2,i}$ (Equation 2.36):

$$N_i = 5.54\left(\frac{t_{R,i}}{w_{1/2,i}}\right)^2 \tag{2.36}$$

To compare the plate numbers of columns with different lengths an efficiency per meter (N_L) is often used (Equation 2.37):

$$N_{L,i} = \frac{N_i}{L_c} = \frac{1}{\text{HETP}_i} \tag{2.37}$$

For asymmetrical peaks the mean retention time of the chromatogram is more precisely calculated by the first absolute moment μ_t (Equation 2.38).

Accordingly, the variance σ_t^2 is determined by the second central moment applying Equation 2.39:

$$\mu_t = \frac{\int_0^\infty t \cdot c(t) \cdot dt}{\int_0^\infty c(t) \cdot dt} \qquad (2.38)$$

$$\sigma_t^2 = \frac{\int_0^\infty (t-\mu_t)^2 \cdot c(t) \cdot dt}{\int_0^\infty c(t) \cdot dt} \qquad (2.39)$$

The plate number for a certain asymmetrical peak is then calculated by Equation 2.40:

$$N_i = \frac{\mu_{t,i}^2}{\sigma_{t,i}^2} \qquad (2.40)$$

For asymmetrical peaks another empirically derived equation has been introduced by Bidlingmeyer and Waren (1984). According to Equation 2.41 and the parameters illustrated in Figure 2.7b, a plate number of nonsymmetrical peaks can be estimated as

$$N_i = 41.7 \left[\frac{(t_{R,i}/w_{t0.1})^2}{1.25 + (b_{0.1i}/a_{0.1i})} \right] \qquad (2.41)$$

The influence of different band broadening parameters on the overall efficiency of a column is shown in Figure 2.11, where the plate height is plotted versus the mobile phase velocity. This famous relationship can be described well by the Van Deemter equation (Van Deemter, Zuiderweg, and Klinkenberg, 1956) (Equation 2.42):

$$\text{HETP}_i = A_i + \frac{B_i}{u_{\text{int}}} + C_i u_{\text{int}} \qquad (2.42)$$

The three terms of Equation 2.42 describe different effects that have to be taken into account when selecting a combination of stationary and mobile phases in

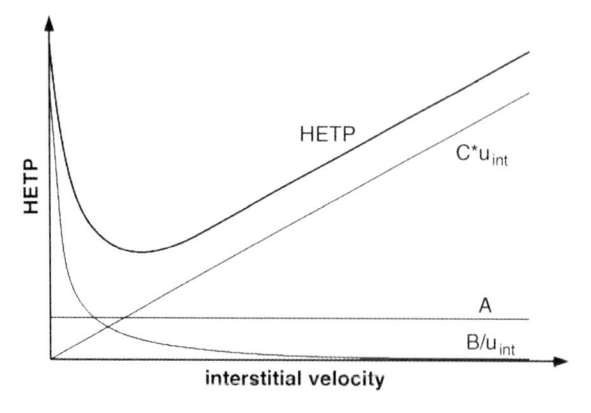

Figure 2.11 Relationship between HETP and interstitial phase velocity.

preparative chromatography. The A term, which is almost constant over the whole velocity range, is mainly governed by eddy diffusion. It results from packing imperfections as well as adsorbents with very broad particle size distributions. The absolute height of the A term is proportional to the mean particle diameter. The plate height can therefore be decreased by using smaller particles.

The hyperbolic B term, B_i/u_{int}, expresses the influence of axial diffusion of the solute molecules in the fluid phase. It is relevant only in preparative systems with large adsorbent particles operated at very low flow rates. In most cases of preparative chromatography it can be neglected, as the velocities of the mobile phase are sufficiently large. As the longitudinal diffusion depends on the diffusion coefficient of the solute, it can be influenced by changing the mobile phase composition. To achieve higher diffusion coefficients, low-viscosity solvents should be preferred, which will, in addition, result in lower column pressure drops.

The linear increase of the plate height at high velocities is captured by the C term ($C_i u_{int}$). It is caused by the increasing influence of mass transfer resistances at higher velocities. Mass transfer resistances inside the pores cannot be reduced by increasing the fluid velocity. Their relative contribution to overall band broadening increases linearly with an increase of the flow velocity. The slope of the C term depends on the nature of the packing material. The more optimized an adsorbent is in terms of pore accessibility and minimal diffusion path lengths, the lower the contribution of the C term and the higher the column efficiency at high mobile phase velocities.

It should be mentioned that in addition to Equation 2.42 several more sophisticated plate height equations have been suggested, for instance, by Knox (2002).

2.4.3
Resolution

In contrast to analytical chromatography, where it is possible to deconvolute overlapping peaks and to obtain even quantitative information from nonresolved peaks, preparative chromatography typically requires complete peak resolution if the components of interest are to be isolated with 100% purity and yield. Chromatographic resolution R_S is a measure of how well two adjacent peak profiles of similar area are separated. R_S is mostly defined as (Equation 2.43)

$$R_S = \frac{2 \cdot (t_{R,j} - t_{R,i})}{w_i + w_j} \tag{2.43}$$

where $t_{R,j}$ and $t_{R,i}$ are the retention times of the two neighboring components j and i, of which component i elutes first ($t_{R,j} > t_{R,i}$); w_j and w_i are the peak widths for components j and i at their baseline ($w = 4\sigma$).

A resolution of 1.5 corresponds to a baseline separation at touching band situation. At a resolution of 1.0 there is still an overlap of 3% of the peaks.

The assumption of equivalent peaks, where the peak width is equal for both components, leads to the following equation:

$$R_S = \left(\frac{\alpha - 1}{\alpha}\right)\left(\frac{k'_j}{k'_j + 1}\right)\frac{\sqrt{N_j}}{4} \qquad (2.44)$$

Note that the index j belongs to the component with the longer retention time. The equivalent expression for resolution based on the less retained component i is

$$R_S = (\alpha - 1)\left(\frac{k'_i}{k'_i + 1}\right)\frac{\sqrt{N_i}}{4} \qquad (2.45)$$

The resolution of a separation system depends on three important parameters: the selectivity, the capacity factors, and the column efficiency (Equations 2.44 and 2.45).

In theory, the above equation offers three possibilities for increasing the resolution. Increasing the capacity factor k' is only of limited use because it is linked to a higher retention time and thus to an increase of the cycle time. This leads not only to lower throughput but also to more band broadening, which reduces the column efficiency.

The two real options for resolution optimization are illustrated in Figure 2.12a–c.

Figure 2.12a shows a separation with a resolution factor of 0.9. The column efficiency is given by a plate number of 1000 N m^{-1}, while the selectivity is equal to 1.5. The first approach optimizes the column fluid dynamics, for example, by improving the column packing quality, reducing the particle diameter of the adsorbent, or optimizing the mobile phase flow rate. The resulting chromatogram is shown in Figure 2.12b. The resolution has been increased to 1.5 due to the higher plate number of 2000 N m^{-1} while the selectivity is kept constant at 1.5. Conversely, the system thermodynamics can be changed by optimizing the temperature or altering the chromatographic system by changing the mobile phase and/or the adsorbent. Figure 2.12c shows the effect of the thermodynamic approach. The plate number is still constant at 1000 N m^{-1}, but the selectivity has increased to 2.0, resulting in an increased resolution of 1.5.

The above two optimization approaches have different effects on the cost of preparative chromatography. Efficiency is increased by decreasing the particle diameter of the stationary phase. However, then the costs of the phase rapidly increase, along with the costs of the equipment, which needs to be more pressure-stable for operation with smaller particle diameters. It is, therefore, worth looking for systems with optimized selectivities.

With Equations 2.44 and 2.45 it is also possible to calculate the plate number necessary for a given resolution (Equation 2.46):

$$N_j = 16 \cdot R_S^2 \left(\frac{\alpha}{\alpha - 1}\right)^2 \left(\frac{1 + k'_j}{k'_j}\right)^2 \qquad (2.46)$$

Figure 2.13 shows the necessary plate numbers for a resolution of 1.5 and 1.0 versus the selectivity. For separation systems with selectivities <1.2 the necessary plate numbers are very high. As higher plate numbers are linked to high-pressure

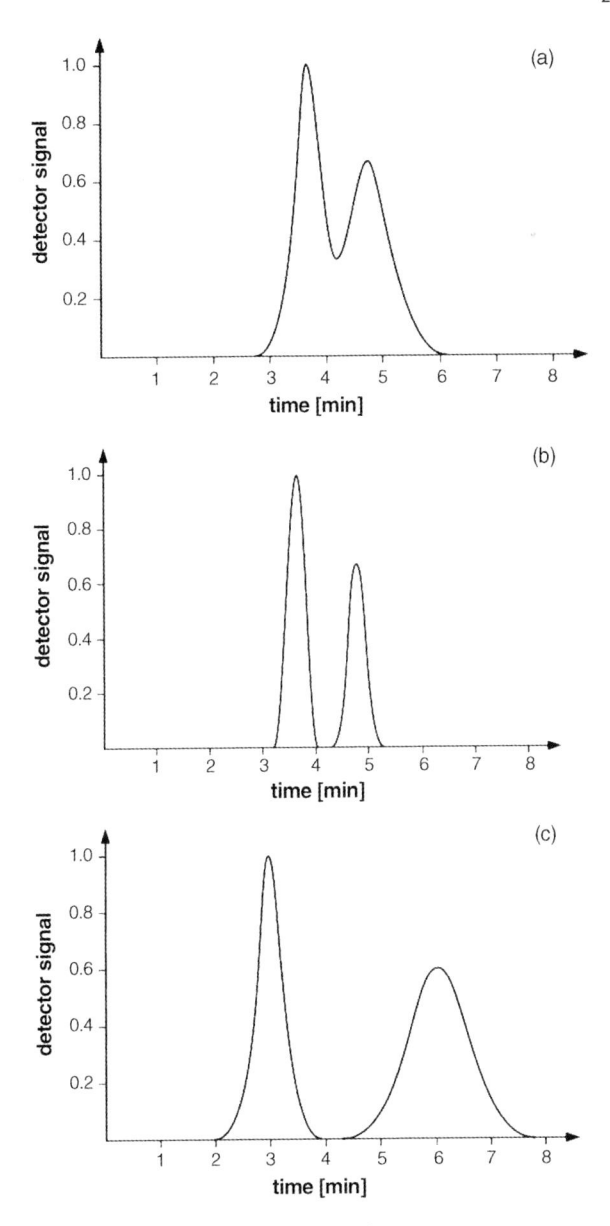

Figure 2.12 Optimization of peak resolution.

systems with small particles, these separations obviously cause higher costs compared to separations which can exploit higher selectivities. This emphasizes that rigorous screening for the most selective separation system is necessary.

In summary, for optimized preparative chromatographic systems some consideration should be taken into account right from the beginning:

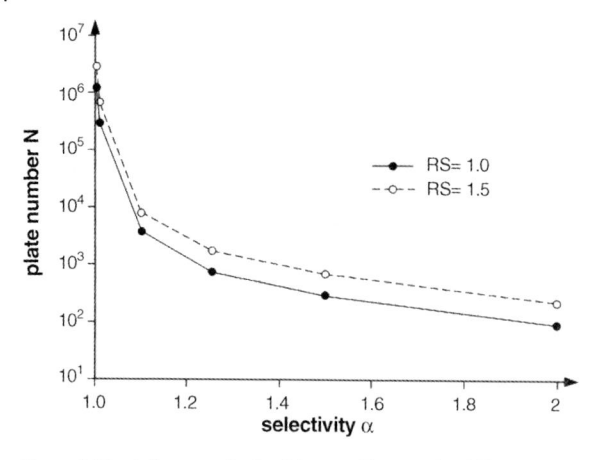

Figure 2.13 Influence of selectivity on efficiency for different resolutions.

1) **System efficiency:** good packing quality and low band broadening by limiting axial dispersion and mass transfer resistances.
2) **Resolution versus load:** good resolution in the nonlinear range of the adsorption isotherm (see below) has to be achieved.
3) **Pressure drop:** should be as low as possible to allow operation at maximum linear velocity.

2.5
Equilibrium Thermodynamics

2.5.1
Definition of Isotherms

The most important difference between analytical and preparative chromatography is the extension of the working range into the nonlinear region of the adsorption isotherms. Therefore, the behavior of single components as well as their mixtures has to be known over a wide concentration range.

As with any other phase equilibrium the adsorption equilibrium is defined by the equality of the chemical potentials of all interacting components in all phases. A more detailed description of the thermodynamic fundamentals can be found in the literature (Ruthven, 1984; Guiochon et al. 2006).

The adsorption of molecules on a solid surface can reach significant loads. The surface concentration may be defined as the quotient of the adsorbed amount of substance and the surface of the adsorbent in mol m^{-2}. However, the effective internal surface area of an adsorbent is difficult to determine because it depends on the nature and size of the solutes. It is therefore advisable to use the mass or the volume of the adsorbent instead of the internal surface. The loading is then expressed as mol g^{-1} or mol l^{-1} adsorbent. Adsorbent volume can be expressed as

2.5 Equilibrium Thermodynamics

total adsorbent volume V_{ads} or as solid-phase volume ($V_{ads} - V_{pore}$). According to Equation 2.47 both values can be transformed into each other. In Equation 2.47 $c_{p,i}$ represents the concentration of component i within the pore system. The total load of the adsorbent is given as q_i^* and the pure solid load as q_i:

$$q_i^* = \varepsilon_P \cdot c_{P,i} + (1 - \varepsilon_P) \cdot q_i \tag{2.47}$$

Plotting the solid load q_i or the total adsorbent load q_i^* versus the concentration of the solute in the fluid phase c_i at constant temperature illustrates the course of the adsorption isotherm. In the following the pure solid load, q_i, will be typically applied.

In the literature (e.g., Brunauer, Emmett, and Teller, 1938; Giles, Smith, and Huitson, 1974; Kümmel and Worch, 1990) different types of possible courses of isotherms are reported (Figure 2.14).

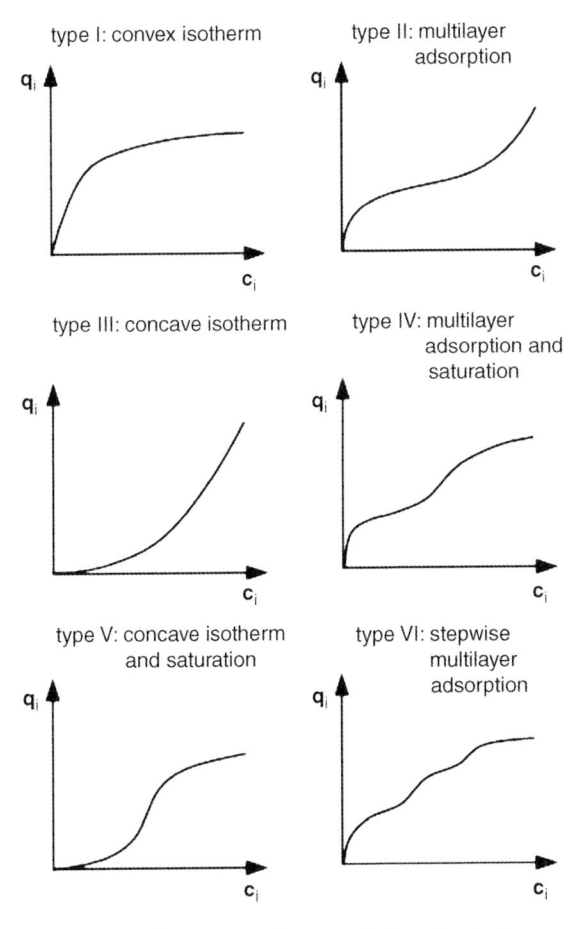

Figure 2.14 Different types of courses of adsorption isotherms.

In contrast to well-developed thermodynamic methods for determining and correlating gas/liquid equilibria, the reliable theoretical prediction of adsorption isotherms is not yet state of the art. A few approaches for determining multi-component isotherms from experimentally determined single-component isotherms are known (see Section 2.5.2.3). Careful experimental determination of the single-component adsorption isotherms is therefore necessary. Different approaches for isotherm determination are discussed in Section 6.5.7.

2.5.2
Models of Isotherms

For modeling and simulation of preparative chromatography experimentally determined adsorption equilibrium data have to be represented by suitable mathematical equations. From the literature a multitude of different isotherm equations is known. Many of these equations are derived from equations developed for gas phase adsorption. Detailed literature can be found, for example, in the textbooks of Guiochon et al. (2006), Everett (1984), and Ruthven (1984) or articles of Seidel-Morgenstern and Nicoud (1996) or Bellot and Condoret (1993).

Isotherm models have to be divided into single-component and multicomponent models. Because most of the multicomponent models are directly derived from single-component models, the latter will be presented first.

2.5.2.1 Single-Component Isotherms

Figure 2.15 illustrates the course of a single-component adsorption isotherm of the Langmuir type, the most common type describing situations encountered in preparative chromatography. Increasing the concentration of the solute in the mobile phase above certain limits does not anymore cause an increase of the amount adsorbed onto the stationary phase. Only in the very first low mobile phase concentration region there is a linear relationship. This region is used for quantitative analysis in analytical chromatography, because only in this region it is ensured that no retention time shifts take place if different amounts are injected.

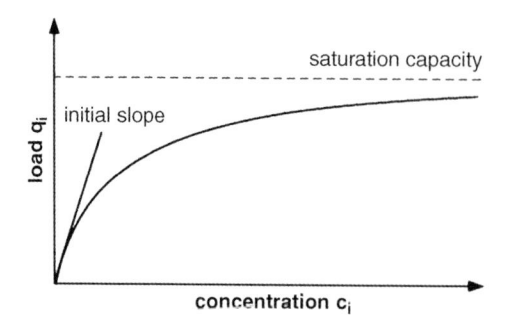

Figure 2.15 Single-component Langmuir isotherm.

2.5 Equilibrium Thermodynamics

In the initially linear range of the adsorption isotherm the relationship between the mobile and the stationary phase concentrations c and q is expressed by Equation 2.48:

$$q = H \cdot c \tag{2.48}$$

For the determination of the Henry coefficient H the retention time of a substance, the column dead time, and the total porosity are necessary (Equation 2.16):

$$H = \left(\frac{t_R}{t_0} - 1\right) \frac{\varepsilon_t}{1 - \varepsilon_t} \tag{2.49}$$

The higher the Henry coefficient for a substance, the longer is its retention time under analytical conditions. This definition shows that for two components to be separated their Henry coefficients have to differ. According to Equations 2.3 and 2.49, the ratio of the Henry coefficients specifies the selectivity of a separation system:

$$\alpha = \frac{H_j}{H_i} \tag{2.50}$$

Figure 2.16 illustrates the isotherms of two different components and their Henry coefficients.

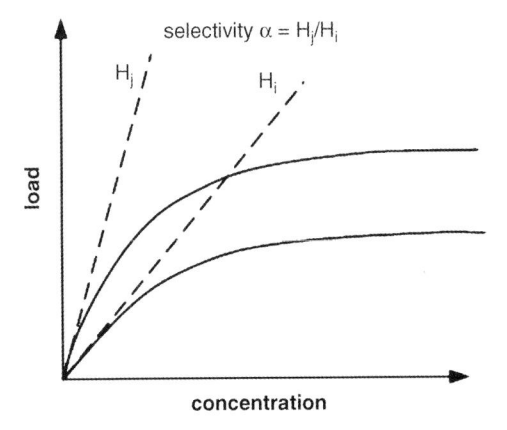

Figure 2.16 Initial slopes of the isotherms for two different components.

The Langmuir isotherm model is represented by Equation 2.51. The following assumptions are the theoretical basis of Langmuir-type isotherms:

- All adsorption sites are considered energetically equal (homogeneous surface).
- Each adsorption site can only adsorb one solute molecule.
- Only a single layer of adsorbed solute molecules can be formed.
- There are no lateral interferences between the adsorbed molecules.

$$q = q_\text{sat} \cdot \frac{b \cdot c}{1 + b \cdot c} = \frac{H \cdot c}{1 + b \cdot c} \qquad (2.51)$$

In the above q_sat is the saturation capacity of the monolayer and the parameter b quantifies the adsorption energy:

$$b = \frac{H}{q_\text{sat}} \qquad (2.52)$$

Experimental data can be often fitted more precisely to an adsorption isotherm model when an additional linear term is introduced that covers the nonspecific adsorption of the solute to the adsorbent. This modifies the single-component Langmuir isotherm (Equation 2.53) (Seidel-Morgenstern and Nicoud, 1996):

$$q = \lambda \cdot c + q_\text{sat} \frac{b \cdot c}{1 + b \cdot c} = \lambda \cdot c + \frac{H \cdot c}{1 + b \cdot c} \qquad (2.53)$$

An even more flexible equation is given by the bi-Langmuir isotherm (Graham, 1953) (Equation 2.54):

$$q = q_{\text{sat},1} \frac{b_1 \cdot c}{1 + b_1 \cdot c} + q_{\text{sat},2} \frac{b_2 \cdot c}{1 + b_2 \cdot c} = \frac{H_1 \cdot c}{1 + b_1 \cdot c} + \frac{H_2 \cdot c}{1 + b_2 \cdot c} \qquad (2.54)$$

The second Langmuir term covers the adsorption of the solute molecules on a second, completely independent group of adsorption sites of the adsorbent. Those completely independent adsorption sites can occur on reversed phase adsorbents with remaining silanol groups or on chiral stationary phases with a chiral selector coated or bonded to a silica surface (Guiochon et al. 2006).

Limitations of the Langmuir model can also be overcome adding a third model parameter for example, by the Toth isotherm model (Toth, 1971):

$$q = q_\text{sat} \cdot \frac{c}{((1/b) + c^e)^{1/e}} \qquad (2.55)$$

The model shown in Equation 2.55 has three independent fitting parameters, q_sat, e, and b, which allow independent control of slope and curvature. For $e = 1$ it approaches the Langmuir model.

Other flexible isotherm models extending the Langmuir model are based on the theory of heterogeneous surfaces (Jaroniec and Madey, 1988) and on concepts provided by statistical thermodynamics (Hill, 1960). The latter approach allows deriving the following second-order isotherm model that is capable to describe inflection points in the isotherm courses:

$$q = q_\text{sat} \frac{c(b_1 + 2 \cdot b_2 \cdot c)}{1 + b_1 \cdot c + b_2 \cdot c^2} \qquad (2.56)$$

2.5.2.2 Multicomponent Isotherms Based on the Langmuir Model

If mixtures of solutes are injected into a chromatographic system, not only interferences between the amount of each component and the adsorbent but also interferences between the molecules of different solutes occur. The resulting competition and displacement effects cannot be described with independent single-component isotherms. Therefore, an extension of single-component isotherms that also takes into account the interference is necessary. The competition can be included in the denominator of the multicomponent Langmuir isotherm equation (Equation 2.57):

$$q_i = q_{sat} \frac{b_i c_i}{1 + \sum_{j=1}^{n} b_j c_j} = \frac{H_i c_i}{1 + \sum_{j=1}^{n} b_j c_j} \tag{2.57}$$

A prerequisite for Equation 2.57 is the equality of the maximum loadability q_{sat} of all n solutes. In case of different loadabilities for the different solutes ($q_{sat,i} \neq q_{sat,j}$), Equation 2.57 no longer fulfils the Gibbs–Duhem equation and is thus thermodynamically inconsistent (Broughton, 1948). This is, for example, the case if solutes with substantially different molecular masses are separated on sorbents where the pore accessibility is hindered for large molecules.

The multicomponent Langmuir isotherm based on Equation 2.53 is shown in Equation 2.58 (Charton and Nicoud, 1995):

$$q_i = \lambda_i c_i + \frac{H_i c_i}{1 + \sum_{j=1}^{n} b_j c_j} \tag{2.58}$$

The bi-Langmuir isotherm (Equation 2.54) can be extended in the same way to give the multicomponent bi-Langmuir isotherm (Equation 2.59) (Guiochon et al. 2006):

$$q_i = q_{sat,1} \frac{b_{1,i} c_i}{1 + \sum_{j=1}^{n} b_{1,j} c_j} + q_{sat,2} \frac{b_{2,i} c_i}{1 + \sum_{j=1}^{n} b_{2,j} c_j} = \frac{H_{1,i} c_i}{1 + \sum_{j=1}^{n} b_{1,j} c_j} + \frac{H_{2,i} c_i}{1 + \sum_{j=1}^{n} b_{2,j} c_j} \tag{2.59}$$

Modified multicomponent Langmuir and multicomponent bi-Langmuir isotherms offer a large flexibility for adjustment to measured data if all coefficients are chosen individually. To limit the number of free parameters that need to be specified it is sometimes possible to use in Equations 2.58 and 2.59 constant adjustment terms ($\lambda_i = \lambda$) or equal saturation capacities ($q_{sat,1,i} = q_{sat,2,i} = q_{sat,i}$).

2.5.2.3 Competitive Isotherms Based on the Ideal Adsorbed Solution Theory

There are several methods available that allow the prediction of mixture isotherms based on general single-component information. An application can significantly reduce the necessary number of experiments. The most successful approach is the ideal adsorbed solution (IAS) theory initially developed by Myers and Prausnitz (1965) to describe competitive gas phase adsorption. This theory was subsequently extended by Radke and Prausnitz (1972) to quantify adsorption from dilute (i.e., also ideal) solutions.

Adsorption occurs in a three-phase system consisting of the solid (stationary) phase (adsorbent), the fluid (mobile) phase, and the adsorbed phase (Figure 2.1). According to Gibbs' classical theory, the adsorbed phase is viewed as a two-dimensional boundary layer between the other two phases. This allows describing adsorption as a two-phase process based on interactions between the adsorbed and the fluid phases. It is further assumed that the surface of the adsorbent, A, is identical for all components. Based on the fundamental property relation (Myers and Prausnitz, 1965), the starting point for the derivation of the IAS theory is the following expression for the free Gibbs' energy of the adsorbed phase:

$$dG_{ads} = -S_{ads}\,dT + A\,d\pi + \sum_{i=1}^{n} \mu_{i,ads}\,dq_i, \quad i = 1, \ldots, n \tag{2.60}$$

The two-dimensional spreading pressure π is an intensive property of the adsorbed phase. Considering the adsorption of just a single (superscript 0) component i and respecting the Gibbs–Duhem equation provides for constant temperature and pressure and equilibrium conditions the well-known *Gibbs' adsorption isotherm* (Myers and Prausnitz, 1965):

$$A\,d\pi_i = q_i^0\,d\mu_{i,ads} \tag{2.61}$$

In order to apply this expression, the better accessible identical chemical potentials in the fluid phase, μ_i, can be used to substitute the corresponding adsorbed phase chemical potentials:

$$\mu_{i,ads} = \mu_i = \mu_i^{ref} + RT\ln\frac{c_i^0}{c_i^{ref}} \quad \text{or} \quad d\mu_{i,ads} = RT\,d\ln c_i^0 \tag{2.62}$$

Integration of the *Gibbs' adsorption isotherm* over a running variable ξ leads to

$$\pi_i(c_i^0) = \frac{RT}{A}\int_0^{c_i^0} q_i^0(\xi)\,d\ln\xi = \frac{RT}{A}\int_0^{c_i^0}\frac{q_i^0(\xi)}{\xi}\,d\xi \tag{2.63}$$

The requirement of phase equilibrium in mixtures is established by stating that all the components in the adsorbed phase exert the same spreading pressure, that is:

$$\pi_j = \cdots = \pi_N, \quad j = 1, \ldots, n \tag{2.64}$$

or

$$\int_0^{c_1^0}\frac{q_1^0(\xi)}{\xi}\,d\xi = \int_0^{c_j^0}\frac{q_j^0(\xi)}{\xi}\,d\xi, \quad j = 2, \ldots, n \tag{2.65}$$

The single-component isotherm equations forming the input information for the IAS theory are the functions $q^0(\xi = c^0)$. Under ideal conditions the following analogy to Raoult's law exploiting the fractions of the components in the adsorbed phase x_i holds for each component in a mixture:

$$c_i = c_i^0 x_i, \quad i = 1, \ldots, n \tag{2.66}$$

Equations 2.65 and 2.66 contain the c_i^0 and the x_i as $2n$ unknowns. An important drawback of the IAS theory is related to the fact that these unknowns can often not be determined analytically and the application of numerical methods is needed.

Once the c_i^0 are known, the remaining explicit equations allowing to calculate the loadings q_i as a function of given fluid phase concentration c_i are

$$q_i = \left[\sum_{j=1}^{n} \frac{c_j}{c_j^0 \cdot q_j^0(c_j^0)}\right]^{-1} \frac{c_i}{c_i^0}, \quad i = 1, \ldots, n \tag{2.67}$$

Only for a small number of relatively simple single-component adsorption isotherm models analytical solutions of the set of IAS theory equations can be derived. This is possible, for example, for the Langmuir model. If the saturation capacities of all components in the corresponding single-component isotherm equations (Equation 2.51) are identical, the IAS theory generates the same competitive isotherm model as given by Equation 2.57.

Recently, Ilić, Flockerzi, and Seidel-Morgenstern (2010) derived a useful new explicit solution of the IAS theory equations for the calculation of adsorbed phase concentrations of binary systems whose individual component behavior can be represented by flexible second-order (quadratic) adsorption isotherms (Equation 2.56). In this case Equation 2.65 provides

$$\int_0^{c_1^0} \frac{b_{11} + 2 \cdot b_{12} \cdot \xi}{1 + b_{11} \cdot \xi + b_{12} \cdot \xi^2} d\xi = \int_0^{c_2^0} \frac{b_{21} + 2 \cdot b_{22} \cdot \xi}{1 + b_{21} \cdot \xi + b_{22} \cdot \xi^2} d\xi \tag{2.68}$$

After integration and exploitation of Equation 2.66 results:

$$1 + b_{11} \cdot \frac{c_1}{x_1} + b_{12} \cdot \left[\frac{c_1}{x_1}\right]^2 = 1 + b_{21} \cdot \frac{c_2}{1 - x_1} + b_{22} \cdot \left[\frac{c_2}{1 - x_1}\right]^2 \tag{2.69}$$

This equation provides the following third-order polynomial in terms of the adsorbed phase fraction of the first component x_1:

$$a_1(c_1, c_2) \cdot x_1^3 + a_2(c_1, c_2) \cdot x_1^2 + a_3(c_1, c_2) \cdot x_1 + a_4 = 0 \tag{2.70}$$

with

$$a_1(c_1, c_2) = \alpha \cdot c_1 + c_2$$

$$a_2(c_1, c_2) = c_1 \cdot [\beta \cdot c_1 - 2 \cdot \alpha] - c_2 \cdot [1 + \gamma \cdot c_2]$$

$$a_3(c_1, c_2) = c_1 \cdot [\alpha - 2 \cdot \beta \cdot c_1]$$

$$a_4(c_1, c_2) = \beta \cdot c_1^2$$

$$\alpha = \frac{b_{11}}{b_{21}} > 0, \quad \beta = \frac{b_{12}}{b_{21}} \geq 0, \quad \gamma = \frac{b_{22}}{b_{21}} \geq 0$$

The only physically meaningful solution for x_1 in the interval [0, 1] can be identified (Ilić, Flockerzi, and Seidel-Morgenstern, 2010) and the IAS theory delivers the following flexible and widely applicable explicit isotherm model:

$$q_i(c_1, c_2) = \left[\frac{x_1}{q_1^0(c_1^0)} + \frac{x_2}{q_2^0(c_2^0)}\right]^{-1} \cdot x_i \tag{2.71}$$

with

$$x_2 = 1 - x_1, \quad c_i^0 = \frac{c_i}{x_i}, \quad i = 1, 2$$

Numerous further attempts were undertaken in order to facilitate the solution of the set of IAS equations. Typically, iterative Newton-type methods have been applied to solve the equations in their original form or after suitable transformations. Hereby, a crucial point is always the selection of initial estimates leading to convergence of the iteration schemes applied. A new approach, based on reformulating the equations into a dynamic problem, avoids this drawback (Rubiera et al., 2012). The following simple decoupled system of ordinary differential equations describes the orbit along a running variable φ building up the spreading pressure:

$$\frac{dc_1^0}{d\varphi} = 1$$

$$\frac{dc_i^0}{d\varphi} = \frac{q_1^0(\varphi)/c_1^0}{q_i^0(\varphi)/c_i^0}, \quad i = 2, \ldots, n \qquad (2.72)$$

The integration starts at well-defined initial values:

$$c_i^0(\varphi = 0) = 0, \quad i = 1, \ldots, n \qquad (2.73)$$

and stops when the closure condition originating from Equation 2.66 is fulfilled, that is, if the following holds:

$$\sum_i^n \frac{c_i}{c_i^0(\varphi^{\text{stop}})} = 1 \qquad (2.74)$$

After specifying the c_i^0, it remains again just solving the explicit equations (2.67). The described algorithm can be efficiently applied to solve the set of IAS equations for arbitrary increasing single-component adsorption isotherm models and any component number n.

2.5.2.4 Steric Mass Action Isotherms for Ion Exchange Equilibria

The state of equilibrium in ion exchange chromatography is currently described by stoichiometric models where the solute, for example a protein, displaces a stoichiometric number of salt ions bound on the ion exchanger. A basic concept is the stoichiometric displacement model developed by Kopaciewicz et al. (1983). For monovalent counterions the reaction is described as follows:

$$c_i + \nu_i q_S = q_i + \nu_i c_S \qquad (2.75)$$

where q_S is the salt load of the adsorbent, q_i the solute load of the adsorbent, c_i the concentration of the solute, and c_S the salt concentration in the mobile phase. The characteristic charge representing the number of sites on the resin surface the solute interacts with is given by ν_i. Due to the mass action law the

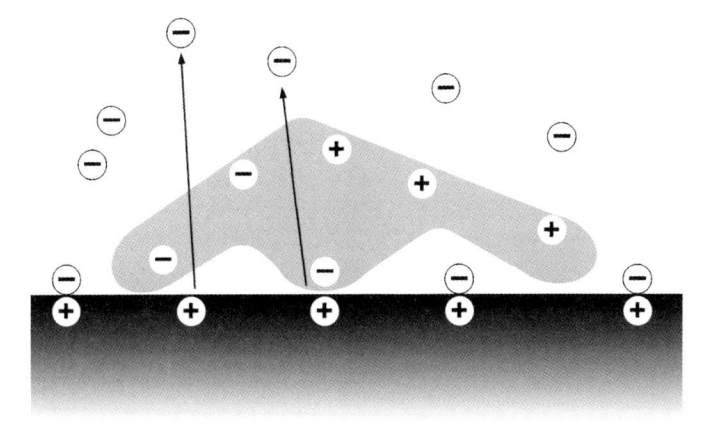

Figure 2.17 Protein adsorption on ion exchange resin with steric shielding.

equilibrium constant is

$$K_{M,i} = \frac{q_i}{c_i} \left(\frac{c_S}{q_S}\right)^{\nu_i} \tag{2.76}$$

Brooks and Cramer (1992) extended this model by the steric shielding of ion exchange groups by large solutes such as proteins as illustrated in Figure 2.17.
Because of the shielding, the equilibrium constant is given as

$$K_{M,i} = \frac{q_i}{c_i} \left(\frac{c_S}{\hat{q}_S}\right)^{\nu_i} \tag{2.77}$$

with the free accessible salt load $\hat{q}_S$. The total ion exchange capacity, that is, the total number of binding sites on the resin surface, is

$$\Lambda = \hat{q}_S + \sum_{i=2}^{N_{comp}} (\nu_i + \sigma_i) q_i \tag{2.78}$$

where σ_i is the steric shielding parameter and ν_i the characteristic protein charge. The component index for the salt is $i = 1$. From Equations 2.77 and 2.78 results the implicit form of the steric mass action (SMA) isotherm that has the same concave curvature as Langmuir isotherms for high concentrations:

$$c_i = \frac{1}{K_{M,i}} \frac{q_i (c_S)^{\nu_i}}{\left(\Lambda - \sum_{i=2}^{N_{comp}} (\nu_i + \sigma_i) q_i\right)^{\nu_i}} \tag{2.79}$$

For dilute protein solutions or at high salt concentration the protein load is very small so that

$$\Lambda \gg \sum_{i=2}^{N_{comp}} (\nu_i + \sigma_i) q_i \tag{2.80}$$

and similarly to Langmuir isotherms the SMA isotherm reduces for $c_i \to 0$ to a linear function:

$$q_i = K_{M,i} \left(\frac{\Lambda}{c_S}\right)^{\nu_i} c_i \tag{2.81}$$

Brooks and Cramer (1992) recommend experimental procedures such as using isocratic/gradient elution chromatography to determine the parameters $K_{M,i}$, ν_i, and σ_i. SMA isotherms do not take into account an influence of the pH value. Corresponding extensions have been developed by Gerstner, Bell, and Cramer (1994) and Bosma and Wesselingh (1998).

2.6
Thermodynamic Effects on Mass Separation

2.6.1
Mass Load

In analytical chromatography typically very dilute samples are injected into the column. Therefore, one stays within the linear range of isotherms and retention times are independent of the amount injected. The concentration profiles are symmetrical and Gaussian (Figure 2.18a). The sample mass for a certain component i, m_i, is given by Equation 2.82:

$$m_{i,inj} = c_{i,inj} V_{inj} \tag{2.82}$$

where c_{inj} is the sample concentration and V_{inj} the sample volume. Typically, 10–100 µl of a sample mixture is injected into an analytical column (length: 100–250 mm; diameter: approximately 4 mm). The concentrations of solutes are often around 1 mg ml^{-1}. When one increases the sample concentrations at constant sample volume, the peak profiles change as indicated in Figure 2.18b. A differently shaped profile is obtained when the sample volume is increased at constant sample concentration (Figure 2.18c). The first case is called concentration overload or mass overload and the second case is called volume overload. In a third case of overloading both sample concentrations and sample volumes are changed.

The term "overload" has been introduced by analytical chemists. For analysis a column should not be overloaded in order to achieve constant retention times for reproducible analytical detection of each component peak. Preparative chromatography has a different aim, which is typically to achieve a high "productivity." To achieve this goal the columns are operated under so-called "overloaded" conditions. From the engineering viewpoint overloaded systems are nonlinear because nonlinear isotherms connect the phases involved. A suitable measure of overloading is provided by the loading factor, which is the ratio of injected sample amount to the corresponding column saturation capacity (Guiochon et al., 2006). A column is sometimes called overloaded by of analytical chemists when the injection causes a

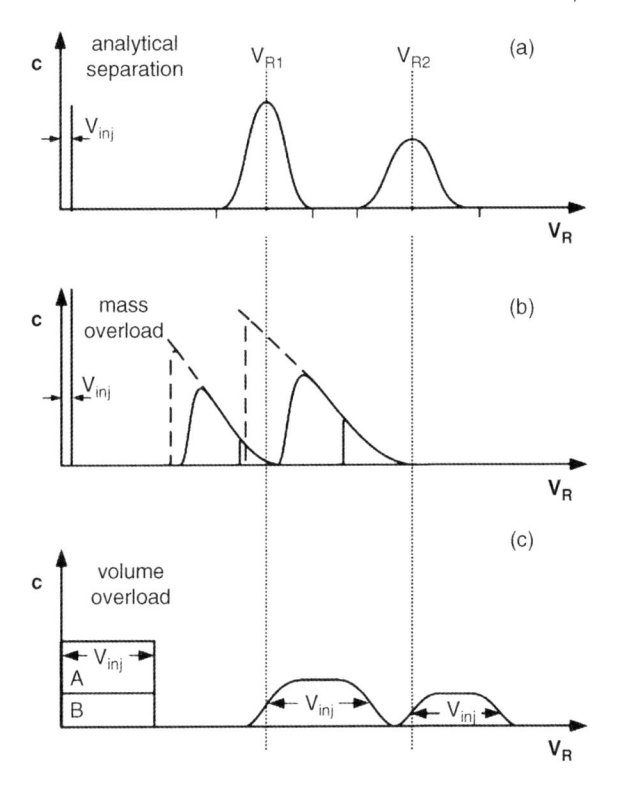

Figure 2.18 Overloaded chromatographic columns.

10% decrease in the retention factor k' or a 50% decrease of the plate number N compared to analytical conditions.

2.6.2
Linear and Nonlinear Isotherms

In many cases the adsorption isotherms are of the Langmuir type with a steep slope at the initial part and reaching a saturation value at higher amounts of adsorbed solute. The local slopes of the isotherms correspond to distribution coefficients K, discriminating two parts of the isotherm: the linear and nonlinear parts. In the linear range K is equal to the Henry constant. For the nonlinear part K becomes smaller and reaches a limiting value at high solute concentrations.

Analytical chromatography aims to achieve an adequate, not necessarily a maximum, resolution of solute bands to identify and to quantify the analytes based on their retention coefficient, peak height, and peak area. Information on the analytes is the target.

In preparative chromatography the goal is to purify and to isolate compounds at a high yield, high purity, or high productivity. Productivity is the major goal, being the mass of isolate (target compound) per unit mass of packing and per unit time. To increase productivity samples are applied on the column with much higher concentrations than in analytical chromatography. One should keep in mind that at higher column loading the selectivity coefficients as well as the column plate numbers change. Typically, the gain in selectivity is the major objective, which is controlled by the choice of eluent and packing.

There are additional effects occurring during the elution of a binary mixture at higher concentration depending on the specific thermodynamics described by the isotherms and their coefficients as well as the relative mass ratio of the two components in the mixture. Two of the most prominent effects are the displacement effect and the tag-along effect (Guiochon et al., 2006). Both are ruled by the competition of the more and the less retained molecules with the interaction sites of the adsorbent. Depending on the mass load and the composition of the feed, it results in quite different elution profiles.

To explain these effects Figure 2.20 depicts elution profiles that have been theoretically calculated based on the transport dispersive model (Chapter 6). The isotherms are described by a multicomponent Langmuir isotherm and the mass ratio of the feed concentration is varied from 1:9 to 9:1.

In the first case a small amount of the earlier eluting component A (dashed line) is displaced by a large amount of the second eluting component B (Figure 2.20a). By reducing the retention time of this low-affinity component the elution profile is sharpened and productivity increased. Characteristic for displacement is the strong reduction of the isotherm of the first eluting component in the mixture situation compared to the single-component behavior, while the isotherms of the late eluting component B in the mixture and for the pure component are nearly the same (Figure 2.19a). To exploit the displacement effect the chromatographic systems can also be operated by adding an additional displacer with well-defined adsorption characteristics. An easier and more economic way is to take advantage of the described self-displacement effect.

If the two components are present in similar amounts in the feed, both equilibrium loadings are reduced compared to the corresponding single-component isotherms. The corresponding elution profiles are deformed in a similar way and have an enhanced tendency to overlap (Figures 2.19b and 2.20b).

The second, so-called tag-along effect can occur if the feed composition is dominated by the first eluting component. Under these conditions the highly concentrated molecules of the first component A desorb the molecules of the second component, thus forcing them into the mobile phase and resulting in shorter retention times of the second component. Analogous to case (a) the isotherm for the late eluting component B is strongly influenced by the presence of large amounts of A crossing the mixture isotherm of this component (Figure 2.19c). The tag-along effect causes a long and undesired plateau of the second component under the elution profile of the first eluting component (Figure 2.20c).

The two described effects origin from nonlinear thermodynamic interactions between the components. If in a concrete separation the peak resolution is not high

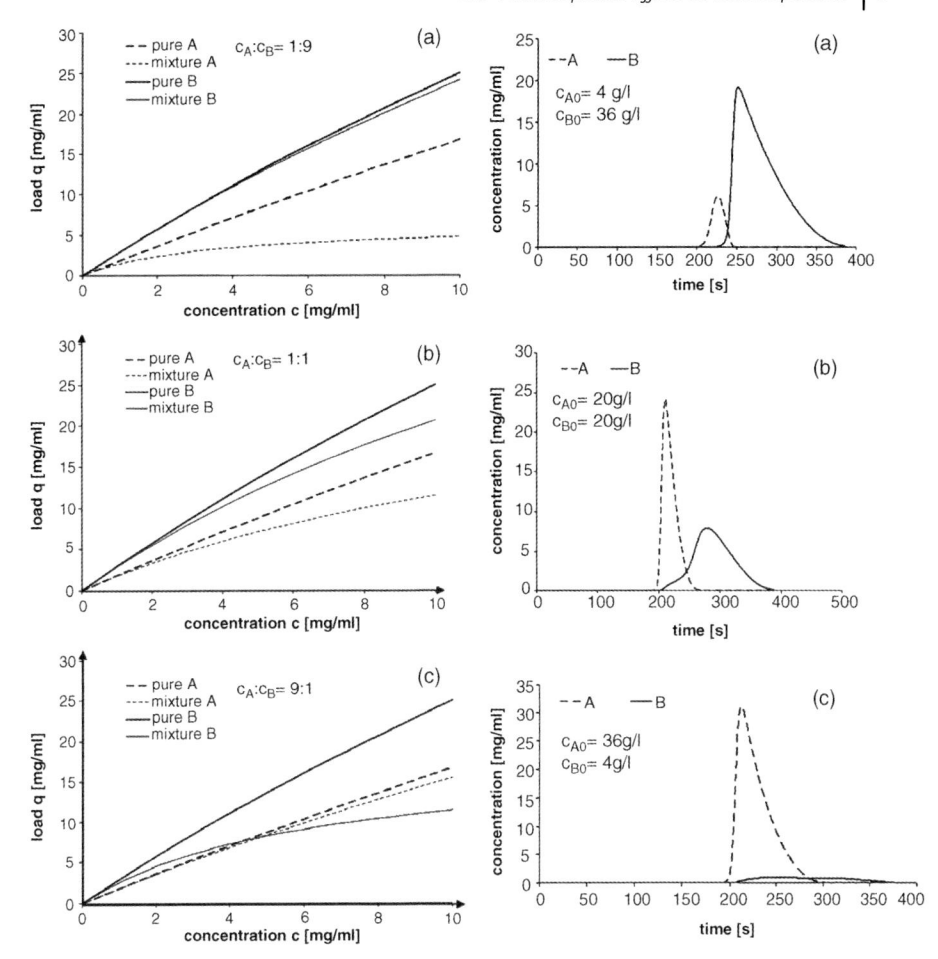

Figure 2.19 Single component Langmuir isotherms and multicomponent Langmuir isotherm for different mixture compositions.

Figure 2.20 Elution profiles for different mass ratios of the feed mixtures.

enough, there is no way to overcome this problem by increasing column efficiency. If a tag-along effect occurs, the use of a smaller particle diameter has no effect, as long as the surface chemistry of the stationary phase and the mobile phase composition are kept constant. In such cases, only a systematic screening for an improved chromatographic system can lead to a more productive separation process.

2.6.3
Elution Modes

Two basic elution modes can be applied in operating column chromatography: isocratic elution and gradient elution (Guiochon et al., 2006). In isocratic elution the composition of the eluent is kept constant throughout the operation. In (solvent) gradient elution the eluent composition is changed during the separation, such that one starts with a weak solvent A and constantly adds volumes of a stronger solvent B. Frequently a linear gradient is applied. Alternatively, a stepwise gradient can be used.

At isocratic elution the peaks broaden and become flatter with increasing retention, and might even disappear in the baseline noise. As a rule of thumb, the retention factor k' of the last eluting solute in isocratic elution should not be larger than 10. This corresponds to an elution volume of less than 10 column volumes. Very strongly retained solutes may not be eluted and, hence, the column has to be washed with a strong solvent from time to time.

At gradient elution the solvent strength is changed continuously or stepwise. As a result of the gradient, the peaks become sharper and higher and one can cover several orders of magnitude of the retention coefficient. Gradient elution is commonly applied at the screening of a complex mixture to receive information on the number and polarity of the components. A gradient accelerates the migration speed of the late-eluting solutes. An example is given in Figure 2.21. After running the gradient one has to go back to the initial conditions, which means that the column has to be washed and reconditioned. Parameters to adjust a gradient with respect to optimum resolution are the starting and final eluent composition and the gradient time and steepness. Problems may occur with respect to the reproducibility of the eluent composition at low amounts (volumes) of solvents A and B,

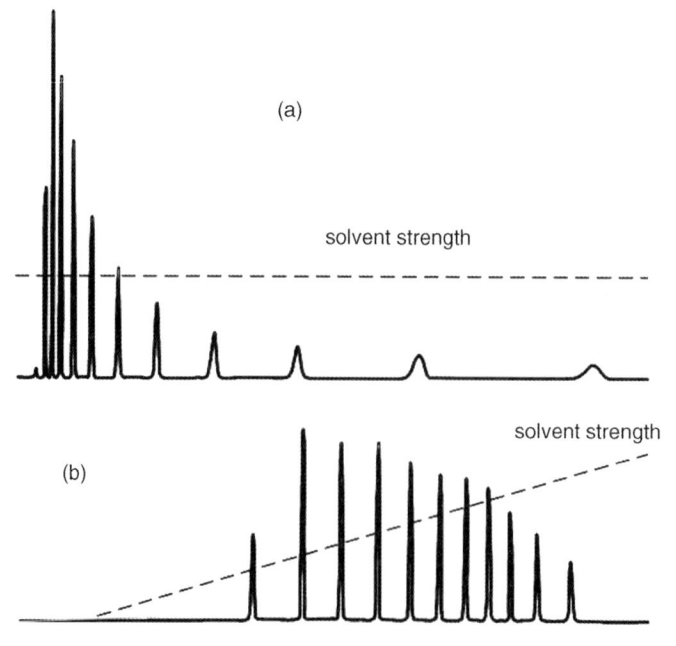

Figure 2.21 Chromatograms for (a) isocratic elution and (b) gradient elution.

respectively, in a binary mixture. Gradient elution is a preferred technique in analytical HPLC and is less frequently used at preparative scale. One disadvantage is that one ends up with solvent mixtures, which makes distillation and reuse difficult and expensive.

References

Atkins, P.W. (1990) *Physikalische Chemie*, Wiley-VCH Verlag GmbH, Weinheim.

Bellot, J.C. and Condoret, J.S. (1993) Modelling of liquid chromatography equilibria. *Process Biochem.*, **28**, 365–376.

Bidlingmeyer, B.A. and Waren, F.V. (1984) Column efficient measurement. *Anal. Chem.*, **56**, 1588–1595.

Bosma, J. and Wesselingh, J. (1998) pH dependence of ion-exchange equilibrium of proteins. *AIChE J.*, **50**, 848–853.

Brauer, H. (1971) *Grundlagen der Einphasen- und Mehrphasenströmungen*, Verlag Sauerländer, Aarau, Frankfurt am Main.

Brooks, C. and Cramer, S. (1992) Steric-mass-action ion exchange: displacement profiles and induced salt gradients. *AIChE J.*, **38**, 1968–1978.

Broughton, D.B. (1948) Adsorption isotherms for binary gas mixtures. *Ind. Eng. Chem.*, **40** (8), 1506–1508.

Brunauer, S., Emmett, P.H., and Teller, E. (1938) Adsorption of gases on multimolecular layers. *J. Am. Chem. Soc.*, **60** (2), 309–319.

Charton, F. and Nicoud, R.-M. (1995) Complete design of a simulated moving bed. *J. Chromatogr. A*, **702**, 97–112.

Everett, D.H. (1984) Thermodynamics of adsorption from solutions, in *Fundamentals of Adsorption Processes* (eds A.L. Meyers and G. Belfort), Engineering Foundation, New York.

Gerstner, J., Bell, J., and Cramer, S. (1994) Gibbs free energy of adsorption for bio molecules in ion exchange systems. *Biophys. Chem.*, **52**, 97–106.

Giles, C.H., Smith, D., and Huitson, A.J. (1974) General treatment and classification of the solute adsorption isotherm. I. Theoretical. *Colloid Interface Science.*, **47**, 755–765.

Graham, D. (1953) The characterization of physical adsorption systems. I. The equilibrium function and standard free energy of adsorption. *J. Phys. Chem.*, **57**, 665–669.

Guiochon, G., Felinger, A., Shirazi, D.G., and Katti, A.M. (2006) *Fundamentals of Preparative and Nonlinear Chromatography*, Elsevier/Academic Press, Amsterdam.

Guiochon, G., Golshan-Shirazi, S., and Katti, A. (1994) *Fundamentals of Preparative and Nonlinear Chromatography*, Academic Press, Boston.

Hill, T.L. (1960) *An Introduction to Statistical Thermodynamics*, Addison-Wesley, Reading, MA.

Ilić, M., Flockerzi, D., and Seidel-Morgenstern, A. (2010) A thermodynamically consistent explicit competitive adsorption isotherm model based on second-order single component behaviour. *J. Chromatogr. A*, **1217**, 2132–2137.

IUPAC (1972) Manual of symbols and terminology, appendix 2, Pt. 1, colloid and surface chemistry. *Pure Appl. Chem.*, **31**, 578.

Jaroniec, M. and Madey, R. (1988) *Physical Adsorption on Heterogeneous Solids*, Elsevier, Amsterdam.

Knox, J.H. (2002) Band dispersion in chromatography – a universal expression for the contribution from the mobile zone. *J. Chromatogr. A*, **960**, 7–18.

Knox, J.H., Laird, G., and Raven, P. (1976) Interaction of radial and axial dispersion in liquid chromatography in relation to the "infinite diameter effect". *J. Chromatogr.*, **122**, 129–145.

Kopaciewicz, W., Rounds, M., Fausnaugh, J., and Regnier, F. (1983) Retention model for high performance ion-exchange chromatography. *J. Chromatogr.*, **266**, 3–21.

Kümmel, R. and Worch, E. (1990) *Adsorption aus Wäßrigen Lösungen*, VEB Deutscher Verlag für Grundstoffindustrie, Leipzig.

Ludemann-Hombourger, O., Bailly, M., and Nicoud, R.-M. (2000a) Design of a simulated moving bed: optimal size of the stationary phase. *Sep. Sci. Technol.*, **35** (9), 1285–1305.

Ludemann-Hombourger, O., Bailly, M., and Nicoud, R.-M. (2000b) The VARICOL-process: a new multicolumn continuous chromatographic process. *Sep. Sci. Technol.*, **35** (12), 1829.

Myers, A.L. and Prausnitz, J.M. (1965) Thermodynamics of mixed-gas adsorption. *AIChE J.*, **11** (1), 121–127.

Radke, C.J. and Prausnitz, J.M. (1972) Thermodynamics of multi-solute adsorption from liquid solutions. *AIChE J.*, **18** (1), 761–768.

Rubiera Landa, H.O., Flockerzi, D., and Seidel-Morgenstern, A. (2011) A method for efficiently solving the IAST equations with an application to adsorber dynamics, AIChE Journal, 2012, doi:10.1002/aic.13894.

Ruthven, D.M. (1984) *Principles of Adsorption and Adsorption Processes*, John Wiley & Sons, Inc, New York.

Sattler, K. (1995) *Thermische Trennverfahren – Grundlagen, Auslegung, Apparate*, Wiley-VCH GmbH, Weinheim.

Seidel-Morgenstern, A. and Nicoud, R.-M. (1996) Adsorption isotherms: experimental determination and application to preparative chromatography. *Isolation Purif.*, **2**, 165–200.

Toth, J. (1971) State equations of the solid–gas interface layers. *Acta Chim. Acad. Sci. Hung.*, **69**, 311.

Tsotsas, E. (1987) *Über die Wärme- und Stoffübertragung in durchströmten Festbetten*, VDI Fortschrittsberichte, 3rd Series, No. 223, VDI-Verlag, Düsseldorf.

Van Deemter, J.J., Zuiderweg, F.J., and Klinkenberg, A. (1956) Longitudinal diffusion and resistance to mass transfer as causes of nonideality in chromatography. *Chem. Eng. Sci.*, **5**, 271–289.

3
Stationary Phases and Chromatographic Systems

Michael Schulte, Matthias Jöhnck, Romas Skudas, Klaus K. Unger, Cedric du Fresne von Hohenesche*, Wolfgang Wewers*, Jules Dingenen, and Joachim Kinkel*

3.1
Column Packings

3.1.1
Survey of Packings and Stationary Phases

Adsorbents can be grouped according to their chemical composition and mode of interaction. In general, one distinguishes between inorganic types such as active carbon, zeolites, porous glass, and porous oxides, for example, silica, alumina, titania, zirconia, and magnesia, and the families of cross-linked organic polymers. The former are classical adsorbents and possess a crystalline or amorphous bulk structure. They possess hydrophobic as well as hydrophilic surface properties. The main criterion that distinguishes them from cross-linked polymers is their high bulk density and the porosity, which is permanent except under certain conditions, for example, very high pressures. The texture of inorganic adsorbents resembles a corpuscular structure rather than a cross-linked network more closely.

The mechanical strength of polymers is achieved through extensive cross-linking whereby a three-dimensional network of hydrocarbon chains is formed. Depending on the extent of cross-linking, soft gels, for example, agarose, and highly dense polymer gels are obtained. The bulk density is much lower than that of inorganic adsorbents. The hydrophilicity and hydrophobicity of cross-linked polymer gels are tuned by the chemical composition of the backbone polymer as well as by the surface chemistry. Organic chemistry provides enormous scope in designing the functionality of the surface of polymeric adsorbents. Even at a high degree of cross-linking, polymers, still show a swelling porosity, that is, the porosity of a polymeric adsorbent depends on the type of solvent. For example, the volume of a soft gel can be ten times higher when immersed in a solvent than in the dry state. For both types of adsorbents the porosity and pore structure can be manipulated by additives such as templates, volume modifiers (so-called porogens), and other additives.

*These authors have contributed to the first edition.

Preparative Chromatography, Second Edition. Edited by H. Schmidt-Traub, M. Schulte, and A. Seidel-Morgenstern.
© 2012 Wiley-VCH Verlag GmbH & Co. KGaA. Published 2012 by Wiley-VCH Verlag GmbH & Co. KGaA.

A note associated with hydrophobicity and hydrophilicity: these are frequently employed in characterizing adsorbents. Hydrophilicity/hydrophobicity is a qualitative measure of an adsorbent, characterizing its behavior towards water. The term lipophilicity/lipophobicity is applied to characterize the polarity of a compound. The overall parameter of hydrophobicity and hydrophilicity is a combination of different interaction principles between the sorbent surface and the adsorbed molecule. The most prominent interaction forces are hydrogen bonds, van der Waals forces, ionic and Π–Π-interactions. For the characterization of solvents, Reichardt introduced a betaine dye as a solvatochromic indicator which has since been known as Reichardt's dye (Reichardt, 2003). This principle has been extended to solid surfaces giving a value for the average surface functionality of the sorbent (Macquarrie et al., 1999). With hydroxyl functions dominating the interaction the obtained value can be nevertheless used as a simple correlation to the composition of the sorbent surface. Figure 3.1 gives a summary of adsorbents used in preparative chromatography and their main interaction principles.

3.1.2
Generic, Designed, and Customized Adsorbents

Another classification of adsorbents which considers their use in preparative chromatography is the division into generic, designed, and customized stationary phases. Generic stationary phases include nonsophisticated adsorbents that are made available in large quantities and are used for rather simple purification processes. In contrast, customized stationary phases represent adsorbents with high specificity and are employed in small-scale isolations unless customized for very large-scale at reasonable costs. Designed stationary phases are located in-between the former groups. "Design" usually refers to modification of the adsorbents surface chemistry to match a given isolation process for a specific group of target molecules.

The basic features for preparative chromatography are their applicability, selectivity, and specificity, which affect the column performance, productivity, and cost. Furthermore, it is important to know how flexible the materials have to be with respect to integration into a given process chain and if bulk quantities are available when a scale-up of the purification and isolation is considered.

3.1.2.1 Generic Adsorbents
Generic adsorbents usually represent bulk materials of low cost produced in several thousand metric tons per year. They are employed in industrial purification processes, for example, in cleaning drinking water or in drying air or removing pollutants. Typical examples are active carbons and bentonites. Various types of materials have been surveyed by Nawrocki et al. (1993), Kurganov et al. (1996), and Yang (2003) and are summarized in Table 3.1.

3.1.2.1.1 **Activated Carbons** Activated carbons are the most widely used adsorbents in gas and liquid-adsorption processes. They are manufactured from

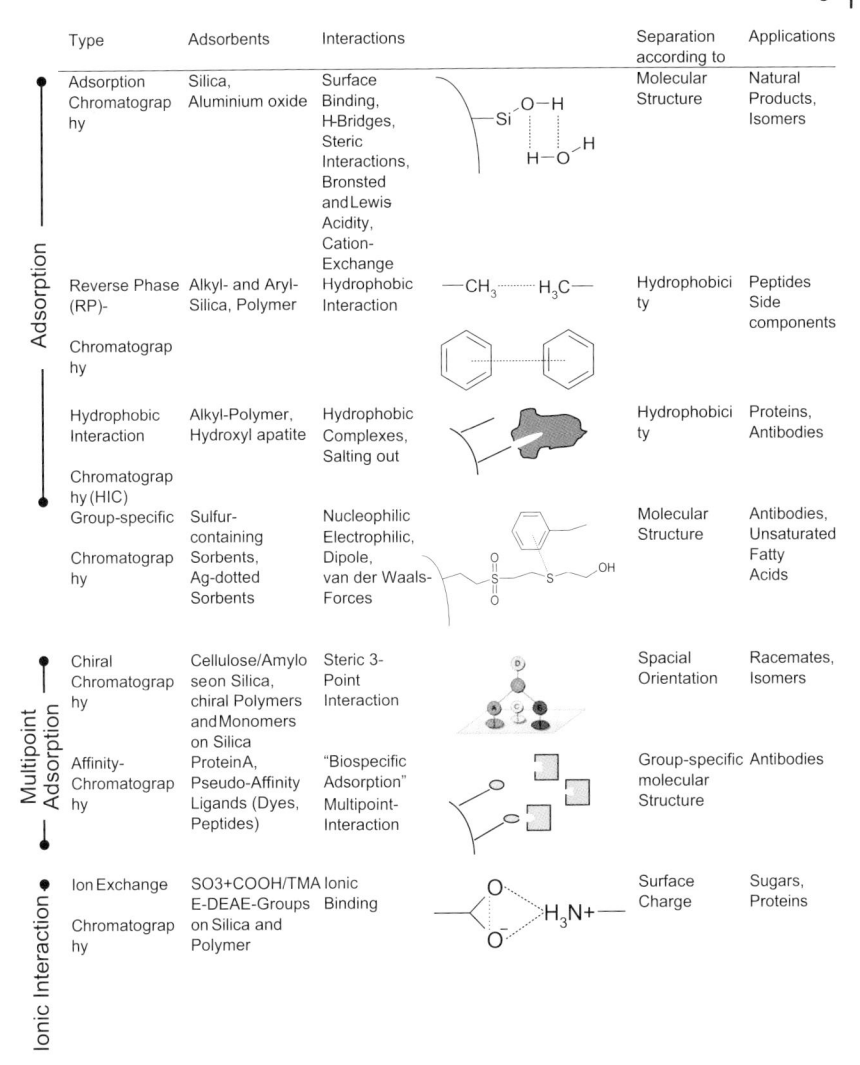

	Type	Adsorbents	Interactions		Separation according to	Applications
Adsorption	Adsorption Chromatography	Silica, Aluminium oxide	Surface Binding, H-Bridges, Steric Interactions, Bronsted and Lewis Acidity, Cation-Exchange		Molecular Structure	Natural Products, Isomers
	Reverse Phase (RP)- Chromatography	Alkyl- and Aryl-Silica, Polymer	Hydrophobic Interaction		Hydrophobicity	Peptides Side components
	Hydrophobic Interaction Chromatography (HIC)	Alkyl-Polymer, Hydroxyl apatite	Hydrophobic Complexes, Salting out		Hydrophobicity	Proteins, Antibodies
	Group-specific Chromatography	Sulfur-containing Sorbents, Ag-dotted Sorbents	Nucleophilic Electrophilic, Dipole, van der Waals-Forces		Molecular Structure	Antibodies, Unsaturated Fatty Acids
Multipoint Adsorption	Chiral Chromatography	Cellulose/Amyloseon Silica, chiral Polymers and Monomers on Silica	Steric 3-Point Interaction		Spacial Orientation	Racemates, Isomers
	Affinity-Chromatography	Protein A, Pseudo-Affinity Ligands (Dyes, Peptides)	"Biospecific Adsorption" Multipoint-Interaction		Group-specific molecular Structure	Antibodies
Ionic Interaction	Ion Exchange Chromatography	SO3+COOH/TMA E-DEAE-Groups on Silica and Polymer	Ionic Binding		Surface Charge	Sugars, Proteins

Figure 3.1 Types of chromatographic sorbents and their main interaction principles.

carbonaceous precursors by a chain of chemical and thermal-activation processes. The temperature for carbonization and activation reaches up to 1100 °C in thermal processes. Activated carbons develop a large surface area, between 500 and 2000 m^2 g^{-1}, and micropores with an average pore diameter <2 nm. Mesoporosity and macroporosity are generated by secondary procedures such as agglomeration. The products are shaped as granules, powders, and pellets depending on their application.

The bulk structure is predominantly amorphous and the surface is hydrophobic. During activation with polar oxygen functional groups such as hydroxyl, both carbonyl and carboxyl functions are formed which act as acidic and basic surface sites.

Table 3.1 Survey of the generic types of inorganic packings for chromatography and their characteristic properties.

Designation	Bulk composition and bulk structure	Surface characteristics
Activated carbons	Microcrystalline carbon	Hydrophobic, terminating polar groups such as hydroxyl, carboxyl, and so on
Zeolites	Crystalline alumosilicates, three-dimensional network of silica and alumina with adjusted silica to alumina ratio	Hydrophobic or hydrophilic depending on the silica/alumina ratio, cation exchanger, Brønsted and Lewis acidity
Porous glass	Silicates	Hydrophilic, cation exchanger, Brønsted acid sites
Porous silica	Amorphous, partially crystalline	Hydrophilic, cation exchanger, Brønsted acid sites, point of zero charge at pH 2–3, pK_a of Brønsted acid sites 7
Porous titania	Anatas	Brønsted and Lewis acid and basic sites, cationic and anionic exchange groups, point of zero charge at pH 5, pK_a of Brønsted acid sites 0.2–0.5
Porous zirconia	Crystalline, monoclinic	Brønsted and Lewis acid and basic sites, cationic and anionic exchange groups, point of zero charge at pH 10–13, pK_a of Brønsted acid sites 7
Porous alumina	γ-alumina	Brønsted and Lewis acid and basic sites, cationic and anionic exchange groups, point of zero charge at pH 7, pK_a of Brønsted acid sites 8.5

As a result, hydrophilicity is introduced to the surface. The world production of activated carbons in 2002 was estimated to be about 750 000 metric tons. There is discrimination between gas- and liquid-phase carbons. Typical liquid-phase applications are: potable water treatment, groundwater remediation, and industrial and municipal waste-water treatment and sweetener decolorization. Gas-adsorption applications are: solvent recovery, gasoline emission control, and protection against atmospheric contaminants.

3.1.2.1.2 **Synthetic Zeolites** Zeolites represent a family of crystalline alumosilicates with a three-dimensional structure. They possess regularly shaped cavities with eight-, ten-, and twelve-membered silicon–oxygen rings. The pore openings range from 0.6 nm (10-ring) to 0.8 nm (12-ring). Zeolites are cation exchangers. The exchange capacity is controlled by the aluminum content. Silica-rich zeolites, for example, MFI-type, possess a hydrophobic surface. Zeolites are made from water-glass solutions under alkaline conditions. An amorphous gel is first formed, which is subjected to hydrothermal treatment. The amorphous gel then converts

Table 3.2 Survey of the type of synthetic zeolites and their applications.

Structure	Cation	Window		Effective channel diameter (nm)	Applications
		Obstructed	Free		
4A	Na^+	8-ring		0.38	Desiccant; CO_2 removal; air separation (N_2)
5A	Ca^{2+}		8-ring	0.44	Linear paraffin separation; air separation (O_2)
3A	K^+	8-ring		0.29	Drying of reactive gases
13X	Na^+		12-ring	0.84	Air separation (O_2); removal of mercaptans
10X	Ca^{2+}	12-ring		0.80	
Silicalite			10-ring	0.60	Removal of organics in aqueous systems

into crystallites 0.1–5 μm in size. Silica-rich zeolites are manufactured in the presence of a structure-directing template, for example, tetraalkylammonium salts. Table 3.2 gives a survey of the type of zeolites and their applications (Ruthven, 1997).

The most important applications are the UOP Sorbex processes (Ruthven, 1997). These are automated continuous processes that separate hydrocarbon mixtures mainly in the liquid phase.

3.1.2.1.3 Porous Oxides: Silica, Activated Alumina, Titania, Thoria, Zirconia

Porous inorganic oxides are made through a sol–gel process. The sol is converted into a hydrogel that is subjected to dehydration to form a porous xerogel. Special techniques have been developed to combine the sol–gel transition with an aim to develop spherical shaped particles (Figure 3.2).

The starting material for processing porous alumina is crystalline gibbsite, $Al(OH)_3$, which is subjected to a heat treatment, under controlled conditions, between 500 and 800 °C. During this process, water is released, leaving crystalline porous alumina (γ-alumina) (Unger, 1990).

3.1.2.1.4 Porous Glasses

Porous glass is a glass manufactured from silica (50–75%), sodium oxide (1–10%), and boric acid (to 100%) by special heat treatment and leaching processes where the porosity and pore size is adjusted and controlled (Janowski and Heyer, 1982; Schnabel and Langer, 1991). The pore size distributions of these controlled pore glasses (CPG) named sorbents is very homogeneous and can be varied from 7.5 to 400 nm. The particle shape due to the manufacturing process is irregular, but special treatments to obtain spherical materials are known. The high mechanical strength and defined chemical composition are the advantages of CPG. Modification chemistry is based on silane linkers comparable to those technologies described for modified silica gels. As silica gels CPG does not remain alkali-resistant above pH 8 but cleaning and sanitization procedures based

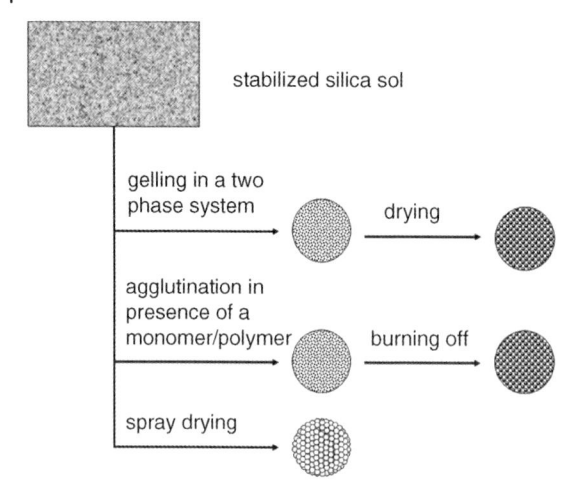

Figure 3.2 Procedures employed to manufacture spherical silica particles.

on acidic solvents are well described and widely used in the biopharmaceutical industry.

3.1.2.1.5 Cellulose/Agarose/Dextrans Between 1956 and 1962 three hydrophilic natural polymers had been introduced as chromatographic sorbents:

- Cellulose by Peterson and Sobers in 1956;
- Cross-linked dextran by Porath and Flodin in 1960; and
- Beaded agarose by Hjerten in 1962.

These polymers show low unspecific binding to proteins due to their hydrophilic nature but are of limited stability when operated in large-scale columns. Due to the weight of the packing material the packed bed is compressed and the linear flow rate has to be drastically reduced. To overcome this drawback, some materials have been cross-linked to stabilize the particles. These classical type materials have been widely reviewed elsewhere (Jungbauer, 2005; Curling and Gottschalk, 2007).

3.1.2.1.6 Hydrophilic Polymers Other hydrophilic base polymers, for example, poly-methacrylates are mainly used as functionalized sorbents and will be discussed in Section 3.1.4.3.

3.1.2.1.7 Hydroxyapatite Hydroxyapatite, which has the formula of $Ca_{10}(PO_4)_6(OH)_2$, is widely used as a sorbent for hydrophobic interaction chromatography. Hydroxyapatite is an underivatized matrix with a specific inherent surface functionality. In the 1980s, it had been introduced as ceramic beaded material with good physical strength (Kato, Nakamura, and Hashimoto, 1987). The interactions of the surface phosphate and calcium-ions with the amino and carboxyl groups of proteins are complex and not fully understood. The elution of proteins is typically performed at a pH of 6.8. Even though it has been reported that lower pH is preferable with respect to separation it has to be kept in mind that the crystalline character of hydroxyapatite is rapidly destroyed at a pH below 5.

3.1.2.1.8 **Styrene–Divinylbenzene Copolymers and Derivatives** The first styrene–divinylbenzene copolymers had been developed by Griesbach at Farbenfabrik Wolfen in 1937. These classical adsorbents are widely used in water deionization as well as in applications such as sugar separations and the separations of different ions, for example, in the nuclear industry and for the separation of rare-earth elements. Classical ion-exchange resins are typically produced in the mm-size and show a strong swelling behavior in different solvents. Due to their low-cross-linked polymer structure they can only withstand a moderate pressure drop and are not intended for use in high-pressure liquid chromatography. They are directly modified mainly with sulfo-groups (cation-exchangers) and quaternary ammonium-groups (anion-exchangers). Today, the main products on the market are Lewatit[R] (Lanxess), Dowex[R] (Dow Chemicals), Amberlite[R] (Rohm and Haas), and Purolite[R] (Purolite).

Purolite has recently introduced a core–shell (or shallow shell) technology resin with the Purolite SST product range. Taking into account that the volume of the inner core of a particle is very small and thus not contributing much to the overall capacity of the resin and on the other hand the diffusion path length towards this inner core is the longest the idea of a core–shell particle is quite convincing. On a hard inner core, which increases the overall particle stability, a porous shell is constructed. Purolite claims that without significantly reducing the resins capacity, the regeneration costs for the resins can be reduced by 20–50% because of the lower amount of regeneration solvent that is needed.

The synthesis of high-pressure stable monodisperse polystyrene-based chromatographic resins is based on the work of Ugelstad (1984). The synthesis starts with a monodisperse, non cross-linked submicron seed particle that is grown by a swelling process up to diameters typically ranging between 5 and 15 μm. The swelling monomer mixture typically consists of styrene and divinylbenzene monomers as well as the porogen solvent mixture determining the later pore structure. Resins of this type are, for example, Source (GE Healthcare) Amberchrom HPR10 (Rohm & Haas), Mitsubishi CHP 5C, Varian PLRP-S. These resins are used mainly in the polishing step of small proteins and peptide-purification processes.

Another type of polystyrene resin has been introduced in 1990 by Perseptive Biosystems with the POROS[R] Perfusion Chromatography[R] resins. They are used as ion-exchange and Protein A-resins for some biopharmaceutical downstream purification processes. It has to be pointed out that due to its hydrophobic nature polystyrene is incompatible with protein purification. Any contact with the hydrophobic surface will lead to severe precipitation of the protein. Therefore the surface has to be first hydrophilized by the application of a hydrophilic network polymer. Due to the particle diameter of around 45 μm, which is significantly smaller compared to other types of ion-exchange resins, these resins show a steep breakthrough curve and high dynamic binding capacity compared to agarose-type resins. The overall

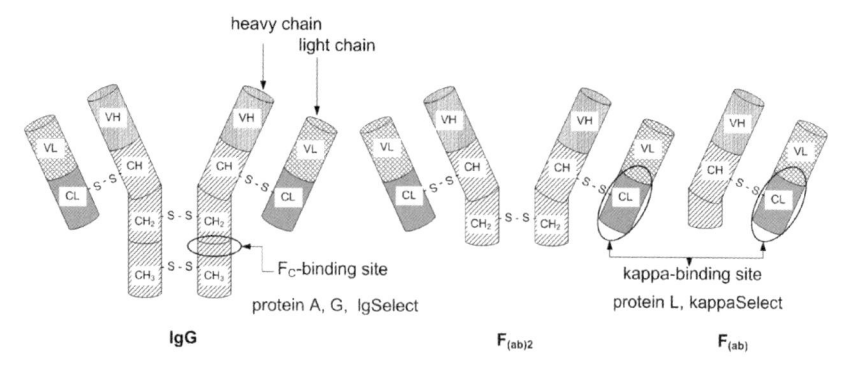

Figure 3.3 Binding sites of affinity selectors.

performance is in accordance with that of other modern-type polymeric resin materials.

3.1.2.2 Designed Adsorbents

In contrast to generic adsorbents, designed adsorbents have been developed for the isolation of specific groups of compounds with common physico-chemical properties. They normally consist of a more or less generic sorbent base material with a surface modification of specific affinity towards the desired chemical function. Their adsorption principle might be based on a multitude of interactions as is the case for the specific proteins, for example, Protein A, G, L or based on single complexing interaction, for example, as in the metal-chelating sorbents.

The use of designed adsorbents is dominated in the area of antibody-capture chromatography as antibodies are by far the largest homogeneous group of molecules of interest to the pharmaceutical industry. Therefore a lot of approaches have been made to develop adsorbents, which can capture the antibodies from the fermentation broth based on a key–keylock (magic bullet) approach, binding to different recognition sites of the antibodies (Figure 3.3). The most successful of these approaches was the development of the Protein A affinity sorbents.

3.1.2.2.1 Affinity Sorbents for Antibodies and Antibody Fragments Protein A – Affinity Sorbents

Protein A affinity sorbents were first introduced in the analytical scale by Hjelm and Kronvall in the 1970s (Hjelm (1972) and Kronvall (1973)). Protein A is a highly selective and specific surface receptor protein originating from *Staphylococcus aureus*, which is capable of binding the Fc portion of immunoglobulins, especially IgGs (IgG1, IgG2 and IgG4, but not IgG3), from a large number of species.

With the tremendous success of monoclonal antibodies in therapy, the development of Protein A adsorbents that are robust enough for a preparative use has been started and has reached a certain level. Today, monoclonal antibodies are captured in the multiton range by the use of Protein A affinity adsorbents.

The advantage of Protein A affinity adsorbents is their ease of use. No feed adjustment is necessary, they are generic for all IgG-antibodies and only a limited method development is needed. In a single capture step they offer a high binding capacity and a good purity of often >95% and the separation of aggregates is possible by the application of linear gradients. Today, the use of Protein A affinity adsorbents is well accepted by the regulative authorities so that they are a gold standard in the majority of current monoclonal antibody purification processes.

The drawbacks of Protein A affinity chromatography are related to the labile chemical nature of Protein A as an affinity selector: due to the necessary low elution pH the target proteins might be denatured or fragmented. The formation of aggregates is a well-known problem and a certain loss of biological activity might occur. Even though the stability of Protein A affinity sorbents could be increased with the upcoming new generations of sorbents, the risk of a certain amount of leaching of the Protein A selector is still present making a subsequent IEX-step that is able to capture the leached Protein A necessary. The biggest challenge in Protein A affinity chromatography is the cost of the sorbent itself and the need for replacement due to limited column lifetime. Table 3.3 lists the most commonly used Protein A affinity sorbents, giving the chemical nature of their base material as well as the type of Protein A selector.

In scientific publications and vendor product data sheets, 10% breakthrough values are typically given. Real processes are never run to such high breakthrough levels to avoid product losses. Therefore more important than 10% breakthrough levels or static binding capacities are 1% breakthrough levels in order to enable a praxis-relevant comparison of different Protein A resins. These 1% breakthrough levels are typically significantly lower than SBC and 10% breakthrough levels.

Starting with the first generation of Protein A affinity sorbents that used native Protein A bound onto an agarose matrix two main development routes have been followed: the increased mechanical stability of the base bead and the increased chemical stability of the Protein A ligand. To achieve an increased mechanical stability of the base bead a very difficult trade-off between the high rigidity of the backbone and still large-enough pore size had to be done. Besides agaroses with different degrees of cross-linker poly-methacrylate beads have been introduced alongside poly-acrylamide gels in a ceramic macrobead shell ("gel in a shell" approach). The large-throughpore material developed by Perseptive is based on polystyrene, which has to be hydrophilized to be used in protein chromatography. Even though this material shows a low static binding capacity it maintains a large percentage of that capacity even at high linear flow rates. A large-pore but rigid base material is controlled pore glass. The Prosep A materials developed by Millipore are available in 700 and 1000 Å pore sizes depending on the total capacity to be achieved.

In the field of Protein A variations, the first products used native Protein A from *S. aureus*, while later products introduced recombinant and thus more controllable Protein A. Repligen announced an increase in their capacity to produce recombinant Protein A significantly so that there is no shortage even for large-scale production and a certain drop in price could be expected. Millipore introduced the first

Table 3.3 Commercially available Protein A affinity sorbents.

Name	Manufacturer	Base material	Protein A selector	*SBC (Q_{max}) **DBC 1% bt or 10% bt
Protein A Sepharose 4 FF	GE Healthcare	Highly cross-linked 4% agarose	Native Protein A	*61.6[a]
rProtein A Sepharose 4 FF	GE Healthcare	Highly cross-linked 4% agarose	Recombinant Protein A	*55.1[a]
MabSelect	GE Healthcare	HF agarose, highly cross-linked	Recombinant Protein A	*67.3[a]
MabSelect Xtra	GE Healthcare	HF agarose, highly cross-linked	Recombinant Protein A	**50.0[b], 10% bt
MabSuRe	GE Healthcare	HF agarose, highly cross-linked	Base-stable Protein A domain	**43[b] 1% bt **51[b] 10% bt
Poros 50 A High Capacity	Applied Biosystems	Poly(styrene-divinyl benzene), hydrophilized	Recombinant Protein A	*42.2[a] **31[b] 1% bt **32[b] 10% bt
Prosep A High Capacity	Merck Millipore	Controlled pore glass, 1000 Å pore	Native Protein A	*37.7[a]
Prosep rA High Capacity	Merck Millipore	Controlled pore glass, 1000 Å pore	Recombinant Protein A	*38.8[a]
Prosep A Ultra Plus	Merck Millipore	Controlled pore glass, 700 Å pore	Recombinant Protein A	**65.0[b] **42[b] 1% bt **46[b] 10% bt
Protein A Ceramic HyperD F	Biosepra	Polyacrylamide gel in ceramic macrobead	Native Protein A	*41.7[a]
AF Protein A Toyopearl 650 M	Tosoh Biosep	Polymethacrylate		**28[b] 10% bt

a) Hahn, Schlegel, and Jungbauer (2003).
b) Merck data at 6 min residence time.

product with a recombinant Protein A produced without any animal-derived component in production ("vegane" Protein A, Prosep vA) following the trend to avoid any mammalian fermentation compound that might bear a risk of a contamination with human pathogens. The latest development is the use of Protein A subunits. The Swedish company Affibody has increased the alkaline stability of Protein A resins by the modification of the amino acid sequence of the natural *staphylococcal* Protein A.

3.1.2.2.2 Other IgG-Receptor Proteins: Protein G and Protein L
Beside the dominating Protein A affinity sorbents, two other IgG-binding receptor proteins have been introduced as affinity chromatography sorbents: Protein G and Protein L. Protein G is a cell surface protein form Group G *streptococcus* species. It exhibits two Fc-binding sites as Protein A and has an overall similar binding spectrum with a clear focus on IgG. Its advantage over Protein A is that it also binds human IgG3-subtype antibodies, which are not bound by Protein A. On the other hand, Protein G in its native form also binds albumin and α2-macroglobulin, so that a recombinant form is developed which avoids the undesirable retention of these proteins. The commercial products on the market differ considerably in their properties. Their price is typically higher than the price for Protein A sorbents. A hybrid Protein A/G sorbent is described in literature.

Protein L from *Peptostreptococcus magnus* is able to bind whole antibodies as well as single-chain variable fragments (scFv) and Fab fragments as long as they contain kappa light chains onto which Protein L binds. One of its prominent features is the binding of IgM, a class of antibodies not well retained by Protein A and G.

3.1.2.2.3 Other IgG-Receptor Proteins: Camelidae Antibodies
The Dutch company BAC BV has developed a technology using Camelidae heavy-chain antibodies (CaptureSelect^R). The technology is based on the identification of stable and specific affinity ligands by using immune llama antibody libraries and the expression of resulting heavy-chain antibody fragments of 129 amino acids (approx. 12 kDa) in the yeast *Saccharomyces cerevisiae*. It is claimed that the ligand has a high affinity, specificity, and capacity in combination with a good caustic stability. In addition, it should bind all IgG-subclasses compared to Protein A that does not bind IgG3.

In Table 3.4 the properties of some affinity-selector sorbents other than Protein A are listed.

Magic bullet resins for Mab-capture have also been developed using small molecular ligands, for example, by Prometic or Xeptagen but so far they are not widely used. Therefore, these approaches are discussed in Section 3.1.2.3 "Customized Adsorbents."

3.1.2.2.4 Sorbents for Derivatized/Tagged Compounds: Immobilized Metal Affinity Chromatography (IMAC)
Certain natural amino acids, especially histidine exhibit a high complexation affinity towards two- and three-valent transition metals, for example, Ni(II), Cu(II), Zn(II), Co(II), and Fe(III). When those transition metals are fixed via complexation onto a solid support these sorbents can be used as affinity resins for proteins naturally rich in histidines or those with genetically engineered oligo-histidine tags (Hochuli, Doebeli, and Schacher, 1987, 1988). When expression vectors became available that attach 6 or 10 histidine units at the N- or C-terminus of a protein IMAC using the metal-iminodiacetic acid (IDA) or metal-nitrilotriacetic acid (NTA) affinity the IMAC technique could be used in a very selective way. Since an oligo-histidine end of a protein is not common in nature the selectivity over host-cell or process-related protein impurities is large, making it

Table 3.4 Commercially available affinity receptor proteins other than Protein A.

Name	Manufacturer	Base material	Affinity selector	Specific binding purposes
Ultralink Immobilized Protein G Plus	Thermo Scientific	Azlactone activated polyacrylamide, 50–80 μm	Recombinant Protein G in E. coli, 22 kDa	> 25 mg of human IgG ml^{-1} of gel
Prosep Protein G	Millipore	Controlled Pore Glass, 100 μm, 1000 Å pore size	Recombinant Protein G in E. coli, 22 kDa	20–24 mg ml^{-1} for human polyclonal IgG
Protein G Sepharose FastFlow	GE Healthcare	Highly-cross-linked Agarose, 4%, 45–165 μm	Recombinant streptococcal Protein G lacking the albumin-binding region produced in E. coli, 17 kDa	Multipoint attachment to minimize leakage, Protein G lacks albumin binding site, 17 mg ml^{-1} human IgG
Protein G Sepharose HighPerformance	GE Healthcare	Highly-cross-linked Agarose, 6%, 34 μm		
POROS Protein G	Applied Biosystems	Polystyrene divinylbenzene coated with hydroxylated polymer, 20 μm, 800–1500 Å pore size	Recombinant Protein G	Binding capacity 15–30 mg ml^{-1} IgG
CIM^R r-Protein G 8f ml Tube Monolithic Column	BIA Separations	Poly(glycidyl methacrylate–co ethylene glycol dimethacrylate) monolith, 8000 Å pore size	Recombinant Protein G	Max 1340 cm h^{-1}
Protein A/G recombinant fusion protein	Thermo Scientific	Beaded agarose, 6% cross-linked or azlactone activated polyacrylamide, 50–80 μm	Recombinant protein in E. coli, 50 kDa	Max 3000 cm h^{-1} for polyacrylamide
Protein L	Thermo Scientific	Beaded agarose, 6% cross-linked, 45–165 μm	Recombinant protein from Peptostreptococcus magnus in E. coli, 36 kDa	

(continued)

Table 3.4 (Continued)

Name	Manufacturer	Base material	Affinity selector	Specific binding purposes
IgSelect affinity medium	BAC BV, distributed by GE Healthcare	Agarose, highly cross-linked, 75 μm	Recombinant protein in S. cerivisiae, 14 kDa	Max 600 cm h^{-1} (at 20 cm CL), DBC 17 mg ml^{-1} (at 2.4 min residence time, 22 mg ml^{-1} @ 6 min)
KappaSelect affinity medium	BAC BV, distributed by GE Healthcare	Agarose, highly cross-linked, 75 μm	Recombinant protein in S. cerivisiae, 13 kDa	DBC 11 mg Fabml^{-1}, max 600 cm h^{-1}

possible to achieve a protein purity of >90% in a single chromatographic step. Table 3.5 lists commercial expression systems for affinity tags.

To achieve a good retention of the target proteins the imidazole nitrogens in the histidyl residues have to be in their nonprotonated form and therefore the loading buffer has to be neutral to slightly basic. By using a high-ionic-strength buffer of 0.1–1.0 M NaCl nonspecific electrostatic interactions are suppressed. For the elution of the target protein the pH of the elution buffer has to be low. If the target protein is not stable in low pH alternative elution modes can be applied. Ligand exchange with imidazole as a buffer additive at neutral pH can be used as can the application of a strong chelating agent, for example, EDTA. Both methods have their drawbacks. For the imidazole elution, the column has to be saturated and equilibrated with imidazole prior to the separation. After the elution with EDTA, the sorbent has to be recharged with metal ions as the application of EDTA results in an elution of the metal ions as well as the bound protein. IMAC resins are reported to exhibit capacities of 5–10 mg Protein ml^{-1} measured for the purified proteins. Commercially available IMAC-resins are listed in Table 3.6. They have been reviewed by Gabers-Porekar (Gabers-Porekar and Menart, 2001).

The process of tagging a protein is today compatible with most of the used expression systems, for example, prokaryotic and eukaryotic organisms which produce the protein intracellularly or secrete it.

The advantages of IMAC are the good specificity, high binding strength, and simple structure of the selector. The good stability of the chemical ligand allows using IMAC resins under denaturing conditions with urea or Guanidinium–HCl. There are a few examples in literature where His-tagged proteins are bound to the IMAC-resin under denaturing conditions, renaturated while still bound to the solid phase by simply lowering the concentration of the chaotropic agent and later eluted in their active forms (Rogl, 1998). This approach is of importance especially for

Table 3.5 Some commercial expression system encoding various histidine-rich affinity tags.

Expression system	Histidine tag	Cleavage	Immunodetection of histidine tag	Commercial source
QIAexpress systems/*E. coli*	(H)$_6$ extensions at N- or C-terminus	No cleavage at N- or C-terminus	Penta-His^{TM} mAb RGS-His^{TM} mAb Tetra-His^{TM} mAb	Qiagen
pcDNA, pEF, and so on series/mammalian cells	(H)$_6$ extensions at N- or C-terminus	No cleavage	anti-HisG mA anti-His (C-term) mAB	Invitrogen, Carlsbad, CA
pMET and pPICZ series/methylotrophic yeasts	(H)$_6$ extensions at C-terminus	No cleavage	anti-His(C-term) mAb	Invitrogen
pYes series/classical yeast	(H)$_6$ extensions at C-terminus	No cleavage	anti-His(C-term) mAb	Invitrogen
pTtriEx vectors/*E. coli*, baculovirus, and mammalian cells	Protein-(H)$_8$	No cleavage	His-Tag mAb	Novagen, Madison, WI
pET systems/*E. coli*	MG(H)$_{10}$SSGHID*DDK↓* H-Protein MG(H)$_{10}$SSGHIEGR↓ H-Protein MGSS(H)$_6$SSG*LVPRGS↓* H-Protein Protein(H)$_6$	Enterokinase Factor X Thrombin No cleavage	His-Tag mAb against N- or C-terminal His tags	Novagen
pHAT vectors/*E. coli*	MKDHLIHDVHKEEHAHAHNKI-*DDDDK↓*-Protein	Enterokinase	HAT Polyclonal	Clontech
TAGZyme kit/*E. coli*	MK(HQ)$_6$Q↓-Protein and various, other His tags optimized for this kit	Dipeptidyl aminopeptidase I (DPP I) alone or a combination of DPP I, glutamine cyclotransferase (GCT) and pyroglutamyl aminopeptidase (PGAP)		UNYZYME Laboratories, Horsholm

Table 3.6 Commercially available IMAC-resins.

Name	Manufacturer	Base material	Ligand/metals
Fractogel EMD Chelate 650	Merck	Polymethacrylate, M (65 µm) and S (30 µm)	Tentacle-iminodiacetic acid (IDA)
PROSEP Chelating, I, II, III	Millipore	CPG 100 µm; 230, 500, 1000 Å pore size	Iminodiacetic acid (IDA)
Toyopearl AF-Chelate 650 M	Tosoh Bioscience	Toyopearl HW 65 µm	Iminodiacetic acid (IDA)
XpressLine Prep Metal Chelate	UpFront Chromatography A/S	HP 70 µm, 165 µm	Iminodiacetic acid (IDA)
Chelating Sepharose FF, High Performance	GE Healthcare	Sepharose 6 FF 90 µm, Sepharose HP 34 µm	Iminodiacetic acid (IDA)
POROS MC	Applied Biosystems	20 µm	Iminodiacetic acid (IDA)
BD Talon Sepharose 6B	BD Bioscience		Nitrilotriacetic acid (NTA)

Escherichia coli expression systems where the target proteins are secreted into inclusion bodies and have to be refolded to obtain their full biological activity.

IMAC offers the possibility to modulate the binding strength of a protein by varying the metal ion and the selector. The binding strength of different metal chelators is normally of the order Cu(II) > Ni(II) > Zn(II) > Co(II) while tetradentate ligand such as NTA bind stronger compared to tridentate ligands, for example, IDA. The modulation of binding strength can be used in two ways: first to obtain mild binding and thus mild elution conditions. If the binding with the classical His6 or His10-tags is too strong the use of a His2-tag might even be considered. On the other hand, the protein can be immobilized on the IMAC column if the binding is strong enough and a reactive column with a certain enzyme activity can be achieved. For rather stable proteins the strong binding can be used to apply harsh washing conditions using detergents, high imidazol concentrations, and organic solvents to get rid of host cell proteins while the target protein is bound to the sorbent.

It has to be pointed out that IMAC suffers from two severe drawbacks limiting its use especially in the production of therapeutic proteins: two artificial compounds, the His-tag as well as the chelating metal are introduced into the production process and have to be completely removed from the final product.

Nickel, especially, is a metal with high human toxicity and has shown carcinogenicity in animal experiments. The problem of metal contamination of the final product is even more difficult as the metal might be incorporated inside the protein and is not easily removed by any subsequent purification step, for example, dialysis or ion exchange.

The removal of the His-tag after the chromatography is the second big challenge. Chemical cleavage of the tag often leads to loss in protein activity, therefore, only enzymatical cleavage can be applied but these methods are often inefficient with a cleavage rate between 60% and 90% only. Remaining tagged protein has to be regarded as an impurity and has to be removed by chromatographic polishing steps. This is an additional effort and reduces the overall process yield.

The integration of the His-tag might in addition reduce the expression level of the protein and oxidative reductive conditions inside the column might lead to metal-induced cleavage of the protein. The number of large-scale applications of IMAC is therefore limited and the biggest potential should be seen in proteins not intended for human use.

Some proteins might be purified in their native, nontagged forms avoiding at least the problem of the His-tag cleavage. Single-chain antibody Fv fragments could especially benefit from IMAC chromatography as their capture is not possible with standard Protein A affinity resins (Casey, 1995; Freyre et al., 2000).

One potential use of IMAC might arise in the area of vaccine purification. In vaccines, the attached His-tag might add some additional antigenicity and serve as an adjuvant, which is not to be removed from the final product. This approach has been shown by Kaslow for a malaria-transmission-blocking vaccine candidate (Kaslow, 1994). Due to the fact that a large volume of cell-culture supernatant had to be purified rapidly, a Ni–NTA agarose was used as a batch adsorbent as the product offered a too low flow rate in a packed column. By using sorbents with better pressure/flow-characteristics this problem should be overcome in large-scale production.

3.1.2.3 Customized Adsorbents

Customized adsorbents are the resins synthesized for the purification of one specific component. They are usually made by sophisticated chemistry to perfectly fit into the three-dimensional structure of the target molecule. Due to the advent of combinatorial ligand-synthesis techniques and high-throughput screening a variety of different customized synthetic and biologically derived ligands have been introduced in preparative chromatography. For customized adsorbents it is important to consider aspects such as stability, production of the sorbent, costs, patents, and regulatory needs besides the performance.

The main aspect of small chemical ligands is their higher stability especially towards harsh sanitization conditions and the fact that they do not introduce any additional material of biological origin into the production process.

For customized adsorbents it is necessary to first define a separation target. The separation target can be defined in positive mode (the compound of interest is bound) as well as in negative mode (a critical impurity is bound and the compound of interest is eluted in flow-through mode). After the target is fixed, a suitable ligand library is tested with respect to its binding affinity towards the pure compound. Two major problems have to be overcome in this stage:

1) For the screening experiments, sufficient target compound in pure form has to be available. Therefore, the screening procedure should be done in an automated and downscaled way.

2) As the affinity has to be screened against the pure compound it has to be tested if the same binding capacity and strength that can be achieved once the compound of interest is subjected to the resin in the original feed composition. Often, other feed components have a drastic influence on the binding kinetics of the target compound.

Once the selector is identified, the binding chemistry on the base resin has to be optimized and the whole adsorbent scaled-up. Finally, the large-scale production under controlled conditions and the registration of the final adsorbent have to be ensured.

Customized adsorbents are available with low-molecular-weight chemical ligands as well as with natural (proteins, poly-nucleotides) or artificial polymers.

3.1.2.3.1 Low Molecular Weight Ligands

Dyax Phage Display Technology The obvious way to reduce the complexity of affinity ligands and bring them down to the absolute essential structure is the use of small peptide fragments. If the binding region of a target molecule is known it can be modeled by Molecular Dynamic (MD) calculations. The other way is through the use of diverse peptide libraries and identifying the best binding selector. In the 1980s, Dyax developed and patented the phage display technology (Smith, 1985). By the introduction of additional genetic information into the genome of small bacterial viruses (bacteriophages) different peptides with 6–13 amino acids, antibody fragments, or proteins will be displayed on the surface of the bacteriophages. By variations in the sequence of the 13 amino acid positions using 19 different amino acids, a library of approximately 100 million different selectors is achieved. The library of bacteriophages can be subjected to a screening procedure and the best binding structure is identified. The best selector molecule can be easily identified and in the case of peptides the selector is chemically synthesized later. Dyax recently published that the new blood coagulation factor FVIII from Wyeth (Xyntha®) is produced using a Dyax customized affinity sorbent. In FVIII-downstream processing, there is typically an immunoaffinity step with a mouse antibody bound to a sorbent. Wyeth states for Xyntha that the replacement of the mouse antibody affinity sorbent by the synthetic peptide sorbent eliminates any potential risk of murine viral contaminants (www.dyax.com).

One of the oldest and best known small molecular ligands is Cibacron blue, initially a textile dye, which was found to have a certain affinity in staining experiments. Simplification and further development of the aromatic selectors attached to the triazine core element by the Canadian company Prometic lead to the affinity ligand A2P (Lowe, 2001, www.prometic.com, Figure 3.4).

A2P is known to bind IgG especially well when isolated from blood plasma. Different products under the label MAbsorbent™ have been developed, for example, in cooperation with GlaxoSmithKline, Boerhringer Ingelheim, and Abbott for the capture of monoclonal antibodies and antibody fragments (FAB). With the cooperation of Prometic and Octapharma the P-CAPT® Prion capture resin for plasma-derived injectible drugs was developed. Prometic states on its website that one of its customized resins is used in 800 l scale in a column of 1.8 m diameter.

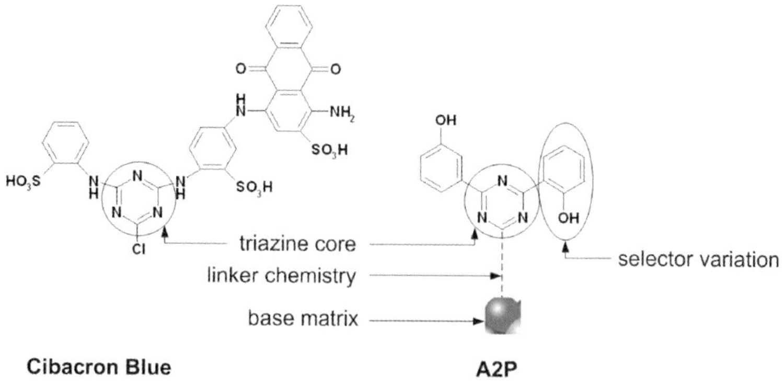

Figure 3.4 Structure of Cibacron Blue and the affinity ligand A2P.

Another approach based on peptides is used by the Italian company Xeptagen (www.xeptagen.com). The D-PAM ligand optimized from a peptide library of 5832 randomized tripeptides is a tetramer of tripeptides linked together by a tetradentate lysine core [structure formula: ([NH$_2$–[R]Arg–(R,S)Thr–[R]Tyr)4–(S)Lys2–(S)Lys–Gly]. All amino acids are in their artificial D-form giving the selector a higher proteolytic stability (Fassina et al., 2001).

The binding affinity is supposed to be especially good for IgM so that under selected conditions the isolation of IgM from IgG-containing feed streams can be achieved. For pure samples, the capacities for different classes of immunoglobulins are claimed to be 10–25 mg ml^{-1} support for IgG, 10 mg ml^{-1} for IgM, and 30–60 mg ml^{-1} for IgY. The high variation in binding capacity for IgG is supposed to be related to the different types of base matrix. The high influence of the support as well as the linker could be verified by molecular dynamic calculations. Within the frame of the EU-funded project Advanced Interactive Materials by Design (AIMs) it could be shown that the two ligands A2P and D-PAM bind to the same region as Protein A between the CH$_2$ and CH$_3$ domains of IgG (Busini et al., 2006; Zamolo et al., 2008; Moiani et al., 2009). This region is not easily accessed and binding needs a high sterical flexibility of the ligand. Binding of the ligand with the target molecule is always in competition with binding of the ligand with its linker or the surface of the base material. For the high flexibility of a small ligand a long linker is preferred although with a too long linker a reorientation of the ligand towards the linker can be observed. For the D-PAM ligand it could be shown that two of the four arms are directed towards the binding site of the protein while the other two are directed more to the linker and the surface. As a conclusion it can be stated that the simple optimization of the ligand is only one part of the design work. Only if the linker as well as the base matrix are carefully adjusted a good customized sorbent can be achieved. Some design criteria can be concluded from the molecular dynamic calculations:

- A too high hydrophobicity of the base material surface leads to a decrease of the interaction between the affinity ligand and the target protein.

- The linker influences the binding as well. A certain length of the linker is needed to give the ligand enough sterical freedom to bind to the target molecule. The hydrophilicity or hydrophobicity of the linker has to be carefully monitored as a back orientation of the ligand towards the base material surface has to be avoided. This phenomenon might occur if the ligand and the base matrix exhibit a certain hydrophobicity and the linker is too hydrophilic.

3.1.2.3.2 Natural Polymers (Proteins, Polynucleotides) Beside the more generic approach to develop an antigen/antibody-based affinity matrix, for example, for the whole class of human immunoglobulins, proteins can be equally well customized towards other protein or virus targets. Two products have been recently developed by BAC BV based on their CaptureSelect® technology. The 14 kDa protein ligands are coupled to 6% cross-linked agarose via a hydrophilic spacer. The base matrix allows a linear velocity of up to $150\,cm\,h^{-1}$ at a bed height of 30 cm. The two available sorbents target alpha-1-antitrypsin (AAT) and an adeno-associated virus (AVB-Sepharose High Performance). AAT is a protease inhibitor protecting body fluids and tissues from the body's own protease enzyme. Diseases related to an AAT-deficit can lead to emphysema, chronic obstructive pulmonary disease (COPD), or liver damage. AAT can be derived from blood plasma and in recombinant form from mammalian cell culture. In addition, it has been the first product expressed in transgenic animals but the project has been stopped; nevertheless, purification strategies for three different production media were needed (www.bacbv.com).

The second very variable type of natural polymers are nucleic acid based. The uniform genetic coding system of DNA or RNA-polymers can be used as well to bind any given target molecule just via its three-dimensional structure. The so-called Aptamers are short-chain oligonucleotides of 25–70 base pairs which are synthesized in large random sequence libraries. Those libraries are subjected to a high-throughput screening to find optimum binding properties for the target molecules. Binding strength can be even tuned by selecting a predefined equilibrium rate and thermodynamic parameters of the interaction between aptamer and target.

The German start-up company AptaRes (www.aptares.net) is offering their MonoLex™ concept as ligands for affinity resins. After the target molecule is bound to the library of random aptamers, those complexes are sorted into different pools. Either the whole pool ("polyclonal aptamers") or the best-binding sequence ("monoclonal aptamer") can be further used for the development of an affinity resin. The aptamers are afterwards chemically synthesized and coupled to the affinity matrix in a site-directed approach. So far no performance data has been reported in the open literature.

3.1.2.3.3 Artificial Polymers A molecular-recognition technology known in the literature for more than 40 years and often supposed to be of great potential for preparative chromatography too is the design of special polymers that function as an artificial key-hole for the target molecule. But Molecular Imprinting introduced by Mosbach and Wulff in the 1970 was never successful in preparative chromatography due to the fact that the polymerization process was rather complex, not scalable, and the resulting chromatographic phases showed a serious diffusion limitation. This diffusion limitation lead to low loadings and long peak tailing reducing the throughput to very low numbers [Alexander et al., 2006].

The only technology on the market today is the so-called Polymer Instruction-Technology offered by the German company InstrAction (www.instraction.com). Their approach uses preformulated linear polymers bearing particle structures of, for example, amino acids or sugars, fixed to a silica support. After the target molecule is added in a physiological environment (water, buffers, or organic solvents) the polymer chains form a three-dimensional network fitting into the target molecule and thus the polymer is "instructed." The polymer instruction is fixed by some type of cross-linking in the linear polymer chains to form a permanent and specific network.

For a lot of the customized adsorbents only very limited or no performance data is available in the open literature. If any of the processes has been scaled-up to a certain size, these projects are highly confidential, due to the fact that the developed materials are target specific for a given and often-patented molecule.

In the following sections, the most widely used generic stationary phases are described:

- reversed phase silicas
- cross-linked organic polymers
  - hydrophobic polymers
  - ion-exchange resins
- chiral stationary phases

3.1.3
Reversed Phase Silicas

Reversed phase silicas are discussed in-depth here as this type of packing materials are predominately applied in preparative chromatography. According to the classification used above, reversed-phase silica is an example of a designed adsorbent.

The term "reversed phase" stands for a hydrophobic packing. Reversed-phase packings are operated with polar mobile phases, typically aqueous mobile phases containing organic solvents such as methanol, acetonitrile, or tetrahydrofurane as a second solvent. For hydrophobic solutes, the retention time increases with the hydrophobic character. The hydrophobic character of a solute is proportional to its carbon content, its number of methylene groups in the case of a homologue series, its number of methyl groups in the case of alkanes, or its number of aryl groups in the case of aromatic compounds.

The term reversed-phase packing is a synonym for a packing with a hydrophobic surface: the most common reversed-phase packings are silicas with surface-bonded long-chain n-alkyl groups, also termed reversed phase silicas. The same term is used to describe silicas with hydrophobic polymer coatings. Reversed-phase packings are also hydrophobic cross-linked organic polymers (cross-linked styrene–divinylbenzene copolymers) and porous graphitized carbons. These reversed-phase packings differ in the degree of hydrophobicity in the relative sequence:

porous graphitized carbon > polymers made from cross-linked styrene/divinylbenzene > n-octadecyl (C18) bonded silicas > n-octyl bonded (C8) silicas > phenyl bonded silicas > n-butyl (C4) bonded silicas > n-propylcyano bonded silicas > diol bonded silicas.

We will demonstrate the synthesis of n-alkyl bonded silicas by chemical surface modification and their properties.

3.1.3.1 Silanisation of the Silica Surface

3.1.3.1.1 **Objectives** Chemical surface modification of the silica serves to:

- Chemically bind desired functional groups (ligands) at the surface, mimicking the structure of solutes and thus achieving retention and selectivity (group-specific approach).
- Deactivate the original heterogeneous surface of the silica surface to avoid matrix effects. As silica has a weak acidic surface basic solutes are strongly adsorbed, which should be minimized by surface modification.
- Enhance the chemical stability of silica, particularly at the pH range above pH 8.

3.1.3.1.2 **Silanisation** The silica surface bears 8–9 µmol m^{-2} of weakly acidic hydroxyl groups (~5 silanol groups per nm^2) when silica is in its fully hydroxylated state. The hydroxyl groups react with halogen groups, OR groups, and other OH groups, leaving acids, alcohols, and water respectively. The most suitable approach is the use of organosilanes with reactive groups X. Thus, the surface reaction can be written as

$$\equiv SiOH + X - SiR_3 \rightarrow \equiv Si - O - SiR_3 + HX. \quad (3.1)$$

As a result, a siloxane bond is formed and the functional group R is introduced by the organosilane. Silanisation can be performed in many ways and thus the products differ in surface chemistry, which is reflected in the chromatographic behavior.

3.1.3.1.3 **Starting Silanes** Silanes differ in the type of the reactive group X and in the type of the organic group R. At constant R and X one can discriminate three types of silanes: monofunctional, bifunctional, and trifunctional. Monofunctional silanes undergo a monodentate reaction, that is, they react with one hydroxyl group, bifunctional silanes may react with one or two X groups. Trifunctional silanes react with a maximum of two X groups per molecule in case of anhydrous reaction conditions. When water is present in the reaction mixture trifunctional silanes hydrolyze and form oligomers by intramolecular condensation. The starting compounds as well as the intermediate then perform a condensation with surface hydroxyl groups. The organic group R is an n-alkyl group with a chain length of 8 or 18. There are also silanes employed with terminal polar groups such as diol, cyano, and amino groups with a short n-alkyl spacer such as n-propyl. The latter are employed for consecutive reactions at the terminal polar group.

Depending on the surface modification, reversed-phase silicas can be grouped into (a) monomeric reversed-phase silicas chemically modified with monofunctional silanes and (b) polymeric reversed-phase silicas with a polymeric layer made by surface reaction with trifunctional silanes.

3.1.3.1.4 Parent Porous Silica

The parent silica is usually subjected to activation prior to silanisation, for example, treatment with diluted acids under reflux. In this way, the heterogeneous surface becomes smoother with homogeneously distributed surface hydroxyl groups to ensure batch-to-batch reproducibility. Simultaneously, the treatment extracts traces of metals that would otherwise affect the chromatographic separation due to a secondary interaction mechanism, for example, ionic interaction of the solute with the adsorbent. Special care has to be devoted to the average pore diameter of the parent silica in relation to the size of the silanes and to the size of solute molecules to be resolved. For example, long-chain silanes drastically reduce the pore opening of the modified adsorbent, which leads to hindered diffusion of the components to be isolated. Peak broadening, reduced capacity, and low resolution are the resulting effects. Commonly, 10 nm pore diameter materials are recommended as starting silicas because a reduction of the specific surface area, a diminution of the specific pore volume, and a decrease of the average pore diameter occur by the silanisation (see also Table 3.16 in Section 3.1.7.1).

3.1.3.1.5 Reaction and Reaction Conditions

Silanisation is a heterogeneous reaction. Silanes can be in the gas or liquid phase or in solution. The reaction is carried out at elevated temperatures, depending on the volatility of the silane and solvent, in a vessel under gentle stirring or in a fluidized bed reactor. To enhance the kinetics, catalysts are added. With chlorosilanes, organic bases are added as acid scavengers; acids are employed in case of alkoxysilanes as reagents. By-products must be carefully removed by extraction with solvents.

3.1.3.1.6 Endcapping

The term endcapping originates from polymerization chemistry, when reactive groups are removed by a specific reaction after polymerization has occurred. After primary silanisation the maximum ligand density amounts to 3.5–4.5 $\mu mol\,m^{-2}$ for monofunctional silanes. As the initial hydroxyl group concentration is about 8 $\mu mol\,m^{-2}$, only half of the hydroxyl groups have reacted. The large size of the silanes makes it almost impossible to convert all hydroxyl groups due to steric reasons. The remaining are still present at the surface and provide the surface with a partially hydrophilic character. As a result, the chromatographic separation will show significant peak tailing due to the weak ion-exchange properties of the hydroxyl groups present. Reversed-phase silicas, even those bonded with C_{18} groups, are operated with aqueous eluent up to approximately 70% v/v water/organic solvent, that is, the C_{18} bonded phases are not completely hydrophobic. To diminish the hydroxyl groups and the so-called silanophilic activity, the silanized materials are subjected to a second silanisation with reactive short-chain silanes. Hexamethyldisilazane (HMDS) and others are the preferred reagent. Figure 3.5a represents the surface of a C_8-modified silica after endcapping.

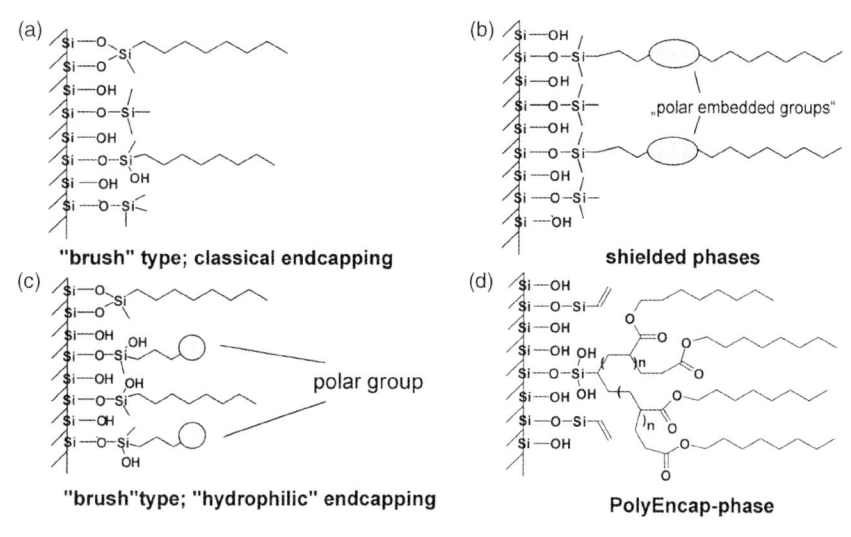

Figure 3.5 Types of RP columns (reprinted from Engelhardt, Grüner, and Scherer (2001) with permission).

Endcapped reversed-phase silica packings exhibit a different selectivity towards polar solutes than nonendcapped materials. Unfortunately, these "base deactivated" phases possess low polarity and therefore similar selectivity towards polar compounds. To overcome the lack of selectivity, a new type of base deactivated stationary phase with polar groups, such as amides or carbamates, "embedded" in the bonded phase (Figure 3.5b) have been developed. These polar embedded phases provide polar selectivity without the poor chromatographic performance associated with stationary phases that have high silanol activity. The use of polar or hydrophilic endcapping (Figure 3.5c) along with bonding of longer alkyl chains such as C_{18} is a successful approach for stationary phases that can retain polar analytes reproducibly under highly aqueous conditions. These polar or hydrophilic endcapping chemicals allow the silica surface to be wetted with water and allow the full interaction with the longer alkyl chains, that is, even 100% water can be applied as solvent.

Depending on the parent silica and the way the reversed-phase silica was modified with silanes, reversed-phase columns exhibit a distinct selectivity towards hydrophobic and polar solutes (Engelhardt, Grüner, and Scherer, 2001).

3.1.3.2 Chromatographic Characterization of Reversed Phase Silicas

Surface-modified silica-based stationary phase packings in chromatography are mostly characterized under isocratic conditions. The employed tests help to assess chromatographic parameters and make it possible to compare different stationary phases. Robustness, reproducibility, and easy handling are the requirements for such tests. It is also important to separate extra-column effects in order to be able to evaluate the column itself rather than the whole HPLC plant system.

The following tests give information on hydrophobic properties (retention of nonpolar solutes), silanol activity (retention of base solutes), performance, purity and shape selectivity towards selected solutes of modified materials in reversed-phase HPLC. It is impossible to find one single suitable test that covers the whole range of chromatographic properties. In addition, the following tests are performed under analytical chromatography conditions.

3.1.3.2.1 Chromatographic Performance The number of theoretical plates N is a measure of the peak broadening of a solute during the separation process (for definitions see Chapter 2). The efficiency of a column can be given for any solute of a test mixture but is strongly dependent on the retention coefficient of the solute.

3.1.3.2.2 Hydrophobic Properties A dependency on the type of ligand, its density, the eluent used, and temperature is found when evaluating hydrophobicities of stationary phases. This property can be assessed by the retention factor of a hydrophobic solute or by the ratio of the retention factors of two nonpolar solutes. The latter is called selectivity; for example, when the components differ only in one methyl group, the term methylene selectivity coefficient is applied. Hence, hydrophobic properties describe the polarity of a column and its selectivity towards solutes with only small differences in polarity. This becomes rather important when endcapped stationary phases are compared (Section 3.1.3.1) as some new types of adsorbents allow separation with 100% water as eluent.

3.1.3.2.3 Shape Selectivity Molecular recognition of the solute by the stationary phase with respect to its geometrical dimension is called shape selectivity. For this test, one can employ aromatic components with identical hydrophobicity that differ only in their three-dimensional shape. The chromatographic selectivity of o-terphenyl/triphenylene or tetra-benzonaphthalene/benzo[a]pyrene are commonly used and show dependencies on several features of the phase, for example, pore structure, ligand type, and density. Figure 3.6 shows a chromatogram of a test mixture of uracil (t_0 marker), n-butylbenzene, and n-pentylbenzene (to assess hydrophobic properties and efficiency), and o-terphenyl and triphenylene (to assess shape selectivity). The test mixture was chosen to provide a short analysis time and to facilitate calculation of parameters from baseline-separated peaks.

3.1.3.2.4 Silanol Activity As already mentioned, a certain amount of silanol groups remain unreacted on the surface after silanisation. To suppress the resulting secondary interactions in HPLC, buffers can be applied. The selectivity of two basic compounds is a measure of silanol activity. Another way to gain information on this property is to assess the peak symmetry of a basic solute, which is defined as the tailing factor by the USP (United States Pharmacopeia) convention. Figure 3.7 shows a chromatogram of a test of uracil (t_0 marker), benzylamine, and phenol using a nonendcapped stationary phase. As can be seen, benzylamine shows peak tailing, indicating strong interaction with residual hydroxyl groups of the silica surface. Some novel adsorbents with hydrophilic endcapping have been

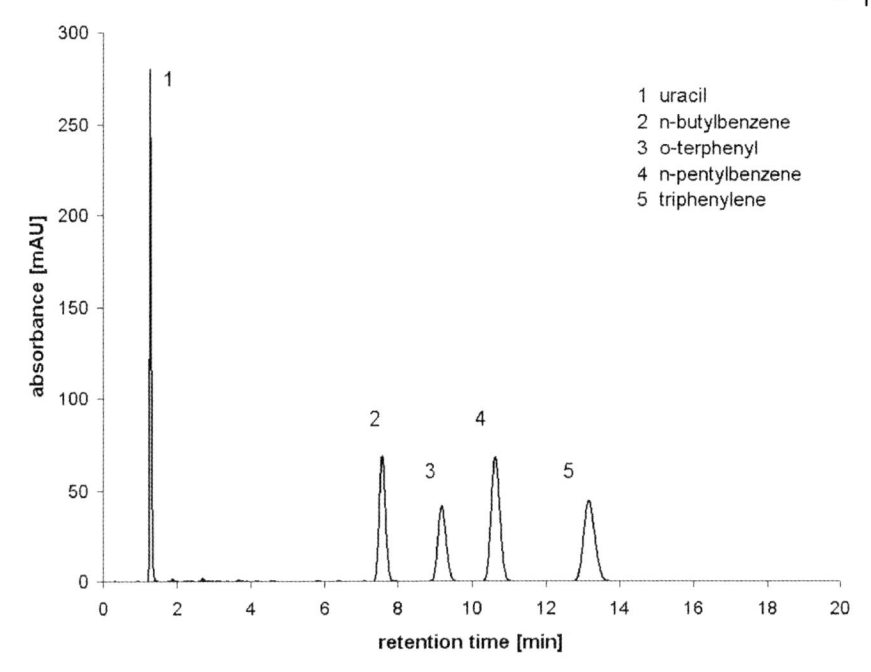

Figure 3.6 Chromatogram of a test mixture to assess hydrophobic properties, efficiency, and shape selectivity of a RP material (eluent: methanol–water, 75:25 v/v).

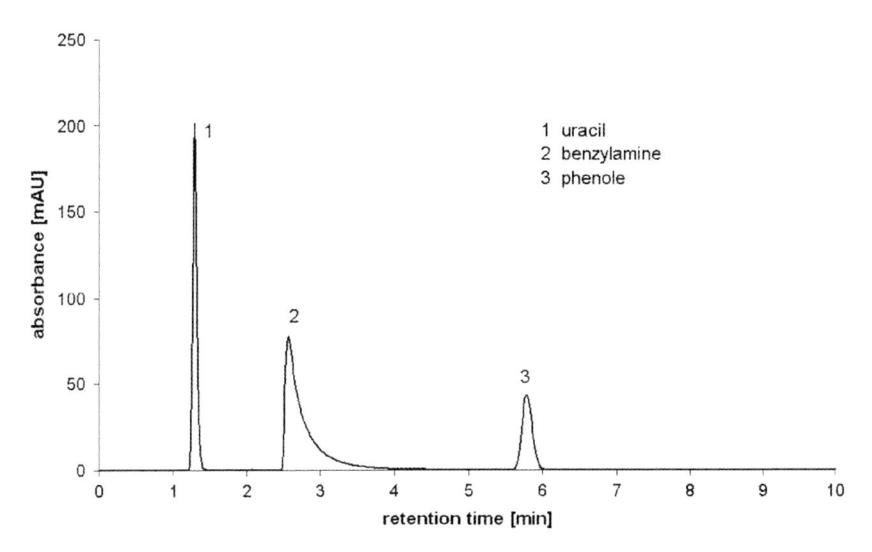

1 uracil
2 n-butylbenzene
3 o-terphenyl
4 n-pentylbenzene
5 triphenylene

Figure 3.7 Chromatogram of a test mixture to assess the silanol activity of a RP material (eluent: methanol–buffer, 30:70 v/v, pH 7.0).

1 uracil
2 benzylamine
3 phenole

developed which reduce peak tailing of base components while retaining high selectivity towards polar and nonpolar solutes.

3.1.3.2.5 **Purity** Metals present at the surface of the phase increase the number of secondary interactions with basic substances. One of the proposed tests using 2.2′-bipyridine (complex forming type) and 4.4′-bipyridine (inactive type) shows a good correlation between the metal content and the peak symmetry of the complexing base, starting at impurity levels around 100 ppm. However, the results change with increasing lifetime of the column as metal ions are accumulated during use.

3.1.4
Cross-Linked Organic Polymers

Organic-based supports for use in liquid chromatography have appeared mostly through applications in biochromatography and in size-exclusion applications for organic polymers. Such applications also represent the two different basic "sources" of organic polymers for separation purposes: natural polymers, such as agaroses and dextranes with varying degrees of cross-linking, and synthetic organic polymers such as hydrophobic styrene divinyl benzene-copolymers as well as more hydrophilic materials such as poly(vinylacetates), synthesized in the late 1960s by Heitz (1970) predominantly for use in size-exclusion chromatography. In biochromatography, research focused on the increase in biocompatibility and alkaline stability of dextrane and agarose gels, allowing the uninhibited use of sodium hydroxide for cleaning and sterilization protocols. However, the relative softness of such gels has always been a limitation in large-scale applications.

Synthetic organic porous polymers in chromatography suffered for a long time from structural problems. One was the diffusion hindrance in the porous structures – mostly traced back to a considerable amount of micropores generated during the synthesis. The other issue interfering with its more widespread use was the compressibility of the beads, limiting their use significantly, especially in high-pressure applications above around 5 MPa. Several approaches have helped to significantly overcome many of these drawbacks, allowing better utilization of their useful properties in loading, selectivity, and cleaning. One approach was the development of more selective synthesis procedures, generating better-defined pore structures with significantly reduced micropores together with the synthesis of macroporous material with much higher rigidity, making the use of porous polymers possible even in large-scale preparative chromatography environments. Another was the control of particle size during synthesis, leading to a much better particle size distribution and even to monodispersity, resulting in reduced backpressure during operation and longer maintenance of the initial system pressure due to lack of generation of fines during operation. These improvements, together with an increasing demand for more selective adsorption systems, have already made organic polymers a valuable asset to modern chromatographic supports, and will do so even more in the future. Among other things, their enhanced properties will also allow their broader use in hybrid materials for upcoming separation tasks, especially in the booming life sciences.

3.1.4.1 General Aspects

Synthetic cross-linked organic polymers were introduced as packings in column liquid chromatography one decade later than oxides. The first organic-based packings were synthetic ion exchangers made by condensation polymerization of phenol and formaldehyde (Adams and Holmers, 1935). In the 1960s, procedures were developed by Moore (1964) to synthesize cross-linked polystyrenes with graduated pore sizes for size-exclusion chromatography. The synthesis of cross-linked dextrane (Porath and Flodin, 1959; Janson,) and agarose (Hjerten, 1964) were milestones in the manufacture of polysaccharide-type packing. At the same time, polyacrylamide packings were synthesized from acrylamide and *N,N*-bismethylene acrylamide by Lea and Sehon (1962).

All the above products, except the first, served as packings in size-exclusion chromatography. The major breakthrough in the synthesis of cross-linked organic polymers with tailor-made properties for column liquid chromatography occurred in the decade 1960–1970 (Seidl *et al.*, 1967). Since then, cross-linked organic polymers have maintained a leading position as adsorbents in ion-exchange and size-exclusion chromatography while their use in column liquid-adsorption chromatography to resolve nonpolar and polar low molecular weight compounds has been rather limited and bare silica has dominated the market. Currently, the situation seems to be changing slightly, but distinctly. It is the authors' belief that organic-based packings will gain greater importance.

Textbooks often treat the structure of organic- and oxide-based materials according to different aspects (Epton, 1978). On viewing a particle of an organic polymer and an oxide, its structure is best described by a coherent system either of three-dimensionally cross-linked chains or of a three-dimensional array of packed colloidal particles as limiting cases. Thus, previous classifications into xerogel, xerogel–aerogel hybrid, and aerogel appear to be inadequate. To provide a sufficient rigidity of particles, the chains or colloidal particles should be linked by chemical bonds rather than by physical attraction forces.

Some examples serve to illustrate the structure of organic-based packings. Figure 3.8 (Hjerten, 1983) shows a scheme of the structure of a cross-linked polyacrylamide gel (right) and a cross-linked agarose gel of an equivalent polymer concentration (left). Both polymers possess a random coil structure. In the agarose, the double-helix-shaped chains are collected in bundles that generate quite an open structure (Arnott *et al.*, 1974). The structure is stabilized by hydrogen bonds between the chains. When agarose is subjected to cross-linking, links are formed among the chains in these bundles (Porath, Laas, and Janson, 1975).

Macroporous, macroreticular, or isoporous polymer packings exhibit another type of structure (Figure 3.9). As the name implies, these polymers contain so-called macropores (>100 nm) and micropores (<2 nm), the latter being inaccessible to large solutes. In other words, macroporous polymer particles constitute an agglomerate made of secondary particles that themselves represent

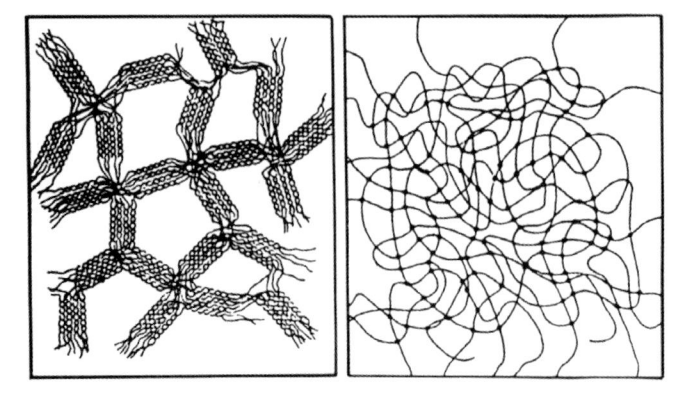

Figure 3.8 Scheme of the agarose gel network (left) compared with a network formed by random chains of Sephadex or Bio-Gel P (right) at similar polymer concentrations (reprinted from Hjerten (1983) with permission).

agglomerates of microspheres. This structure resembles that of porous silica particles, which are composed of agglomerates of spherical colloidal silica particles. With respect to the mechanical rigidity of the polymeric packings, cross-linking becomes an essential means in the synthesis. Other requirements that must be met are insolubility, resistance to oxidation and reduction, and a defined, controllable, and reproducible pore structure.

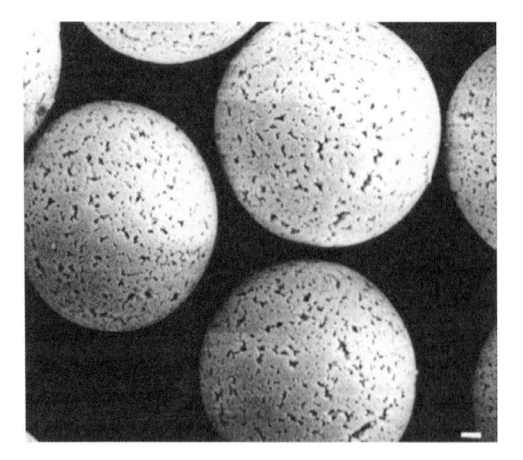

Figure 3.9 SEM picture of a macroporous polymer bead; the bar represents 1 μm.

Polymerization is performed either by condensation or addition polymerization, depending on the type of starting monomer. For cross-linking, comonomers such as divinylbenzene (styrene), ethylene glycol dimethacrylate, epichlorohydrin, 2,3-dibromopropanol, and divinylsulfone (saccharides) are added (Ghethie and Schell, 1967; Porath, Janson, and Laas, 1971; Laas, 1975). The cross-linking reagent can amount to as much as 70 wt.%. Macroporous copolymers are synthesized in the presence of an inert solvent that functions as a volume modifier. Both the cross-linker and the inert solvent have a substantial impact on both the polymerization kinetics and the resulting properties of the copolymer. The decisive parameters relevant for the synthesis of macroporous copolymers have been reviewed by Mikeš et al. (1976). As in the synthesis of silica packings, specific processes must be chosen in polymerization to manufacture polymeric packings with beads of controlled size (Bangs, 1987). Emulsion polymerization starts with a solution of a detergent to which the monomers are added. As a result, micelles swollen with the monomer are formed. After a water-soluble initiator is added (for styrene as a monomer), polymerization leads to particles of exactly the same size as the swollen micelles. Emulsion polymerization processes generate particles of up to 0.5 μm in one step.

Suspension polymerization is usually designed to prepare larger beads, > 5 μm mean particle diameter. The monomer or comonomer solution is vigorously agitated in water in the presence of a colloidal suspending agent. The colloidal agent coats the hydrophobic monomer droplets (in the case of, e.g., styrene or divinylbenzene). Coalescence of the droplets is prevented by their surface. Adding a lipophilic catalyst or initiator starts the polymerization in the droplets and this continues until the beads are solidified in bulk. The size of the beads is thus controlled by the size of the droplets via the stirring speed.

A third variant in polymerization technology is the swollen emulsion polymerization pioneered by Ugelstad et al. (1980). The procedure is performed in two steps. First the polymerization is started by adding a swelling agent, which causes the submicrometre polymer particles to swell by large volumes of the monomer. The increase in volume can reach a factor of 1000. Second, the monomer-swollen beads of defined size are polymerized in a consecutive step.

Having briefly examined the structure of organic packings and the various routes in their manufacture, the most important features may be summarized as follows:

- hydrophilic as well as lipophilic organic packings are synthesized with a controlled pore and surface structure depending on the type of monomer/comonomer and the polymerization reaction. Surface structure can be altered by controlled consecutive surface reactions.
- In accordance with the bulk composition, polymer packings are stable across almost the entire pH range, particularly under strong alkaline conditions.
- Chemical stability is affected by oxidizing and reducing solutions.

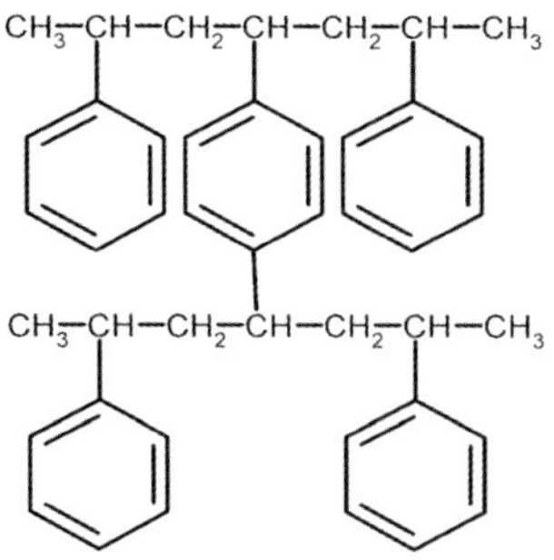

Figure 3.10 Poly(styrenedivinylbenzene).

- Although cross-linking reactions have been optimized in as much as rigid pressure stable particles can be manufactured, some remaining swelling property is often noted when changing the solvent composition in HPLC.
- As in the manufacture of silica, porosity, pore size, and surface area of polymer packings can be adjusted over a wide range, and micro-, meso-, and macro- as well as nonporous beads are synthesized reproducibly.

3.1.4.2 Hydrophobic Polymer Stationary Phases

The synthesis of cross-linked copolymers of styrene and divinylbenzene has been studied intensively and is well documented. The starting monomer is styrene, and divinylbenzene (DVB) is used as cross-linker. The amount of DVB can reach up to 55 wt.%. At 55% DVB, the copolymer shows practically no swelling and possesses a permanent porosity. Figure 3.10 illustrates the network structure of poly(styrenedivinylbenzene) (St-DVB).

Commercial products differ in bead size and pore size. There are even nonporous products on the market, designed for the rapid separation of peptides and proteins by reversed-phase HPLC (Maa and Horvath, 1988). St-DVB copolymers are stable in the pH range 0–14. They find increasing application in the separation of low molecular weight compounds, peptides, and proteins by means of reversed-phase chromatography (Tanaka, Hashizume, and Araki, 1987; Tweeten and Tweeten, 1986) and as parent materials for the synthesis of derivatized packings in interaction chromatography of biopolymers (Unger, Janzen, and Jilge, 1987; Regnier, 1987). Numerous other organophilic polymer packings have been synthesized for size exclusion and interactive HPLC after suitable derivatization. They are reviewed in depth by Mikes (1988).

3.1.4.3 Hydrophilic Polymer Stationary Phases

Hydrophilic polymer sorbents are mainly based on natural polymers, for example, agarose and polymethacrylates. Some newly developed phases use poly-vinylethers or hydrophilized poly-styrene/divinylbenzenze copolymers. These materials will be discussed in the form of their functionalized products. Nevertheless, some of the polymers are used in their native form as resins for size exclusion chromatography (SEC).

3.1.4.4 Ion Exchange (IEX)

Protein purification by ion exchange has been dominated in the past by soft gel matrices based on agarose, cross-linked cellulose, or dextran. These were the only hydrophilic base materials available which could be modified with ionic groups. Coarse and nonhydrophilised ion-exchange materials based on polystyrene can not be used for protein chromatography as the hydrophobic surface will denature the proteins and the pore sizes of PST-based materials are usually too small. Today, a huge variety of hydrophilic polymers based on methacrylate or vinylethers are available. These materials have a much higher rigidity and allow a faster operation even of large-diameter columns, opening a new dimension for the multiton production of modern therapeutic proteins as well as specialized enzymes.

Ion-exchange materials may be easily modified with anionic or cationic groups of different strength. Mixed interaction materials with more than one ionic interaction principle or an ionic active group in combination with, for example, hydrophobic groups are summarized under the group of mixed mode-resins and will be discussed under Section 3.1.4.5. The ionic groups can be divided into strong and weak ionic groups for both cation and anion exchangers. "Strong" ion exchangers exhibit their ionic nature over a broader pH-range compared to "weak" ion exchangers. The binding strength toward a certain protein is not covered by this definition. The most common ionic groups together with their operating pH-ranges are listed in Table 3.7. The two main groups for strong ion exchangers are the Q-group (derived from the quaternary amino function of the trimethylammoniumethyl-group) and the S-group from the sulfo-group.

The type of attachment of the ion group on the surface of the base sorbent is of great importance, as the proteins to be separated have to come into close contact with the ionic group. If the ionic group is directly attached to the sorbent surface (Figure 3.11a) the pore has to be quite wide and open so that the ionic groups are presented on a plain surface able to bind the protein over its full footprint. Ionic groups hidden in small pores or on a rough sorbent surface will not take part in the adsorption of proteins and are therefore useless for the separation. To overcome this problem, sorbents have been developed where the ionic groups are presented on flexible polymer chains comparable to the tentacles of an octopus (Figure 3.11b). The advantage of this set-up can be clearly seen with its flexibility towards the ionic groups of the protein, which can up to a certain extent be caught by the tentacles. Recent development optimized the ratio between charged and thus binding groups with nonbinding spacer groups enhancing the flexibility of the tentacle arms (Figure 3.11c).

The advantages of ion-exchange chromatography are its straightforward adsorption principle that is based on the Coulomb interaction that can be adjusted according to the isoelectric point (pI) of the proteins to be separated. The binding mechanism can be tuned over a wide area of conditions by choosing different types

Table 3.7 Ion-exchange ligands.

Functional group	Description	Operating range pH	pK value/category
TMAE/Q	Trimethylammoniumethyl	6–10	>13/strongly basic
DEAE	Diethylaminoethyl	6–8.5	11/weakly basic
DMAE	Dimethylaminoethyl	6–8.5	8–9/weakly basic
SO_3/SE/SP/S	Sulfoisobutyl/sulfoethyl/sulfopropyl/sulfonate	4–8	<1/strongly acidic
COO	Carboxy	5–8	4.5/weakly acidic
CM	Carboxymethyl	5–8	4.5/weakly acidic

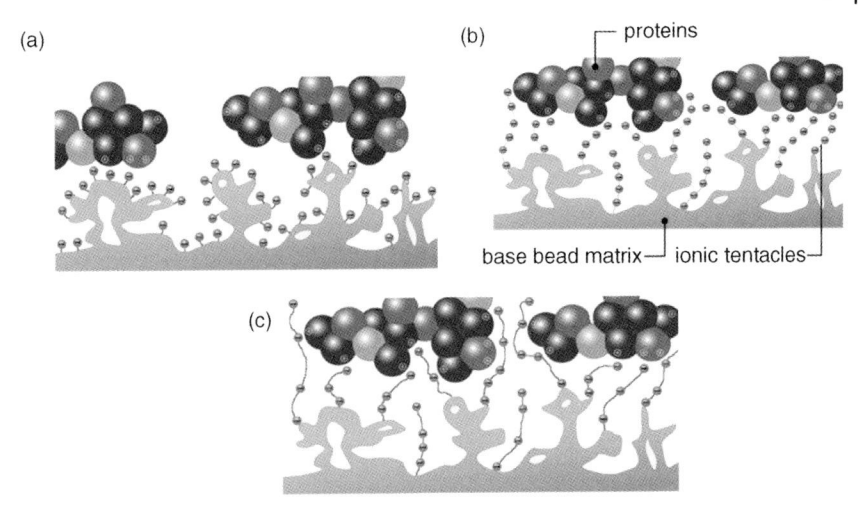

Figure 3.11 Presentation of ionic ligands on different ion-exchange type reins, (a) directly attached to the surface (b) tentacle type ion-exchange resin (c) enhanced flexibility tentacle resins.

of buffers at different pH and salt concentration for the feed, wash, and elution buffer. This high degree of freedom on the other hand leads to a high number of possible process conditions, which should be exploited in a consequential way. Robotic systems are more and more used to systematically screen for the best conditions of a given separation task.

Ion-exchange resins have additionally a relatively high protein-binding capacity and a quite high resolving power, especially if linear gradients are applied. The ease of performance and good packing properties of modern ion exchangers allows to pack large-scale columns of up to 2 m diameter. Tables 3.8 and 3.9 list the properties of the most commonly used ion exchangers

IEX sorbents are densely substituted with ionic groups leading to a total ionic capacity of around 100–500 μmol ml^{-1} gel, corresponding to a concentration of ion exchanging groups of 0.1–0.5 M. The binding capacity for proteins is lower due to their size and it strongly depends on the available surface area for the molecular weight and size of the protein as well as on the chemical nature (surface charge) of the protein.

3.1.4.4.1 **Optimization of Ion-Exchange Resins** To optimize an ion-exchange sorbent several options have to be taken into account:

- The pores should be large enough so that the performance limiting pore diffusivity is maximized.
- On the other hand the number of ionic groups which is in most cases proportional to the surface area should be as high a possible.
- To ensure that the sorbent can be packed in large-diameter columns without self-compression of the packed bed the rigidity of the matrix, governed by the amount of stabilizing cross-linker, should be high.

Table 3.8 Properties of cation exchangers as provided by manufacturers.

Resin	Manufacturer	Base matrix/surface modification	Particle size [μm]	Pore size [nm]	Ionic capacity (μeq ml^{-1} gel)	Static protein-binding capacity (mg protein ml^{-1} gel)	Operation range (cm h^{-1})
SO$_3$/SE/SP/S							
Fractogel EMD SO$_3$	Merck Millipore	Poly-methacrylate/tentacle	65	100	80–100	110–150	400 @ 2 bar
Fractogel EMD SE Hicap	Merck Millipore	Poly-methacrylate/tentacle	65	100	60–90	120–160	400 @ 2 bar
Eshmuno S	Merck Millipore	Poly-vinyl derivative/tentacle optimized	75–95	80	50–100	115–165	1200 @ 2 bar
SP Sepharose FF	GE Healthcare	Agarose/conventional	45–165	—	180–260	70	300–400 @ 1 bar at 15 cm bed height
SP Sepharose XL	GE Healthcare	Agarose/conventional	45–165	—	180–250	160	300–500
SP Sepharose Big Beads	GE Healthcare	Agarose/conventional	100–300	—	180–250	70	1800 @ 2 bar
SP Sepharose HP	GE Healthcare	Agarose/conventional	34	—	150–200	55	—
Capto S	GE Healthcare	Agarose/Three-dimensional network	90	—	110–140	120	700 @ 3 bar (20 cm, water)
MacroPrep High S	Biorad	Poly-methacrylate/conventional	50	100	120–200	55	400 @ 2 bar, 14 * 17 cm
UNOsphere S	Biorad	Poly-methacrylate/conventional	80		280	60	1200 @ 2 bar, 1.1 * 20 cm

Table 3.8 (Continued)

S F Ceramic Hyper D	BioSepra/Pall	Polyacrylamide gel in ceramic macrobead	50	—	>150	75	600 due to diffusion, max $\Delta P = 70$ bar
GigaCap S	Tosoh	Poly-methacrylate/three-dimensional polymer network	50–100	—	100–200	150	1500 @ 2.5 bar at 14 cm bed height
Poros 50 HS	Applied Biosystems	PST/DVB hydrophilized	44[a]	—	—	27[a], mAb 10% bt	—
Toyopearl SP-650 M	Tosoh	Poly-methacrylate/conventional	40–90	100	120–170	39[a] mAb 10% bt	400 cm @ 2 bar

a) Staby et al. (2004), all other data is from the manufacturer's technical data sheets.

Table 3.9 Properties of anion exchangers as provided by manufacturers.

Adsorbent	Base matrix/surface modification	dp (μm)[a]	Mean pore radius (nm)	Ion capacity (μmol ml^{-1})	Dynamic capacity (BSA mg ml^{-1})
Merck Millipore Fractogel EMD TMAE (M)	Methacrylate/tentacle	65	100	—	100 (static)
Merck Millipore Fractogel EMD TMAE Hicap (M)	Methacrylate/tentacle	65	100	—	180 (static)
Merck Millipore Fractogel EMD DEAE (M)	Methacrylate/tentacle	65	100	—	100 (static)
Merck Millipore Fractogel EMD DMAE (M)	Methacrylate/tentacle	65	100	—	100 (static)
Merck Millipore Eshmuno Q	Poly-vinyl derivative/tentacle, optimized	85	80	90–190	>80
Tosoh Bioscience Super Q-650C QAE-550C	Methacrylate/conventional	40–90	~50	200–300	105–155
	Methacrylate	50–150	~25	280–38	60–80
GE Healthcare					
Source 30Q	Polystyrene/divinyl-benzene	30	Not given		>40[b]
Q Sepharose FF	Agarose	45–165	Not given	180–250	120 HAS
Q Sepharose XL	Agarose with bound dextran	45–165	Not given	180–260	>130[c]
Whatman Express-Ion Q	Microgranular cellulose/conventional	60–130[d]	Not applicable	1 meq/dry g	55
Sterogene Bioapplications Q Cellthru Big Bead Plus	4% Agarose/conventional	300–500	Not given	Not given	65
Applied Biosystems POROS 50 HQ	Polystyrene/divinylbenzene, hydrophilized	50	Not given	Not given	60–70

a) Particle diameter.
b) 50% breakthrough.
c) 10% breakthrough.
d) Fiber length.

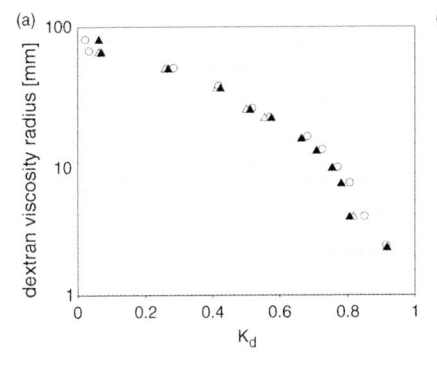

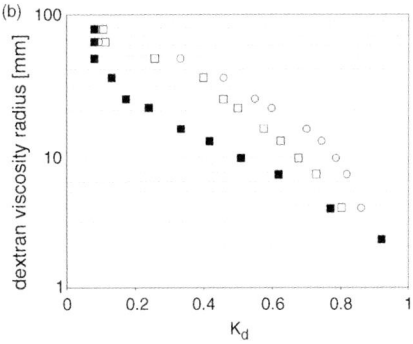

Figure 3.12 Inverse size exclusion data of different ion exchange sorbents, (a) TosoHaas HW 65 F SEC (○) and TosoHaas SP 650 M cation exchanger (▲), TosoHaas CM 650 M (△), (b) TosoHaas HW 65 F SEC (○) and Merck Fractogel SO3 (M) (■), Fractogel COO (M) (□), (reproduced from DePhillips and Lenhoff, 2000).

Pore dimensions and their geometry play a vital role for the adsorption process in protein chromatography on ion exchangers. Different methods have been used to evaluate the pore network of a sorbent, which is always a random arrangement of holes in the matrix with different width and length (Chang and Lenhoff, 1998; Kopaciewicz, 1987). For all these analytical methods it is of utmost importance to control the experimental conditions. The combination of Confocal Laser Scanning Microscopy (CLSM) with Inverse Size Exclusion Chromatography (ISEC) is very well suited for the determination of intraparticle diffusion coefficients and thus the optimization of ion-exchange sorbents.

Especially the tentacle-based ion exchangers make the analytical situation even more complex. The tentacles attached to the matrix surface have a three-dimensional structure stretching out into the free pore space. Even more, their structure depends on the salt concentration in the mobile phase. At high salt concentration, the ionic groups of the tentacles are neutralized and the tentacles are able to pack much more densely. The difference in open pore space can be seen from Figure 3.12 where DePhillips and Lenhoff showed the influence of tentacles for two different ion exchange sorbents. In Figure 3.12, the open pore space is shown as a function of the ion-exchange modification. All four ion-exchange sorbents (TosoHaas SP 650 M, TosoHaas CM 650 M, Merck Fractogel SO3 (M), and Fractogel COO (M)) are based on the matrix HW 65 F. While the first two have an ordinary surface modification the later two exhibit the tentacle principle. The influence of the tentacle modification stretching out into the open pore space can be seen for the uncharged dextran molecules, which have been used as size markers. While the TosoHaas materials show the same pore accessibility for unmodified and modified sorbents, the modified Fractogel tentacle materials have a smaller accessible pore volume.

K_d represents the extent of permeation into the pore volume, a totally excluded molecule would show a K_d equal to 0, a molecule small enough to penetrate all pores shows a K_d of 1.

84 | *3 Stationary Phases and Chromatographic Systems*

| (a) Fractogel ® EMD SO3 | (b) experimental Fractogel® EMD SO3 modification 1 | (c) experimental Fractogel® EMD SO3 modification 2 |

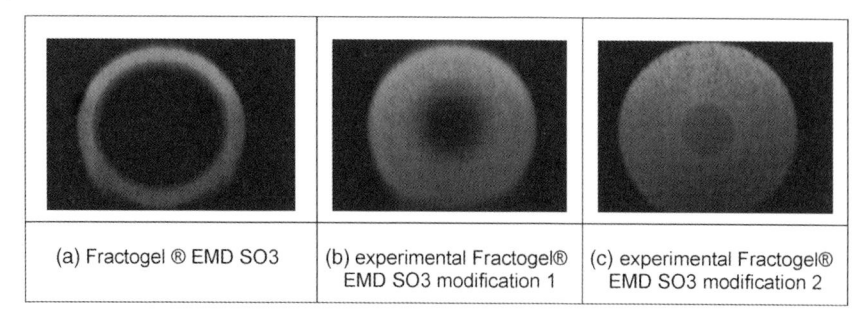

Figure 3.13 Adsorption of dye labeled proteins on different ion-exchange sorbents (a) Fractogel® EMD SO3 (b) and (c) experimental Fractogel® EMD SO3 modifications with advanced tentacle flexibility.

The accessibility of the pore volume is just one side of the medal: even with the lower accessible pore volume the Fractogel tentacle ion exchangers show a very high protein-binding capacity due to the fact that the tentacle network is very well able to bind charged molecules even if it excludes the noncharged marker molecules.

The pore diffusion coefficients in tentacle-based matrices are typically higher as the transport mechanism is not based on free diffusion but most probably on a faster film diffusion principle (Thomas *et al.*, 2012, de Neuville *et al*, 2012).

The development in the imaging technologies especially CLSM enabled to investigate the pore diffusional mechanisms of such tentacle structures in detail (Stanislawski, Schmit, and Ohser, 2010). Proteins marked by fluorophores can be followed in their uptake procedure inside the bead pore system and the spatial distribution of the radial density can be reconstructed. Using this technology, it is possible to determine the influence of tentacle structure modifications on protein uptake clearly demonstrating that an optimum exists between charged groups and noncharged spacers forming the three-dimensional tentacle network inside the pore system. This optimization work led to the development of new sorbents with an optimized combination of protein loading and diffusivity allowing high linear flow rates without compromising protein dynamic binding capacity. Figure 3.13 shows the classical tentacle material Fractogel EMD SO3 (M) in comparison to materials based on the same matrix but with an optimized ratio of ionic and spacer groups. The pictures taken at the identical residence time of 4 min clearly show that the diffusivity under binding conditions using a fluorophore-labeled antibody as the marker molecule is much faster for the optimized materials. Under column chromatographic conditions, it could be shown that dynamic binding capacities as high as 200 mg ml^{-1} can be achieved at a residence time of 2 min under experimental conditions (Urmann *et al.*, 2010).

Besides microscopic technologies, the molecular dynamics simulation using coarse grain models to speed up the simulations improved the understanding of the effects of tentacle surface modifications (Cavalloti, 2008). The tentacular structure of IEX materials is characterized by a macroscopic reorganization caused by van-der-Waals interactions among the tentacle backbones and electrostatic interactions of charged groups with the solvent. Thus, the density of tentacles and their structure in terms of tentacle length and number of charged groups per tentacle are important parameters for the material performance. Salt effects can be observed which indicate that external parameters can trigger an important structural reorganization of the tentacular structure at a mesoscopic level.

The optimization of the tentacle structure still exhibits the possibility to optimize the mass-transfer characteristics and thus develop improved resins for the purification of biopharmaceutical molecules of different shape, size, and ionic properties.

3.1.4.5 Mixed Mode

Over the last few years, new resins have been launched which exhibit more than one mode of interaction. The goal of the combination of hydrophobic interaction principles with ionic interactions is the generation of complementary selectivities for mixtures difficult to separate by just ionic or hydrophobic interactions (Johansson et al., 2003a, 2003b). Especially, the separation of antibody aggregates from the corresponding monomeric form can be achieved with these resins.

The most prominent materials of this interaction type use an aromatic moiety in combination with a quaternary amine function (Capto Adhere) or a sulfonic group (Capto MMC). A slightly different concept is the Hydrophobic Charge Induction chromatography where the resin charge is generated by adjusting the pH accordingly (Pall MEP Hypercell).

All resins exhibit quite remarkable selectivities but suffer from significantly higher efforts for the development of the operating conditions as two interaction principles have to be adjusted and controlled in parallel. Even when operated under optimal conditions the product recovery is lower than that for conventional ion-exchange resins. Table 3.10 lists the commercially available mixed mode resins.

3.1.5
Chiral Stationary Phases

Chiral stationary phases (CSP) for preparative use have been developed tremendously over the past decade. After the introduction of the first commercially available phases in the early 1980s the reproducible synthesis of those phases in large quantities has improved significantly, so that today several multiton productions of enantiopure drugs are operated applying chiral chromatography. CSPs most often used in preparative chromatography are listed in Table 3.11.

The breakthrough in the application of chiral stationary phases for preparative use came with the controlled bonding or coating of the chiral selector onto mesoporous silica offering both a better thermodynamic control due to the optimized silica pore system and much higher pressure stability. With these types of stationary phases the efficient packing of large-scale columns became possible. The only known application for one of the natural or synthetic soft polymers is the preparative separation of a precursor of the cholesterol-lowering drug Lipobay (which was later withdrawn from the market due to side effects) on a polyacrylamide phase

Table 3.10 Properties of mixed-mode sorbents.

Adsorbent	Selector (base matrix)	Particle diameter d_p (μm)	Capacity	Working pH
CaptoAdhere (GE Healthcare)		75 μm		2–14
Capto MMC (GE Healthcare)		75 μm		2–14
MEP hypercell (Pall)		80–100 μm	> 20 mg ml^{-1} hIgG	2–12

Table 3.11 Commercial packings with enantioselective properties.

Name or class	Trade name	Manufacturer
(1) Natural polymers or derivatives		
Cellulose triacetate	CTA	Merck, EMD Chemicals
Cellulose tribenzoate	CTB	Riedel-de Haen
(2) Synthetic polymers		
Poly-(triphenylmethyl methacrylate)	Chiralpak® OP/OT	Daicel, Chiral Technologies
(3) Silica with monomolecular ligands (Brush-type CSP)		
3-[N-(3,5-dinitrobenzoyl) phenylglycin-amido]propyl-silica	ChiraSep® DNBPG	Merck, EMD Chemicals
	Chirex 3001	Phenomenex
3-[1-(3,5-dinitrobenzamido)-1,2,3,4-tetrahydrophenanthrene-2-yl]-propyl-silica	Whelk-O 1®	Regis, Merck
11-[2-(3,5-dinitrobenzamido)-1,2-diphenylethylamino]-11-oxoundecyl-silica	ULMO®	Regis
3-{3-[N-[2-(3,5-dinitrobenzamido-1-cyclohexal]]-3,5-dinitrobenzamido}-2-hydroxy-propoxy}-propyl-silica	DACH-DNB®	Regis
3,5-Dinitrobenzamido 4-phenyl-β-lactam	Pirkle 1-J®	Regis
N-3,5-Dinitrobenzoyl-a-amino-2,2-dimethyl-4-pentenyl phosphonate	α-Burke® 2	Regis
®-3-[N-(3,5-dinitrobenzoyl)-1-naphthylglycine-amido]propyl-silica	Sumichiral 2500	Sumitomo
	Chirex 3005,	Phenomenex
(S)-*tert*-Leucine and ®-1-(α-naphthyl)ethylamine urea linkage	Sumichiral® OA-4700	Sumitomo
	Chirex 3020	Phenomenex
(S)-indoline-2-carboxylic acid and ®-1-(α-naphthyl) ethylamine urea linkage	Sumichiral® OA-4900	Sumitomo
	Chirex 3022	Phenomenex

(continued)

Table 3.11 (Continued)

Name or class	Trade name	Manufacturer
O-9-(*tert*-Butyl carbamoyl) (8R,9S)-Quinine	Chiralpak QN-AX	Daicel (Chiral Technologies)
O-9-(*tert*-Butyl carbamoyl) (8R,9S)-Quinidine	Chiralpak QD-AX	Daicel (Chiral Technologies)
(4) Silica with natural selector derivatives		
Cellulose- and amylose-derivatives, see Table 3.12		
β-Cyclodextrin	ChiraDex®, Cyclobond®	Merck, EMD Chemicals Astec
Vancomycin	Chirobiotic® V1 and V2	Astec
Teicoplanin	Chirobiotic® T1 and T2	Astec
Teicoplanin-aglykon	Chirobiotic® TAG	Astec
(5) Silica with synthetic polymer		
Poly[(S)-N-acryloylphenylalanine ethyl ester]	Chiraspher®	Merck, EMD Chemicals
Cross-linked-O,O'-bis-(3,5-dimethylbenzoyl)-L-diallyltartaramide	Kromasil®-CHI-DMB	Eka Chemicals
Cross-linked- O,O'-bis-(4-tert-butylbenzoyl)-L-diallyltartaramide	Kromasil®-CHI-TTB	Eka Chemicals
(R,R) or (S,S) Poly-(trans-1,2-diamino-1,2-diphenylethane)-N,N-diacrylamide, *trans*-1,2-diaminocyclohexane)-N,N-diacrylamide	P-CAP DP, P-CAP	Astec

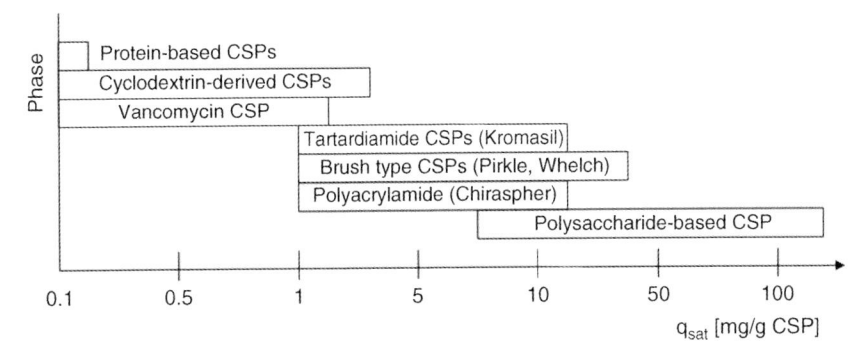

Figure 3.14 Typical saturation capacity of the most used commercially available CSPs (reproduced from Franco et al., 2001).

produced by Bayer (Angerbauer et al., 1993). The other type of full polymer phases are the cellulose-fibers Cellulose-triacetate (CTA) and –tribenzoate (CTB) and the later developed cross-linked cellulosic beads (Francotte and Wolf, 1991). Even though they offer in principle the advantage of high loadability they are not easy to handle in large columns and thus have never been commercialized (Franco et al., 2001; Ikai, 2007).

The CSP based on cellulose and amylose-derivatives coated onto silica, which have been developed and patented by the pioneering work of Okamoto (1984) are by far the most often used ones in preparative chromatography and, especially, SMB applications. These adsorbents offer good productivities because of their high loadabilities (Figure 3.14). In addition, the four most commonly used CSPs of this type separate a broad range of different racemates. (Borman et al., 2003) The major problem of these adsorbents is their limited solvent stability, especially towards medium-polar solvents such as acetone, ethyl acetate, or dioxane. In the past, their use in conjunction with aqueous mobile phases was also not recommended by the manufacturer. This problem had been overcome when Ishikawa and Shibata (1993) as well as McCarthy (1994) reported good stabilities for the use of polysaccharide-type phase with aqueous mobile phase systems. The stability of the adsorbent after switching to RP conditions has been reported by Kummer and Palme (1996) to be at least 11 months and by Ning (1998) to be 3 years. No peak deviation is observed after switching to RP mode.

A novel development is the chemical bonding of the chiral selector onto the silica backbone of the CSP by either cross-linking of the chiral polymer in the presence of a supporting silica bead (Franco et al., 2001) or single-point attachment onto a silica surface (Franco et al., 2001; Ikai et al., 2008). The covalent bonding of the cellulose or amylose-polymer opens the door to the use of medium polar solvents, which are otherwise widely used in preparative chromatography and which offer good solubilities for a lot of important racemates. However, compared to the coated phases where only a limited number of solvent combinations could be used, the mobile-phase selection process for a separation on the immobilized phases could be more challenging.

Table 3.12 Cellulose- and amylase-based chiral stationary phases.

Selector	Coated	Immobilized
Cellulose		
tris(4-methyl-benzoate)	Chiralcel OJ	
tris(3,5-dimethyl-phenylcarbamate)	Chrialcel OD, Kromasil CelluCoat, Lux Cellulose-1, RegisCell, Nucleocell delta, Eurocel 01	Chiralpak IB
tris(3,5-dichloro-phenylcarbamate)		Chiralpak IC
tris(4-chloro-3-methyl-phenylcarbamate)	Chiralcel OZ	
tris(3-chloro-4-methyl-phenylcarbamate)	Lux Cellulose-2	
Amylose-		
tris(3,5-dimethyl-phenylcarbamate)	Chiralpak AD, Kromasil AmyCoat, RegisPack, Nucleocel alpha, Europak 01	Chiralpak IA
tris([S]-á-methylbenzylcarbamate)	Chiralpak AS	
tris(2-chloro-5-methyl-phenylcarbamate)	Chiralpak AY	
tris(5-chloro-2-methyl-phenylcarbamate)	Lux Amylose-2	

The new possibilities of the solvent-resistant CSPs were shown in an impressive way by Zhang (2005a) for the Ca-sensitizing drug candidate EMD 53986. The major obstacle for the development of an enantiopure form of this drug substance was always its limited solubility in solvents that could be used in preparative chromatography. Even though good selectivities were found on different CSPs, none of them could be used in preparative mode with sufficient economy (Schulte, 1997). By using the new chemically bonded immobilized cellulose-CSP dichloromethane/THF as the solvent increased the solubility tremendously and thus improved the productivity for the preparative separation 6.5 fold from 430 g enantiomer kg^{-1} CSP/day to 2834 g $(kg\,day)^{-1}$ (Zhang, Schaeffer, and Franco, 2005). Descriptions of the immobilized CSPs as well as recommendations for their use and special screening protocols can be found in Cox and Amoss (1980) and Zhang (2005b, 2006, 2008). Table 3.12 lists the commercially available cellulose- and amylose based chiral stationary phases.

Chiralcel and Chiralpak are trade names of Daicel (Chiral Technologies), Kromasil of EKA, RegisCell and RegisPak of Regis Technologies, Lux of Phenomenex, Nucelocel of Macherey & Nagel, and Eurocel of Knauer.

After the first patents of the coated cellulose- and amylose-type CSPs expired in 2006 a series of generic CSPs have been introduced into the market. The main arguments for choosing one of these phases for a preparative separation process should be the loadability and stability of the stationary phase beside the price per kg and also the batch-to-batch reproducibility records and the availability in large batch sizes.

For method development and separation optimization, straightforward methods are described in the literature mainly for analytical purposes (Perrin, 2002a, 2002b, Matthijs, 2006; Ates, 2008; Zhou, 2009). For preparative separations, it should be always kept in mind that some process parameters have a tremendous influence on the economy of the separation compared to analytical separations, for example, the cost, reusability, and toxicity of the mobile phase.

3.1.5.1 Antibiotic CSP

Antibiotics have been introduced as natural chiral selectors in a variety of commercially available chiral stationary phases (Armstrong et al., 1994). Initially, they have been critically reviewed due to the fact that they are potent drug molecules, which are introduced as part of the production process of other pharmaceutical compounds and leaching of the antibiotic selector might occur. The manufacturers of the chiral phases have reacted by optimizing the binding chemistry in the phases. The CSPs can be operated in normal as well as reversed-phase mode and additionally in SFC mode. Their main field of application is the separation of underivatized amino acids (Berthod et al., 1996) and small peptides (Ilisz, Berkecz, and Peter, 2006). As especially underivatized amino acids and small chiral acids are difficult to be separated there has been some effort put into an integrated separation approach by first using SMB-chromatography to obtain an enantiomerically enriched fraction and later obtaining the final purity by fractionated crystallization. An example for the separation of L,D-threonine and L,D-methionine on a Teicoplanin-CSP is given by Gedicke et al. (2007).

3.1.5.2 Synthetic Polymers

The synthetic polymers based on N-acryloyl amino acid-derivatives developed by Blaschke in the 1970 and transferred to silica-bonded phases in the 1980 are especially useful for the separation of 5- and 6-membered N- and O-heterocycles with chiral centers (Review in Kinkel, 1994). Their wide chemical variety has been intensively exploited by Bayer Healthcare for their portfolio of chiral molecules. One example of this approach has been published in a joint work of Merck and Bayer (Schulte, 2002). This work explicitly shows how important it is to screen different intermediates in addition to the final drug compound. Due to different selectivities and solubilities, the productivity for the preparative separation can be dramatically different.

New types of synthetic polymeric CSP are those based on tartaramide as the selector (Andersson et al., 1996) and those based on (R,R) or (S,S) trans-1,2-diaminocyclohexane (Zhong, 2005). The first one shows its best performance for strongly hydrophobic molecules, when a high alkane content in the mobile phase can be used. The second one has moderate selectivities and resolutions and a comparable loading to the cellulose-based CSP (Barnhart, 2008). It offers the possibility to reverse the elution order by choosing the other enantiomer of the chiral selector.

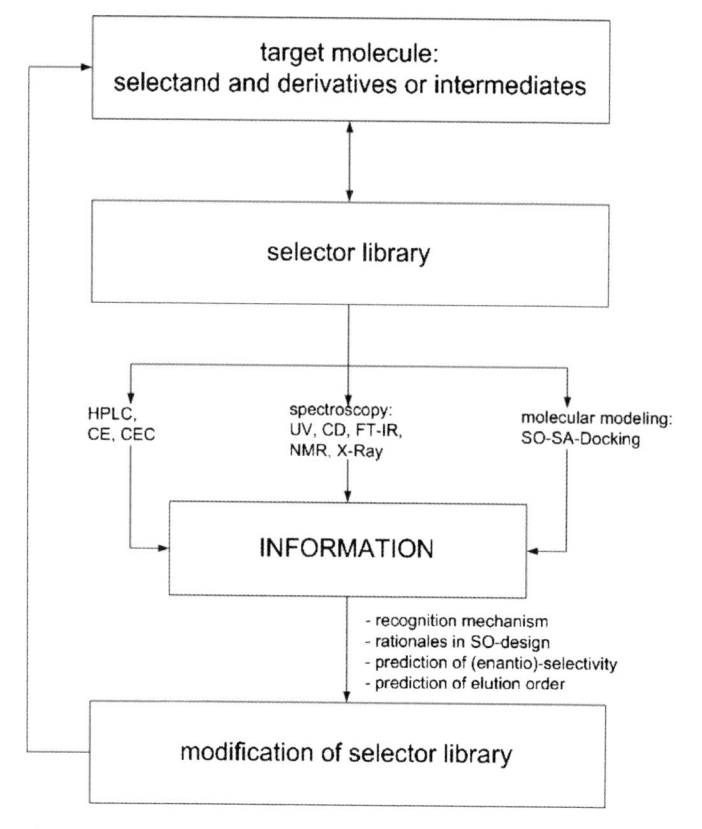

Figure 3.15 Rational screening process for enantioseparation by applying targeted selector design.

3.1.5.3 Targeted Selector Design

Besides the general approach of polysaccharide-type phases more and more chiral phases are designed for special separation problems (Figure 3.15). The targeted selector or reciprocal design approach first used by Pirkle and Däppen (1987) can offer good productivities for the desired enantiomers. For the reciprocal-design approach one enantiomer of the target molecule is fixed via a spacer to a silica support. A series of racemic compounds, which can be used as chiral selectors (the Selector library), are subjected onto a column packed with the target enantiomer CSP. The chiral selector, which is separated on that column with the highest selectivity, is then used to prepare the actual phase for the preparative separation of the target enantiomer. Screening for useful selectors can be automated in a combinatorial high-throughput approach (Wu et al., 1999).

The classical Pirkle-type phases are the most stable chiral stationary phases. They are mainly operated in normal phase mode and can be used to their full extent with scCO$_2$ (Macaudiere, 1986; Kraml, 2005). As the chiral selector is rather small and synthetically available, chiral phases with both enantiomers of the selector are often available offering the possibility to choose the elution order. (Welch 1993, Dungelova 2004) If this possibility is given it is always advisable to elute the target molecule first. If a displacement effect can be observed in the separation this choice will lead to an enormous gain in productivity.

A special subclass of brush-type chiral stationary phases are the ion-exchange CSPs developed by Lämmerhofer and Lindner (1996). By introducing an additional ionic moiety, a strong interaction between the selector and the selectand can be achieved. The first commercial available phases are based on the Cinchona alkaloids Quinine and Quinidine with an additional weak anion-exchange function offering good separation possibilities for chiral acids. A strong dependence of the capacity from the counter ion of the mobile phase could be demonstrated by Arnell (2009).

In a complementary approach, chiral cation exchangers as well as zwitterionic phases have been developed but not yet commercialized. The good separation properties for a chiral amine with only small sidegroups next to the chiral center ($R1 = -H$, $R2 = -CH_3$) has been recently published by Merck (Helmreich et al., 2010). By screening different selectors in a reciprocal approach an optimized chiral stationary phase could be developed to separate a molecule that has so far not been separated by all other known and commercially available chiral phases.

It has to be once again stressed that it is not always the best choice to separate the final chiral molecule. By screening intermediates as well as easy obtainable derivatives, a dramatically better productivity might be achieved. A better solubility of the feed can especially lead to a much higher overall process performance. The approach of derivative screening is shown in more detail in Strube (2006).

As a summary, Table 3.13 lists the features of the main groups of commercially available chiral stationary phases used for preparative chromatography.

3.1.5.4 Further Developments

The development of new chiral phases for preparative use is still an important topic even though more than a hundred phases are commercially available (recent review in Lämmerhofer, 2010). The main features to be still improved are selectivity, capacity, and mechanical as well as chemical stability, especially against a wide variety of different solvents. Optimization of the selectivity for chromatographic phases is necessary only up to a certain value. Process SMB applications are usually operated with selectivities between 1.3 and 2.0. Using too high a selectivity results in high eluent consumption needed to desorb the extract compound from the stationary phase. Conversely, phases with much higher selectivities would make it possible to use easy adsorption/desorption procedures.

The increase in capacity should always be combined with a high chemical stability against a wide variety of different solvents. The high capacities of chiral stationary phases can be used to their full extent only if a good solubility of the racemate can be found.

A third field of improvement would be the development of chiral stationary phases that combine a low pressure drop with a good efficiency and stability, for example, the pressure stability requirements for the SMB systems could be

Table 3.13 Product features of different classes of chiral stationary phases.

CSP	Application range	Solvent use	Loadability (mg g^{-1} CSP)	Stability, restrictions for use	Availability (>10 kg)
Cellulose-fibres CTA, CTB	Broad	Alkanes, alcohols, water	10–100	Good, large column packing difficult	Yes
Brush-type columns	Small, but focused	Unlimited including scCO$_2$	1–40	Very good	Major types: yes, several manufacturers
Cellulose-, amylose-derivatives coated on silica	Very broad	Alkanes, alcohols, acetonitrile, scCO$_2$	5–100	Good, but solvent restrictions	Yes, several manufacturers
Cellulose-, amylose-derivatives immobilized on silica	Very broad	Unrestricted	5–100	Good	Yes
Cyclodextrin-CSP	Broad	Reversed phase, alcohols	0.1–3	Limited	No
Antibictic-CSP	Broad	Mainly polar organic mode	0.1–1	Limited	Yes
Polymer-silica composites (Chiraspher, Kromasil)	Limited	Mainly alkanes + alcohols	1–20	Very good	Yes

Source: Loadability data from Franco et al. (2001).

reduced. In that context, monolithic stationary phases may offer some possibilities.

3.1.6
Properties of Packings and their Relevance to Chromatographic Performance

The relevant packing properties are divided into bulk and column properties. The former pertain to the bulk powder before it is packed into the column, the latter characterize the chromatographic properties of the packed column.

3.1.6.1 Chemical and Physical Bulk Properties

The bulk composition and the bulk structure depend on the type and chemical composition of the adsorbent and are largely determined by the manufacturing process. Parameters that characterize the bulk structure are the phase composition, phase purity, degree of crystallinity, long- and short-range order, and defect sites, and so on. Special care has to be taken with regard to the purity of the adsorbents. Metals incorporated in the bulk and present at the surface of oxides often give rise to additional and undesired retention of solutes. Remaining traces of monomers and polymerization catalysts in cross-linked polymers are leached during chromatographic operation and may affect column performance. Thus, high-purity adsorbents are aimed for during the manufacturing processes for the isolation of value-added compounds, for example, pharmaceuticals.

Among the physical properties of stationary phases the skeleton density and the bulk density of an adsorbent are of major interest, in particular when the column packing is considered. Skeleton density is assessed as the apparent density due to helium from helium-penetration measurements. For silica, the skeleton density varies between 2.2 and 2.6 g cm^{-3} depending on the bulk structure. The bulk density of a powder is simply determined by filling the powder in a cylinder under tapping until a dense bed is obtained. The bulk density is inversely proportional to the particle porosity: the higher the specific pore volume of the particles the lower is the bulk density. In column chromatography, the packing density is a commonly used parameter, expressed in gram packing per unit column volume. The packing density varies between 0.2 and 0.8 g cm^{-3}. The low value holds for highly porous particles, the high value for particles with a low porosity.

3.1.6.1.1 Morphology

The morphology of the adsorbent controls the hydrodynamic and kinetic properties in the separation process, for example, the pressure drop, mechanical stability, and column performance in terms of number of theoretical plates. In classical applications, granules of technical products were applied with irregularly shaped particles. The desired particle size was achieved by milling and grinding and subsequent sieving. Irregular chips were successively replaced by spherical particles. The latter generate a more stable column bed with adequate permeability. For crude separations, low-cost irregular particles can be used, for example, in flash chromatography.

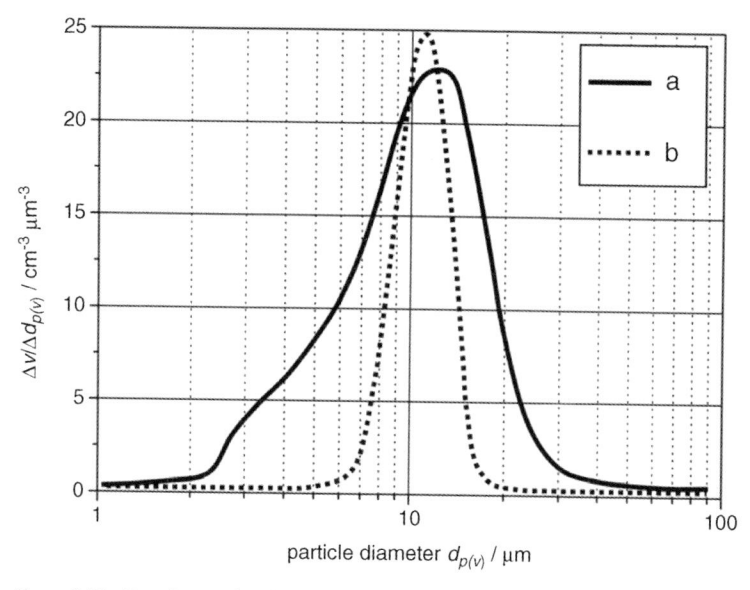

Figure 3.16 Particle-size distribution of silica spheres before (a) and after (b) size classification.

Although columns packed with beads can be supplied by manufacturers with highly reproducible properties, the mechanical stability of columns needs to be improved, particularly in terms of the maintenance of a constant flow rate in automated chromatographic systems such as simulated moving bed processes or high-throughput separations. This problem is solved by the design of monolithic columns composed of polymeric materials or silica. Monoliths are constituted of a bimodal, highly interconnected pore system: flow-through pores in the size range of a few micrometres and mesopores in the 10 nm pore size range. Monoliths consist of highly interconnected macropore systems, providing a much higher permeability and lower pressure drop than particulate columns. Monoliths are mechanically stable and show neither bed-settling nor channeling if the correct cladding is performed. One of the greatest challenges in monolith production is to increase the diameter. Presently, silica monoliths with a diameter of up to 25 mm can be produced. Commercial monolithic silica columns, for example, Chromolith columns of Merck KGaA possess a column permeability equivalent to a column packed with 15 µm particles, but show a column performance of a column packed with 5 µm beads. The major benefit of monolithic columns, however, is their robustness in use, their tailored pore structure, and the tunable surface chemistry.

3.1.6.1.2 Particulate Adsorbents: Particle Size and Size Distribution

Especially when beading processes are applied, the final material has to be subjected to a sizing process to obtain a narrow size distribution. The sized factions are characterized by the average particle diameter and the size distribution. Figure 3.16 shows a differential size distribution based on the volume-averaged particle diameter of the as-made product (Figure 3.16a) and the sized product (Figure 3.16b). As expected,

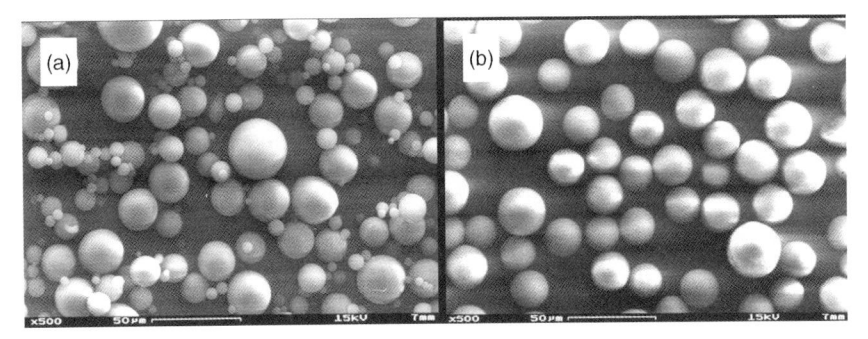

Figure 3.17 SEM pictures of silica spheres before (a) and after (b) size classification.

the amount of small and very large particles is reduced by the sizing process. The particle-size distribution of the sized material has a near Gaussian appearance.

Usually, the width of the distribution is expressed by the ratio of d_{p90}/d_{p10}, which should be lower than 2.5 for the chromatographic application; d_{p90} (d_{p10}) equals to the average particle diameter at 90% (10%) of the cumulative size distribution. Figure 3.17a and b visualize the effect of size classification and efficient removal of fine particles.

The particle size is given by an average value: d_p. The average particle diameter can be expressed as a number average, $d_{p(n)}$, a surface area average, $d_{p(s)}$, a weight average, $d_{p(w)}$, and a volume average, $d_{p(v)}$. Based on statistics the following sequence is achieved $d_{p(n)} > d_{p(s)} > d_{p(w)} < d_{p(v)}$.

Usually, in chromatography the volume-average particle diameter is employed. For comparison, the particle-size distribution based on the number and the volume average is shown for the same silica measured by the same technique (Table 3.14).

Apart from the respective value the distribution expressed as the ratio of $d_{p(90)}/d_{p(10)}$ is different. Based on the volume average, $d_{p(90)}/d_{p(10)}$ is calculated to be 1.66, for the number average it is 3.52.

The particle size of a packing affects two major chromatographic properties: the column pressure drop and the column performance in terms of plate number per column length. For simplicity:

- Δp is inversely proportional to the square of the average particle diameter of the packing and
- the plate number is inversely proportional to the particle diameter.

The optimum average particle size of preparative stationary phases with respect to pressure drop, plate number, and mass loadability is between 10 and 15 μm.

3.1.6.1.3 Pore Texture The pore texture of an adsorbent is a measure of how the pore system is built. The pore texture of a monolith is a coherent macropore system with mesopores as primary pores that are highly connected or accessible through the macropores. Inorganic adsorbents often show a corpuscular structure; cross-linked polymers exhibit a network structure of interlinked hydrocarbon chains

Table 3.14 Comparison of the particle size distribution data of a LiChrospherSi 100 silica.

	Volume statistics	Number statistics
Calculation range	0.96–33.5%	0.96–33.5%
Volume	100%	100%
Mean	8.267 µm	6.240 µm
Median (d_{p50})	8.236 µm	7.138 µm
Mean/median ratio	1.04	0.874
Mode	8.089 µm	7.341 µm
Standard deviation	0.091	0.221
Variance	0.0082	0.049
Skewness	462.5 left skewed	−2.010 left skewed
Size analysis		
d_{p5}	6.067	1.545
d_{p10}	6.475	2.865
d_{p50}	8.236	7.138
d_{p90}	10.740	9.344
d_{p95}	11.540	10.090
d_{p90}/d_{p10}	1.66	3.52

Source: Kindly supplied by Dr. K.-F. Krebs, Merck KGaA, Darmstadt, Germany.

with distinct domain sizes. Porous silicas made by agglutination or solidification of silica sols in a two-phase system are aggregates of chemically bound colloidal particles (Figure 3.18).

The size of the colloidal nonporous particles determines the specific surface area of an adsorbent (Equation 3.2) and the porosity of particles is controlled by the average contact number of nonporous primary particles.

$$a_s = \frac{6}{d_p \cdot \varrho_{app(He)}}. \tag{3.2}$$

The average pore diameter is given by Equation 3.3 and shows that pore structural parameters such as the specific surface area, pore diameter, and specific pore volume are interrelated and, therefore, can be varied independently only to a certain extent.

$$d_{pore} = \gamma \left(\frac{v_{sp}}{a_s} \right). \tag{3.3}$$

As an example, spray drying of nanosized porous particles yields spherical agglomerates with a bimodal pore size distribution: the pores according to the primary particles (intraparticle pores) and the secondary pores (interparticle pores) made by the void fraction of the agglomerated nanobeads. Figure 3.19 shows spherically agglomerated nanoparticles that build up a porous bead. For a given size d_p of the nanoparticles the pore diameter of the interstitial pores

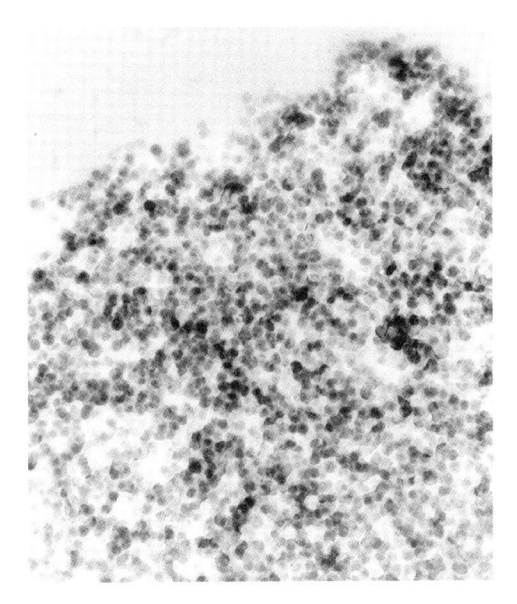

Figure 3.18 TEM image of a silica xerogel (diameter of the primary particles approx. 10 nm).

corresponds to about 40% of d_p. According to porosimetry experiments, this rule of thumb was verified, showing a mean pore size at 300 nm in the case of 750 nm nanoparticles.

3.1.6.1.4 Pore Structural Parameters The pore structure of an adsorbent is characterized by the dimensionality of the pore size (unidimensional channels,

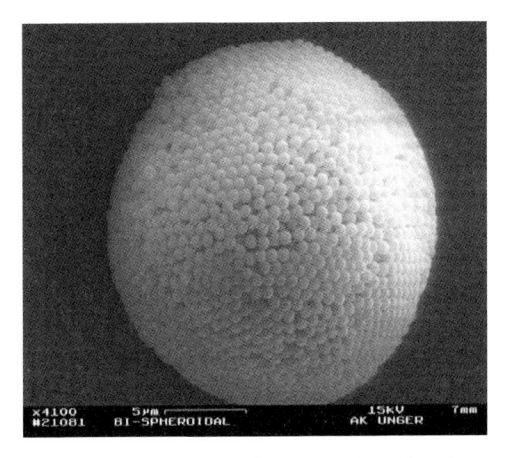

Figure 3.19 SEM picture of a 20 μm spherical agglomerate consisting of 750 nm particles.

three-dimensional pore system), pore size distribution, pore shape, pore connectivity, and porosity. The specific surface area is related to the average pore size. Micropores generate high specific surface areas in excess of $500\,m^2\,g^{-1}$, mesopores have values between 100 and $500\,m^2\,g^{-1}$, and macropores of d_{pore} larger than 50 nm possess values of less than $50\,m^2\,g^{-1}$. Micropores of d_{pore} smaller than 2 nm possess low specific pore volumes of smaller than $0.2\,cm^3\,g^{-1}$. Mesopores have specific pore volumes of the order of 0.5–$1.5\,cm^3\,g^{-1}$ whereas macropores generate much higher porosities. Porosity can change due to swelling with cross-linked organic polymers. Porous oxides exhibit a permanent porosity at common conditions. The pore size distribution of adsorbents with a permanent porosity is determined either by nitrogen sorption at 77 K (micropore size and mesopore size range) or by mercury intrusion (macropore size and mesopore size range). The pore-size distribution can be expressed as a number or volume distribution. However, the pore size distribution does not give any indication of how the pores are interconnected across a porous particle. The most useful parameter in this context is the (dimensionless) pore connectivity n_T, which is derived from the pore network model applied to the experimental nitrogen sorption isotherm (Meyers and Liapis, 1998, 1999). Pore connectivity describes the number of pore channels meeting at a node at a fixed lattice size. Low n_T values indicate a low interconnectivity; n_T can be as high as 18 for highly interconnected pore systems. Such a high interconnectivity leads to fast mass-transfer kinetics and favorable mass transfer coefficients (Unger et al., 2002).

3.1.6.1.5 Surface Chemistry
In *normal-phase* chromatography the native, non-modified adsorbent is employed with organic solvent mixtures as eluents. Normal phase chromatography was the classical chromatography mode performed with native silica or alumina, that is, the adsorbent's surface is hydrophilic and the interaction with the solutes takes place via the hydroxyl groups on the surface. As an example, the surface of silica consists of silanol groups of different types. One can distinguish between free (isolated, nonhydrogen bonded), terminal, vicinal (hydrogen-bonded), geminal, and internal hydroxyl groups (Figure 3.20).

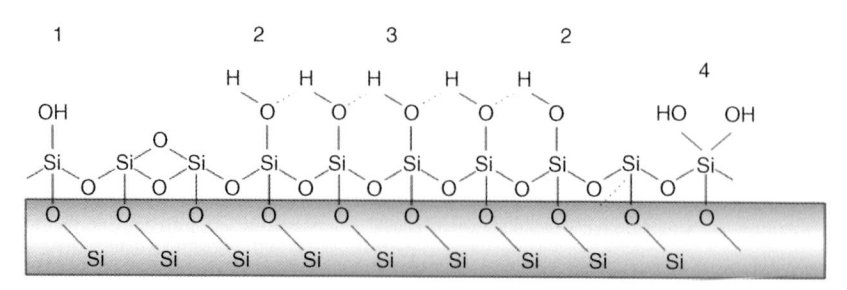

Figure 3.20 Types of silanol groups: isolated (1), terminal (2), vicinal (3), and geminal (4) species.

Table 3.15 Interrelationship between adsorbent characteristics and chromatographic properties.

Method	Parameter derived	Chromatographic properties affected
Light microscopy, scanning electron microscopy	Particle morphology, particle size distribution	Stability of packed bed, column performance
Light-scattering coulter counter	Particle size distribution, volume average particle diameter, number average particle diameter	Stability of packed bed, hydrodynamic column properties, column performance
Gas sorption (nitrogen at 77 K), mercury intrusion/extrusion (mercury porosimetry)	Specific surface area (BET), pore size distribution, average pore diameter, specific pore volume, particle porosity	Retention of solutes, mass loadability, column regeneration, column performance, pore and surface accessibility for solutes of given molecular weight, mechanical stability, column pressure drop, pore connectivity
Atomic absorption spectroscopy, neutron activation analysis, ICP-optical emission spectroscopy	Inorganic bulk and surface impurities of packings	Chemical stability of packing, retention of solutes, peak tailing
Inverse size exclusion chromatography iSEC	Accessibility of ligands in the porous network	Mass loadability, adsorption, and desorption kinetics

Knowledge of the materials surface chemistry is crucial. The required adsorbents are silicas with a chemically modified surface carrying bonded n-octadecyl or n-octyl groups. The long n-alkyl ligands invert the surface polarity from hydrophilic (native adsorbent) to hydrophobic. Silica-based reversed-phase columns are operated with aqueous/organic eluents and inversion of the elution order takes place (hydrophilic components elute first). More details are given in Table 3.15.

3.1.6.2 Mass Loadability

For lab-scale purifications as well as a quick adsorbent-screening method the determination of the resolution as a function of the load is used. Increasing amounts of the feed mixture are injected onto the column and the corresponding resolution of the compounds of interest is determined. With this methodology, the maximum load for 100% recovery and yield can be extrapolated from the graph at the point where the resolution of 1.5 (baseline resolution) cuts the curves for the single adsorbents at a given load. Even if no interactions between the substances in the nonlinear range of the isotherm are taken into account, this approach is useful for the near-linear range with resolution factors down to 1.0–0.8. Figure 3.21 shows the resolution versus load for different adsorbents. All adsorbents are normal phase silica with similar particle sizes. A mixture of two tocopherol isomers from a

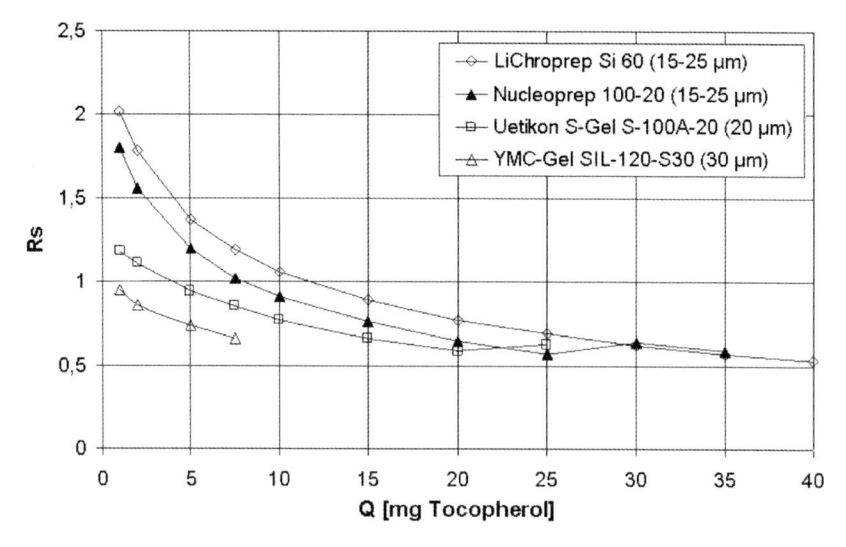

Figure 3.21 Determination of the resolution factor at different loads for four different silicas.

sunflower oil source was injected and the resolution between the two main peaks calculated. The graph shows that the starting resolution is quite different for each adsorbent, which depends on the selectivity α and the efficiency of the column. As all resolution factors decline with increasing amounts of tocopherol loaded onto the column, the higher the load on the column the smaller the differences between the columns. In that region, interactions between the compounds in the feed mixture have to be taken into account, and the complete isotherm should be determined. Nevertheless from Figure 3.21 a good solution for the isolation of amounts of the target compound in milligrams can be easily obtained.

3.1.6.3 Comparative Rating of Columns

The rating of columns is performed according to two aspects:

1) Assessment of column performance data obtained under analytical conditions. This includes:
    - Plate height of the column: The plate height should correspond to two to three times the average particle diameter of the packing.
    - Plate height of the column as a function of the linear velocity of mobile phase: the flow rate should have an optimum where the plate height is at a minimum.
    - Number of plates per column length: Commonly, the plate number is between 5000 and 10 000 to sufficiently separate mixtures.
    - Retention coefficient of a solute.

    All these parameters depend on the mass loadability of the column and change significantly when a critical loadability is reached. The critical mass

loadability of analytical columns is usually reached at a 10% reduction of the retention coefficient or at 50% decrease of column plate number. At higher values, the column is, in chromatographic terms, overloaded.
2) Assessment of parameters under preparative conditions.

Preparative columns are compared with regard to the column productivity, yield, purity, and cost. Productivity is defined as the mass of isolate per kg of packing per day at a set purity and yield. It depends, essentially, on the applied technical chromatographic process. For example, the productivity for the separation of racemates into enantiomers by the simulated bed technology has increased from 2 kg per kg packing per day to about 10 kg of product per kg packing per day. The yield can be low provided the purity is high. Desired purities are sometimes 99.9% and higher. The overall cost is the only measure that is taken for chromatographic production processes. The main cost contributions come from costs related to the productivity (column size, amount of stationary and mobile phase), yield losses of the product, and work-up costs linked to the product dilution.

3.1.7
Sorbent Maintenance and Regeneration

Preparative chromatography in the nonlinear range of the isotherm means applying a high mass load of substance onto the adsorbent. Some compounds from the feed mixture may alter the surface of the adsorbent and change its separation activity. Therefore, regular washing procedures and reactivation should be performed to ensure a long column lifetime. Different goals for washing procedures can be distinguished.

Removal of highly adsorptive compounds (cleaning in place, CIP).
Reconditioning of silica surfaces.
Sanitization in place (SIP) of adsorbents.
Column and adsorbent storage conditions.

3.1.7.1 Cleaning in Place (CIP)
Highly adsorptive compounds should always be removed before subjecting the feed onto the main column. This can be achieved by a precolumn, which has to be changed at regular intervals. Sometimes, simple filtration of the feed over a small layer of the adsorbent is sufficient to remove sticking compounds. Nevertheless, it has to be checked that the feed composition is not drastically changed after the prepurification step so that the simulation of the chromatogram and thus the process parameters are still valid. If compounds stick to the adsorbent (irreversible adsorption) and if the performance of the column in terms of efficiency and loadability decreases, a washing procedure has to be carried out (CIP). For this washing procedure, two general considerations have to be kept in mind:

- The flow direction should be always reversed so that the sticking compounds are not pushed through the whole column.

Table 3.16 CIP procedures for silica-based adsorbents after Majors (2003a) and Rabel (1980).

	Reversed phase		Normal phase	Ion-exchange	Protein removal
	C18, C8, C4, Phenyl, CN		Si, NH$_2$, CN, Diol	SAX, SCX, DEAE, NH$_2$, CM on silica	C18, C8, C4, Phenyl
	Distilled water (up to 55°C)	Distilled water (up to 55°C)	Heptane–methylene chloride	Distilled water (up to 55°C)	Distilled water (up to 55°C)
	Methanol	Methanol	Methylene chloride	Methanol	0.1% trifluoroacetic acid
	Acetonitrile	Isopropanol	Isopropanol	Acetonitrile	Isopropanol
	THF	Heptane	Methylene chloride	Methylene chloride	Acetonitrile
	Methanol	Isopropanol	Mobile phase	Methanol	Distilled water
	Mobile phase with buffer removed	Mobile phase with buffer removed		Mobile phase with buffer removed	Mobile phase

- As the adsorbent has to be washed with solvents of very different polarity, care has to be taken that all solvents used in a row are miscible. Especially when a solvent with a different viscosity is introduced into the column or the mixture of two solvents results in a higher viscosity than the single solvents, for example, methanol and water, the pressure drop can rapidly increase and damage the packing. Therefore, a maximum pressure should be set in the system control and the flow rate for the washing procedure should be reduced if the resulting pressure drop is unknown.

If the column can be opened at the feed injection side, which is not always possible for technical or regulatory (GMP) reasons, the first one or two centimetres of packing can be removed. The top of the bed is then reslurried, fresh adsorbent added, and the column recompressed. Up to a column diameter of 200–300 mm this procedure might make it possible to avoid repacking of a column and ensures the further use of the packed column for some time. Table 3.16 shows some CIP-regimes for silica-based adsorbents with different functionalities.

As a rule of thumb at least 10 to 15 column volumes of each washing solvent should be pumped through the column. No general rules can be given as to whether column washing and regeneration is more economic than replacing the packing material as this strongly depends on the nature of the components to be removed and the type of the packing material. Naturally, if coarse, cheap packing materials are used the solvent and time consumption for washing and regenerating a column is more expensive than unpacking and repacking with new adsorbent.

The cleaning in place protocols for ion-exchange and affinity resins are of even greater importance due to the risk of contamination from the fermentation broth to be purified.

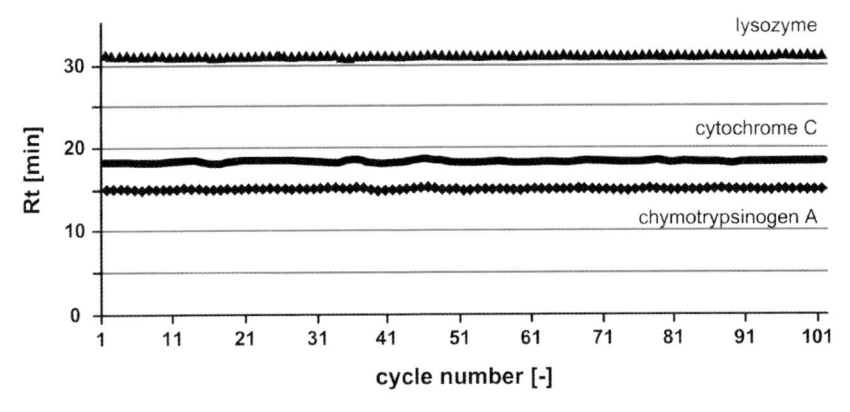

Figure 3.22 Variation of retention time for test protein mixture after 100 cycles; each cycle consisting of 1 h 1 M-NaOH-treatment.

3.1.7.1.1 CIP for IEX The classical CIP-solution for ion-exchange resins typically consists of up to 1 M NaOH-solution which is also accepted by the regulatory authorities. Classical resins show only little loss in capacity and performance if the resins are stored in the NaOH solution. Modern type resins can be operated for a longer time even when CIP is done in a flow-through mode. Figure 3.22 shows the selectivity in standard protein separations for Eshmuno S when operated in alternating cycles of 1 M NaOH cleaning solution for 1 h and running buffer for the chromatographic separation. No variations in selectivity or retention time of the eluted proteins are being observed.

In most cases, quaternary ammonium-based anion-exchange resins show a reduced alkaline stability due to the Hoffmann elimination reaction.

Table 3.17 lists some recommended conditions for Fractogel® IEX sorbents.

3.1.7.1.2 CIP of Protein A-Sorbents Stability of the expensive Protein A-resins is a crucial point. Protein A in its native form has only a limited stability towards caustic conditions. Therefore, careful CIP procedures have to be followed.

Table 3.17 Recommended cleaning and regeneration steps.

Reagent	Concentration	Column volumes (CV)
NaCl	Up to 2 M	2
NaOH	0.1–1 N	1–5
HCl	1–2 N	1–5
Pepsin/HCl	0.1%/0.01 N HCl	1–2
Sodium Laurylsulfate (SLS)/NaCl	2%/0.25 M NaCl	1–2
Isopropanol, acetonitril or ethanol	20%	1–2
Urea	6 M	1–2

Füglistaller (1989) reported Protein A leakage of 1.8 up to 88 ppm (weight/-weight). Caustic and acidic steps as well as treatments with detergents are described in the literature for column hygiene, for example, 0.4 M acetic acid, 0.5 M NaCl, and 0.1% Tween 20. For cleaning the affinity resin, short washing times (contact time max 15 min) with 50 mM NaOH and 1 M NaCl are recommended with a cleaning run every 1–10 separation runs and a sanitization run in-between different product batches.

O'Leary et al. (2001) estimated the column lifetime of carefully maintained Protein A-columns with 225 cycles at production scale. Even though the capacity of the resin was significantly reduced it was still able to reduce the HCP and DNA content to the required minimum (O'Leary, 2001). Jiang et al. (2009) reported the same behavior for the purification of an Fc fusion protein I on either rProtein A Sepharose or MabSelect. Sanitization with 3 CV of 0.1 N NaOH led to a drop in yield of 30% after 66 cycles for rProtein A Sepharose. Using even milder sanitization conditions (50 mM NaOH + 500 mM NaCl) on MabSelect showed the same trend. When Q_{max} of the isotherm was determined for both resins it declined by 38% (rProtein A Sepharose) and 43% (MabSelect) respectively after 100 cycles of cleaning with 0.1 N NaOH. To improve column lifetime two approaches have been followed: the addition of stabilizing excipients, for example, NaCl or Na_2SO_4 and the integration of a stripping step following column elution with the same elution conditions (low pH in the case of Protein A). Three different regimes have been compared:

Strip	Regeneration	Salt (stabilizing excipient)	% of initial yield (at cycle #)
No	0.1 N NaOH	No	70% (cycle 68)
0.1 M Phosphoric acid	50 mM NaOH	0.5 M NaCl	68% (cycle 95)
No	50 mM NaOH	0.5 M NaCl	88% (cycle 95)

As the addition of a stabilizing agent showed the biggest influence different additional agents were tested. Sucrose, xylitol, and ethylene glycol showed the greatest stabilizing influence, while urea, EDTA, glycerol, and L-arginine had little or no effect. The capacity of Mabselect Protein A with 1 M xylitol and 1 M NaCl in the regeneration solution was still at a resin capacity of 25 mg ml^{-1} after an incubation time of 41 h in 50 mM NaOH compared to 17 mg ml^{-1} for a resin incubated with 0.5 M NaCl as the stabilizing excipient. A similar trend could be observed for rProtein A Sepharose as well (Jiang, 2009).

3.1.7.2 Conditioning of Silica Surfaces

One of the drawbacks of chromatography with normal phase silica is the strong and changing interactions the silica surface can undergo with the mobile phase and feed components. In contrast to reversed-phase adsorbents, where the interaction takes place with the homogeneously distributed alkyl-chains bonded to the surface, the silica surface has a lot of energetically different active groups such as free,

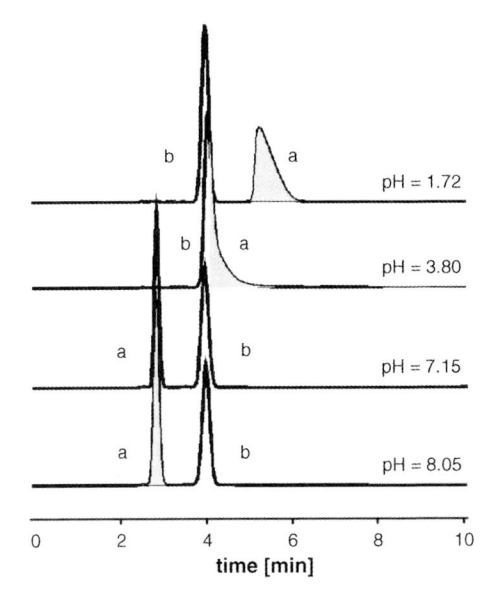

Figure 3.23 Influence of surface pH on the retention time of two test compounds, 2-aminobenzophenone (b) and 4-nitro-3-aminobenzophenone (a).

geminal, and associated silanol groups, incorporated metal ions in the silica surface and water, which might be physisorbed or chemisorbed (Figure 3.37 in Section 3.2.4). It is of great importance to condition the silica before using it in a preparative separation. Especially, the activity, which is determined by the amount of water, and the surface pH have to be controlled to ensure constant conditions from one column packing to another. The surface pH of a silica adsorbent depends heavily on its production process and is especially influenced by the applied washing procedures (Hansen, Helböe, and Thomasen, 1986). The apparent surface pH is measured in a 10% aqueous suspension of the silica. The dramatic influence of different surface pH is illustrated in Figure 3.23 where the surface pH of a LiChroprep Si adsorbent has been adjusted with sulfuric acid and ammonia respectively and two test compounds (2-aminobenzophenone and 4-nitro-3-aminobenzophenone) have been injected onto the columns. The retention time of 4-nitro-3-aminobenzophenone is strongly influenced by the surface pH. Even a reversal of the elution order can be achieved at low pH. The adjustment of surface pH is possible even in large batches by washing the silica with diluted acids, removal of the water, and subsequent operation of the column with nonpolar solvent systems. Under these conditions, the apparent surface pH is remarkably stable during the operation of the column.

A second adsorbent characteristic that should be controlled is the water content of the silica. The water content can vary from 0 to 10 wt.%. As silica is a well-known desiccant (drying agent) care has to be taken with adsorbent types that are adjusted with low amounts of water. Here, significant uptake of water from the air might

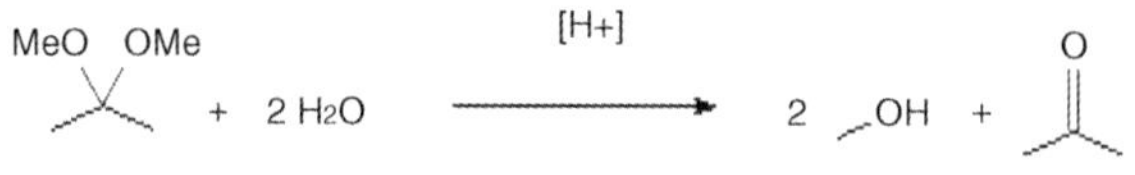

Figure 3.24 Reaction of dimethoxypropane for chemical removal of water from silica adsorbents.

influence the chromatographic conditions. Silica should therefore generally be stored in well-closed containers. The adjustment of silica to certain water contents is a necessary standard procedure in large batch sizes for most large-scale silica manufacturers. If water has to be removed from a column a chemical drying procedure can be applied. Dimethoxypropane reacts with water, resulting in methanol and acetone (Figure 3.24).

For this procedure, the column should be flushed with 10 column volumes of a mixture of dichloromethane, glacial acetic acid, and 2,2-dimethoxypropane (68:2:30 v/v/v). Later, the column should be subsequently washed with 20 column volumes of dichloromethane and 20 column volumes of *n*-heptane.

3.1.7.3 Sanitization in Place (SIP)

For purification of recombinant products from fermentation broth, care has to be taken to prevent microbial growth on adsorbents. Therefore, regular SIP procedures have to be applied. When adsorbents based on organic polymers or sepharose are used (for ion-exchange and adsorption chromatography), decontamination is performed at high pH, for example, by using 0.1–1 M NaOH. With silica-based adsorbents procedures based on salts, solvents, or acids are commonly used due to the chemical instability of silica above pH 8–9. Some recommended washing procedures are listed in Table 3.18.

3.1.7.4 Column and Adsorbent Storage

If packed columns or adsorbents removed from columns are to be stored for a period of time, the adsorbent should be carefully washed and then equilibrated

Table 3.18 SIP procedures.

Solvent	Composition
Acetic acid	1% in water
Trifluoroacetic acid	1% in water
Trifluoroacetic acid + isopropanol	(0.1% TFA)/IPA 40:60 (v/v) viscous → reduce flow rate
Triethylamine + isopropanol	40:60 (v/v), adjust 0.25 N phosphoric acid to pH 2.5 with triethylamine before mixing
Aqueous urea or guanidine	5–8 M (adjusted to pH 6–8)
Aqueous sodium chloride, sodium phosphate or sodium sulphate	0.5–1.0 M (sodium phosphate pH 7)
DMSO–water or DMF–water	50:50 (v/v)

Table 3.19 Recommended solvents for storage and flushing.

Adsorbent type	Flushing solvent	Storage solvent
Normal phase	Methanol or other polar solvent	Heptane + 10% polar compound (isopropanol, ethyl acetate)
Reversed phase	Methanol or acetonitrile, THF, or dichloromethane for oily samples	>50% organic in water
IEX (silica based)	Methanol	>50% organic in water

with a storage solvent. Compounds strongly adsorbed to the adsorbent and, especially, any source of microbial growth have to be carefully removed before the adsorbent is stored. In addition, any buffer substances or other additives, which might precipitate during storage, have to be washed out. For flushing and storage different solvents are recommend (Table 3.19).

Ion-exchange resins are typically stored in either diluted NaOH-solution or ethanolic salt solutions (e.g., 150 mM NaCl, 20% EtOH). Cool storage is an additional requirement for the storage or Protein A-resins.

3.2
Selection of Chromatographic Systems

This chapter aims to provide guidelines for the selection of chromatographic systems related to different key issues of the whole process in preparative chromatography. These key issues do not only focus on the engineer's point of view. Along with engineering parameters such as productivity, yield, and recovery we have to take into account economy, scale of the separation, speed, time pressure, hardware requirements and availability, automation and legal aspects towards documentation, safety, and others. Obviously, rules of thumb related to these criteria may not cover all possible practical scenarios, but they may be useful in avoiding pitfalls.

The selection of chromatographic systems is critical for process productivity and thus also impacts process economy. While the selection of the chromatographic system must be regarded as holding the biggest potential for optimizing a separation process it is also a source of severe errors. The choice of the chromatographic system includes the selection of the adsorbent and the mobile phase for a given sample.

Some general recommendations can be given for the selection of a chromatographic system:

1) Choose a stationary phase that is manufactured reproducibly in large (kg) quantities and is allowed to be used under regulatory issues.
2) Prefer native and pure adsorbents over those with a sophisticated surface chemistry.

3) Use stationary phases whose retention and selectivity can be adjusted by simple solvents. Solvent removal and solvent recovery can become an expensive step.
4) Follow carefully the recommendations of the packing manufacturer with respect to solvent use, washing, and regeneration.
5) Apply pre-columns packed with the same material or with one showing a high adsorptivity, for example, zeolites, towards undesired impurities or by-products. Or use a separate pre-purification (polishing) step of the crude sample prior to chromatography because, very often pre-columns are not as well packed as the main column, which can result in a significant loss of the main column performance.

Besides the pure chromatographic behavior of the stationary phase some further features with regard to the sorbent performance as well as the supplier's reliability should be taken into account especially for a large-scale separation project:

Sorbent related features	Supplier related features
High dynamic capacity	Good documentation (regulatory support file)
Good recovery	Lot-to-lot consistency
Straight scalability	Long-term supply issues, for example, financial stability
Cleanability	Large batch size
Long shelf life	Certificates of analysis
Longevity Mechanical and chemical stability	Possibility to do vendor audits (initially and periodically)
Sufficient selectivity	Reasonable costs
Appropriate pore size	Technical support

The selection of a chromatographic system can be based on a systematic optimization of the system through extensive studies of solubility, retention, and selectivity, but, sometimes, the use of generic gradient runs with standard systems is sufficient.

One source of severe mistakes in system development is a common misunderstanding about preparative chromatography. The classical differentiation between analytical and preparative chromatography is based on the size of the equipment, the size of adsorbent particles, or the amount of sample to be separated. This old view of preparative chromatography leads to a geometric scale-up of systems that are dedicated for analytical applications. Scale-up from analytical to preparative systems is problematic because the goal and restrictions for the two operation modes are different (Table 3.20) and thus the parameters to be optimized are also different. The goal of analytical separations is to generate information whereas preparative applications aim to isolate a certain amount of purified products. For these reasons, preparative and analytical chromatography should be differentiated on the basis of the goal and not the geometric size of the system.

Table 3.20 Differentiation between analytical and preparative chromatography.

Goal	Analytical Information	Preparative Material
Restrictions	Fast information generation All separation modes allowed	Optimal product recovery Recovery of unchanged solutes must be guaranteed
Optimization parameters	• Necessary separation sensitivity • Peak capacity ($1 < k < 5$) • Speed • Selectivity	• Purity • Yield • Productivity • Economics

For analytical separations, the sample can be processed, handled, and modified in any way suitable to generate the required information (including degradation, labeling, or otherwise changing the nature of the compounds under investigation). In preparative mode, the product has to be recovered in the same condition that it was in before undergoing the separation (i.e., no degrading elution conditions, no reaction etc.). This determines the chromatographic process and chromatographic system far more than any consideration of process size.

Preparative HPLC separations are used in all stages of process development. The three main stages in process development as well as examples for the tasks and corresponding criteria for process development are described in Figure 3.25. They differ in time pressure, frequency of the separations, and amount of target products, which dominate the economic structure of the process. At the laboratory and technical scale the development of the separation is dominated by the time pressure and the separation problem includes only a limited amount of sample and a limited number of repetitions. Process optimization in this small- to semi-scale is not helpful, because the effort for optimization will not be repaid in reduced operational costs.

In the early stages of process development, too little sample is available for extensive optimization studies. These studies are not needed and are counterproductive to economic success because separation costs in these early stages are dominated by the time spent developing the separation and not by the operational costs. If purified products are needed for further research (e.g., lead finding) time delays are problematic. Furthermore, the number of different separation problems at this early stage is quite high, while the number of repetitions of one separation is low. In this case, generic gradient runs, which can separate a broad range of substances, provide a quick and effective means of separation.

In production-scale processes economic pressure comes to the fore. Next to economic pressure it is very often not possible to separate the huge amounts involved with noneffective chromatographic separations. Although the time pressure is still present, an optimization of the chromatographic system with extensive studies is required. The investment in an optimized system will be repaid though by the reduced operation costs of a subsequent production process.

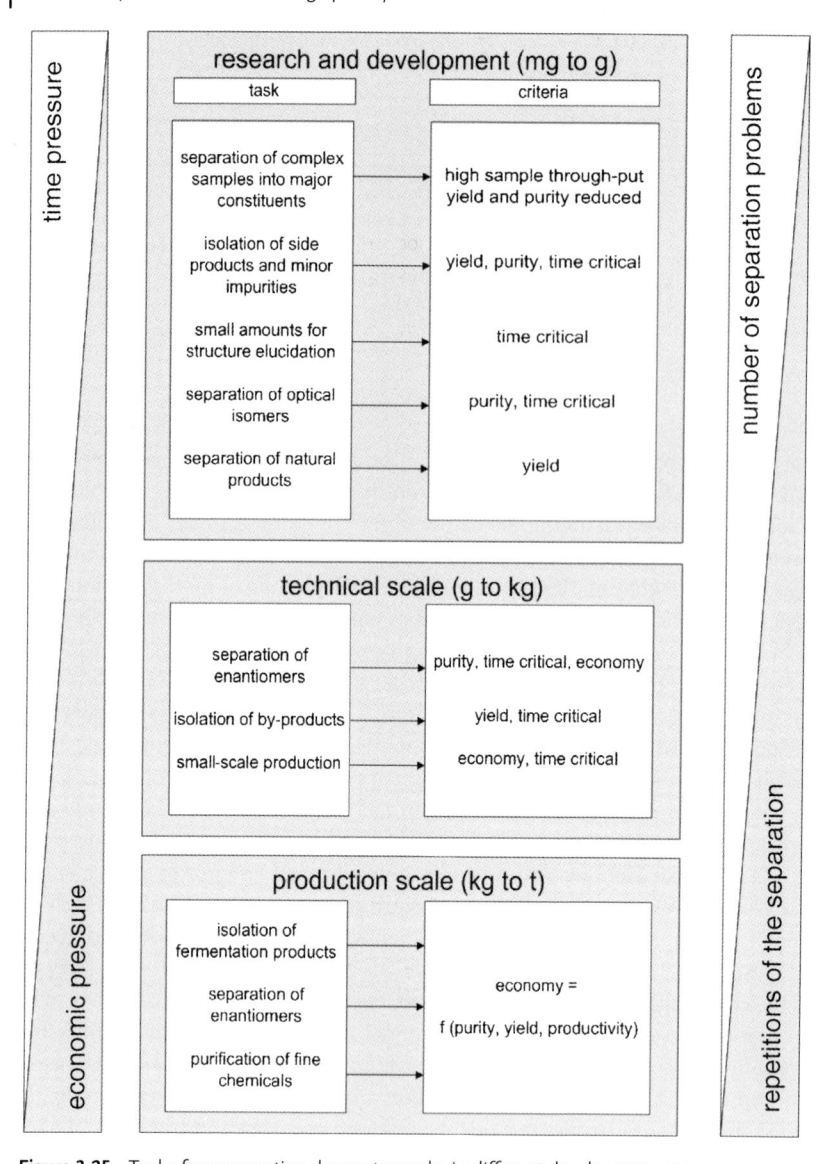

Figure 3.25 Tasks for preparative chromatography in different development stages.

research and development (mg to g)

task	criteria
separation of complex samples into major constituents	high sample through-put yield and purity reduced
isolation of side products and minor impurities	yield, purity, time critical
small amounts for structure elucidation	time critical
separation of optical isomers	purity, time critical
separation of natural products	yield

technical scale (g to kg)

task	criteria
separation of enantiomers	purity, time critical, economy
isolation of by-products	yield, time critical
small-scale production	economy, time critical

production scale (kg to t)

task	criteria
isolation of fermentation products	
separation of enantiomers	economy = f (purity, yield, productivity)
purification of fine chemicals	

Axes: time pressure / economic pressure (left); number of separation problems / repetitions of the separation (right).

However, due to the vast number of variables and the restrictions regarding the environment of the preparative separation one may say: *"There is no general chromatographic system to meet all requirements."*

Comparable to the different scales of processes (Figure 3.25) the users of preparative separations can also be distinguished by their tasks, and differentiated into the following groups.

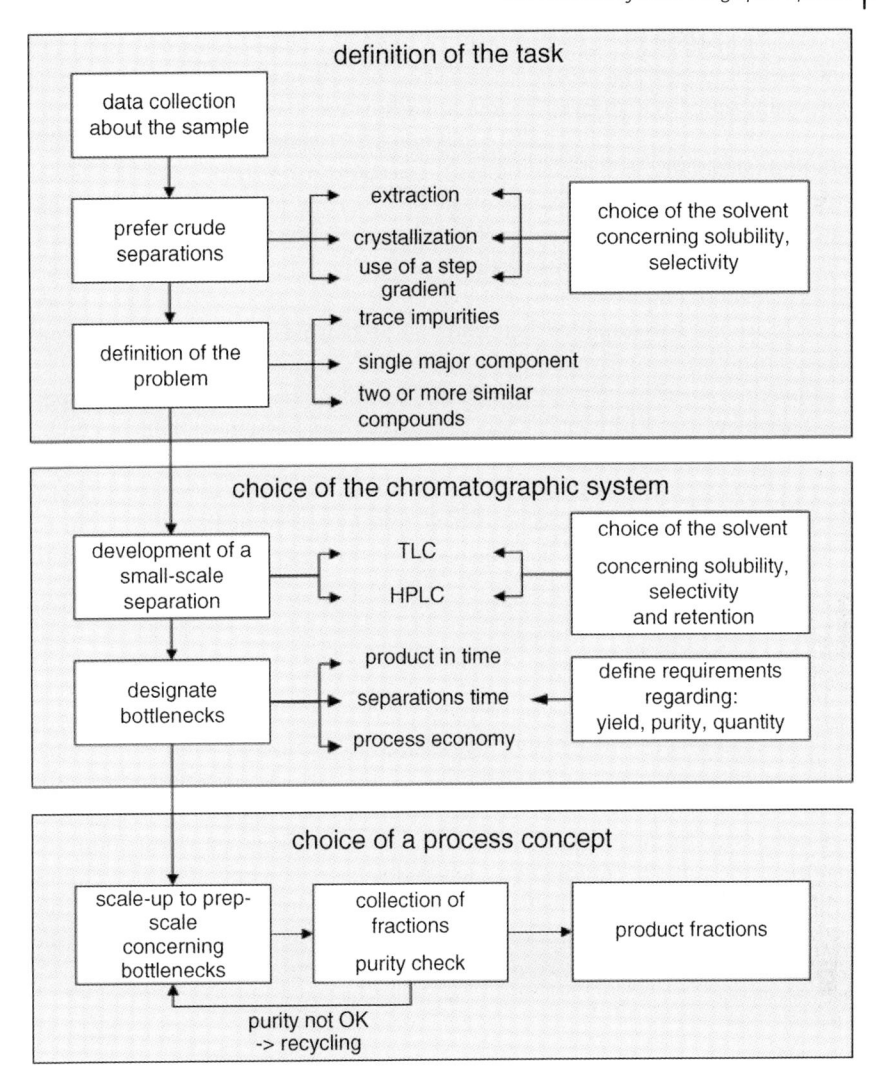

Figure 3.26 Development of a chromatographic method.

There are users, who have to solve many different separation problems in a short time. These users represent the laboratory-scale production of chromatography. The frequency of changes in their tasks and thus the time pressure is high and rules of thumb for a fast selection of suitable chromatographic systems are needed. Other users have to develop the separation for one production-scale process. This group is interested in guidelines for the systematic optimization of the chromatographic system and for the choice of the process concept. Rules of thumb are only used for a first guess of the system.

The development of a chromatographic separation can be divided into three stages, which are discussed here and in Chapter 5. As shown in Figure 3.26, the first step is to define the task (Section 3.2.1), meaning nothing else than a reflection and classification of the separation problem. After the task is defined, a suitable chromatographic system must be found. For this purpose, the properties of chromatographic systems are discussed in Section 3.2.2, while criteria for the selection of chromatographic systems are given in Sections 3.2.3–3.2.8. The last step of method development is the choice of the process concept, which depends on the given task and chosen chromatographic system as well as the available equipment. Different process concepts are introduced in Sections 5.1 and 5.2, while Section 5.3 gives guidelines for the selection of the chromatographic method. The whole development process is exemplified there and thus Section 5.3 can be regarded as a conclusion resulting from Chapters 3 and 5.

Numerous strategies to solve these tasks have been presented and there is nothing really new to say. The main areas of progress have been in equipment design, modeling, and simulation tools. Nowadays, it is possible to predict the chromatographic behavior and economy of a separation task. Next to these predictive procedures the operator of a plant is also aided during process operation by advanced process control systems and documentation of the separation, for example, with respect to GMP regulations.

3.2.1
Definition of the Task

Any selection of a chromatographic system for preparative isolation of individual components should start with a definition of the task. The starting point can be described as careful reflection of the main criteria influencing the separation.

Good advice for primary actions is to collect all available data describing the nature of the sample. Information exchange is especially important if synthesis and separation of the products are done by different departments. For this purpose, a rough characterization of the sample and the preparative task made by following the questionnaire of Table 3.21 eases the planning of experimental procedures.

Table 3.21 Questionnaire for the collection of the information.

- What is the consistency of the sample (solid, liquid, suspension, etc.)?
- What is the source of the sample (chemical synthesis, fermentation, extraction from natural products, etc.)?
- Which possible solvents for the sample as well as solubilities are known?
- Is the sample a complex mixture of unknown composition?
- Which chemical or physical properties of the sample are available (stability (chemical, oxidative), structures, UV spectra as well as toxic and biohazard data)?
- Is the structure, for example, physical, or chemical behavior of the target compound(s) known?
- Is the structure, for example, the physical, or chemical behavior of the impurities known?
- Is any kind of chromatographic data available?

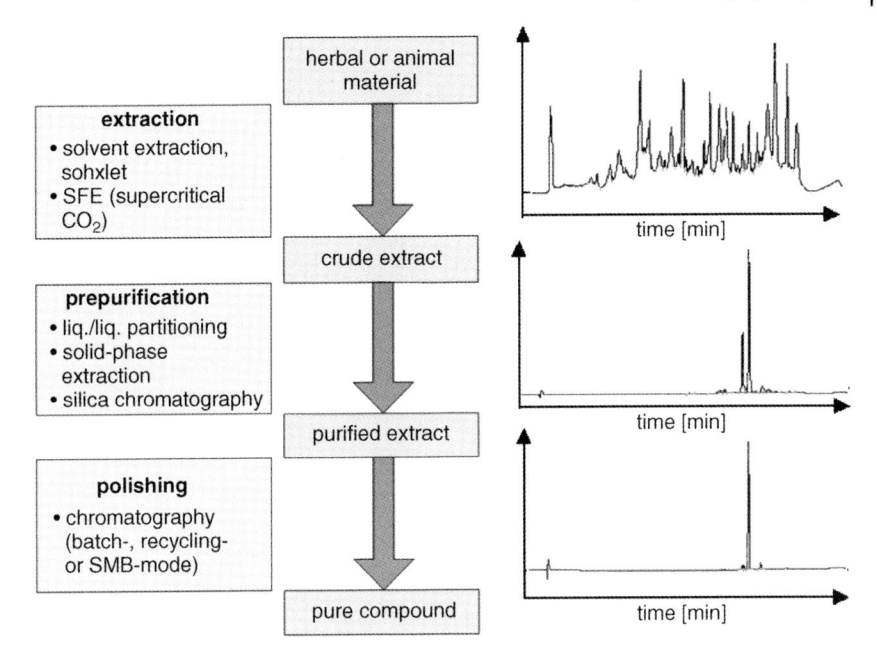

Figure 3.27 Purification of Paclitaxel.

Before starting the selection of any chromatographic system the use of crude separation steps such as extraction, crystallization, or flash chromatography should be checked. The separation task can be simplified by the use of these crude separations if impurities or the target product are removed in the first capture steps.

An example of these steps of downstream processing is the isolation of Paclitaxel, an anticancer drug that is extracted for example from the pacific yew tree. This is a very challenging task due to the low concentration (0.0004–0.08%) in the bark (Pandey, 1997; Pandey et al., 1998). The target product is extracted from the herbal material by Soxhlet extraction. The concentrated extract contains a large range of impurities, which is shown in the first chromatogram of Figure 3.27. Many prepurification steps are necessary before the separation is downsized to a separation problem of paclitaxel and closely related taxanes. Owing to the difficulty of this last separation step, paclitaxel is isolated by preparative chromatography.

The isolation of paclitaxel exemplifies that most preparative separations must be downsized to a level where a limited number of individual compounds are present to ease the final purification steps. This downsizing of the separation problem can be done by crude separations or by a cascade of consecutive chromatographic separation steps. One finally ends up at a point where a multicomponent mixture with a broad concentration range of the different substances has to be fractionated to a series of mixtures. This approach is described in Figure 3.28. In general, a mixture can be split into three types of fractions, where each represent a specific separation

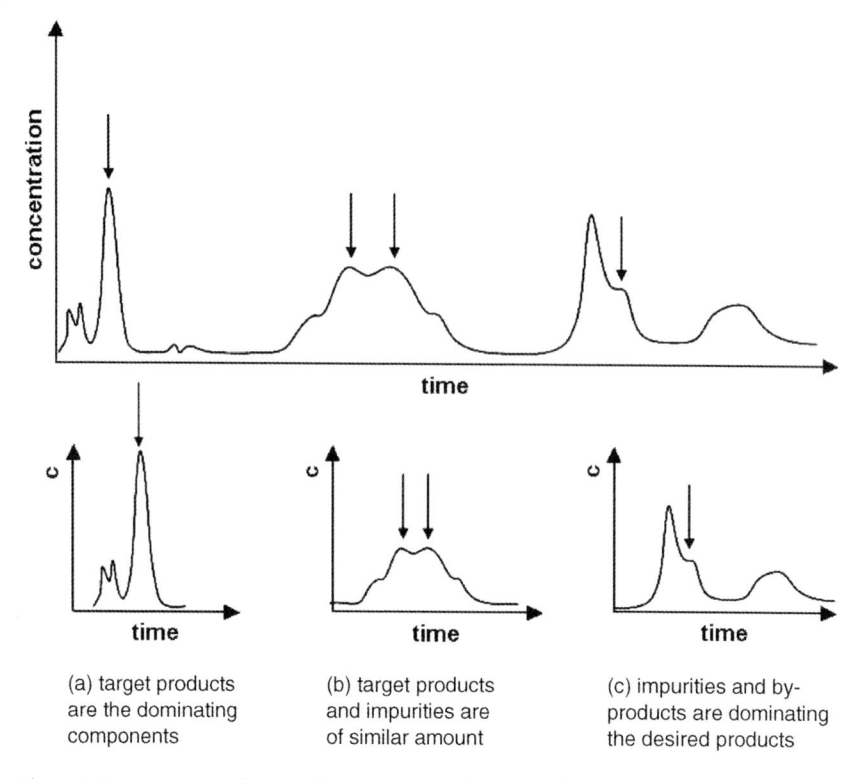

Figure 3.28 Main types of preparative chromatographic-separation scenarios.

(a) target products are the dominating components

(b) target products and impurities are of similar amount

(c) impurities and by-products are dominating the desired products

problem. These three fractions exemplify possible separation scenarios that differ with regard to the ratio of target products and impurities.

The first separation scenario is the most comfortable problem. The target product is the major component in the mixture and thus it is easy to reach a yield and purity in the desired range. Thermodynamic effects can be used to displace impurities during the separation process.

In the second scenario, target products and impurities are present in similar amounts. The separation of racemic mixtures exemplifies this scenario. The problem of racemic mixture formation often occurs during chemical synthesis, where 50% of the mixture consists of the wanted enantiomer (eutomer) and 50% is the unwanted enantiomer (distomer). In this case, competitive adsorption and the elution order of the enantiomers are of special interest (Section 3.2.6).

The last scenario represents the most difficult separation problem. The concentration of the target product is quite small and thus it is very important to reach, simultaneously, high yield and high purity. Because of these special constraints the chromatographic system must be selected carefully. An impropriate system may result in a wrong elution order and thus prevent separation of the products.

Solute
- structure
- separation problem (lipophilic/steric)
- Impurities

solubility
stability

spec. adsorption
(selectivity)

Process Performance
- retention
- selectivity
- resolution
- productivity
- eluent consumption

Mobile phase
- viscosity
- purity
- modifier
- temperature
- composition

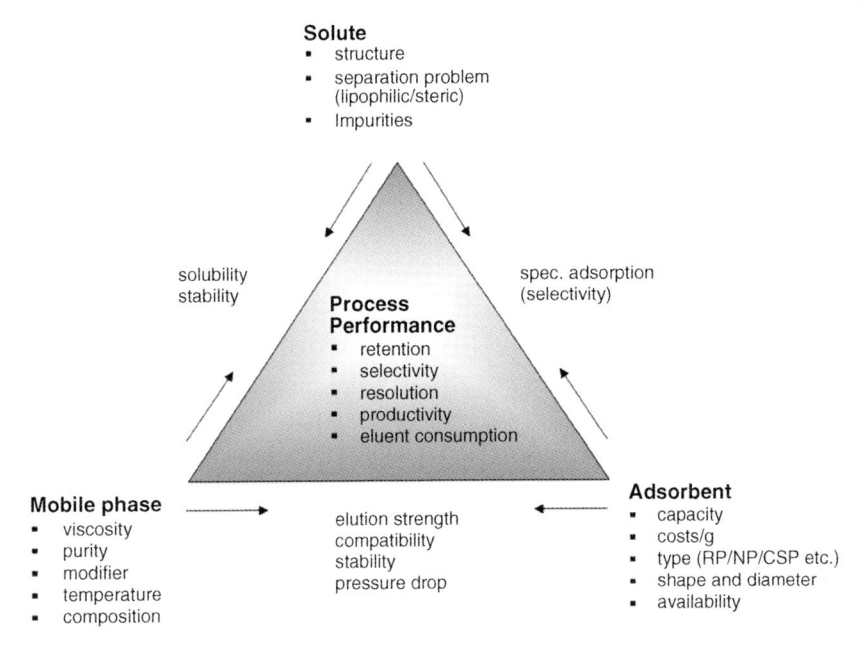

elution strength
compatibility
stability
pressure drop

Adsorbent
- capacity
- costs/g
- type (RP/NP/CSP etc.)
- shape and diameter
- availability

Figure 3.29 Elements of the chromatographic system.

The choice of chromatographic system and the process concept are influenced by the classification of the separation problem into one of the three scenarios of Figure 3.28. This section focuses on the influence of the chromatographic system, while the influence on the process concept is explained in Section 5.3. Here, it should be kept in mind that the elution order of the components is essential for the whole process and the elution order is determined by the chromatographic system. Especially if one component is in excess, as in scenarios (a) and (c) in Figure 3.28, the use of thermodynamic effects such as displacement or tag along are a special source for optimization as well as for severe errors and mistakes (Section 2.6).

Before a chromatographic system is selected, its attributes have to be defined. As described in Section 2.1 a chromatographic system consists of the adsorbent (stationary phase), mobile phase (eluent, solvent, desorbent), and solutes (samples, analyte, etc.). Figure 3.29 illustrates the interrelationship of these three constituents. The selection of the chromatographic system is influenced by their properties and their interaction. These properties are described in this section, while rules and criteria for the selection of the chromatographic system are explained in Sections 3.2.3–3.2.8.

The solute is the product of the prepurification steps. It contains the target product and the impurities. Although the solute is a part of the chromatographic system it is not a free parameter like the eluent or the adsorbent. Subsequent to previous crude separations the composition of the feedstock is fixed and the chromatographic system is completed by the choice of mobile phase and adsorbent.

3.2.2
Mobile Phases for Liquid Chromatography

The first step of system development is the choice of the mobile phase. While the use of neat solvents is preferred, very often, a mixture of solvents must be used to obtain good results. The selection of the mobile phase should be based on:

- Throughput
- Stability
- Safety concerns
- Operating conditions.

The most important criterion for solvent selection is throughput, which mainly depends on a sufficient solubility of the solutes and the corresponding selectivity of the separation. Because solubility and selectivity depend on the interaction between the three elements of the chromatographic system, the selection of the mobile phase dependent on these parameters is further discussed in Sections 3.2.3–3.2.8.

Along with these interactions there are a lot of properties of the neat solvents, with regard to stability, safety, and operation conditions, which help to decrease the number of suitable solvents for preparative chromatography. In contrast to analytical separations, the number of applicable solvents for preparative use is decreased by these limitations. Snyder has examined the behavior of 74 different solvents for general use in liquid chromatography (Snyder and Kirkland, 1979). Only 27 solvents have been proven for use in preparative separations. Solvents that are said to be a good starting point for the evaluation are marked in gray in Table 3.22.

3.2.2.1 Stability

The stability of all components of the chromatographic system must be assured for the complete operation time. The solvent must be chemically inert towards all kinds of reaction. Neither an instable solvent, which for example tends to form peroxides, nor a solvent that reacts with the sample or the adsorbent is suitable for an economically successful solution of the separation problem. Of course, corrosion of the HPLC unit must be prevented, too.

In chromatographic systems operated with a high percentage of water or buffer the microbial contamination of the solvent has to be carefully monitored.

3.2.2.2 Safety Concerns

The safety of preparative processes depends on the flammability and toxicity of the solvent. Owing to the huge amounts of solvents handled in preparative chromatography, low flammability should be preferred. Flammability is described by the vapor pressure, the explosion limits, or the temperature class of the solvent. Generally, the use of flammable solvents can not be avoided and thus the risk must be minimized by good ventilation and other precautions in the laboratory.

Table 3.22 Properties of solvents for preparative chromatography.

Solvent	Solvent strength	Boiling point (°C)	Viscosity$_{25°C}$ (cP)	RI (−)	UV cut-off (nm)	LD 50 ($\mu g\,kg^{-1}$)	ICH-class	Vapor pressure 25 °C (bar)	Explosion limit (vol. %)
n-Hexane	~0	69	0.31[a]	1.375	195	28710	2	0.202	1.2–7.8
n-Butyl ether	2.1	142	0.64	1.397	220	7400		—	0.9–8.5
di-Isopropyl ether	2.4	68	0.38	1.365	220	8470		0.200	1–21
Methyl tertiary butyl ether	2.7	55	0.27[a]	1.369	210	3870	3		
di-Ethyl ether	2.8	35	0.24	1.35	218	1213		0.669	1.8–
n-Butanol	3.9	118	2.6	1.397	210	790	3	0.009	1.4–11.3
2-Propanol	3.9	82	2.4[a]	1.377	205	5045	3	0.05	
1-Propanol	4	97	1.9	1.385	240	1870	3	0.028	
Ethanol	4.3	78	1.2[a]	1.361	195	7060	3	0.079	3.4–19
Methanol	5.1	65	0.55[a]	1.328	205	5628	2	0.169	5.4–44
Tetrahydrofuran	4	66	0.55[a]	1.407	212	1650	2	0.216	
Pyridine	5.3	115	0.88	1.507		891	2	0.028	
Methoxy ethanol	5.5	125	1.6	1.4	210	2370	2	0.013	
Dimethylformamide	6.4	153	0.8	1.428	268	2800	2	0.005	
Acetic acid	6	118	3.3	1.447		5800	3	—	4–17
Formamide	9.6	210	3.3	1.447	233	1600	2	—	
Dichloromethane	3.1	40	0.44[a]	1.424	233	1600	2	0.573	—
1,2-Dichloroethane	3.5	83	0.78	1.442	228	670	2	0.105	
Ethyl acetate	4.4	77	0.45[a]	1.372	256	5620	3	0.126	2–12
Methyl ethyl ketone	4.7	80	0.38	1.376	329	2600	3	—	
1,4-Dioxane	4.8	101	1.2	1.42	215	5200	2	0.050	2.2–13
Acetone	5.1	56	0.36[a]	1.359	330	5800	3	0.306	

(continued)

Table 3.22 (Continued)

Solvent	Solvent strength	Boiling point (°C)	Viscosity$_{25°C}$ (cP)	RI (–)	UV cut-off (nm)	LD 50 ($\mu g\,kg^{-1}$)	ICH-class	Vapor pressure 25 °C (bar)	Explosion limit (vol. %)
Acetonitrile	5.8	82	0.38[a]	1.344	190	2460	2	0.118	3–16
Toluene	2.4	111	0.59[a]	1.497	284	636	2	0.038	1.4–8
Benzene	2.7	80	0.6	1.498	280	930		—	
Chloroform	4.1	61	0.53	1.443	245	908	2	0.259	
Water	10.2	100	1[a]	1.333	<190			0.032	

a) Value for 20 °C.

Using higher amounts of alcohols in aqueous buffers may require specific attention regarding explosion protective equipment and environment. The regulations given by the authorities differ from country to country.

The toxicity of the solvent is important with reference to safety at work and product safety. Toxicity is classified, for example, by the LD_{50} value. If no alternative solvent is available any danger of toxification of the employees during the production process has to be avoided by precautions in the laboratory. The safety of the product is another concern linked to the choice of certain eluents. A contamination of the product might occur through the inclusion of solvent in the solid product due to inadequate drying of the product. The residual amount of solvents in the product is regulated by the different pharmacopoeia as well as the International Guidelines for Harmonization (ICH-Guideline Q3C Impurities: Residual Solvents www.fda.gov/cber). Class 1 solvents should be avoided by any means, while the use of Class 2 solvents can not be avoided and, therefore, has to be carefully optimized. If possible, Class 3 solvents with low toxic potential should be used. The ICH classification as well as the classification limit of the single solvents in pharmaceutical products is given in Table 3.22.

3.2.2.3 Operating Conditions

For operational reasons it is important to look at detection properties, purity, recycling ability, and viscosity of the mobile phase. If solvent mixtures are used as mobile phase, the miscibility of the solvents is an absolute condition for their use, which has to be guaranteed for the whole concentration range of the solute as well (Figure 3.30). For example, acetonitrile and water are miscible, but when sugars are added at high concentrations (e.g., fructose) the system demixes.

The detection system is another boundary condition for the choice of the mobile phase. In most cases ultraviolet (UV) absorbance or refractive index (RI) detectors are used. An imprecise detection leads to insufficient recognition of the target substances as well as the impurities, thus causing purity problems. To avoid these problems the UV cut-off value and the RI are given in Table 3.22 for different pure solvents.

UV detection is based on the UV absorption of the solutes and, thus, the detection wavelength is determined by the UV spectra of the solutes. The mobile phase must be highly transparent at the detection wavelength. This is most important for solutes with low UV activity, due to missing double bonds. To detect these solutes it might be necessary to choose a wavelength below 220 nm. In this range, most solvents absorb UV light significantly as well. The UV absorbance of the solvent is described by the UV cut-off value or by plots and tables of absorbance at different wavelengths, which are documented in the technical data sheet of the solvent. The UV cut-off, which is given in Table 3.22 for the most common solvents, is defined as the wavelength at which the absorbance of the pure solvent measured in a 1 cm cell against air reaches 1 absorbance unit (Poole, 2003). It represents the lower limit for the wavelength, at which the solvent absorbance is getting problematic. Due to this definition, the solute should have a significant UV absorbance at wavelengths higher than the UV cut-off value. Otherwise only the absorbance of the solvent is

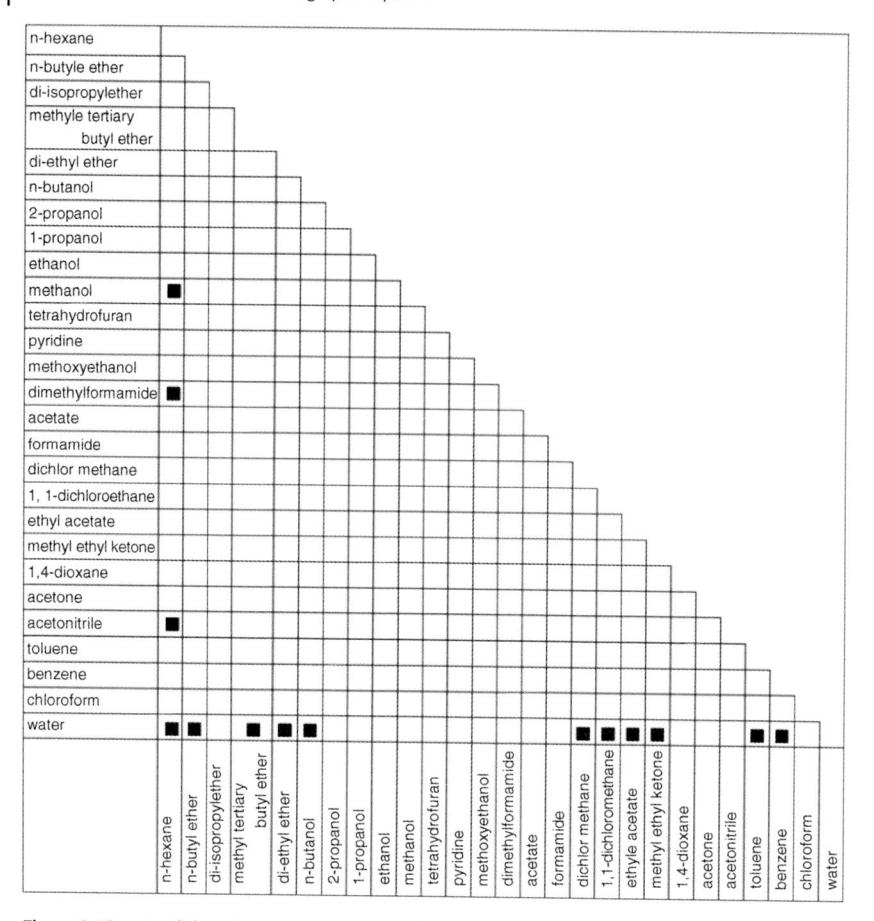

Figure 3.30 Miscibility of preparative HPLC solvents.

measured while the solutes cannot be detected. Since the UV cut-off values in Table 3.22 are measured for pure solvents these data should be regarded as a criterion for the exclusion of UV detection if certain solvents, for example, acetone, are used. For technical solvents it is advisable to rely on absorbance data in the technical data sheet, because the detection limits of different solvents are influenced by impurities within the solvents. Traces of impurities can increase the UV cut-off by 50–100 nm. For example, saturated hydrocarbons such as n-hexane are reasonably transparent down to 190 nm, but even "research grade" solvents often contain a few tenths of a percent of olefins of similar boiling point (e.g., 1-hexene). This contamination is already sufficient to make the solvent opaque at wavelengths below 260 nm (Snyder and Kirkland, 1979).

RI detection depends on the difference in refractive index of the solute and the solvent. If the refractive index of the solute is known, a solvent with a significantly different refractive index, and thus high detection selectivity, should be chosen.

As mentioned before, detection problems can result from impurities in the solvents. For the production of pure substances, nonvolatile impurities in the solvent are also problematic. In most cases, the solvent is removed after separation by evaporation and thus nonvolatile impurities accumulate in the product fraction. For these reasons, the use of HPLC grade solvents is recommended and the availability of appropriate amounts has to be assured.

Dependent on economic pressure, the recycling ability of the solvent is becoming ever more important. Especially in case of production-scale processes the recycling of the mobile phase is essential for economic success. For solvent recycling the following rules should be taken into account:

Prefer pure solvents to binary or ternary mixtures.
With binary mixtures prefer azeotropic mixtures.
Prefer solvents with low boiling points and enthalpies of evaporation.

The last decision parameter concerning process operation is the viscosity. Process economy is influenced by the viscosity in two ways. The pressure drop of the system is proportional to the viscosity of the eluent and, in most cases, the pressure drop is the limiting factor for the flow rate and, thus, productivity. High viscosity mobile phases also show low diffusion of the solutes and thus a worse mass transfer inside the pore system.

Table 3.22 lists the viscosities of pure solvents. For nonaqueous solvent mixtures they can be calculated with Equation 3.4 from the viscosities of the pure solvents and their mole fractions X_A and X_B (Snyder, Kirkland, and Glajch, 1997).

$$\log(\eta) = X_A \cdot \log(\eta_A) + X_B \cdot \log(\eta_B). \tag{3.4}$$

Especially with strongly associating solvent mixtures (alcohol–water, acetonitrile–water) the viscosity shows anomalous variations with composition. The viscosity of the mixture is very often larger than the viscosity of the pure solvents (Figure 3.31). For example, the maximum viscosity of an ethanol–water mixture is more than twofold higher compared with the pure solvents. Next to the composition, the viscosity is strongly influenced by temperature. At higher temperatures the viscosity reduces dramatically. However, chromatographic processes are very often operated at ambient temperature, because it is difficult and expensive to guarantee a constant elevated process temperature. Further information for a few solvent mixtures is given by Gant, Dolan, and Snyder (1979).

3.2.2.4 Aqueous Buffer Systems

The choice of aqueous buffer systems depends on different factors: protein stability, resin performance, waste water treatment, and standard operation conditions, for example, standardization of solvents in a production facility.

The most common buffer systems used in biochromatography are listed in Table 3.23.

Positively charged buffering ions should be used on anion exchangers to avoid an interaction or binding to the functional group. Therefore Tris (pK_a 8.2) is preferred

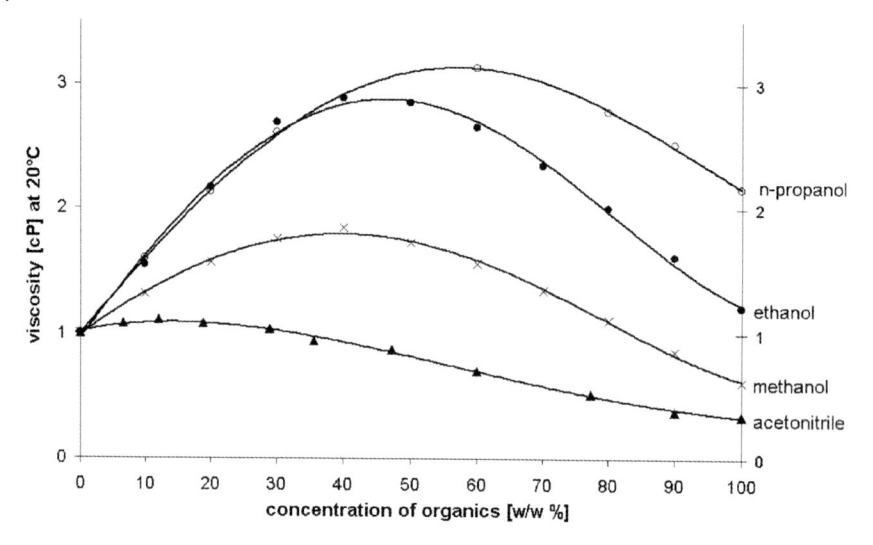

Figure 3.31 Viscosity of different mixtures of water and organic solvents.

with Cl⁻ as counterion. For cation exchangers, the buffering ion should be negatively charged, for example, carbonate, acetate, or MES, and the counterion is K^+ or Na^+. Phosphate buffers are generally used on both exchanger types. The buffer concentration is in the range of 10–50 mM.

Table 3.23 Common buffer compounds used in biochromatography.

Common name	pK_a at 25 °C	Buffer range	Molecular weight	Full compound name
Phosphate	2.15	1.7–2.9	94.9	
Glycil-glycine	3.14	2.5–3.8	132.1	2-[(2-aminoacetyl)amino]acetic acid
Acetate	4.76	3.6–5.6	60.05	
Maleate	5.13	4.0–6.0	116.05	
MES	6.15	5.5–6.7	195.2	2-(N-morpholino)ethanesulfonic acid
Citrate	6.40	5.5–7.2	192.1	
PIPES	6.76	6.1–7.5	302.4	Piperazine-N,N'-bis(2-ethanesulfonic acid)
SSC	7.0	6.5–7.5	189.1	Saline sodium citrate
MOPS	7.20	6.5–7.9	209.3	3-(N-morpholino)propanesulfonic acid
HEPES	7.55	6.8–8.2	238.3	4-2-Hydroxyethyl-1piperazineethanesulfonic acid
Tricine	8.05	7.4–8.8	179.2	N-Tris(hydrocymethyl)methylglycine
Tris	8.06	7.5–9.0	121.14	Tris(hydroxymethyl)methylamine
Bicine	8.35	7.6–9.0	163.2	N,N-bis(2-hydroxyethyl)glycine
TAPS	8.43	7.7–9.1	234.3	3-{[(tris(hydroxymethyl)methyl]amino} propanesulfonic acid

Table 3.24 Buffer additives.

Additive	Typical concentration	Effect
2-Mercaptoethanol	10 Mm	Prevent protein oxidation
Dithiothreitol (DTT),	1 mM	Prevent protein oxidation
Glycerol	20%	Polarity reductant, reduces hydrophobic interactions
$MgCl_2$	5 mM	Chaotropic salt, enhances solubility for hydrophobic proteins
Benzamidine-HCl	1 mM	Serine protease inhibitor
Chymostatin	10 µg ml^{-1}	Peptide protease inhibitor
Ethylenedimaintetraacetic acid EDTA	2 mM	Bind heavy metals, avoid poisoning of sensitive proteins, deactivate metalloproteases, reduce sulfhydryl oxidation
Triton	0.5%	Nonionic detergent, relaxing hydrophobic interactions
Sodium cholate	0.25%	Ionic detergent, relaxing hydrophobic interactions
α-Methyl-mannoside	0.25 M	Elution buffer, esp for Con A affinity chromatography
Sodium azide	0.02%	Prevent bacterial growth (not to be used with peroxidises and heme proteins)

In addition to the buffering salts, special additives are used in some cases to stabilize the target protein or to enhance the elution and selectivity. Table 3.24 lists some of these additives together with the effect they are offering. As any additive is increasing the complexity of the chromatographic system it should be carefully considered if the result is justifying its introduction.

3.2.3
Adsorbent and Phase Systems

The last two sections characterized the elements of a chromatographic system and discussed different separation tasks. This section provides guidelines to develop the chromatographic system for a given separation problem. The classical recommendation for preparative separations performed by elution chromatography is as follows (Figure 3.32):

Find a mobile phase with high solubility, a compatible stationary phase, and then optimize selectivity, capacity, and efficiency followed by a systematic increase in column load.

This is certainly true for the technical and production scale domain, but for high-throughput separations or other projects, where time dominates productivity, other solutions have to be applied based on limited or even not optimized methods.

These two main fields of application for chromatography result of course in a dissimilar approach to selecting the chromatographic system. If time pressure

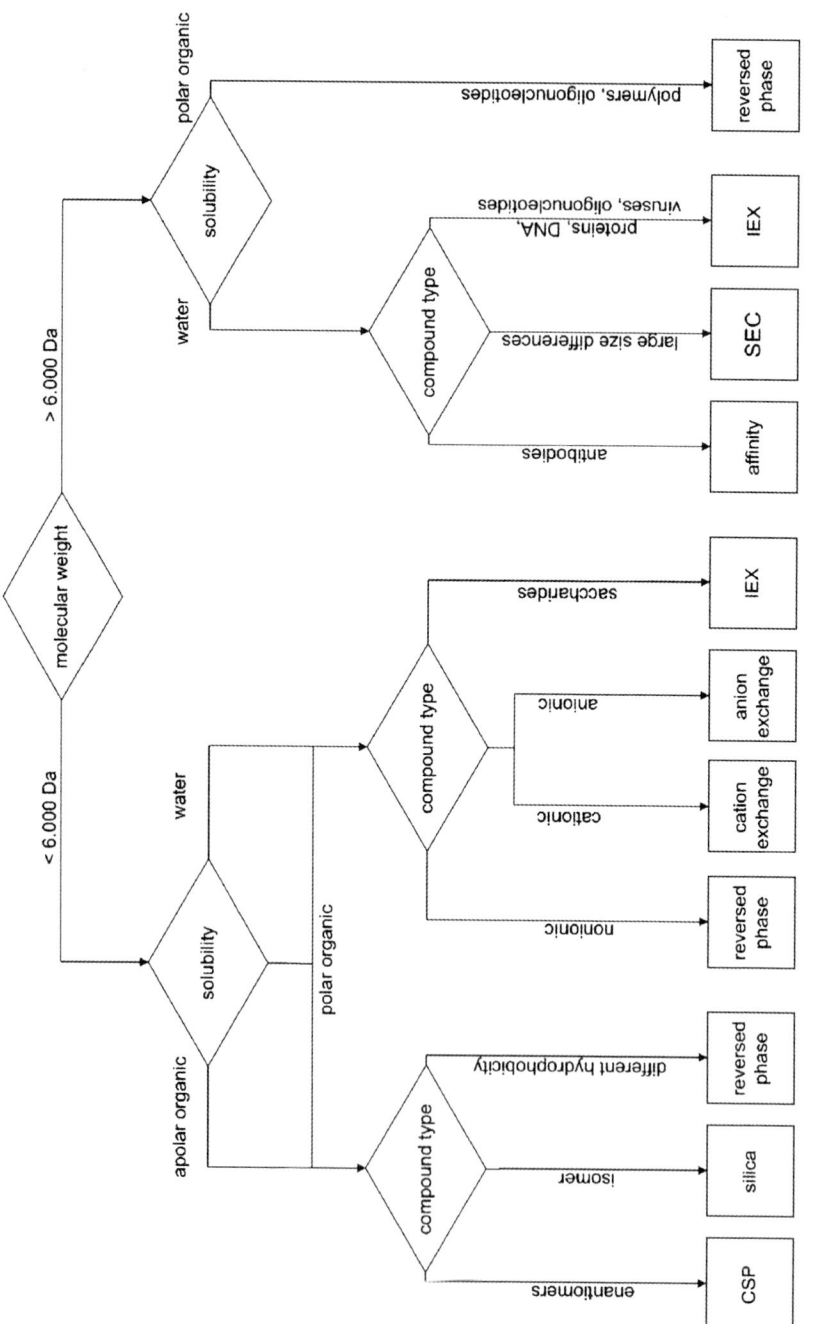

Figure 3.32 Selection of phase systems dependent on the separation problem.

minimizes the possible number of experiments the use of generic gradient systems (Section 3.2.3.4) is recommended. In such cases, solubility problems are also important, but they should be solved quickly (Section 3.2.3.2) and not by extensive solubility studies.

For production-scale processes, economic pressure arises and thus a successive optimization of the chromatographic system will pay off by a reduction in operating costs. Therefore, the development of the chromatographic system has to start with the search for a suitable mobile phase with high solubility for the given solute (Section 3.2.3.1) and a compatible adsorbent. Dependent on the nature of the phase system the separation is optimized by adjusting the mobile phase composition (Sections 3.2.4–3.2.6).

3.2.3.1 Choice of Phase System Dependent on Solubility

The most important step in developing a chromatographic system is the choice of mobile phase. Although the mobile phase influences the separation process in many ways, the main decision parameters for the choice of the mobile phase are the maximum solubilities c_s of the target components and the selectivity of the separation. Solubility is essential for the productivity of the chromatographic separation process, because it is a precondition for exploiting the whole range of column loadability.

An old alchemist maxim, "similia similibus solvuntur" ("like dissolves like"), is the oldest rule for selecting suitable solvents, meaning that the nature of the solute determines the nature of the solvent. Due to the classification of phase systems in Section 3.2.3, organic and aqueous solvents are distinguished. Non-polar to medium-polar substances show best solubility in typical organic solvents while medium-polar to polar substances show best solubility in aqueous solvents.

Referring to this differentiation in the nature and the polarity of suitable solvents a first classification concerning the adsorbent and thus the phase system can be made. Figure 3.32 describes how the solvent and the nature of the separation problem influence the choice of adsorbent. Separation of enantiomers needs CSP, which will be discussed in Section 3.2.6.

NP phase systems are used for the separation of nonionic, apolar to medium-polar samples. If the sample is soluble in organic solvents an NP phase system should be applied as a first test system (e.g., silica and organic mobile phase). However, if the sample shows best solubility in aqueous solvents the sample is quite polar. In this case, RP phase systems with semi-polar (cyano, amino or diol) or apolar packings such as alkylsilica should be used.

Solvent mixtures are often used to obtain a desirable chromatographic separation. For this reason, the elution strength of the strong solvent with high solubility is adjusted by a weaker solvent. Water is used in RP chromatography, while heptane, hexane, or cyclohexane is used for NP systems. For solvent mixtures the dependency of the solubility on the solvent composition is very important. Good solubility results, in most cases, in low retention and thus poor separation. Usually, the weaker solvent according to the elution strength possesses less solubility and thus the choice of the solvent is a compromise between retention and solubility.

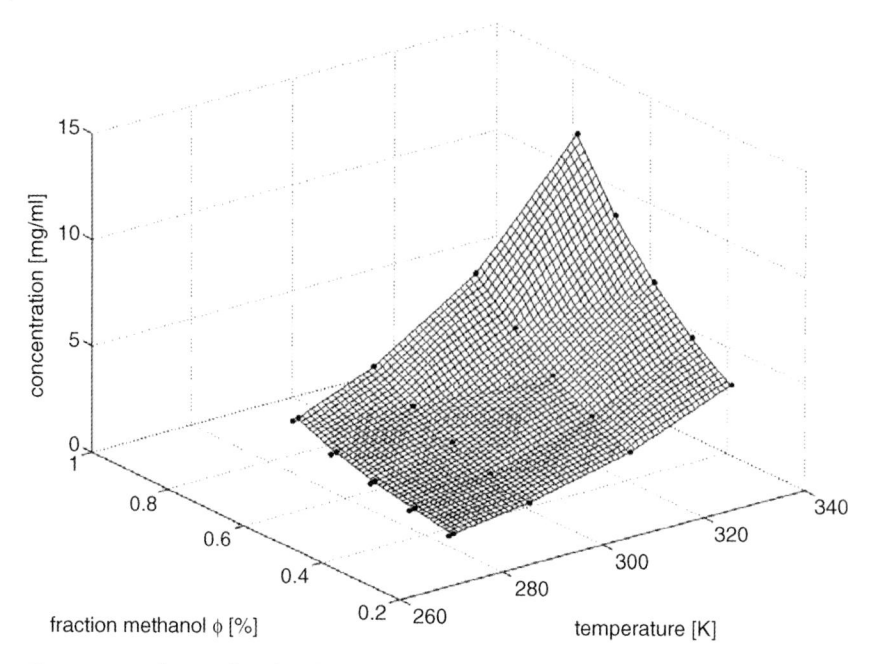

Figure 3.33 Influence of methanol content and temperature on solubility.

For example, in RP chromatography very often organic solvents and water are used. The solubility in the organic solvent is higher and water is used to adjust retention and selectivity. Figure 3.33 shows a typical plot for the dependence of the solubility of a pharmaceutical intermediate on both the volume fraction ϕ of methanol in a mixture with water and the temperature. Generally, solubility increases with temperature and, in most cases, it increases for high fractions of methanol.

3.2.3.2 Improving Loadability for Poor Solubilities

Although high solubility should be preferred for good productivities, time pressure and thermodynamics can require the acceptance of poor solubilities. Nearly half of R&D laboratories involved in preparative chromatography examine 10–100 samples per day and the number of candidates for preparative separation is still increasing (Neue et al., 2003). Obviously, time and large amounts of sample can not be spent for systematic evaluations and tedious solubility studies. Very often the amount of sample is also too small to carry out extensive experiments.

Cases of severe solubility limitations ($<5\,\mathrm{g\,l^{-1}}$) are handled in different ways. First, the sample may be dissolved in a solvent different from the mobile phase, for example, the sample solvent may be 100% organic and quite different from the mobile phase. In most cases, this solvent is, in addition to its high solubility, also a much stronger chromatographic solvent than the mobile phase. In such a case, large sample injections into the mobile phase may have an adverse effect on the

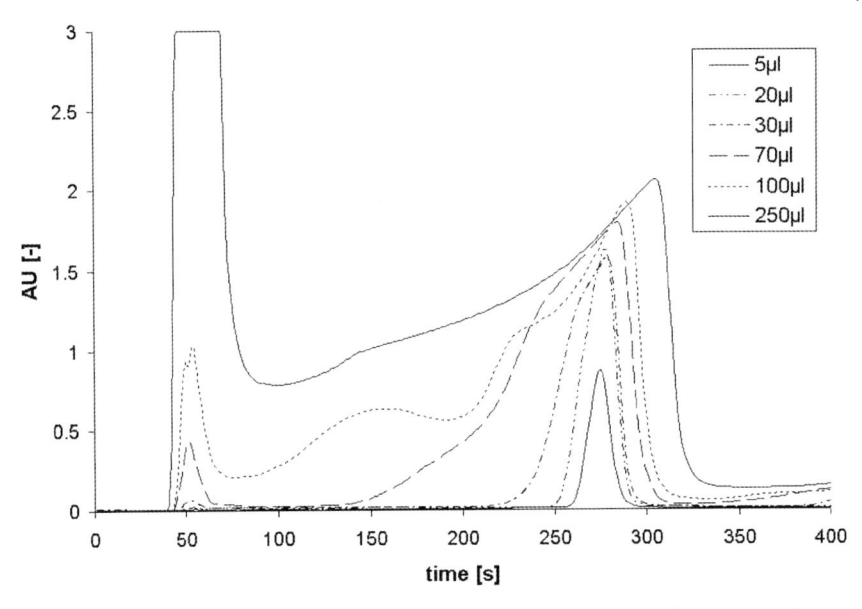

Figure 3.34 Peak distortion due to injection of sample dissolved in pure methanol (Purospher Star RP18; methanol–water, 60:40).

separation process. Peak shapes may be badly distorted and the sample may even precipitate in the mobile phase, if it is only slightly soluble in the mobile phase.

Figure 3.34 shows the separation of a pharmaceutical agent on an analytical alkyl-silica column (Purospher® Star RP18e, 100 mm long, internal diameter 4 mm). The sample was dissolved in pure methanol and between 5 and 250 μl was injected, while the mobile phase consists of methanol and water (60:40 volumetric ratio). At small injection volumes no peak distortion is observed, but with 30 μl the chromatogram shows a distortion of the main peak and band splitting. A small part of the sample elutes unretained with the dead time of the column. The strong injection solvent displaces the sample until the sample and the solvent are sufficiently diluted with the mobile phase solvent. At high injection volumes, methanol and the sample are no longer diluted and, thus, a front of both migrates through the bed. This behavior is exemplified in the left-hand part of Figure 3.34.

These problems are intensified if the sample is dissolved in a solvent completely dissimilar to the mobile-phase solvents. Many samples delivered for preparative separations are dissolved in common universal organic solvents such as dimethyl sulfoxide, dimethylformamide, or other polar organic solvents (Neue et al., 2003). Guiochon and Jandera have examined the influence of sample solvent and advise avoiding injections with a solvent of higher elution strength than the mobile phase, especially at high sample loads (Jandera and Guiochon, 1991). Injection of a large volume of the saturated mobile phase is preferred.

Peak distortion and precipitation problems due to injection of solvents of high solubility can be defused by changing the injection technique. The right-hand part

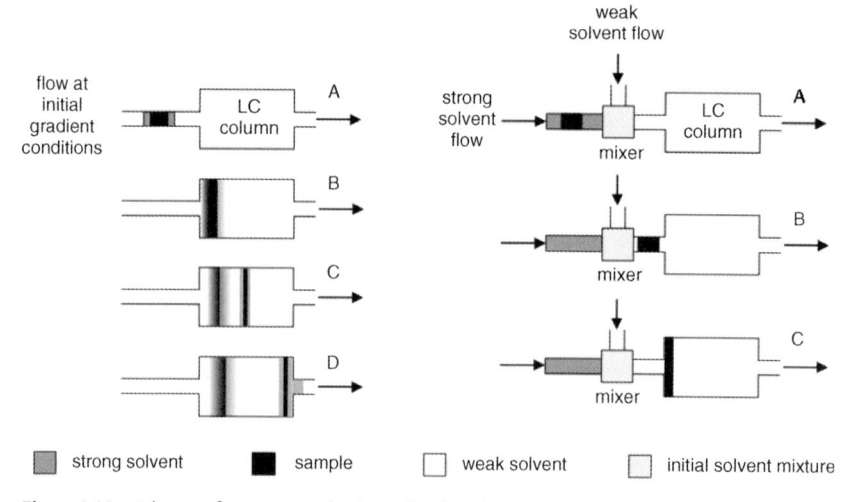

Figure 3.35 Scheme of conventional column loading (left) and at-column-dilution (right).

of Figure 3.35 shows the principle of the at-column-dilution (ACD) technique. Sample is dissolved in a strong solvent but, before it is injected into the column, it is mixed with the weak solvent to adjust the solvent strength to the mobile phase. The solubility of the sample in the mixture is of course decreased but, due to the short path between the mixing point and the packed bed, the growth of particles large enough to clog the frits can be excluded, especially since precipitate formation is preceded by a supersaturated solution (Neue et al., 2003). ACD significantly increases the amount of sample that can be purified relative to conventional injection methods. A desirable peak shape is obtained for the ACD technique although the separation is performed at mass loadings more than ten times greater than conventional loading where the conventional chromatographic separation already fails (Szanya et al., 2001).

Another method for separating low-solubility samples, which is well known in the downstream processing of biopolymers, starts with a large volume of sample dissolved in a chromatographically weak solvent. This solution is fed to the column. As the solute has a relatively strong affinity to the stationary phase, for example, silica, it is adsorbed and concentrated at the column inlet. This has the effect of introducing the sample as a concentrated plug.

When enough sample solution has been injected, the solvent is changed to a chromatographically stronger mobile phase. The sample is thus eluted with a high concentration in a step-gradient-type procedure. When this technique is used, coarse filtration of the original sample solution is advisable to prevent solid material entering the system.

3.2.3.3 Dependency of Solubility on Sample Purity

During scale-up of a multistep chemical synthesis the chemical purity of samples is often very variable. This is a major issue for process optimization of

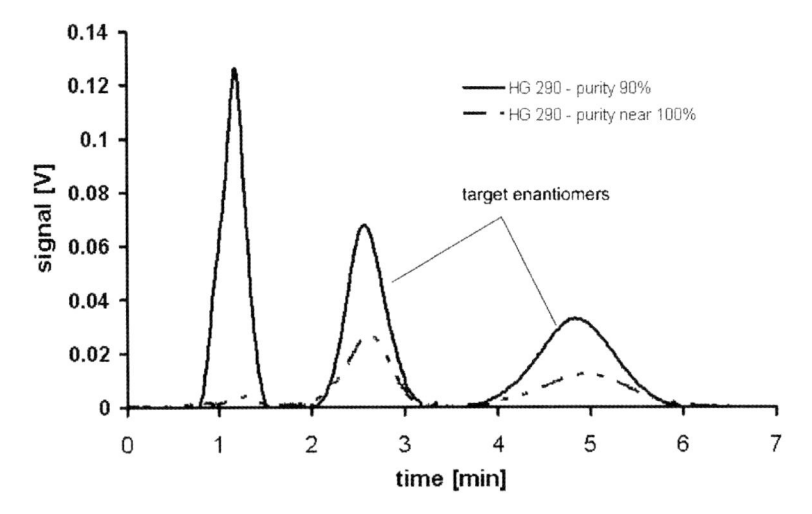

Figure 3.36 Maximum solubility of a sample for two different purities; sample volume 20 µl (reproduced with permission of H. Gillandt SiChem GmbH, Bremen).

chromatographic processes as chemical impurities can strongly influence a sample's solubility. The solubility of the sample can decrease dramatically when it becomes purer.

Figure 3.36 shows the quantitative analysis of the racemic pharmaceutical intermediate HG290 to determine its maximum solubility. Two different samples, with 90 and 100% purity, have been available. For both samples, a big excess of the solid compound was dissolved under slight agitation for 24 h in the solvent. It has to be assured that a residue remains at the bottom of the sample vial to achieve a saturated solution. After 24 h a sample of the clear supernatant is taken and analyzed by HPLC. The peak areas in Figure 3.36 show that the solubility is about two times higher for the sample with 90% purity.

In production-scale processes, the purity of the samples also depends on the quality of the raw material (e.g., educts for chemical reactions) as well as on the equipment used. To avoid problems in the subsequent separation process the influence of purity on solubility must be checked after the first amounts of the actual sample are available.

3.2.3.4 Generic Gradients for Fast Separations

For components to be isolated in the mg to g range, tedious method development is not advisable as these components have to be delivered under time pressure and not with maximum economy. Specialized laboratory units that perform high-throughput purification are often operated within the pharmaceutical industry.

A fast solution for these separation problems is the use of generic gradients. Table 3.25 lists typical solvents and conditions for gradient runs for RP and NP systems.

Table 3.25 Typical gradient runs for reversed and normal phase systems.

Phase system	Adsorbent	Weak solvent	Strong solvent
Reversed phase	RP-18, RP-8	Water	Acetonitrile
Reversed phase	RP-18, RP-8	Water	Methanol
Reversed phase	RP-18, RP-8	Dichloromethane	Acetonitrile
Normal phase	Silica	Heptane	Ethyl acetate
Normal phase	Silica	Dichloromethane	Methanol

Gradient separations always start with a mobile phase composition of 90–100% of the solvent with low elution strength (weak solvent). This leads to the highest possible retention at the beginning. For instance, RP systems start with 100% water, if the stationary phase can tolerate this high amount of water without any collapse of the alkyl chains. Otherwise, water has to be mixed with a small amount of an organic solvent. Section 3.2.5.2 describes the development of a typical RP gradient. The sample should be injected in a solvent composition with weak elution strength. Otherwise the sample breaks through with the front of the injection solvent. Solvent strength is now increased linearly by increasing the volume fraction of the solvent with high elution strength. The mobile phase composition is linearly changed to 100% of the strong solvent during the so-called gradient time, t_g. This gradient time should be adjusted to 10- to 15-fold of the dead time of the column.

3.2.4
Criteria for Choosing NP Systems

Normal phase systems consist of a polar adsorbent and a less polar mobile phase. Because these were the first available chromatographic systems they were named normal phase systems. They are for instance silica gels or other oxides in conjunction with a nonpolar solvent such as heptane, hexane, or some slightly polar solvents such as dioxane. Semipolar adsorbents such as cyano or diol phases can be operated in the normal phase mode as well. NP-systems have excellent separating power characteristics for samples with low to moderate polarity and intermediate molecular weight (<1000). As long as the sample to be separated does not contain extremely polar or dissociating functional groups, it is the method of choice for preparative chromatographic applications. In general, small organic molecules have better solubility in organic solvents than in the water-based mobile phases used in reversed-phase chromatography.

In addition, more product can be loaded on a silica gel than on reversed-phase materials. Furthermore the separated products are readily recovered by simple evaporation procedures, while the isolation of products from an RP separation process often requires additional work-up steps. These steps include the removal of salts or other aqueous-phase additives and extraction of the target product after evaporation of the organic phase (Section 3.2.5.5).

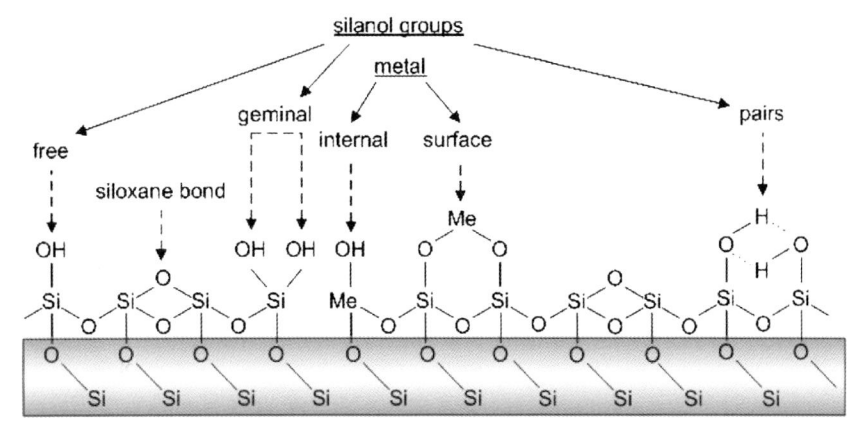

Figure 3.37 Surface groups on silica.

The combination of water and silica is not recommended due to the strong interaction between water and adsorbent. Furthermore silica is slightly soluble in water (strongly depending on the pH value !), which results in a shortened lifetime of the adsorbent. Normal phase systems are limited to organic solvents and thus the solutes have to be soluble in these solvents.

The interaction mechanism in NP chromatography is adsorption on the polar surface of the adsorbent. Because silica is the most widespread NP adsorbent the following focuses on the use of silica. As mentioned in Section 3.1.3 active centers for adsorption are statistically spread over the surface of the silica. Generally, these are siloxane groups and different kinds of silanol groups (Figure 3.37). The silanol groups appear as single (free) or as pairs. Single silanol groups have the strongest polarity while the polarity of silanol pairs is reduced. Impurities such as metal atoms (Ca, Al, or Fe) produce a disturbed surface chemistry and the formation of polarity hot spots, resulting in unwanted strong adsorption on these sites.

Obviously, due to the polar surface, only polar parts of solutes interact with the adsorbent, while nonpolar groups such as alkyls are not adsorbed. This behavior determines the range of applications for NP chromatography. For instance, it is not suitable for the separation of homologues that only differ in the length of an alkyl group. Conversely, solutes that differ in the number, nature, or configuration of their substituents are mainly separated by NP chromatography. This includes the separation of stereoisomers, if they are not enantiomeric. This stereoselectivity can be explained by the isotropic nature of the adsorption on silica. Silanol groups are rigidly fixed on the surface and, due to the short bond length of the silanol groups, the interaction between the solute and surface is very isotropic.

Aqueous mobile phases are not used because the interaction between water and the polar stationary phase would be too strong. However, a minimal amount of water is needed to saturate the above-mentioned hot spots of polarity due to metal atoms. For these two reasons the water content of the solvents should be controlled very carefully (Section 3.2.4.2). Changes in the range of a few ppm, resulting for example from air humidity, can cause significant changes in retention time and peak shape. As a result of this sensitivity to changes in water content the equilibration time for the system is very high, particularly when nonpolar solvents such as n-hexane are employed. A constant baseline and system behavior is reached after about 50 column volumes, while Poole (2003) remarked that column equilibration for mobile phases of different hydration level takes at least 20 column volumes.

3.2.4.1 Pilot Technique Thin-layer Chromatography

Thin-layer chromatography is a simple, fast technique that often provides complete separation of fairly complex mixtures. Any sample type that has been successfully separated by thin-layer chromatography on silica or alumina plates can be separated by means of liquid chromatography.

Mobile phase optimization using HPLC can easily be carried out using fully automated instruments, but this approach is extremely time-consuming. Conversely, modern thin-layer chromatographic techniques allow several mobile phases to be tested in parallel within a very short period of time. Furthermore, different methods are available to visualize the sample components, and the thin-layer chromatogram immediately provides information about the presence of products in the sample that remain at the point of application.

Taking some basic rules into account, the mobile phase composition established by means of TLC can be used directly for preparative chromatographic applications. A pragmatic approach using TLC as a pilot technique for normal-phase preparative chromatographic separations is elucidated in this chapter.

Similar to the retention factor in HPLC the retardation factor R_f describes retention behavior during TLC experiments. It represents the ratio of the distance migrated by the sample to the distance traveled by the solvent front. Boundaries are $0 < R_f < 1$. For $R_f = 0$, the product does not migrate from the origin and for $R_f = 1$, the product is not retained. R_f values are calculated to two decimal places, while some authors prefer to tabulate values as whole numbers, as hR_f values (equivalent to $100 R_f$) (Poole, 2003).

3.2.4.2 Retention in NP Systems

Different models have been developed to describe the retention of substances in adsorption chromatography. The Snyder model (Figure 3.38) assumes that in liquid–solid chromatography the whole adsorbent surface is covered with a monolayer of solvent molecules and the adsorbent together with the adsorbed monolayer has to be considered as the stationary phase (Snyder, 1968; Snyder and Kirkland, 1979; Snyder, Kirkland, and Glajch, 1997). Adsorption of the sample occurs by the displacement of a certain volume of solvent molecules from the monolayer by an approximately equal volume of sample, by which these molecules can adsorb on the adsorbent and become part of the monolayer. In this model, product retention is always caused by the displacement of solvent molecules from the monolayer.

Snyder's model provides a good understanding of separations on alumina as the adsorbent and is fairly good at explaining separations on silica gel using weak

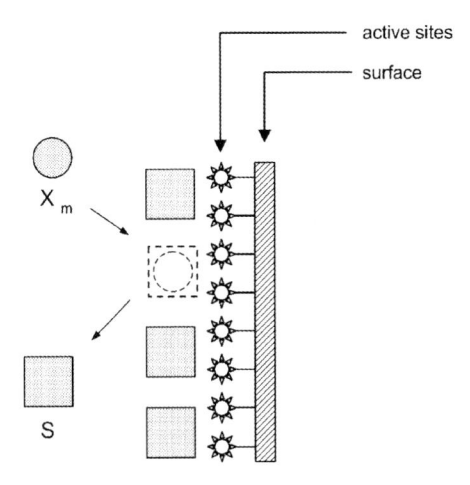

Figure 3.38 Graphical representation of the adsorption process.

solvents. An almost similar model developed by Soczewinski (1969) is more suitable for separations on silica gel using strong eluents.

In contrast, the Scott–Kucera model considers a solvent system composed of an apolar solvent A and a polar solvent B (Scott and Kucera, 1975). When this mixture is pumped through a column, a monolayer of the most polar solvent B is formed by adsorption of B on the adsorbent. Sample molecules are adsorbed on this monolayer instead of on the adsorbent surface. In other words, there is no displacement of adsorbed solvent molecules, and interaction between the molecules of the monolayer and the sample molecules determines the retention of the component. This theory has been adapted by saying that the model is only valid for medium polar mobile phases and solutes with a polarity lower than the most polar solvent in the eluent. These medium polar solvents are called hydrogen-bonding solvents (esters, ethers, ketones). A monolayer of these solvents behaves as a hydrogen-bonding phase. Interaction between this monolayer and the product molecules takes place mainly by means of hydrogen bonding. When the monolayer of strong solvent B is completely formed, a second layer comes into being. Advancement in the build-up of this second layer determines whether the sample can displace solvent molecules. However, product molecules will never displace solvent molecules from the monolayer as long as the sample molecules are less polar than the strongest solvent in the eluent.

As already described, molecules of the mobile phase and solute compete for the active sites of the adsorbent. The competition is represented by Equation 3.5 (Snyder, 1968).

$$X_m + nS_{ads} = X_{ads} + nS_m. \tag{3.5}$$

X_m and X_{ads} represent the sample molecule in the mobile phase and the adsorbed state, respectively. The corresponding states of the mobile-phase molecules are represented by S_m and S_{ads}; n is the number of solvent molecules that have to be displaced to accommodate the sample molecule.

Equation 3.6 gives the equilibrium, which shows that the relative interaction strength of the solvent and the sample molecules for the active sites of the adsorbent, determine the retention of a product.

$$K = \frac{(X_{ads}) \cdot (S_m)^n}{(X_m) \cdot (S_{ads})^n}. \tag{3.6}$$

Both silica and alumina contain surface hydroxyl groups and possess some possibilities of Lewis acid–base type interaction, which determine their adsorption characteristics. The more the hydroxyl groups, the stronger a solute molecule will be retained. The number and the topographical arrangement of these groups determine the activity of the adsorbent.

Silica and alumina have the highest surface activity when the adsorbents are free of physisorbed water. Addition of water blocks the most active sites on the surface since water, as a polar adsorptive, is preferentially adsorbed. Other polar compounds such as alcohols can also adsorb irreversibly at the surface. Consecutive adsorption of water deactivates the adsorbent surface and the solute retention will decrease concurrently. For this reason, the water content of the mobile-phase solvents should be controlled carefully if an apolar mobile phase is used (Unger, 1999). The water content only influences retention when apolar eluents such as hexane or heptane are used. When these solvents are mixed with 10% or more of a moderately polar solvent (e.g., acetone, ethyl acetate) the dependency disappears. The influence can be decreased by the addition of a small amount of acetonitrile to the mobile phase.

The steric orientation of the functional groups in the solute molecule in relation to the spatial arrangement of the hydroxyl groups on the surface enable silica and alumina to separate isomeric components (e.g., *cis–trans*/positional isomers).

3.2.4.3 Solvent Strength in Liquid–Solid Chromatography

Knowing that the solvent and the solute compete for active sites of the adsorbent, it is easily understood that the more the mobile phase interacts with the adsorbent the less a solute molecule is retained. Therefore, the major factor determining product retention in adsorption chromatography is the relative polarity of the solvent compared with the solute.

It is possible to set up a polarity scale by empirically rating solvents in order of their strength of adsorption on a specific adsorbent. A solvent of higher polarity will displace a solvent with a lower rank on the polarity scale. Solvents ranked according to such a polarity scale form an eluotropic series. Table 3.26 gives an eluotropic series developed for alumina as adsorbent. It can be transferred to silica by a simple factor. The ε^0 value in Table 3.26, called the "solvent strength parameter," is a quantitative representation of the solvent strength and can be used for the calculation of retention factors for different solvent compositions. Assuming that the adsorbent surface is energetically homogeneous and the solute–solvent interaction mechanism in the mobile phase does not influence interactions in the

Table 3.26 Eluotropic series for alumina.

Solvent	$\varepsilon^0_{Al_2O_3}$
n-Pentane	0.00
Isooctane	0.01
Cyclohexane	0.04
iso-Propyl ether	0.28
Toluene	0.29
Dichloromethane	0.42
Methyl iso-butyl ketone	0.43
Methyl ethyl ketone	0.51
Triethylamine	0.54
Acetone	0.56
Tetrahydrofuran	0.57
Ethyl acetate	0.58
Methyl acetate	0.60
Diethyl amine	0.63
Acetonitrile	0.65
1-Propanol	0.82
2-Propanol	0.82
Ethanol	0.88
Methanol	0.95
Acetic acid	Large

adsorbent phase, Poole (2003) recommends the following empirical equation (Equation 3.7) for binary mixtures:

$$\log k'_2 = \log k'_1 + aA_s \left(\varepsilon^0_1 - \varepsilon^0_2\right). \tag{3.7}$$

ε^0_1 and ε^0_2 are the solvent strength parameters for the mixture, while k'_1 and k'_2 represent the retention factors for the solute S, which occupies the surface cross-section A_s at the stationary phase; a stands for the activity parameter of the adsorbent phase (see below).

Values in the table only serve as a guide because the relative polarity can change, depending on the type of sample and adsorbent.

An eluotropic series is a tool to adjust the retention of sample components. If the retention of a product has to be reduced, a solvent of higher ranking in the eluotropic series is chosen. Due to increased competition between product and solvent molecules for the active sites on the adsorbent, the retention factor of the product will automatically decrease. The same holds true for the opposite situation.

According to Snyder (1968) it is possible to calculate the solvent strength of binary mixtures using Equation 3.8.

$$\varepsilon^0_{ab} = \varepsilon^0_a + \frac{\log[N_b \cdot 10^{a \cdot n_b \cdot (\varepsilon^0_b - \varepsilon^0_a)} + 1 - N_b]}{a \cdot n_b}. \tag{3.8}$$

n_b describes the molecular cross-section occupied on the adsorbent surface by the solvent molecule b in units of $0.085\,nm^2$ (which is 1/6 of the area of an adsorbed benzene molecule, corresponding to the effective area of an aromatic carbon atom on the adsorbent surface), while the parameter a describes the activity parameter of the adsorbent surface. The adsorbent activity parameter is a measure of the adsorbent's ability to interact with adjacent molecules of solute or solvent and is constant for a given adsorbent. For silica gel a value of 0.57 is given by Poole (2003). The molar solvent ratio is calculated by Equation 3.9 with the molar volumes of the molecules (Equation 3.10).

$$N_b = \text{molar solvent ratio} = \frac{\text{Volume }\%B \cdot \left(\frac{1}{V_b}\right)}{\text{Volume}\% B \cdot \left(\frac{1}{V_b}\right) + \text{Volume }\%A \cdot \left(\frac{1}{V_a}\right)} \quad (3.9)$$

$$\left(\frac{1}{V_a}\right) \text{ and } \left(\frac{1}{V_b}\right) = \frac{\varrho(\text{density})}{MW(\text{molecular} - \text{weight})}. \quad (3.10)$$

In the following example, the solvent strength of a mixture of dichloromethane and methanol with 90% (v/v) dichloromethane is calculated.

Dichloromethane
$\varepsilon_a^0 = 0.42$
$n_a = 4.1$
$a = 0.7$

Methanol
and $\varepsilon_b^0 = 0.95$
and $n_b = 8$

$$\left(\frac{1}{V_a}\right) = \frac{1.325(\text{g ml}^{-1})}{84.93\,\text{g mol}^{-1}} = 0.0157\,\text{mol ml}^{-1} \text{ and } \left(\frac{1}{V_b}\right) = \frac{0.79\,\text{g ml}^{-1}}{32\,\text{g mol}^{-1}} = 0.0249\,\text{mol ml}^{-1}$$

$$N_b = \frac{10 \cdot 0.0249}{10 \cdot 0.0249 + 90 \cdot 0.00157} = 0.1489$$

$$\varepsilon_{ab}^0 = 0.42 + \frac{\log\left(0.1498 \cdot 10^{0.78(0.95-0.42)} + 1 - 0.1498\right)}{0 \cdot 7.8} = 0.42 + 0.38325 = 0.80325$$

3.2.4.4 Selectivity in NP Systems

Separations by means of liquid–solid chromatography are carried out on polar adsorbents. The primary factor that determines the adsorption of a product is the functional groups present in the sample. Relative adsorption increases as the polarity and number of these functional groups increase, because the total interaction between the solute and the polar adsorbent surface is increased.

Uniqueness of retention and selectivity in NP-LC arise from two characteristic phenomena:

- Competition between sample and solvent molecules for the active sites on the adsorbent surface.

- Multiple interactions between functional groups of the sample and the rigidly fixed active sites on the adsorbent surface.

The broad selectivity potential of different solvent systems in NP-LC results from the adjustable concentration of the polar components that are adsorbed into the monolayer. This leads to an enormous variety of possible stationary phases (adsorbed solvent plus adsorbent) and a corresponding broad range of potential selectivities. Due to differences in the content of the polar modifier, different stationary phases are created.

For both TLC and HPLC many mobile-phase optimization procedures and criteria have been described in the literature. Mainly, two strategies are followed:

Keeping the solvent mixture simple by considering binary mixtures only.
Tuning selectivity by using more solvents, which can, of course, result in very complex mixtures.

For binary mixtures, a limited number of experiments are required and this approach is, naturally, the preferred method for preparative chromatographic applications.

3.2.4.5 Mobile-Phase Optimization by TLC Following the PRISMA Model

Before starting extensive experiments, a procedure recommended by Kaiser and Oelrich (1981) to rule out adsorbents by fast experiments should be employed. Each elution experiment takes about 20 s. For this purpose, samples are applied on a 50×50 mm TLC plate at nine points, which are exactly 10 mm apart. Five microlitres of methanol are drawn into a microcapillary with a platinum–iridium point. By applying the point of the filled capillary on one of the sample points on the plate, methanol is introduced onto the plate. A miniature radial chromatogram of about 7 mm diameter is produced. If the sample components remain at the point of application, the use of this adsorbent type is ruled out for HPLC usage. To make sure, the procedure is repeated with 5 µl of acetonitrile and tetrahydrofuran, respectively. If the products still remain at the point of application, the situation will not be changed by using any other mobile phase that is suitable for preparative chromatography work.

However, if the entire sample migrated to the outer ring with methanol, the test is repeated with *n*-heptane. If the entire sample still migrates to the outer ring, then the used adsorbent is again ruled out as stationary phase. If, with *n*-heptane, the entire sample or part of it remains at the starting point, the unlimited possibilities of combining solvents can be exploited to further optimize the method (Kaiser and Oelrich, 1981).

The PRISMA model developed by Nyiredy and coworkers (Nyiredy *et al.*, 1985; Dallenbach-Tölke *et al.*, 1986; Nyiredy and Fater, 1995; Nyiredy, 2002) for use in Over Pressured Layer Chromatography is a three-dimensional model that correlates solvent strength and the selectivity of different mobile phases. Silica is used as the stationary phase and solvent selection is performed according to Snyder's solvent classification (Table 3.27).

Table 3.27 Classification of the solvent properties of common liquids (Snyder and Kirkland, 1979).

Solvent	S_t	x_e	x_d	Group
n-Hexane	0	—	—	
n-Butyl ether	2.1	0.44	0.18	I
DIPE	2.4	0.48	0.14	I
MTBE	2.7	0.49	0.14	I
Diethyl ether	2.8	0.53	0.13	I
iso-Pentanol	3.7	0.56	0.19	II
n-Butanol	3.9	0.56	0.19	II
2-Propanol	3.9	0.55	0.19	II
n-Propanol	4	0.54	0.19	II
Ethanol	4.3	0.52	0.19	II
Methanol	5.1	0.48	0.22	II
Tetrahydrofuran	4	0.38	0.2	III
Pyridine	5.3	0.41	0.22	III
Methoxyethanol	5.5	0.38	0.24	III
Methylformamide	6	0.41	0.23	III
Dimethylformamide	6.4	0.39	0.21	III
Dimethyl sulfoxide	7.2	0.39	0.23	III
Acetic acid	6	0.39	0.31	IV
Formamide	9.6	0.36	0.23	IV
Dichloromethane	3.1	0.29	0.18	V
Benzylalcohol	5.7	0.4	0.3	V
Ethyl acetate	4.4	0.34	0.23	VI
Methyl ethyl ketone	4.7	0.35	0.22	VI
Dioxane	4.8	0.36	0.24	VI
Acetone	5.1	0.35	0.23	VI

The PRISMA model is a structured trial-and-error method that covers solvent combinations for the separation of compounds from low to high polarity. Initial experiments are done with neat solvents, covering the eight groups of the Snyder solvent classification triangle.

The PRISMA model (Figure 3.39a) has three parts: an irregular truncated top prism, a regular middle part with congruent base and top surfaces, and a platform symbolizing the modifier. The model neglects the contribution of the modifier to the solvent strength, because modifiers are usually only present in a low and constant concentration. In NP chromatography, the upper frustum is used in optimizing mobile phases for polar and/or semipolar substances. The regular central part of the prism is used in solvent optimization for apolar and semipolar components.

The three top corners of the model represent the selected undiluted neat solvents. The corner to the longest edge of the prism (A) represents the solvent with the highest strength, while the solvent with the lowest strength (C) corresponds to the corner of the shortest edge. Because the three selected solvents have unequal solvent strengths, the length of the edges of the prism are unequal and the top plane of the prism will not be parallel and congruent with its base. If the prism is

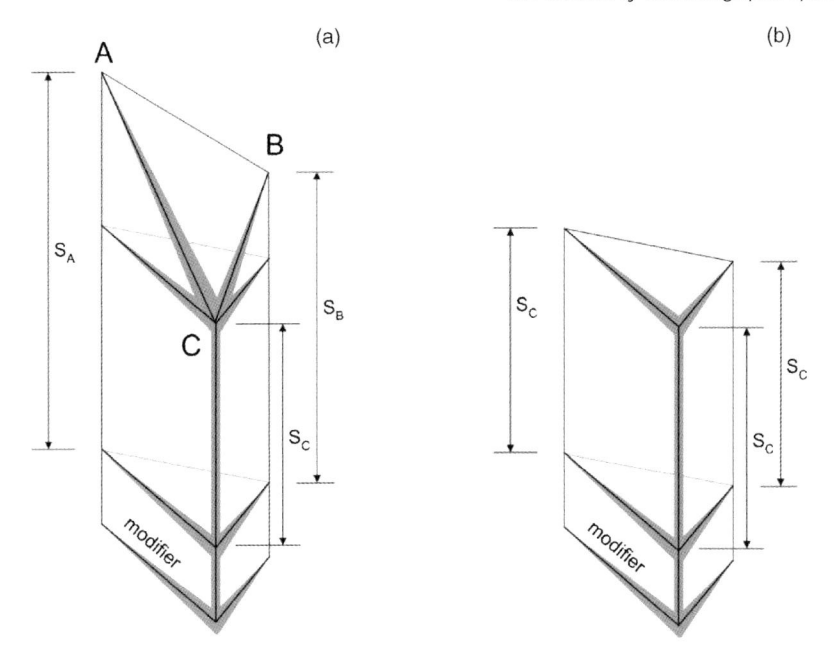

Figure 3.39 Graphic representation of the PRISMA model: (a) complete model (b) regular part of the model.

intersected at the height of the shortest edge and the upper frustum is removed, a regular prism is obtained (Figure 3.39b). The height of this prism corresponds to the solvent strength of the weakest solvent. All the points of the equilateral triangle formed by the cover plate of the prism have equal solvent strength. One of the corners (C) represents the neat solvent with the lowest strength. The composition of the two other corners can be obtained by diluting these solvents with a solvent of zero strength (*n*-hexane for NP and water for RP applications).

In the regular prism, horizontal intersections parallel to the base can be prepared by further diluting the selected solvents with solvent of zero strength (represented by the line from P_s^* to P_s^{**} in Figure 3.40). All points on the obtained triangles represent the same solvent strength, while all points on a vertical straight line correspond to the same selectivity points (P_s).

Points along the edges of the irregular part of the model, which represent mixtures of two solvents (AB, BC, and AC), are shown in the right-hand part of Figure 3.41, while the left-hand part describes the selectivity points of mixtures of the three solvents.

Each selectivity point can be characterized by a coordinate, defined by the volume ratio of the three solvents. The ratio is written in the order of decreasing solvent strength of the undiluted solvent (A, B, C). 100% solvent A with the highest solvent strength is represented by $P_s = 10-0-0$ and the solvent with the lowest strength (C) by $P_s = 0-0-10$.

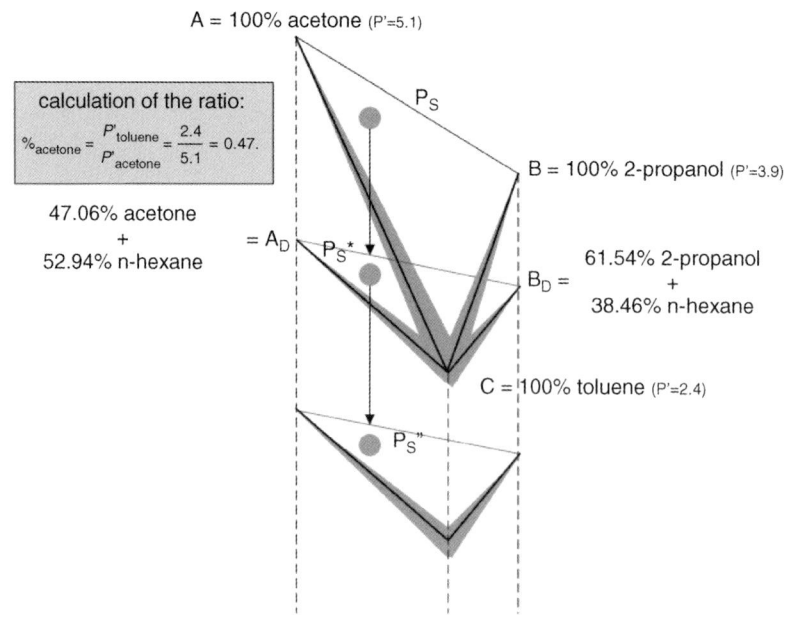

Figure 3.40 Example calculation of mobile-phase composition.

From these data, all other basic selectivity points can be defined by a three-digit number. For example, the selectivity point with coordinates 325 means 30 vol.% A, 20 vol.% B, and 50 vol.% C (Figure 3.42). Finer adjustment of the volume fractions can be expressed by three two-digit numbers ($P_s = 55–25–20$).

The whole strategy of solvent optimization via the PRISMA model includes the following steps:

- Start – selection of neat solvents.
- Step 1 – solvent strength adjustment.

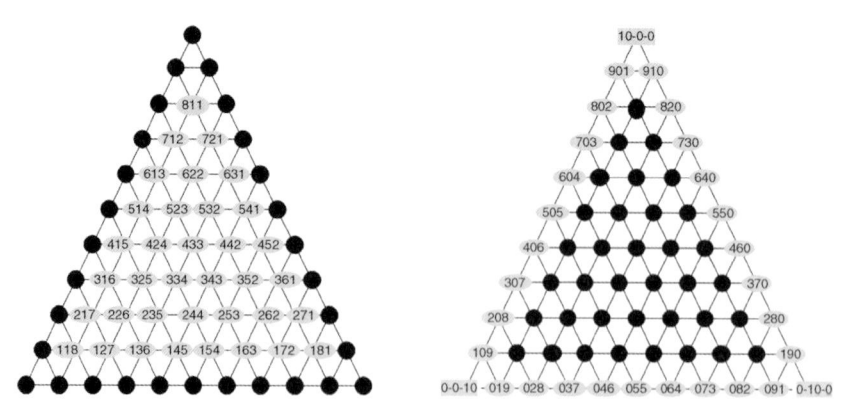

Figure 3.41 Basic selectivity points of the PRISMA model.

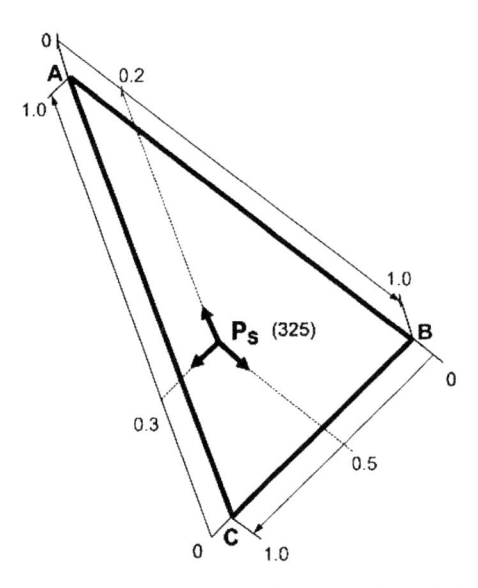

Figure 3.42 Combination of three neat solvents in the irregular part of the model.

- Step 2 – optimization of selectivity.
- Step 3 – final optimization of solvent strength.
- Step 4 – determination of optimal mobile phase composition.

The optimization strategy will be exemplified by a rather difficult isomer separation. The product to be separated was a relatively strong base and first experiments using neat solvents on different types of stationary phases (silica gel, amino-, cyano-, and diol-modified silica gel) revealed that it would be very difficult to separate the two isomers on a larger scale. Therefore, synthetic chemists were contacted to modify one of the secondary base functions with an easily removable BOC (tert-butoxycarbonyl) group.

Preliminary TLC experiments in unsaturated chromatographic chambers were performed using ten neat solvents. Solvents listed in Table 3.28 represent the most common starting solvents from each group of the Snyder model. For the example

Table 3.28 Solvents for preliminary TLC experiments.

Solvent	Group	Solvent	Group
Diethyl ether	I	Dichloromethane	V
2-Propanol	II	Ethyl acetate	VI
Ethanol	II	Dioxane	VI
Tetrahydrofuran	III	Benzene	VII
Acetic acid	IV	Chloroform	VIII

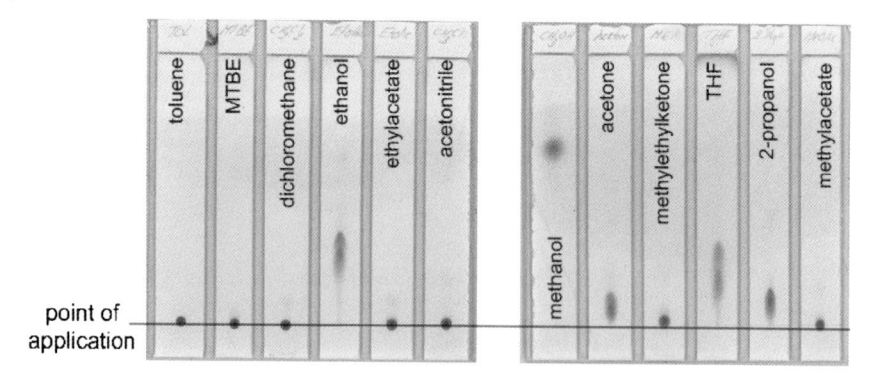

Figure 3.43 TLC experiments for the separation of two isomers using neat solvents.

of the isomers this list was modified slightly to fit the experience of previous experiments. Figure 3.43 shows the results of the solvent-selection experiments.

Based on experience (R_f values, selectivity, and spot shape), one to three solvents were selected. Preferentially, solvents are chosen that demonstrate small, well-defined spot shapes for the title components. Binary mixtures or single solvents that result in round finite spots have to be selected for the first set of experiments.

Figure 3.43 clearly demonstrates that only ethanol, acetone, and tetrahydrofuran showed some selectivity between the two isomers, while only methanol gave a nice round spot. This example shows the practical usefulness of screening with TLC experiments. Because the experiment with each plate took roughly 20 min, the use of 12 different solvents could be examined in 40 min.

As a standard approach for preparative chromatography, binary solvent combinations are always investigated before considering ternary mixtures. Figure 3.44 shows the results of experiments with binary mixtures where selectivity was obtained. Other binary combinations were tested, but none resulted in a good spot

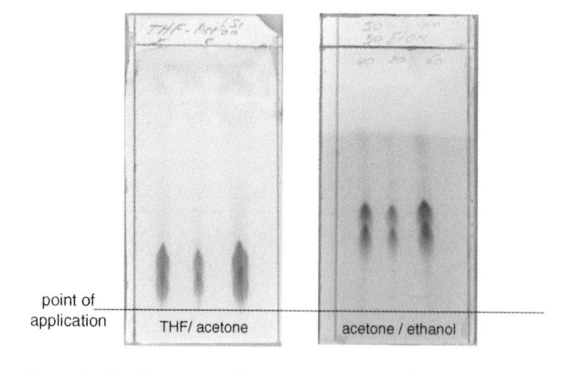

Figure 3.44 Screening of binary mixtures (50:50 v/v) of selected neat solvents.

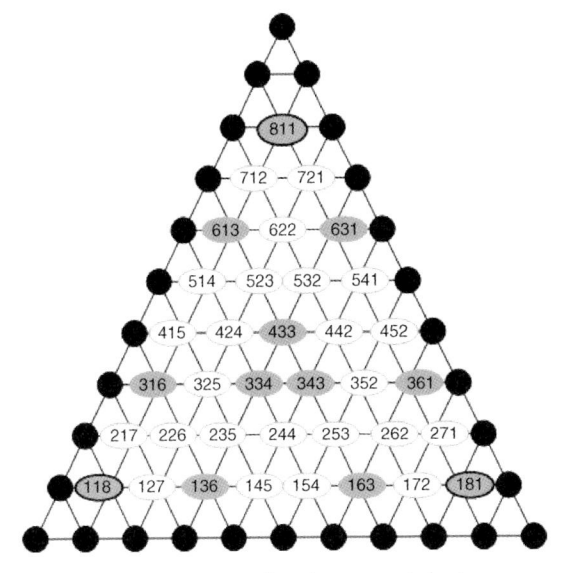

Figure 3.45 Starting points for selectivity optimization.

shape and an acceptable separation between the isomers. Therefore, some additional tests were performed with ternary solvent combinations.

3.2.4.5.1 Step (1): Solvent Strength Adjustment Once the different solvents have been selected, a first experiment is done using the mobile-phase composition corresponding to the center of the cover triangle ($1/3$ solvent A, $1/3$ solvent B, and $1/3$ solvent C). If the observed R_f values are too high, additional experiments are performed to adjust the solvent strength by the addition of zero strength solvent.

3.2.4.5.2 Step (2): Optimization of Selectivity In the triangle with the desired solvent strength, three selectivity points near the edges of the triangle, [**811**(80% A–10% B–10% C)] – [**181** (10% A–80% B–10% C)] – and [**118** (10% A–10% B–80% C)], are the mobile-phase compositions for the following three experiments (Figure 3.45).

For binary mixtures the selectivity points (75% A–25% B–0% C), (25% A–75% B–0% C), (75% A–0% B–25% C), (25% A–0% B–75% C), (0% A–75% B–25% C), (0% A–25% B–75% C) were investigated.

The obtained chromatograms and chromatograms of step 1 serve as guidelines for the next experiments. If necessary, selectivity is further optimized by choosing new points in the triangle near the coordinate or between coordinates that gave the best resolution. Generally, a limited number of additional experiments are required to obtain optimum conditions for separation. The resulting selectivity point (P_s) is used in the final optimization step. To separate the isomers the following solvent combinations were chosen for steps 1 and 2:

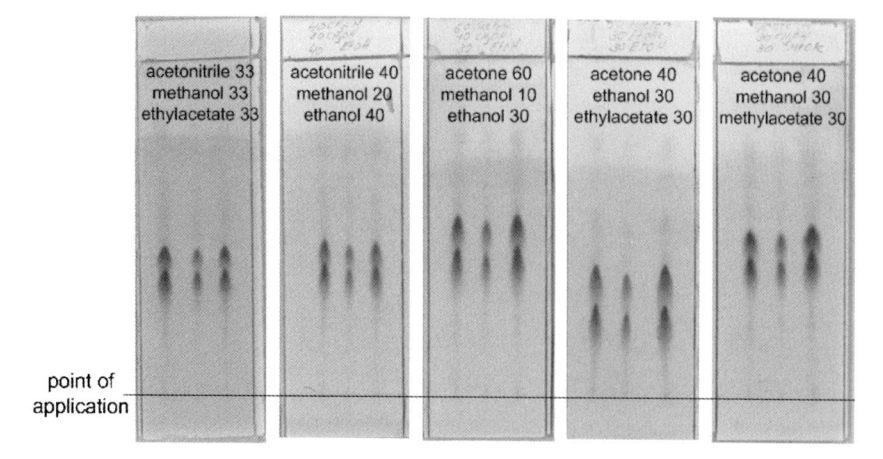

Figure 3.46 Screening of ternary solvent mixtures.

- Acetonitrile, methanol, and ethanol.
- Acetone, methanol, and ethanol.
- Acetone, ethyl acetate, and ethanol.
- Acetone, methyl acetate, and methanol.
- Acetonitrile, methyl acetate, and methanol.

First experiments were done with equal volumes of the three neat solvents as described in step 1. Based on this data some additional experiments were performed using different volume ratios of the most promising solvent mixtures.

Figure 3.46 shows the results of this optimization. The R_f values are in a promising range and thus a zero strength solvent is not needed. The TLC chromatograms clearly illustrate that different ternary solvent combinations separate the isomers well. Both spot shape and resolution are better than for the binary solvent mixtures (Figure 3.44).

3.2.4.5.3 Step (3): Final Optimization of the Solvent Strength

This step is used for the final optimization of solvent strength. It corresponds to a vertical shift in the regular part of the prism, starting from the optimal selectivity point (P_s) established in step 2. If all products of interest are sufficiently separated, the solvent strength can be increased or decreased to reach the desired goal.

3.2.4.5.4 Step (4): Determination of the Optimum Mobile-Phase Composition

After the basic tests, where suitable solvents are selected and the desired solvent strength is established, the effect of eluent changes at a constant solvent strength are further investigated to find the optimum mobile-phase composition.

To obtain reliable results the TLC measurements have to be carried out in saturated chambers, because in unsaturated chambers the reproducibility of R_f may be poor and the formation of secondary and tertiary solvent fronts can affect the interpretation of the generated results. The most versatile TLC chamber in which to perform such experiments is the Vario-KS chamber developed by Geiss and Schlitt (1965). This chamber is suitable for evaluating the effect of different solvents, solvent vapors, and relative humidity. In this type of chamber, the chromatoplate is placed face down over a conditioning tray containing several compartments to hold the required conditioning solvents. The design of the chamber ensures that saturating and developing solvents are completely separated. The chamber's major advantage is that, for the purpose of solvent optimization, up to ten activity and/or saturating conditions can be compared on the same plate using the same eluent.

To formulate a mathematical model for the dependence of hR_f on mobile phase composition, the obtained hR_f are displayed against the selectivity points, which symbolize the composition of the eluent.

Correlations were tested along the axes of the triangle and from the three basic selectivity points (118–811–181) across the middle point to the opposite side of the triangle.

A minimum of 12 experiments are required to determine a local optimum. By measuring hR_f for the selectivity points 118–316–613–811–631–361–181–163–136–433–334 and 343 (Figure 3.45), hR_f for the selectivity points 217–415–514–712–721–541–451–271–172–154–145–127–235–253–325–352, 523, and 532 can be calculated using the mathematical functions obtained from the measured values.

Functions along the axes:

- 811–613–316–118 (solvent A) (calculation of 217–415–514–712)
- 181–361–631–811 (solvent B) (calculation of 271–451–541–721)
- 118–136–163–181 (solvent C) (calculation of 127–145–154–172)

In general the correlation between hR_f and the selectivity points at a constant solvent strength level can be expressed by Equation 3.11.

$$hR_f = a \cdot (P_s)^2 + b \cdot (P_s) + c. \tag{3.11}$$

$$\ln hR_f = a(S_t) + b. \tag{3.12}$$

To obtain a global solvent optimum, the vertical relationship between the solvent strength and hR_f values has also to be investigated and, therefore, some additional tests are required. Using quaternary solvents, the correlation was found to be described by Equation 3.12. Because a linear mathematical function requires a minimum of three measured data points, the vertical relationship between solvent strength and hR_f values has to be tested at three different solvent strength levels. To collect accurate data, the chosen solvent strength levels have to differ individually by 5–10%.

To determine accurately the vertical relationship function, the following selectivity points were investigated:

- Solvent strength level 1: 811–631–118–343–334–181
- Solvent strength level 2: 811–433–118–316–361–181
- Solvent strength level 3: 811–613–118–334–163–181

3.2.4.6 Strategy for an Industrial Preparative Chromatography Laboratory

The effort required to develop a separation method strongly depends on several factors:

1) Recurrence of a separation problem (repeatedly or only once)
2) Amount of product to be separated
3) Do we have to isolate a single target component or all relevant peaks in the mixture?
4) Quality requirements
5) Purpose of the purification process
    - Purification of key intermediates or separation of a final product
    - Preparation of reference standards for analytical purposes
    - Isolation and purification of impurities in the 0.05–0.1 mg ml^{-1} concentration range
    - cGMP related investigations

Obviously, when separating small quantities of many different products, standard separation methods using gradient elution are preferred. Conversely, for larger scale separations it is important to design a robust separation process that can run fully automated 24 h a day. Clearly, for a production-scale process that has to be performed under cGMP conditions. Many experiments are required to be able to design a reproducible, robust process that can be transferred from one production location to another without the need for substantial modifications. For this reason two examples, covering both separations are exemplified.

■ **Example 3.1: Separation of Large Numbers of Different Products in the Range 1–10 g**

As mentioned before, for this type of work, standard procedures using gradient elution are the method of choice and there are two possible approaches:

- Normal phase chromatography on silica.
- Reversed phase chromatography (Section 3.2.5.2).

For such products, HPLC chromatograms using a standardized method and mass spectrometry data are usually available and, therefore, it is very tempting to perform a preparative chromatographic purification using these experimental conditions.

The major advantage of this approach is that for mixtures containing several peaks of approximately equal size, the analytical HPLC-MS data allow immediate recognition of the desired product in the preparative run and, therefore, no additional structural analyzes are required to identify the isolated product. However, the major disadvantage of reversed-phase applications is often the limited solubility of the samples in eluents having high water content. Furthermore, due to the relatively high viscosity of the eluents used in reversed-phase chromatography, the flow rate that can be pumped through a preparative chromatography column is generally lower than that for the pure organic solvent mixtures used in normal phase chromatography. Another point that certainly has to be considered is the loading capacity of a C18 reversed phase material, which is generally only $1/2-1/3$ of the loading capacity of bare silica gel.

However, is it possible to run, reproducibly, solvent gradients on bare silica? Many people have their doubts, especially concerning the necessity of very long equilibration times when returning to the initial solvent composition. Furthermore, it was impossible to perform gradient elutions on the first generation of silica using very polar solvents, for example, methanol, because these solvents washed out the silica dust originating from the production process, which was electrostatically attached onto the particles. This phenomenon certainly is not very convincing. Nonetheless, new spherical materials prepared from ultrapure silica no longer demonstrate this pronounced problem. On these materials, gradient runs ending up with 100% of very polar organic solvent are easily performed. The problem of activity changes when going from an intermediate polarity to a relatively polar solvent is also nonexistent when, for two or more injections in succession, exactly the same procedure is followed, that is, same gradient profile, same rinsing time with polar solvent, same time to go back to the start solvent composition, and, very important, same time at the initial composition of the gradient to start the next injection. Only when, for the next series of samples, the gradient has to start with a zero strength solvent (n-heptane) equilibration problems can be expected and then some measures may have to be taken to reactivate the stationary phase.

3.2.4.6.1 Standard Gradient Elution Method on Silica

To select suitable gradient conditions, TLC experiments are performed using an eluotropic series based on dichloromethane–methanol mixtures. By means of a Camag Vario chamber (Figure 3.47), the eluent combinations following Table 3.29 have been tested.

The mixture dichloromethane–methanol (50:50 v/v) has been chosen on the basis of experience that products remaining in the point of application with solvents of lower strength often can be moved more easily to higher R_f values with this mixture than with pure methanol as the eluent. Probably, this phenomenon is due to the very good solubility characteristics of this solvent mixture for a broad variety of products.

Based on the observed chromatographic profile, the start and end conditions for a preparative gradient run are selected. The solvent composition that corresponds to the solvent mixture where the product of interest just leaves the point of application is used as the starting composition for the gradient. The solvent mixture moving the product of interest closely to the solvent front is the final composition of the gradient.

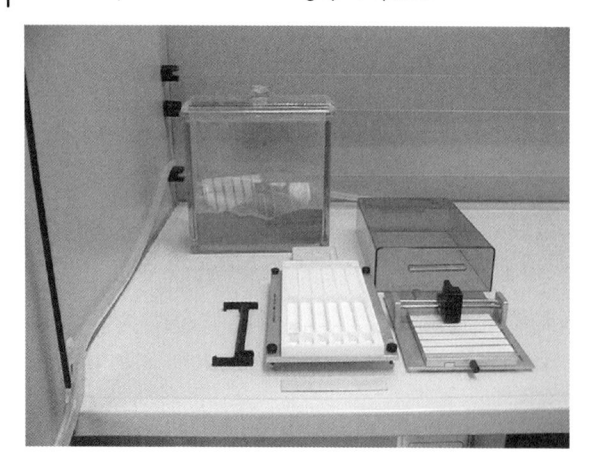

Figure 3.47 Vario chamber with accessories.

3.2.4.6.2 **Simplified Procedure** Instead of testing the whole dichloromethane–methanol eluotropic series, only one TLC run is performed, using a 90:10 volume ratio mixture of dichloromethane–methanol. The location of the product of interest on this plate is then used to establish the preparative gradient conditions. This procedure is shown in Figure 3.48.

The type of silica selected to perform this type of preparative gradient elutions is a 5–10 μm good quality spherical material with excellent mechanical stability (Kromasil®, Akzo – Nobel, Sweden). About 200–250 g are packed at 100–150 bar in a 50-mm internal diameter dynamic axial compression column. Chromatography is executed at a flow rate of 110–125 ml min^{-1} and the average sample amount applied is about 3 g. This gradient approach on native silica has proven to be a very versatile and robust method as long as the sample and eluent are carefully filtered prior to their use, thus avoiding partial blockage of the porous metal plates in the top flange and the piston of the column.

■ **Example 3.2: Pilot Chemistry Laboratories and Pilot Plant Scale Separations (A few Hundred Grams to Tenth'of Kilograms Scale)**

For this type of problem, time constraints generally dominate. Very often, the scale-up of a chemical reaction without extensive optimization

Table 3.29 Solvent series based on dichloromethane–methanol mixtures.

Dichloromethane–n-heptane (50–50 v/v) ($S_T = 1.55$)
Dichloromethane ($S_T = 3.1$)
Dichloromethane–methanol (98–2 v/v) ($S_T = 3.14$)
Dichloromethane–methanol (95–5 v/v) ($S_T = 3.195$)
Dichloromethane–methanol (90–10 v/v) ($S_T = 3.3$)
Dichloromethane–methanol (50–50 v/v) ($S_T = 4.1$)

Desaga chamber for 10 x 10 cm plates (dichloromethane / methanol 90/10$_{v/v}$)	preparative gradient % methanol	
	start	end
$hR_f > 50$	0 %	5 %
$30 < hR_f < 50$	0 %	10 %
$hR_f < 30$	0 %	20 %

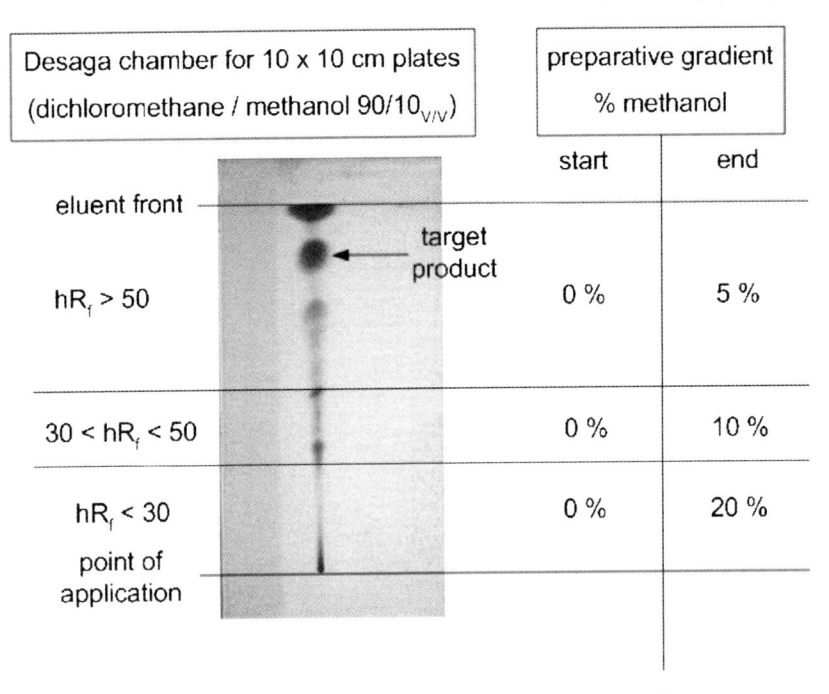

Figure 3.48 Simplified procedure to establish the start and end composition for a preparative gradient elution on native silica.

work results in low yields and bad qualities. The necessity to deliver a certain amount of product within a very narrow time frame, in general does not allow preparative chromatography workers to thoroughly optimize the separation method. Therefore, a standard approach is required.

At first the polarity of the product is investigated by means of a TLC analysis using an eluotropic series based on dichloromethane–methanol mixtures such as described in Table 3.29.

Very often, one of the tested mixtures can be used to perform the separation, or a few additional experiments using the same solvent combination have to be performed to fine-tune the chromatography process.

Simultaneously, two additional silica TLC plates are made to test 12 neat solvents that, from a safety, health, environmental and practical point of view, are acceptable for use on a larger scale.

From the series of solvents proposed in Table 3.28, diethyl ether, dioxane, benzene, and chloroform are excluded. Instead, the following solvents are used:

Plate 1: toluene–*tert*-butyl methyl ether (TBME)–dichloromethane–ethanol–ethyl acetate–acetonitrile.

Plate 1: methanol–acetone–methyl ethyl ketone–tetrahydrofuran–2-propanol–acetic acid.

Toluene, acetone, and methyl ethyl ketone are, due to their cut-off values, not the first choice if UV detection is used to control the process. Tetrahydrofuran has to be avoided because for larger scale use it has to be stabilized with an antioxidant (butylated hydroxytoluene) that will finally end up in the product, often making an additional purification step necessary.

However, for a broad variety of substances, some solvent mixtures based on toluene or acetone or THF have proven to be universally applicable.

- Toluene–2-propanol
- Toluene–ethyl acetate–ethanol
- Toluene–2-propanol–25% ammonia in water
- *n*-Heptane–methyl ethyl ketone (or acetone)
- Acetone–ethanol
- Acetone–ethanol–ethyl acetate
- Dichloromethane (or acetonitrile)–THF

Based on the chromatographic profiles observed on both plates, suitable binary solvent mixtures are chosen. Using another TLC plate, an eluotropic series of these solvents is prepared and tested. The start and end composition of this eluotropic series depends on the R_f values measured for the strongest solvent of the mixture.

If on silica, no satisfactory results are obtained with the neat solvents, the two solvent series (except acetic acid) are, respectively, tested on an amino-cyano and diol modified TLC plate.

Most problems can be solved on silica gel as the stationary phase. If either some specific selectivities or the addition of a basic modifier is required, amino-modified silica gel (90:10 v/v) often brings the solution.

Clearly, this solvent-selection procedure will not always result in optimal preparative chromatographic conditions, but it offers a method that can be applied immediately.

Direct application of a TLC method for preparative chromatography requires consideration of some basic differences between the techniques:

1) Driving force for solvent transport:
    - Capillary forces (TLC)
    - External force (pump) in HPLC
2) Specific surface area of silica gel used to prepare a TLC plate can differ from the packing material in the column
3) Organic or inorganic binders are used to fix the silica layer on the surface of a glass plate. Furthermore, the layer generally contains an inorganic fluorescence indicator
4) TLC starts from a dry layer and we have to deal with a mobile-phase gradient over the whole migration distance

> Preferential adsorption of the most polar component of the solvent mixture in TLC means that the use of the same mobile-phase composition in an HPLC experiment will always result in shorter retention times. (In HPLC the solvent mixture is continuously pumped through the column and, after some time, the active surface spots are occupied with polar solvent molecules.)
>
> Therefore, it is advisable to perform a test injection of the sample on the HPLC column using a mobile phase containing a somewhat smaller amount of the polar component.

3.2.5
Criteria for Choosing RP Systems

The demand for separation of water-soluble solutes that differ in their nonpolar part led to the development of adsorbents for reversed phase (RP) chromatography. In RP systems, the polarities of the adsorbent and mobile phase are inverted compared with NP systems. Therefore, reversed-phase chromatography should be better termed as reversed polarity chromatography. Nonpolar or weak polar adsorbents are used together with polar solvents. The transition to polar mobile phases affords the use of water. A typical solvent for RP systems consists of a mixture of water and a miscible organic solvent, such as methanol or tetrahydrofuran (THF). Due to the aqueous solvents, precise control of the water content as in normal phase chromatography working with nonpolar solvents is not necessary. The nonpolar packing material consists, for instance, of a porous silica support coated with a monolayer of alkyl groups (C_2–C_{18}), aryl groups, or hydrophobic polymers. Alternatives to these reversed-phase silicas are hydrophobic cross-linked organic polymers or porous graphitized carbons. One important property is the hydrophobicity of this surface. It is mainly afforded by the carbon content and the endcapping of the silica (see also Section 3.1.3.1).

Adsorption–desorption mechanisms of RP systems are a turn around from NP systems (reversed polarity). In this case, the nonpolar, or hydrophobic, portion of solute molecules adsorbs to the surface of the stationary phase, while the polar part of the molecule is solvated by the mobile-phase solvent. The result is a reversed elution order – polar before less polar solutes (Figure 3.49).

Reversed phase silicas are called brush type, because of their structure. Numerous alkyl groups point from the surface into the mobile phase and thus the surface is similar to a brush, with bristles of alkyl groups. To describe the adsorption process a simple two-layer model of the system has been proposed (Galushko, 1991), which means that the surface layer of the alkyl groups is assumed to be quasi-liquid. The retained solutes penetrate into the surface layer and retention can be regarded as partitioning between hydrophobic stationary phase and mobile phase, similar to liquid–liquid chromatography.

RP chromatography is mainly used to separate substances that differ in their lipophilic part or in the number of C atoms, for example, the separation of homologues. However, due to the anisotropic nature of the interaction between surface

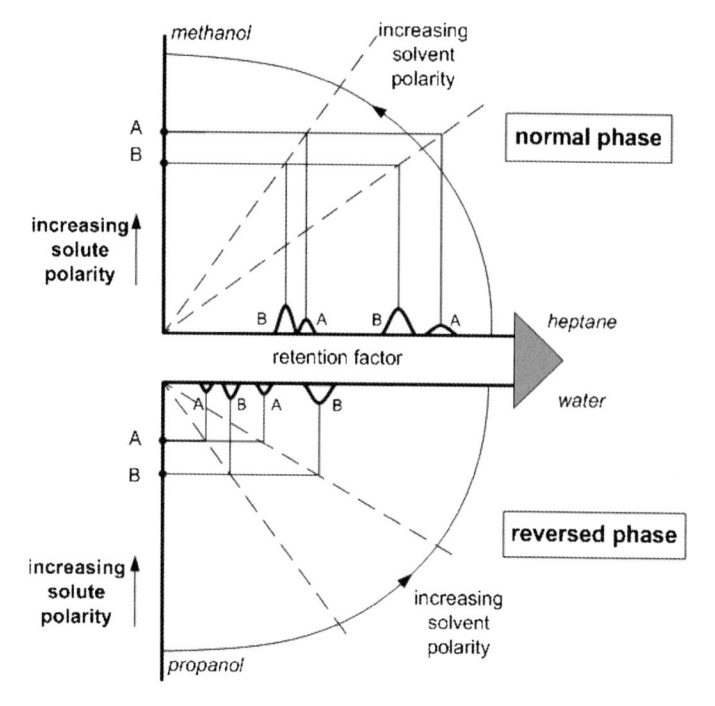

Figure 3.49 Retention behavior in both normal and reversed-phase chromatography.

layer and solute, RP phase systems reach lower selectivities for the separation of stereoisomers than NP phase systems.

The elution order in NP and RP systems is described in Figure 3.49. The retention time and thus the elution order for two solutes of different polarity is exemplified for RP in the lower part of Figure 3.49 and for NP in the upper part. Component A is more polar than B, which results in A eluting before B in RP systems and an inverted elution order for NP systems. The influence of solvent polarity is shown by the two diagonal lines from the origin. Each line illustrates a solvent with a certain polarity. Retention times are given by the point of intersection with the horizontal line, which describes the solute polarity.

Most analytical separations are performed by reversed-phase systems while most preparative separations are performed using an NP system. For example, 84% of the analytical separations at the former Schering AG now part of Bayer Healthcare are done in RP mode while 85% of the preparative separations are performed on normal phase silica gels (Brandt and Kueppers, 2002).

RP systems dominate analytical applications because of their robustness and reduced equilibration time. This fast column equilibration makes RP phases applicable for gradient operation. For preparative chromatography, normal phase systems are preferred because of some disadvantages of the RP systems. Although water is a very cheap eluent it is not an ideal mobile phase for chromatography because it evaporates at higher temperatures, the enthalpy for evaporation is quite high and the viscosity of aqueous mixtures is much higher than the viscosity of mixtures of organic solvents.

Since the polarity range of adsorbents is bordered by silica on the polar side and RP-18 or hydrophobic polymeric phases on the nonpolar side it is easy to assign these absorbents to normal or reversed-phase systems. Medium polar packings (Section 3.1.1) possess polar properties because of the functional group as well as hydrophobic properties contributed by the spacer (Unger and Weber, 1999). Owing to this mixed nature, these phases can not be directly assigned to a certain type of phase system.

3.2.5.1 Retention and Selectivity in RP Systems

Reversed-phase liquid chromatography is used in separating polar to medium polar components. Their separation is based on the interaction of the lipophilic part of the solutes with the nonpolar surface groups. Retention depends on the nature of the active groups bonded on the silica surface as well as the functional groups of the solute. The hydrophobicity of reversed-phase packings differs in the relative sequence:

Porous graphitized carbon > polymers made from cross-linked styrene/divinylbenzene > n-octadecyl (C18) bonded silicas > n-octyl (C8) bonded silicas > phenyl-bonded silicas > n-butyl (C4) bonded silicas > n-propylcyano-bonded silicas > diol-bonded silicas.

The retention time in RP chromatography increases with the surface hydrophobicity. Therefore, for instance, C18 groups cause higher retention times than C8 groups at constant ligand density – as demonstrated in Figure 3.50, which compares the retention on RP-8 and RP-18 phases. Apart from a higher retention time, RP-18 silicas also exhibit different selectivities.

In addition to the surface groups of the adsorbent the nature of the functional groups of the solutes determines the interaction and, thus, the retention time as

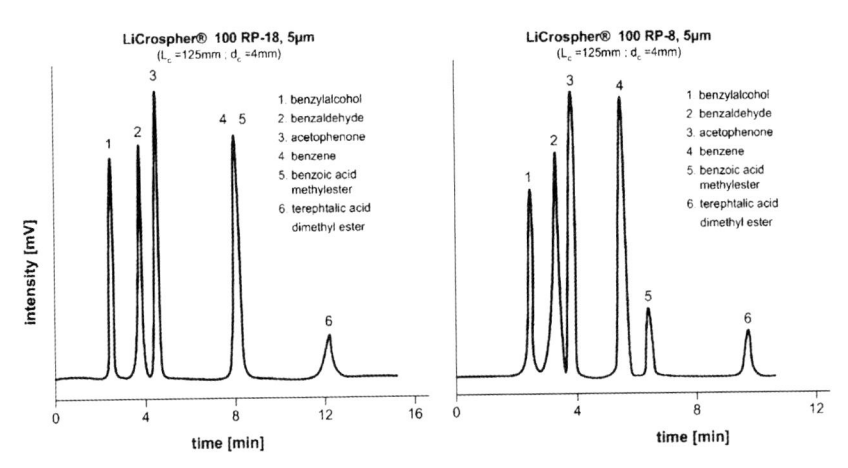

Figure 3.50 Retention of aromatic components on RP-8 and RP-18 columns with water–methanol (50:50 v/v); flow rate 1 ml min^{-1} (reproduced from Unger and Weber, 1999).

well. The following sequence shows the elution order for solutes of similar size with different functional groups (Unger and Weber, 1999):

Amine/alcohols/phenols < acids < esters < ethers < aldehydes/ketones < aromatic hydrocarbons < aliphatic hydrocarbons.

Figure 3.50 presents the influence of functional groups on the retention time. Benzene and some aromatic derivates are separated by RP chromatography. Compared with benzene the retention time is decreased for derivates with polar aldehyde or hydroxyl groups while it is increased for the more bulky apolar esters, for example, benzoic or terephthalic methyl esters.

In RP chromatography, mixtures of organic solvents such as methanol, acetonitrile, and tetrahydrofuran with water are used as mobile phases. The organic (more hydrophobic) solvent also interacts with the nonpolar surface groups of the packing and thus is in competition with the solutes for the nonpolar adsorption sites on the adsorbent. Consequently, the retention time of the solutes decreases with increasing fraction of the organic solvent in the mobile phase. The dependency of the retention factor on the volume fraction ϕ of the organic solvent is described by Equation 3.13 for medium values of ϕ. In this equation, $k'_{0,S}$ is the retention factor for pure water and S is the elution strength of the organic modifier (Snyder and Kirkland, 1979; Snyder, Kirkland, and Glajch, 1997; Snyder and Dolan, 1998).

$$\ln k' = \ln k'_{0,S} - S \cdot \phi. \tag{3.13}$$

Owing to the exothermic behavior of the adsorption, the retention time increases at lower temperatures. An Arrhenius-type equation describes the retention factor's dependence on temperature (Equation 3.14)

$$\ln k' = \ln k'_{0,T} - \frac{b}{T}. \tag{3.14}$$

The impact of the mobile-phase composition and the temperature is unique for each component and thus the selectivity can be optimized by these parameters.

3.2.5.2 Gradient Elution for Small amounts of Product on RP Columns

RP chromatography is widely used to isolate small amounts of target molecules in automated high-throughput systems. Due to good reproducibility, stability, and a broad application range, RP instead of NP chromatography is preferred for this task. For reversed-phase separations, TLC is seldom used as a pilot technique, mainly due to the large difference in the degree of derivatization between RP materials used in column chromatography and the materials that can be used for TLC applications. With the now ready availability of mass spectrometry (MS) detection, and since MS is coupled to columns with diameters of up 25 mm by splitting systems, compound isolation is often directly combined with its identification. The structure of the molecules to be separated largely determines the type of mobile phases that can be used in RP chromatography. For some product types, mixtures of water and organic solvents can be used as the mobile phase. However, in this case one has to deal with basic or acidic molecules and so, in general, the aqueous

3.2 Selection of Chromatographic Systems | 157

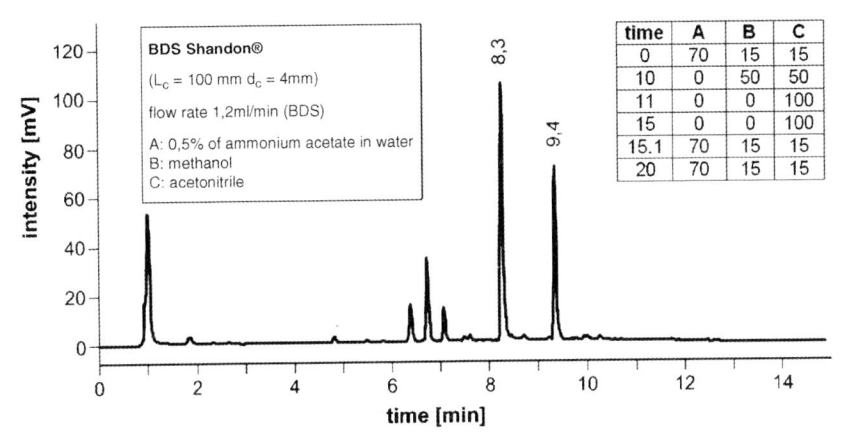

Figure 3.51 Standard HPLC-Mass spectrometric reversed-phase analysis procedure.

time	A	B	C
0	70	15	15
10	0	50	50
11	0	0	100
15	0	0	100
15.1	70	15	15
20	70	15	15

BDS Shandon®
(L_c = 100 mm d_c = 4mm)
flow rate 1,2ml/min (BDS)
A: 0,5% of ammonium acetate in water
B: methanol
C: acetonitrile

phase has to be buffered to obtain acceptable peak shapes. It is common practice in HPLC-MS identification to add an ammonium salt to the aqueous phase.

Figure 3.51 gives experimental conditions that can be used as a standard HPLC-MS analysis procedure. After the column has been equilibrated, the operation starts with a 10 min gradient separation (from 70% A, 15% B, 15% C to 50% B, 50% C) followed by 5 min washing and 5 min re-equilibration.

The obtained analytical results can be directly transferred to a preparative chromatography column. For the standard method depicted in Figure 3.51, a combination of acetonitrile and methanol is used as the organic modifier. Therefore, some additional experiments, investigating the effect of each type of organic modifier individually, are advisable because, very often, large differences in selectivity can be observed.

Owing to microbiological growth, a solution of ammonium acetate in water cannot be stored for a long period. In other words, when for preparative chromatographic applications larger volumes of an ammonium acetate solution have to be stored, some measures have to be taken to avoid this microbiological growth. Different remedies are possible, one of which is to add 5–10 vol.% acetonitrile to the mixture of ammonium acetate and water. Obviously, the presence of this amount of organic modifier has to be taken into account when transferring data from an analytical to a preparative column.

Therefore, the standard gradient for HPLC-MS analysis is modified by using an aqueous ammonium acetate solution containing 10 vol.% acetonitrile. Care has to be taken that the polarity at the start of the gradient remains approximately equal. Chromatograms in Figure 3.52 depict the separation using this modified weak solvent.

3.2.5.3 Rigorous Optimization for Isocratic Runs

After developing a first gradient method, the optimization of an isocratic method should be considered for process-scale separations. Gradient runs help to find

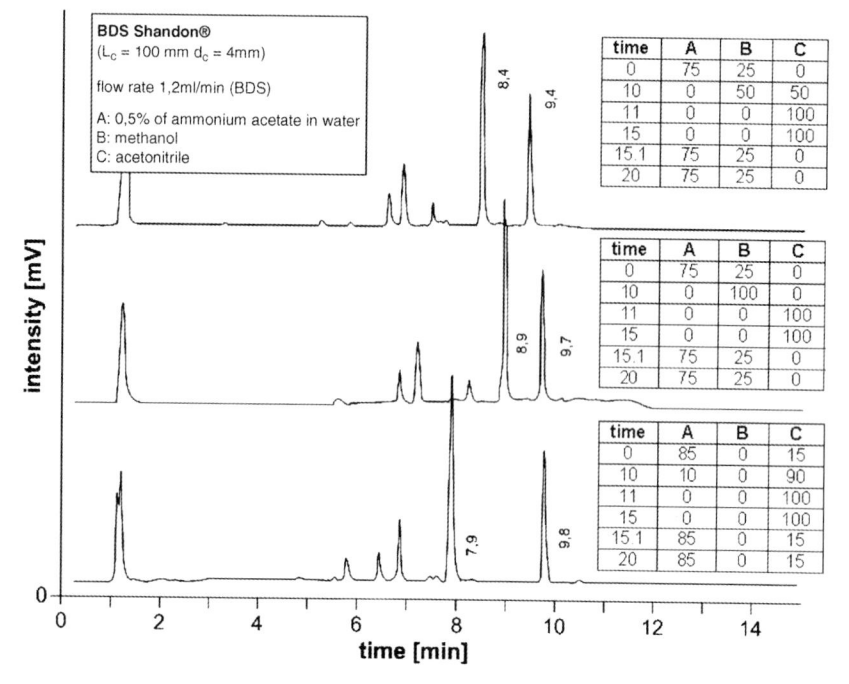

Figure 3.52 Standard gradient with modified weak solvent.

time	A	B	C
0	75	25	0
10	0	50	50
11	0	0	100
15	0	0	100
15.1	75	25	0
20	75	25	0

time	A	B	C
0	75	25	0
10	0	100	0
11	0	0	100
15	0	0	100
15.1	75	25	0
20	75	25	0

time	A	B	C
0	85	0	15
10	10	0	90
11	0	0	100
15	0	0	100
15.1	85	0	15
20	85	0	15

BDS Shandon®
(L_c = 100 mm d_c = 4mm)
flow rate 1,2ml/min (BDS)
A: 0,5% of ammonium acetate in water
B: methanol
C: acetonitrile

optimal conditions for isocratic operation, which should be preferred for preparative purposes as the advantages for large-scale separations are obvious:

Easier equipment can be used (only one pump, lower number of eluent tanks).
No re-equilibration time necessary (touching band-methods as well as closed-loop recycling and SMB-chromatography can be applied).
Easier eluent recycling due to constant eluent composition.

In the following paragraph the development of a separation method is explained in detail by the separation of a pharmaceutical sample consisting of three intermediates.

The general tasks during optimization are:

Minimize total elution time
Minimize cycle time (= time between first and last peak)
Optimize selectivity
Use isocratic conditions

Minimization of the total elution and cycle time results in higher productivity and lower eluent consumption. Optimizing the selectivity can be contradictory to these parameters. However, if the cycle time is minimized the productivity for touching band operation increases.

Figure 3.53 presents a procedure for optimizing RP chromatography by isocratic experiments. It usually begins with gathering solubility data for the sample, using

3.2 Selection of Chromatographic Systems | 159

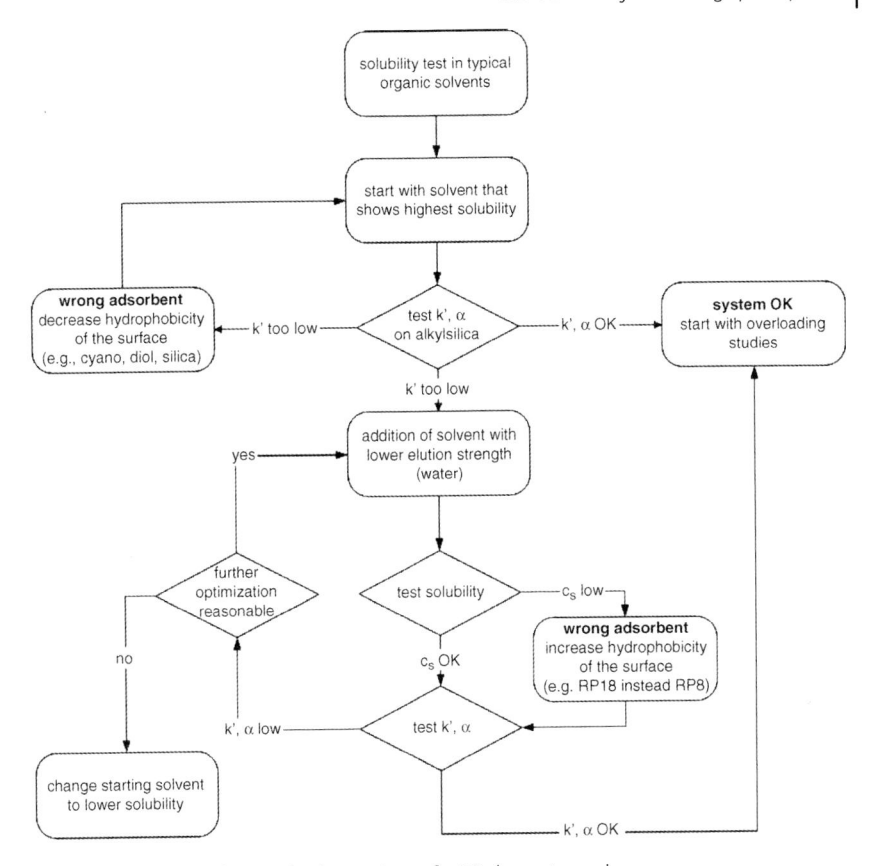

Figure 3.53 Choice of optimal solvent mixture for RP chromatography.

water-miscible solvents. The solvent with highest solubility for the sample is the first mobile phase. A test run is made on an alkyl silica and the retention factor k as well as the selectivity are determined. In most cases, this solvent (e.g., methanol) will not separate the solutes and they elute unretained within the dead time of the column. Solvent strength is then decreased by successively adding water to the mobile phase. As the polarity of the mobile-phase solvent is increased, the retention factors also increase.

When the selectivity α is inadequate for preparative purposes, the stronger solvent, in this case methanol is substituted by another water-miscible organic solvent, such as THF or acetonitrile. The optimization of k is then repeated by adjusting the water:organic ratio. This approach works well for most polar to moderately polar compounds that remain nonionic in the solution.

As an example, Figure 3.54 shows the separation of three pharmaceutical intermediates on LiChrospher® RP-18 with a mobile-phase composition of acetonitrile–water (80:20). Intermediate 1 elutes at 2.28 min while intermediates 3 and 2 coelute at 3.16 min. Obviously, the separation problem is not solved and, thus, the

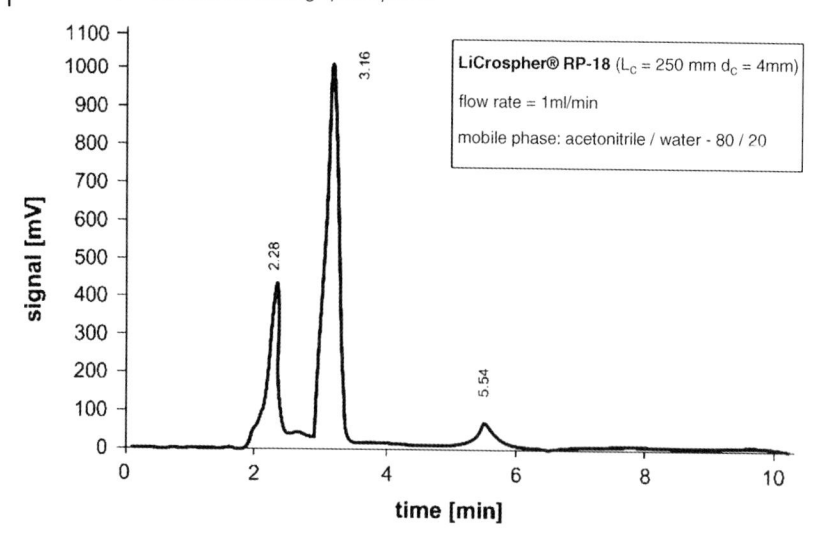

Figure 3.54 Separation with LiChrospher® RP-18 with acetonitrile–water (80:20).

mobile-phase composition must be adjusted in subsequent batch runs with higher water fractions.

Table 3.30 gives the selectivities and retention factors for increased water fractions, while Figure 3.55 shows the plot of ln k' versus ϕ. Corresponding to Equation 3.13, the influence of ϕ on the retention factor of the intermediates is described by the slopes of these straight lines.

The parameters of Equation 3.13 are determined from Figure 3.55, and thus the selectivity can be calculated for the whole range of mobile-phase composition by Equation 3.15. The influence of ϕ on selectivity is mainly determined by the difference between the two slopes S_1 and S_2. If the difference is positive ($S_2 < S_1$) the selectivity increases for higher fractions of the organic solvent.

$$\alpha_{2,1} = \frac{k_2'}{k_1'} = \frac{e^{(\ln k_{0,S2} - S_2 \cdot \phi)}}{e^{(\ln k_{0,S1} - S_1 \cdot \phi)}} = \frac{k_{0,S2}}{k_{0,S1}} e^{((S_1 - S_2) \cdot \phi)}. \tag{3.15}$$

The mobile-phase composition can now be adjusted to the needs of the separation. For example, if intermediate 1 must be isolated with high purity its retention

Table 3.30 Optimization of an RP separation with LiChrospher® RP 18 and different volume fractions ϕ of acetonitrile.

ϕ (%)	k_1	k_2	k_3	$\alpha_{2,1}$	$\alpha_{3,2}$
40	1.34	4.80	10.23	3.58	2.13
50	0.46	2.22	3.51	4.78	1.58
60	0.19	1.23	1.63	6.42	1.33
80	0.04	0.45	0.45	12.13	1.00

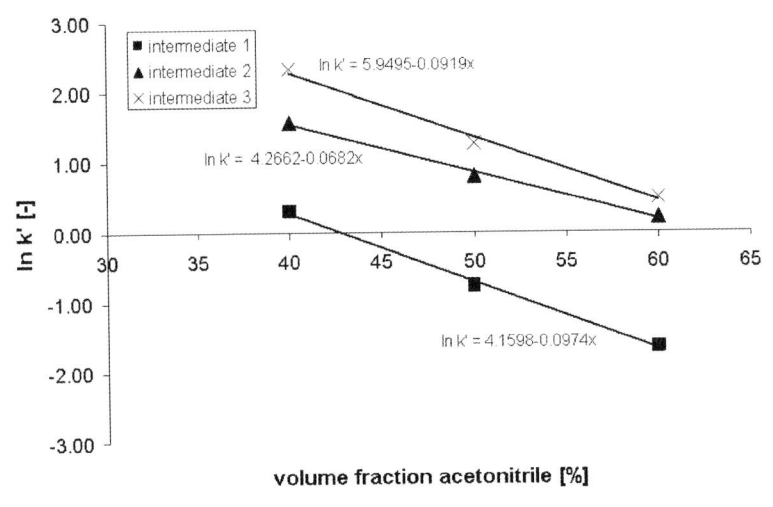

Figure 3.55 Dependency of retention factor on mobile phase composition.

factor should not be too small ($k > 1$; $\ln k > 0$), and thus the volume fraction should not exceed 43% acetonitrile.

The selectivities $\alpha_{2,1}$ and $\alpha_{3,2}$ show a contrary behavior: $\alpha_{2,1}$ increases with ϕ while $\alpha_{3,2}$ decreases at higher volume fractions of acetonitrile. This is due to the differences in slopes in the plot of Figure 3.55; $S_1 > S_2$ and thus $S_1 - S_2$ is positive, while $S_3 > S_2$ and thus $S_2 - S_3$ is negative.

Selection of the optimal mobile phase composition has to take into account the desired product components. If there are more than two product components in the sample a crucial separation must be defined. In the example case a volume fraction of 60% acetonitrile shows best results for the three-component separation because a good selectivity $\alpha_{2,1}$ is observed at low retention times and the reduced k'_1 is still acceptable (Table 3.30).

3.2.5.4 Rigorous Optimization for Gradient Runs

The optimization of reversed-phase separation by gradient runs is sometimes unavoidable even in process scale. Mobile-phase gradients can reduce the development time dramatically as the solvent strength is varied during the chromatographic separation by changing the mobile-phase composition. Figure 3.56 shows the development scheme for optimizing mobile-phase composition by gradient operation.

The optimization by gradient operation starts with a linear gradient over the whole range of 100% water to 100% organic. If the solubility in different solvents is unknown, methanol or acetonitrile can be taken as starting solvents. Gradient time is adjusted to 10–15 fold the dead time of the column. If the target solutes elute at retention factors <1 the solutes are either polar or they ionize in aqueous solvents. Here, the interaction of polar solutes with the nonpolar adsorbent is very low and a polar solvent with low interaction (e.g., 100% water) is recommended. Ionization

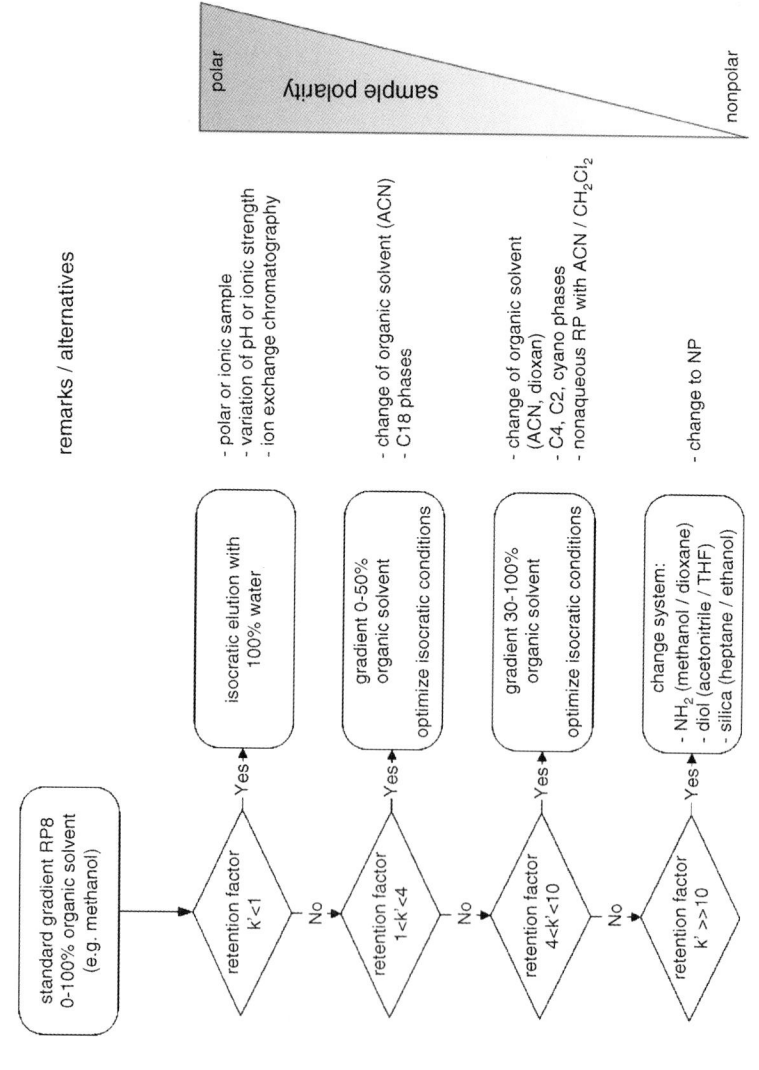

Figure 3.56 Optimization of mobile-phase composition by gradient operation.

must be suitably suppressed (Section 3.2.5.5) or the solutes should be separated by ion-exchange chromatography.

If the retention time is much higher than the time of the initial gradient (retention factor $\gg 10$) the substances seem to be too lipophilic for RP separation on the initial adsorbent. The polarity of the adsorbent should be increased by the use of amino or diol phases. If the adsorbent is already polar the phase system must be changed to NP-chromatography (Section 3.2.4).

If the retention times are in a suitable range ($1 < k < 10$–15), solvent composition is adjusted by further gradient runs. Conditions for subsequent gradient runs are derived from the retention times of the solutes during the previous run. Mobile-phase composition at the elution point of the first target solute is a good starting point for the mobile-phase composition of the following gradient run, while the mobile phase composition at the elution point of the last peak can be taken as the final composition of the mobile phase.

The following example for the optimization of mobile-phase composition by gradient runs refers to a separation of the same pharmaceutical mixture as specified in Figure 3.54. LiChrospher® RP-18 ($L_c = 250$; $d_c = 4$ mm) is used as adsorbent and the mobile phase is a mixture of acetonitrile and water. Figure 3.57 shows the initial

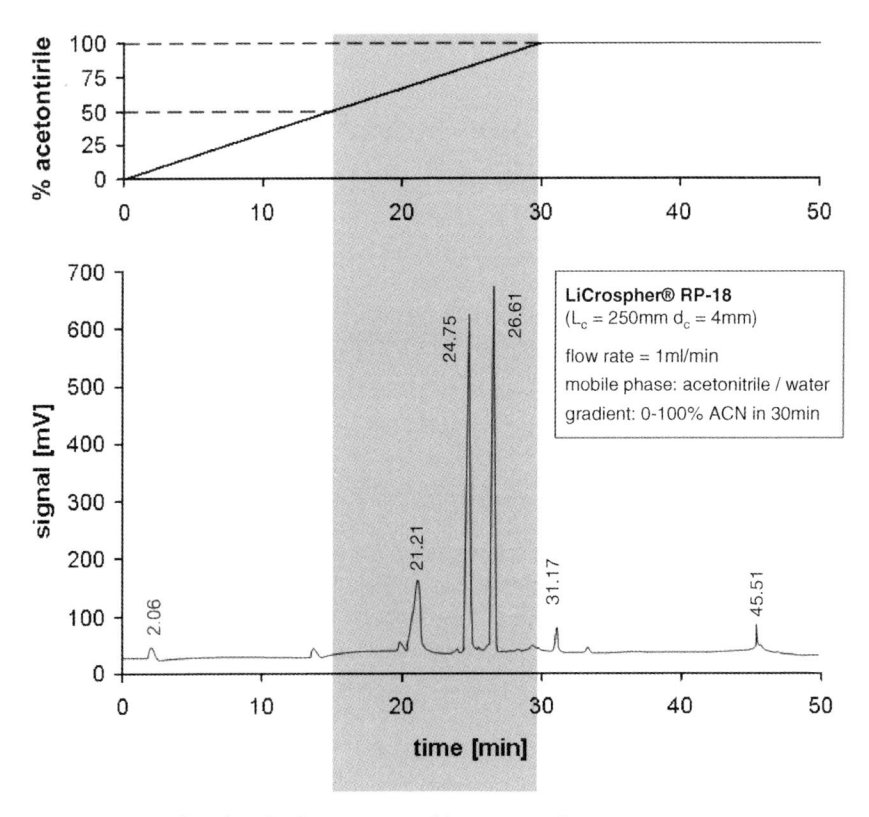

Figure 3.57 Initial gradient for the separation of three intermediates.

gradient from 0 to 100% acetonitrile over 30 min and a flow rate of 1 ml min^{-1}. The three target intermediates elute at 21.21, 24.75 and 26.61 min. The range of the elution times of the three intermediates is marked by the gray box in the chromatogram. It exceeds the above recommended starting and ending points by a few minutes. Corresponding to this time range, the gradient conditions for further optimization are determined according to the upper part of Figure 3.57.

The first mobile-phase composition for the subsequent gradient run corresponds to ϕ at the beginning of the gray marked time range (50%), while the final composition is equal to ϕ at the end of this range (100%). By this strategy the gradient (and thus the mobile phase) is optimized step by step. Table 3.31 shows the gradient optimization. The initial gradient is followed by a second gradient from 50–100% acetonitrile. A third gradient from 70–100% has been examined, but the retention times are too low and, therefore, the components were not separated.

To check the selectivity with different chromatographic systems the separation with LiChrospher® RP-18 is compared with the gradient separation with two different adsorbents. For LiChrospher® RP-8 the separation with methanol–water is shown in Figure 3.58a while Figure 3.58b shows the separation with acetonitrile–water. With methanol the elution order of the first two peaks was inverted compared with the separation with acetonitrile. However, this change of the elution order results in decreased selectivities and thus only acetonitrile is further tested as organic solvent. The third tested adsorbent is LiChrospher® CN. Retention factors and selectivities are shown in Table 3.31. For the RP-8 phase no suitable system was found while the selectivity for the RP-18 and CN phase was satisfying and appropriate gradient conditions could be identified. Based on the good selectivities on these two adsorbents isocratic conditions in the medium polarity region of the gradient composition were also tested and gave good results. The two isocratic separations on RP-18 and CN-silica would be preferred for large-scale separations of these feed mixtures as they show short cycle-times and still have good separation factors. Isocratic mobile-phase compositions for both phases are also presented in Table 3.31.

3.2.5.5 Practical Recommendations

Although RP systems are quite robust, different sources of practical problems during the operation exist:

- Solutes might ionize in aqueous mobile phases.
- Effects of residual silanol groups of the RP packing
- Operational problems due to high fractions of water in the mobile phase

Since retention in RP systems depends on the interaction between hydrophobic groups of the solutes and the adsorbent surface, ionization of the solutes can result in severe peak distortion. Ions are very hydrophilic and are thus poorly retained in RP systems even if the nonionized solute shows good retention. Because ionization reactions (e.g., dissociation of organic acids) are determined by their chemical

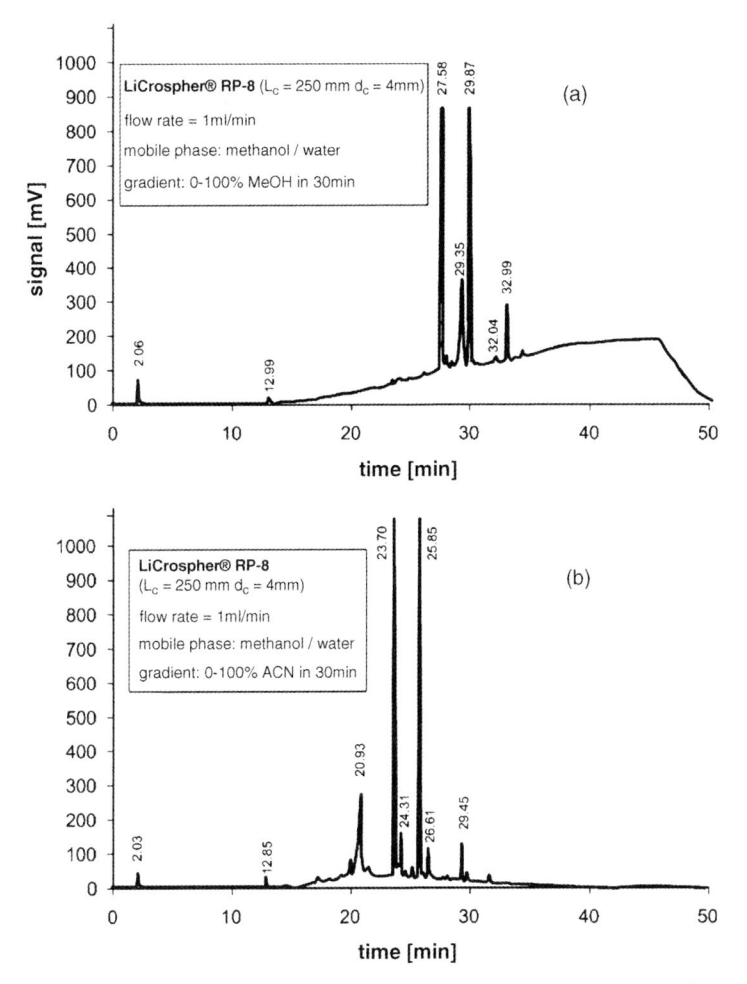

Figure 3.58 Gradient runs for two different organic solvents with LiChrospher® RP-8: (a) methanol and (b) acetonitrile.

equilibrium, solutes are present in both forms. This leads to broad peaks with poor retention. The ionization reaction must be suppressed by buffering the system. Thus, the stability of the adsorbent for the pH range must be checked. Especially for alkaline mobile phases, silica-based materials are problematic and the use of polymeric phases should be checked. If ionization of the solutes is suppressed by adding salts to the aqueous mobile phases a second effect occurs since the salt increases the polarity of the mobile phase, thereby increasing retention times.

Residual silanol groups on the adsorbent surface are a source of secondary interaction different from the bonded groups. Next to hydrogen bonding and

Table 3.31 Optimization of RP separation of three intermediates by gradient operation and resulting isocratic conditions.

Adsorbent	Organic solvent	Gradient	t_G (min)	Cycle time (min)	k_1 (−)	k_2 (−)	k_3 (−)	$a_{2/1}$	$a_{3/2}$
RP-18	Acetonitrile	0–100	30	5.40	8.64	10.26	11.10	1.19	1.08
RP-18	Acetonitrile	50–100	30	5.99	0.46	2.00	3.18	4.35	1.59
RP-18	Acetonitrile	70–100	30	1.24	0.07	0.07	0.64	1.00	8.71
RP-18	Acetonitrile	60	Isocratic	3.16	0.19	1.23	1.63	6.42	1.33
RP-8	Methanol	0–100	30	2.29	$11.54^{a)}$	$12.35^{a)}$	12.58	1.07	1.02
RP-8	Methanol	50–100	30	6.27	$4.17^{a)}$	$6.73^{a)}$	7.02	1.62	1.04
RP-8	Methanol	70–100	30	0.99	$0.81^{a)}$	$1.12^{a)}$	1.26	1.39	1.12
RP-8	Acetonitrile	0–100	30	5.02	8.52	9.78	10.80	1.15	1.10
CN	Acetonitrile	0–100	30	7.58	5.83	8.44	9.28	1.45	1.10
CN	Acetonitrile	30–100	30	8.49	0.56	3.45	4.42	6.22	1.28
CN	Acetonitrile	60	Isocratic	1.77	0.06	0.63	0.87	9.80	1.38

a) Elution order between intermediates 1 and 2 inverted due to the use of methanol.

dipole interactions, silanol groups can be regarded as weak ion exchangers (Arangio, 1998). Especially for basic solutes, the ion-exchange effect dominates. Secondary interactions result in significant changes in retention time and, very often, peak tailing. The ion-exchange mechanism can be suppressed by protonation of the silanol groups at low pH. At pH 3 roughly all residual silanols are protonated and no ionic interactions are observed. At pH > 8 nearly all residual silanol groups are present in ionic form. Along with suppression of ionization by adding buffer, chromatographic efficiency is increased by reducing the number of residual silanol groups by endcapping.

Endcapping of adsorbents is very important if high volume fractions (even 100%) of water are used as mobile phase. Due to the hydrophobicity of the adsorbent surface, problems with moistening of the adsorbent can occur and, as a result, the surface groups can collapse. Standard RP-18 alkyl-chains are not wetted and are thus not stable at 100% water. The alkyl chains, which should point into the mobile phases like bristles from a brush, will collapse and stick together on the surface. As a result, solutes in the mobile phase will not migrate into the surface of the adsorbent and the retention time will change dramatically. This behavior depends strongly on the endcapping of the surface (Section 3.1.3.1). If nonpolar groups are used for endcapping, the hydrophobicity of the alkylsilica increases and thus the stability for 100% water mobile phases gets worse. For this reason, special adsorbents with polar groups for the endcapping or embedded polar groups in the alkyl chain have been designed for use in 100% water. Nevertheless, wherever possible nonendcapped adsorbents should be used in process-scale applications, because they are cheaper and less sophisticated.

Owing to the practical problems with RP chromatography, very often aqueous solvents with buffers are used as mobile phase. Because of the high boiling point of water and the presence of salts in the mobile phase, recovery with high purity of the target solutes from the fractions can be simplified with one more adsorption step. A short RP column with a large diameter (so-called pancake column) can be used for a solid-phase extraction step. The target fraction is diluted with water to decrease the solvent strength of the mobile phase. The diluted fraction is then pumped onto the column and the sample is strongly adsorbed at the column inlet. After the complete fraction is collected in the column, it is washed with a mobile phase of low solvent strength to remove the salts. In the last step, the sample is eluted with a low volume of a solvent with high elution strength (e.g., methanol). The concentration of the sample is now much higher, the salt is removed and the more volatile strong solvent is easier to evaporate.

3.2.6
Criteria for Choosing CSP Systems

In contrast to normal and reversed-phase separations, one critical difference with chromatographic enantio separations on chiral stationary phases (CSP) is that rational development of selectivities on one given stationary phase is nearly impossible. Optimization of the chromatographic parameters for enantio separation is much

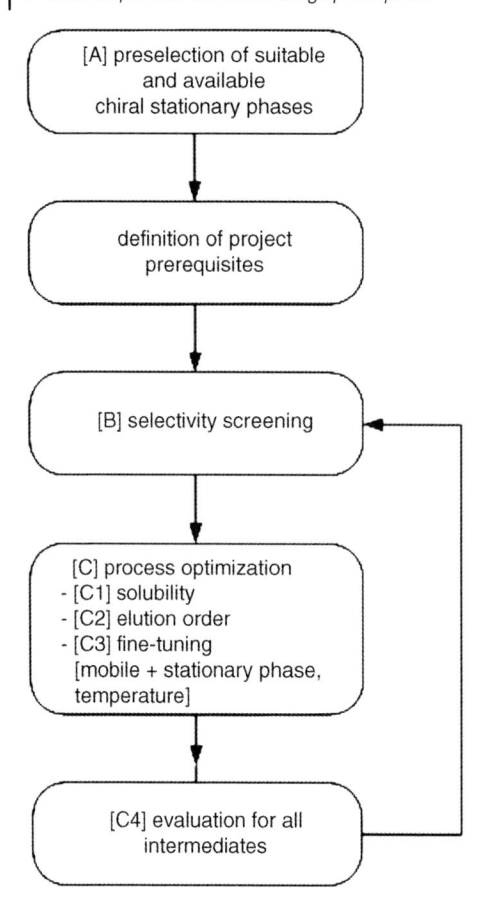

```
┌─────────────────────────────┐
│ [A] preselection of suitable │
│      and available           │
│   chiral stationary phases   │
└─────────────────────────────┘
              │
              ▼
┌─────────────────────────────┐
│   definition of project      │
│       prerequisites          │
└─────────────────────────────┘
              │
              ▼
┌─────────────────────────────┐
│  [B] selectivity screening   │◄──┐
└─────────────────────────────┘   │
              │                    │
              ▼                    │
┌─────────────────────────────┐   │
│ [C] process optimization     │   │
│  - [C1] solubility           │   │
│  - [C2] elution order        │   │
│  - [C3] fine-tuning          │   │
│    [mobile + stationary      │   │
│    phase, temperature]       │   │
└─────────────────────────────┘   │
              │                    │
              ▼                    │
┌─────────────────────────────┐   │
│   [C4] evaluation for all    │───┘
│       intermediates          │
└─────────────────────────────┘
```

Figure 3.59 Screening strategy for chiral separations.

more difficult than for nonchiral molecules and involves the screening of several mobile–stationary phase combinations. As a general guide to successful enantio separation the process should be developed using the following steps (Figure 3.59):

- Suitability of preparative CSP.
- Development of enantio selectivity.
- Optimization of separation conditions.

3.2.6.1 Suitability of Preparative CSP

Before starting to screen for a suitable CSP some general considerations have to be taken into account to optimize the separation in the best direction right from the beginning. The required purity and amount and time frame should be known alongwith some limitations with regard to solvents or other process conditions, for example, temperature range. In addition, it should be known if only one or both enantiomers have to be delivered in purified form.

Over 100 chiral stationary phases are now commercially available (Section 3.1.5). Not all of them are designed and suited for preparative purposes. Certainly, every CSP might be used to isolate some mg amounts of an enantiomer in case of urgent need, but to develop a production process the CSP should fulfill some critical requirements. The main parameters are:

- Selectivity (range).
- Reproducibility.
- Stability.
- Availability.
- Price.

A certain selectivity range should be given for the individual CSP. The selectivity range is of great importance for laboratories where moderate amounts (max. in the gram-range) of a multitude of different racemates are separated in a short time. Chiral phases with the broadest selectivity range are the cellulose and amylose derivatives. With the four chiral phases Chiralcel® OD and OJ and Chiralpak® AD and AS, selectivities for about 70% of all screened racemates are found (Daicel Application Guide 2003). For production purposes, a broad selectivity range is not necessary if only the desired racemate is well separated. Even tailor-made adsorbents could be used, if the production is well controlled and the regulatory and reproducibility issues of the phase are well addressed. The main parameters for overall suitability are related to the mechanical and chemical stability of the adsorbents. Chemical stability is mainly linked to the range of solvents that can be used. As solubility is often an issue for pharmaceutical compounds, the chosen CSP should have a good stability against solvents with different polarities so that the whole spectrum of retention adjustment can be used. For production processes the availability of the CSP in bulk quantities at a reasonable price should also be taken into account. Different CSP groups are characterized according to these parameters in Table 3.13 (Section 3.1.5.3).

3.2.6.2 Development of Enantioselectivity

For chiral stationary phases, the mobile phase selection has to aim first for enantioselectivity α (D, L). The molecules to be separated show identical chemical (same functional groups) and chromatographic behavior (same retention on the bare stationary phase; chiral recognition of enantiomers only by the bonded chiral selector). Therefore, the separation is based on very small differences in complexation energies of the transient diastereomer complexes formed, and similar strategies as for the selections of solvents for liquid–solid chromatography are helpful. Take a good solvent, in which the sample readily dissolves (e.g., ethanol for cellulose or amylose-based packings) and increase retention until enantioseparation may be observed.

As pointed out above, no selectivity guarantee can be given for the separation on a certain chiral stationary phase. Therefore, screening routines have to be followed to obtain the appropriate adsorbent. The first possibility is the knowledge-based approach. Some CSP exhibit a certain group-specificity, for example, the quinine-based CSPs for N-derivatized amino acids or the poly-(N-acryloyl amide derivatives) for five-membered heterocycles. The most advanced statistical and knowledge-based approach is the chiral separation database Chirbase®, developed by Christian Roussel at the University of Marseille (http://chirbase.u-3mrs.fr/). Based on more than 20 000 separations integrated into the database, software has been developed by Chiral Technologies in cooperation with the University of Marseille to suggest the most successful starting conditions for a racemate based on its functional groups (ChiralTOOL, Chiral Technologies Application Guide 2003, www.chiral.fr). More general predictions have been made by the Marseille group based on correspondence analysis of the graphical representation of single CSPs and 15 empirical molecular descriptors of the racemates included in the database (Roussel, Piras, and Heitmann, 1997).

If no rational starting point for the separation of a racemate can be found, a random screening procedure has to be applied. In most cases 250 × 4 or 4.6 mm columns are used for this purpose. Modern HPLC systems often include a column switching device with up to 12 columns. Most systems have programmable software options for the set of different mobile phase compositions. Combination of UV and polarimetric detection allows even very small selectivities to be found and used as a starting point for further optimization. Commercial systems with deconvolution and optimization options are available, together with a given set of 12 different chiral stationary phases (www.pdr-chiral.com). After an initial detection of selectivity on a CSP, the mobile-phase composition and additives as well as temperature have to be varied either manually or by means of an automated system.

For large-scale separations even the design of a new chiral stationary phase might be economically advisable. The most prominent CSP design approach is the reciprocal approach developed by Pirkle and Däppen (1987). One pure enantiomer of the racemate to be separated is bound via a spacer to silica as chiral selector. Onto this chiral stationary phase various racemates are subjected whose pure enantiomers are readily available in sufficient amounts. One enantiomer of the racemate with the highest selectivity is chosen as the chiral selector for the CSP. After manufacturing this reciprocal CSP the production of the desired enantiomer can take place. This approach has been used to obtain a very good stationary phase for the separation of non-steroidal anti-inflammatory drugs, for example, ibuprofen and naproxen (Pirkle, Welch, and Lamm, 1992).

Now that combinatorial and parallel syntheses are available, the number of possible selectors for chiral stationary phases can be drastically increased. Selector synthesis on solid phases and the testing in 96-well plate format have been used to make the CSP-screening process more efficient (Welch, 1998; Murer, 1999; Wang, Bluhm, and Li, 2000; Bluhm, Wang, and Li, 2000; Svec, Wulff, and Fréchet, 2001).

The history of columns used for screening racemates for large-scale production purposes should be carefully documented. Cellulose and amylose derivative CSPs, especially, sometimes show strong reactions on small amounts of acids and bases. These compounds might not only be mobile-phase additives but also racemates bearing acidic or basic functions that have been previously injected. For large-scale projects it is advisable to use new columns to avoid disappointments if selectivities cannot be reproduced on other or larger columns.

3.2.6.3 Optimization of Separation Conditions

3.2.6.3.1 Determination of Racemate Solubility The solubility of the racemate in the mobile phase is the first parameter to be optimized. Several solvents should be screened for solubility and selectivity at the chiral stationary phase that showed the highest selectivity. Notably, once again, not all chiral stationary phases can be operated with all solvents. The instability of the coated cellulose- and amylose-derivative CSPs against medium polar solvents such as dichloromethane, ethyl acetate, and acetone is widely known. Improved phases with immobilized chiral selectors have been recently developed and made available in bulk quantities (Zhang, 2005b). Conversely, some CSP tend to low retention factors if polar solvents are used. Especially, tartar-diamide-based CSPs (Kromasil® CHI) need a certain amount of apolar alkane solvent to achieve sufficient retention. For this type of stationary phase, the use of supercritical carbon dioxide in combination with a polar modifier might be an alternative option. As mentioned in Section 3.2.3.1, productivity can be increased by changing the injection solvent if the solubility is not sufficient in the mobile phase (Dingenen, 1994). This can be realized only in noncontinuous modes, especially in closed-loop recycling chromatography (Section 5.1.3).

3.2.6.3.2 Selection of Elution Order With chromatographic production processes the elution order of the enantiomers is of importance. In SMB processes the raffinate enantiomer can often be obtained with better economics as it is recovered at higher purities and concentrations. If the CSP offers the possibility of choosing one of the two optically active forms of the selector, the adsorbent on which the desired enantiomer elutes first should be chosen. This option can be used especially with the brush-type phases with monomolecular chiral selectors. Even if the CSP is not available in both forms, the elution order should be checked carefully as the elution order might be reversed on two very similar adsorbents or with two similar mobile-phase combinations. Okamoto and Nakazawa (1991) and Dingenen (1994) have shown that by changing only from 1-propanol to 2-propanol, respectively with 1-butanol, the elution order on a cellulose-based CSP might reverse.

3.2.6.3.3 Optimization of Mobile/Stationary-Phase Composition, Including Temperature It should be taken into account that the highest enantioselectivity is observed at the lowest degree of nonchiral interactions, that is, at the level of a nearly non-retained first enantiomer. Moreover, enantioselectivity increases with lower temperature according to Equation 3.16.

$$\alpha = \frac{1}{e^{\left(\frac{\Delta\Delta G_{D,L}}{RT}\right)}}. \tag{3.16}$$

This effect on resolution may be counterbalanced by increased viscosity, leading to lower efficiency of the system. Therefore, fine tuning of mobile-phase composition and temperature should be carefully taken into account for production-scale systems as some economic benefits have to be considered against a higher complexity of the separation system, for example, in terms of controlling temperature and small amounts of modifier.

3.2.6.3.4 Determination of Optimum Separation Step Notably, in the synthetic route towards the final enantiomerically pure compound all intermediates should be taken into account for a chromatographic separation step. After introducing the chiral center the intermediates might differ substantially in solubility and selectivity. No general guidelines can be given as to whether separation on an early or final stage is better economically. Good project coordination involving synthetic chemists and chromatography specialists is the best way to ensure that the optimum separation stage is found.

3.2.6.4 Practical Recommendations

When optimizing the temperature of a chiral separation it should be noted that some chiral compounds tend to racemize at elevated temperatures. On-column racemization might be seen as a typical plateau between the two peaks of single enantiomers (Trapp and Schurig, 2002). Acidic or basic mobile-phase additives might even catalyze such racemization. Conversely, on-column racemization can be used to optimize the yield of a chiral SMB separation. If only one enantiomer is of interest and the other is considered as "isomeric ballast," on-column racemization could be used to enhance the separation yield by transferring the undesired enantiomer into the racemate again. This might be done in an on- or off-line chromatographic reactor (Section 5.2.9).

In chiral chromatography, especially in SMB production systems, the quite sensitive chiral stationary phases are subjected to a high mass load of compounds to be separated – which might cause some conformational changes of the chiral selector. Once the column is rinsed with pure solvent and checked by a pulse injection, severe changes might be observed. Figure 3.60a shows enantioseparation on a column continuously loaded with feed mixture for several hours. Afterwards, the column was washed with pure eluent and checked with a pulse injection (Figure 3.60b).

Clearly, some severe changes have occurred during the 139 h of operation. The retention time of both peaks has shifted and, especially, the tailing of the second peak has increased tremendously with time. After washing the chiral phase with a solvent of high polarity the changes are only partly reversible. This behavior is shown in Figure 3.60b, where the chromatograms at the beginning 0 h and after 139 h of separation are compared with that after washing. Obviously, after the washing step the column does not reach the original condition at the beginning of the separation. For SMB chromatography, the increased tailing had to be adjusted by changing the flow rates. However, finally, the separation had to be stopped because no stable conditions were found under which the change in stationary

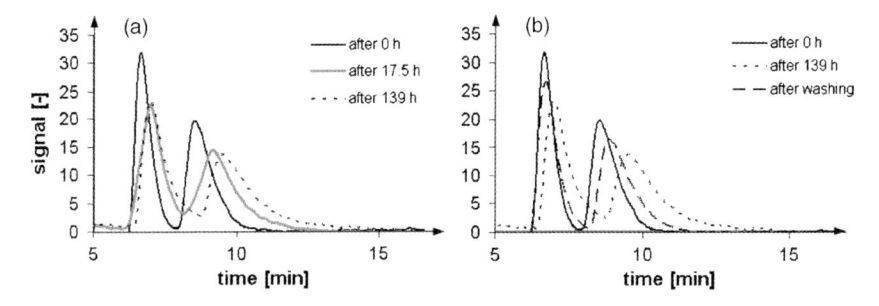

Figure 3.60 Effect of high feed loading on a chiral stationary phase.

phase behavior could be compensated by small flow-rate changes or short washing procedures.

Chiral stationary phases are also used for the separation of positional isomers. Isomers are normally well separated on straight-phase silica or medium polar silica phases, for example, cyano-modified silica. However, if two or more chiral centers are a certain distance apart, they might behave on the columns as if they were pairs of enantiomers. Therefore it is also worth testing CSPs if difficult isomer separations have to be performed. Figure 3.61a–c shows optimized separations of a pair of diastereomers on a straight-phase silica (LiChrospher® Si60), a reversed-phase silica (Purospher RP-18e), and a chiral stationary phase (Chiralpak® AD), respectively. Clearly, the separation with the CSP is much better and simplifies the

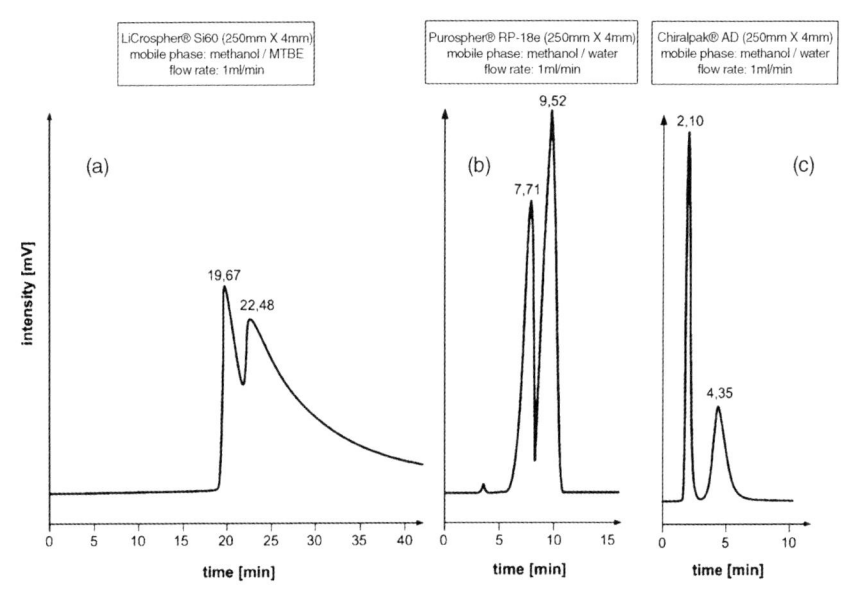

Figure 3.61 Separation of a pair of diastereomers on different nonchiral and chiral stationary phases (a–c).

preparative separation so that the higher costs of the stationary phase are compensated by improved overall process economics.

Derivatization of racemates is widely used for analytical purposes to enhance detection or reduce the interference of the target compound with the matrix (Schulte, 2001), but derivatization might also be used for preparative purposes. Francotte (1998) showed that the reaction of different chiral alcohols with a benzoyl ester functionality results in a series of racemates that are well separated on benzoylcellulose-type CSPs. With difficult enantioseparations for a given target molecule it can be worthwhile derivatizing the racemate with different achiral side groups and testing the series of racemates for optimum selectivity. Table 3.32 shows the results of this approach for a series of derivatives of a chiral C3-building block. This building block is widely used in medicinal chemistry. As the side chain is cleaved off during the following synthesis step the derivatization group might be chosen from a wide range of available components. Table 3.32 shows that substantial differences in enantioselectivity can be observed for the single compounds. Nevertheless, it should be taken into account that the derivatization group is ballast in terms of the mass balance of the total synthetic route. The molecular mass of the group should be kept as low as possible to keep the total amount of components to be separated small. This strategy has, for example, been successfully implemented by Dingenen (1994) for α-(2,4-dichlorophenyl)-1H-imidazole-1-ethanol-derivatives.

3.2.7
Downstream Processing of Mabs using Protein A and IEX

The separation tasks for ion-exchange chromatography differ mainly by the type of expression system used. The most abundant systems are mammalian cells, microbial cells (yeast), and the bacterial organism *E. coli*. All expression systems have a common feature that they not only generate the target protein but also process related impurities, such as host cell proteins, DNA, viruses, endotoxins. In addition the downstream process has to isolate the active form of the target molecule from all its derivatives, for example, oxidated, deamidated, acetylated forms, dimers, aggregates, and unfolded proteins.

Impurities and excipients stemming from the production process are another group of molecules to be removed during the purification process. These process-related molecules can be of low molecular weight, for example, antibiotics, antifoaming agents, inducing and refolding agents, sugars, solvents, salts, and water or molecules of high molecular weight, for example, insulin or albumin.

It is clear that this difficult separation task cannot be done with a single chromatographic step. Therefore, the typical downstream process for example, a monoclonal antibody consists of three distinct chromatographic steps: capturing – intermediate purification and polishing.

Figure 3.62 shows one very typical mAb downstream process, which is focused here on the chromatography steps. The assumption for this downstream process is a fermentation titer of $2\,\text{g}\,\text{l}^{-1}$ mAb in a $12\,\text{m}^3$ bioreactor. Following centrifugation

Table 3.32 Derivatization of a chiral C3-building block and corresponding chromatographic results.

C3 moiety: epoxide-CH₂-O-R3

X	R3	CSP	k′1	α
—O—	—H	Several	n.a.	1.0
—O—	—CH₂—phenyl	Several	n.a.	1.0
—O—	—SO₂—(p-tolyl)	Chiralpak® AS	0.52	1.08
—O—	2-methoxyphenyl	Whelk—O—1®	1.19	1.08
—O—	4-methoxyphenyl	Whelk—O—1®	1.39	1.11
—O—	1-naphthyl	Chiralcel® OJ	1.26	1.21
—O—	diphenylmethyl (CH(Ph)₂)	Whelk—O—1®	0.69	1.05
—O—	trityl (C(Ph)₃)	Whelk—O—1®	2.27	1.07

(continued)

Table 3.32 (Continued)

C3 moiety	X△—O—R3			
X	R3	CSP	k′1	α
—O—	biphenyl group	Whelk—O—1®	1.17	1.24
—O—	fluorenyl group	Exp. Poly-(N-acryloyl amide)	1.26	1.11
—OH, —OH	phenyl group	Chiralcel® OD	3.69	1.25

and filtration the so called clarified cell culture supernatant is loaded to a Protein A affinity resin. The mAb in the Protein A eluted fraction is concentrated by about a factor of 4, and the total volume to be processed is also significantly reduced to almost the same factor. The following steps are virus removal by a low pH step in the virus hold tank, followed by a cation-exchange step, performed in a bind and elute modus, followed by a hold tank where the cation elution pool is adjusted to be pumped over an anion-exchange column in a flow-through or nonbinding mode, respectively. Volumes to be processed and concentration of the target molecule are given in Figure 3.62.

For ion-exchange chromatography the main criteria to be optimized include type of ion-exchange group (weak or strong, cation or anion) and on the mobile phase side the pH, nature, and concentration of the buffer. Some additives might be required in addition.

In finding the right process conditions it should be furthermore distinguished between simple capturing processes, often operated in an on/off-mode with the goal to maximize recovery and intermediate or polishing steps where the separation capability between closely related and often labile compounds is the key point. The method development for chromatography steps can be done in batch mode and in an automated way using multiwell plates or miniaturized multicolumn systems to determine the capacity and the recovery. For more sophisticated processes the chromatographic behavior should be elaborated using linear gradient elution experiments (LGE) in scout columns. The methodology of LGE has been

3.2 Selection of Chromatographic Systems | 177

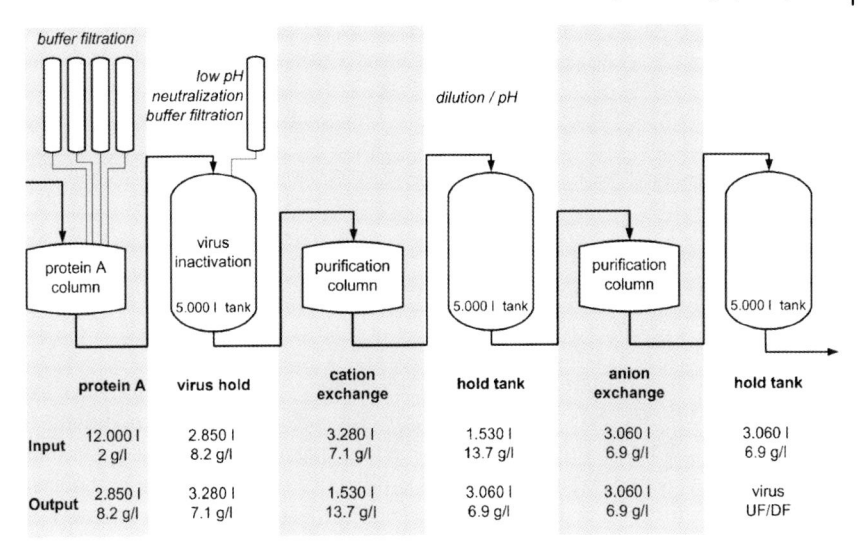

	protein A	virus hold	cation exchange	hold tank	anion exchange	hold tank
Input	12.000 l 2 g/l	2.850 l 8.2 g/l	3.280 l 7.1 g/l	1.530 l 13.7 g/l	3.060 l 6.9 g/l	3.060 l 6.9 g/l
Output	2.850 l 8.2 g/l	3.280 l 7.1 g/l	1.530 l 13.7 g/l	3.060 l 6.9 g/l	3.060 l 6.9 g/l	virus UF/DF

Figure 3.62 mAb purification platform.

introduced into ion-exchange chromatography by Yamamoto in 1988 (Yamamoto 1988). It is the goal of the LGE to obtain plots of the Peak salt concentration I_R for the different proteins versus the normalized gradient slope GH. The GH-value is representing the gradient slope times the stationary-phase volume and is determined at different pH-values. From the GH versus I_R plot the relation between the protein distribution coefficient K versus the ionic strength can be derived. Using Equation 3.17 the dimensionless number O representing the resolution can be introduced.

$$O = (ZI_a)/(G(\text{HETP})_{\text{LGE}}), \tag{3.17}$$

where Z is the column length in cm, I_a a dimensionless constant having a numerical value of 1 (M), G the gradient slope normalized with respect to column void volume and HETP_{LGE} is the plate height in the LGE. Identical elution curves can be achieved with different column geometries, gradient volumes, and linear velocities as shown in Table 3.33.

It can be seen from the table that completely different processes with respect to the amount of stationary and mobile phase and the column dimension have to be compared to obtain the process with the best economy. By calculating iso-resolution curves (Yamamoto and Kita, 2005) the optimization of protein separations in the linear range of the adsorption isotherm can be achieved. For nonlinear conditions more sophisticated calculation methods taking into account the competitive effects have to be used.

Due to the high number of ion-exchange materials used for process chromatography and driven by the fact that a multitude of different process conditions have to be screened there have been attempts over the last years to automatize the process development by using robotic high-throughput liquid-handling stations (LHS).

Table 3.33 Process parameters for the separation of egg white proteins on the strong cation exchanger Fractogel SO$_3$ with identical O-values of 10 000 (Jacob and Frech, 2007).

Value	Parameter set 1	Parameter set 2	Parameter set 3
Column length (cm)	5.4	25.0	24.3
Column diameter (cm)	1	1	2.6
Gradient volume	5.3	5.1	5.5
Linear velocity (cm h^{-1})	100	240	150

These systems screen a large variety of static conditions (variation of pH, salt concentration, and buffer and salt types) for binding as well as for elution conditions.

A typical result demonstrating the static binding characteristics of the two strong cation-exchange resins Fractogel® EMD SO3 (M) and Eshmuno® S for a monoclonal antibody as function of pH and salt concentration in the load buffer is shown in Figure 3.63.

For static binding experiments 96-well plates are typically filled with 10–20 µl of settled resin volume and equilibrated. The plates are then incubated with the protein-containing buffers for up to 2 h to reach full saturation of the beads. The robotic system shakes the plates at defined time intervals. At the end of the incubation time the supernatant is separated from the resin by a vacuum suction step. The protein concentration in the supernatant is analyzed by an on-line UV-detection system. Following a washing and elution step and the determination of the protein concentration in the washing and elution fractions also the recovery and the elution-binding capacity can be determined. This method gives results which are in good accordance with column experiments but basically are being used to compare different operating conditions. It has to be pointed out that the measurement of the UV-signal cannot distinguish between the target protein and any impurity. The integration of an off-line analytical system, for example, by HPLC or Elisa-techniques to analyze the impurities is possible but needs a quite substantial investment.

A solution for quasi-dynamic high-throughput analysis has been developed by Boehringer Ingelheim and named Rapptor® (Eckermann et al., 2007; Rathjen, Wenzel and Studts, 2009, Rathjen et al, 2009). Samples are taken in defined incubation intervals representing different residence times and analyzed for their protein content (Figure 3.64). By closing the mass balance of the fractions taken a breakthrough curve can be calculated from the UV signal of the fractions.

More recently also high-throughput screening systems in flow-through mode have been developed also enabling dynamic binding and elution experiments. These systems use prepacked arrays of chromatographic scout columns, for example, MediaScout MiniColumns from Attoll (Schulze-Wierling et al., 2007). The packed columns are operated by the LHS that presses a defined amount of mobile phase under predefined linear velocities through the single scout columns

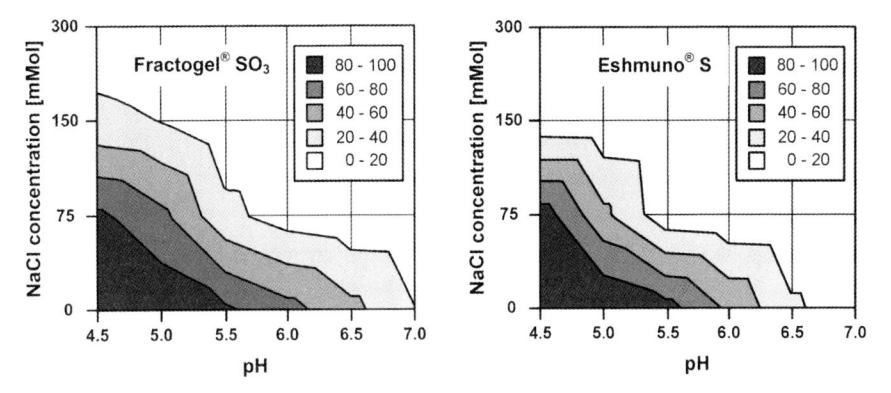

Figure 3.63 Static binding capacity of mAb on different CEX-resins at different pH and NaCl-concentrations.

(Wiendahl et al., 2008). As for each new mobile-phase composition the needle of the roboter has to be replaced and filled again and only step gradients can be applied. Nevertheless the accuracy of the system compared to single-screening columns is sufficient. Using these new screening technologies, the bottleneck in downstream processing development has been shifted to the analytical characterization of the impurity levels in the obtained large numbers of sample fractions.

3.2.8
Size Exclusion (SEC)

Size exclusion resins in their ideal form do not exhibit any adsorption. For protein purification they should be as hydrophilic and inert as possible. The separation principle is a complete partitioning process in between the pore system of the matrix (Yau et al, 1979; Cutler, 2004). Therefore, the separation efficiency is solely based on the column length and for a good resolution of compounds with only minor differences in molecular weight (or better molecular volume) a long column length is needed. As a consequence, the matrix of the sorbent has to be quite rigid, so a high degree of cross-linking is preferred, without reducing the accessible pore volume too much. As no adsorption is taking place the load of a size exclusion column is quite low and the injection volume is restricted to <5% of the column volume if high resolution should be achieved. To maximize throughput, the protein concentration should be high in the range of 2–20 mg ml^{-1} depending on the solubility of the protein and the maximum viscosity the chromatographic system can tolerate.

In addition, the flow rate in SEC is quite low to ensure complete diffusion of the feed into the pore system. From these facts it can be clearly seen that SEC in preparative mode is a quite difficult technology: long columns up to 1 m in length were packed with semirigid sorbents and operated with low flow rates. Nevertheless it is

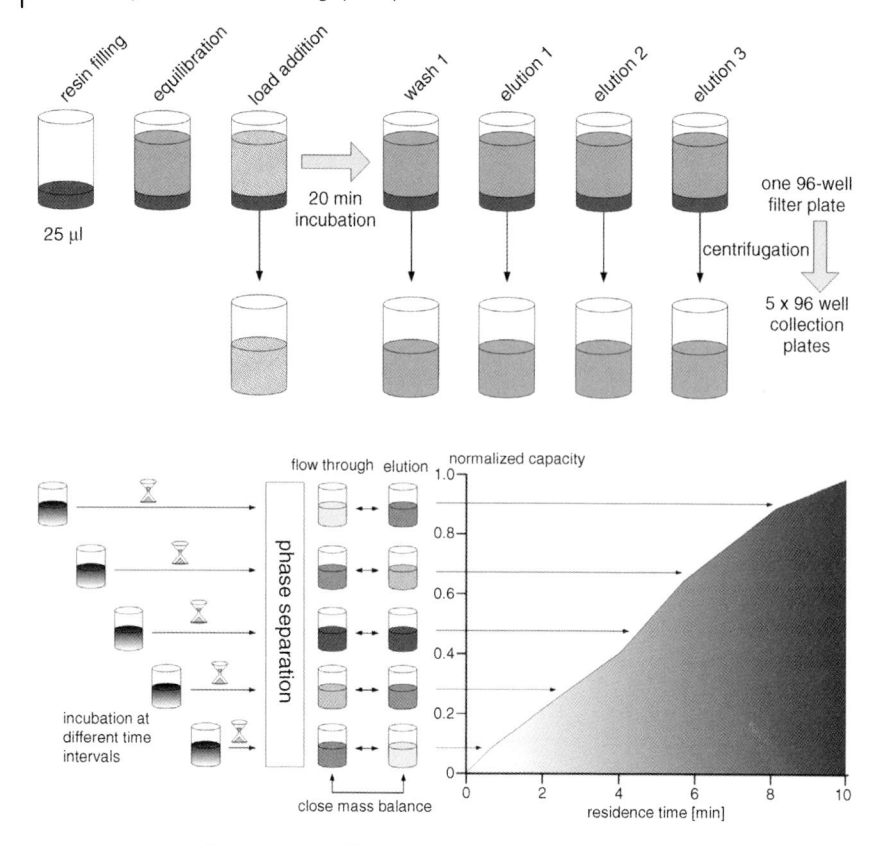

Figure 3.64 Principle of the Rapptor® robotic platform. Samples are incubated for different time intervals, filtered, and the flow-through as well as bounded protein analyzed. (a) robotic examination steps: 1. resin incubation 2. analytical assays; 32 variables in triplicate in a 96-well plate. Four different resins with eight different conditions, 20 min incubation time, load, one washing step, three elution steps, (b) Quasi-dynamic examination of different resins with the robotic system.

quite often used for special purposes, for example, the removal of protein aggregates, vaccine purification, or the fractionation of blood plasma fractions.

It can be observed with certain SEC-matrices that a certain adsorption occurs. Remaining residual charges might come from sulfate groups in agarose gels or carboxyl residues in dextrans. To overcome these adsorptive effects the ionic strength of the buffer should be kept at 0.15–0.2 M to suppress electrostatic or van der Waals forces. Cross-linking agents in polymeric gels may introduce a certain hydrophobicity to the matrix which can especially lead to the adsorption of smaller proteins rich in aromatic amino acids. The additional adsorptive effects can be used to tune the selectivity of the separation but they might be also the reason for incomplete yield due to strong adsorption.

Preparative Size Exclusion matrices are either cross-linked cellulose, agarose or dextranes, or hydrophilic polymers. Table 3.34 lists the most common sorbents with their fractionation range and pH-stability.

3.2.9
Overall Chromatographic System Optimization

3.2.9.1 Conflicts During Optimization of Chromatographic Systems

Up to now many different criteria for the choice of an optimized chromatographic system have been given. Many of these parameters are not independent. For example, a high solubility should always be preferred, but the retention time is very often too low with solvents that provide high solubilities. The most important parameter is high throughput at the desired purity and yield. The resolution gives useful hints for the optimization of a separation, because it merges different effects into one parameter.

$$R_S = \underbrace{\left(\frac{\alpha - 1}{\alpha}\right)}_{\substack{\text{influence} \\ \text{of the} \\ \text{selectivity}}} \cdot \underbrace{\left(\frac{k'_2}{1 + k'_2}\right)}_{\substack{\text{dependent} \\ \text{on the} \\ \text{retention time}}} \cdot \underbrace{\frac{\sqrt{N_2}}{4}}_{\substack{\text{efficiency} \\ \text{of the column}}} \quad (3.18)$$

Resolution can be increased if one of the three terms in Equation 3.18 is increased. As mentioned in Section 2.4.3 the first term describes the influence of selectivity. This term should be maximized by maximizing α in a selectivity screening with different adsorbents and mobile phases. The second term should be kept in a certain range and not be maximized, because the maximum value of 1 is reached for an infinite retention factor. At infinite retention, the productivity would decrease due to the high cycle time. The last term of Equation 3.18 describes the efficiency of the column in terms of the number of plates. Resolution can be increased by selecting efficient adsorbents with small particle size and appropriate narrow particle-size distribution. For these adsorbents, fluid dynamic and mass-transfer resistances are minimized. Conversely, the back-pressure of the column and the stability of the packing must also be taken into account for process-scale separations. Very small particles with a narrow particle-size distribution result in high back-pressure and may show reduced efficiency due to packing imperfections (Section 4.6).

Next to these practical problems of high efficient column packings, a high number of plates is very often not needed if peak broadening results from thermodynamic effects. Seidel-Morgenstern (1995) has discussed the effects of high column loading (Section 2.6) and showed that separation efficiency is not dependent on the number of plates, if isothermal effects come to the fore. A typical isothermal effect is the decreasing retention time for high concentration in the Langmuir range of the isotherm. Highly efficient adsorbents (with high number of plates) are not useful for separations with dominant isothermal effects. Only for

Table 3.34 Commercially available preparative SEC sorbents.

Name (nature)	Manufacturer	Type	Fract range (kDa)	pH-stability	Organic solvent stab.
Sephadex (Dextran)	GE Healthcare	G25	1–5	2.0–10.0	
		G50	1.5–30		
		G75	3–80		
		G100	4–100		
		G150	5–150		
		G200	5–600		
Sepharose (Agarose)	GE Healthcare	6B	10–4000	3.0–13.0	
		4B	60–20 000		
		2B	70–40 000		
Superdex (Agarose/Dextran)	GE Healthcare	30	0–10	3.0–12.0	
		75	3–70		
		200	10–600		
Sephacryl (Dextran/bis-acrylamide)	GE Healthcare	S100HR	1–100	3.0–11.0	
		S200HR	5–250		
		S300HR	10–1500		
Biogel P (Polyacrylamide)	Bio-Rad	P-2	0.1–18	2.0–10.0	
		P-4 gel	0.8–4		
		P-10 gel	1.5–20		
		P-60 gel	3–60		
		P-100 gel	5–100		
Biogel A (Agarose)	Bio-Rad		10–500/100–50.000	4.0–13.0	
Toyopearl (Methacrylate)	Toso Haas		0.1–10/400–30.000	1.0–14.0	Yes
Fractogel EMD BioSEC 650 (Methacrylate)	Merck Millipore		0.1–10/500–50.000	1.0–14.0	Yes
Matrex Cellufine (Cellulose)	Merck Millipore		0.1–3/10–3.000	1.0–14.0	
Ultrogel AcA (Acrylamide/Agarose)	Biosepra		1–15/100–1.200	3.0–10.0	
Trisacryl GF (Acrylamide)	Biosepra		0.3–7.5/10–15.000	1.0–11.0	

3.2 Selection of Chromatographic Systems

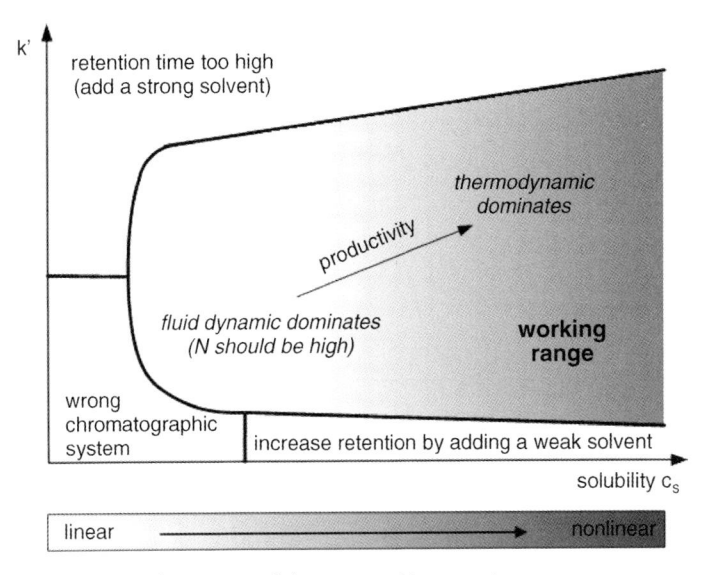

Figure 3.65 Working ranges of chromatographic separations.

separations with low selectivities at low loadings must the number of plates be high to reach a sufficient resolution (Figure 2.12, Section 2.4.3).

This behavior is demonstrated by the different working areas defined in Figure 3.65. Assuming that a certain selectivity can be reached, the influence of the retention factor and the solubility is shown. The working range of process chromatography should have medium retention factors and medium to high solubilities for the feed compounds. For low feed solubilities the influences of thermodynamics and fluid dynamics dominates. In this range, adsorbents with high numbers of plates per meter should be used. At high solubilities, high column loading is possible and thermodynamic effects come to the fore. In this range, less efficient adsorbents for high flow rates can be used.

In preparative chromatography the systems are mainly operated at high feed solubilities and thus thermodynamic effects dominate. If separations with low feed solubilities have to be operated, high-efficiency adsorbents should be used and the ACD injection technique could be applied (Section 3.2.3.2).

Thermodynamic effects are the main source of increasing productivity in preparative chromatography. The most important parameter is the selectivity, which describes the ratio of the initial slopes H of the adsorption isotherms for both components (Equation 2.46).

Especially for feed mixtures with different ratios of the single components, the elution order must be considered. The major component should elute as the second peak, because in this case the displacement effect can be used to ease the separation (Section 2.6.2). If the minor component elutes as the second peak the tag-along effect reduces the purity and loadability of the system.

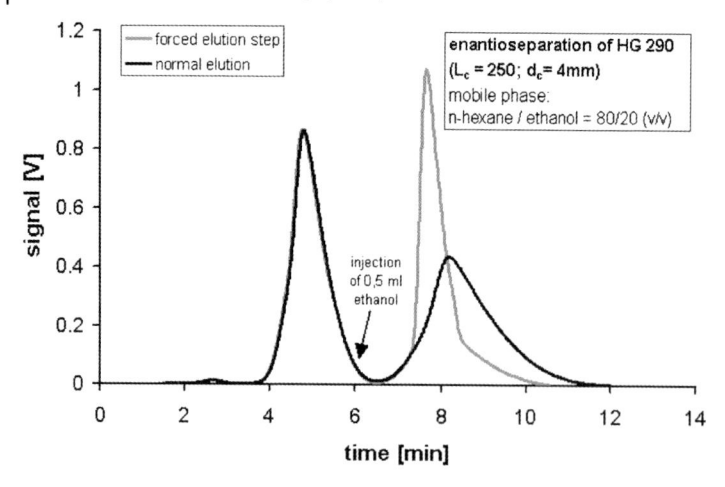

Figure 3.66 Increase in productivity due to forced elution step (injection of a strong eluent) during enantioseparation of HG 290 (reproduced with permission of H. Gillandt, SiChem GmbH Bremen).

If the components to be separated are present in a similar amount (e.g., as for racemic or diastereomeric mixtures) the target component should elute first, because the first peak is normally obtained at higher productivity and concentration.

Very often the cycle time of isocratic batch chromatography is high due to extensive peak tailing of the second peak, which can be explained by a Langmuir-type isotherm. The black line in Figure 3.66 shows a typical chromatogram of an enantioseparation with a polysaccharide adsorbent with n-hexane–ethanol as the mobile phase. Here ethanol is the solvent with higher elution strength. The second peak shows tailing due to the Langmuir shape of the isotherm and high concentrations. The cycle time (time between the starting point of the first peak and the ending point of the last peak) for the isocratic elution mode is 8.2 min.

The gray line shows the separation when a forced elution step is used. After the first peak has eluted from the column, a second injection is made with the strong solvent (in this case ethanol). If the volume of this second injection is large enough the strong solvent displaces the second peak. Owing to this displacement, the second peak elutes much faster and at higher concentrations. The cycle time is decreased from 8.2 to 6.7 min and, thus, productivity is increased by 22%. Eluent consumption is also decreased as well, but recycling of the solvent becomes slightly more problematic because the eluent composition is no longer constant over the whole cycle time. When using the forced elution step-method it should be ensured that the column is in equilibrium at the end of the cycle time. If only small amounts of eluent with high elution strength are used for this method, isocratic process concepts can still be applied, for example, touching–band elution.

3.2.9.2 Stationary Phase Gradients

From the aforementioned optimization strategies it can be seen that the solubility of the feed and the target compound in a suitable solvent is of utmost importance for productive preparative chromatographic separations. Therefore it is of consequence to choose the best solvent for the separation first and later try to optimize the separation by adjusting the selectivity of the stationary phases. This stationary phase selectivity optimization can be achieved by developing a new stationary phase with an optimized selector or simply by mixing or combining stationary phases with different selectors to stationary phase gradients. The concept has been exploited in analytical chromatography by Eppert and Heitmann (2003) as well as Nyiredy, Szucs, and Szepesy [2006, 2007] with the goal to minimize the analysis time and to maximize the peak capacity.

The concept of stationary-phase gradients is similar to 2D-chromatography, which is used in analytical chromatography to increase the peak capacity, but as always in preparative chromatography the concept of stationary-phase gradients tries to keep the system much more simple. In 2D-chromatography, the chromatogram in the first dimension (on the first stationary phase) is cut into fractions and submitted to a second type of stationary phase with an orthogonal selectivity. During the process of fractionation the fractions can be stored and solvents can be adjusted. Using stationary-phase gradients the goal is to end up with a system where simple isocratic chromatography can be applied and gives the best selectivity and capacity for a given separation task.

The optimization procedure follows quite a simple strategy:

1) Setting of boundary conditions, especially in the choice of appropriate solvents.
2) Single runs on the individual stationary phases.
3) Optimization of the combination of stationary phases using the PRISMA model.
4) Combination of the stationary phases and verification of the separation.

For the choice of the mobile phase as a starting point several important criteria should be fulfilled: First of all the solubility of the feed in the solvent should be high enough to ensure a good injection concentration. In addition, the boiling point of the solvent should be in a moderate temperature range (ideally between 60 and 80°C) and the solvent should show a low viscosity. Low UV-adsorption is not as important as in analytical chromatography because detection methods other than UV-adsorption can be used. For reasons of production costs and regulatory constraints the solvent should be available in good quality for a reasonable price and should be of low toxicity. It should definitely fall into the ICH-classes 2 or 3 of residual solvents (ICH, 2009).

By restricting the choice of mobile phases the regulatory impact of chromatographic processes might be significantly reduced. For cosmetic or food ingredients, the restriction to ethanol–water mixtures could be a viable option, as natural extracts from those solvents are generally regarded as safe. If in a second step, only chromatographic adsorption of impurities is applied the whole process is considered as purely "physical" (in contrast to "chemical") and the product is thus a "natural" one.

After the choice of the solvent has been optimized the screening of different stationary phases can start. The phases should be of similar particle size and show similar pore diffusivities to exhibit similar chromatographic properties. The surface modifications should cover different molecular-interaction principles but the

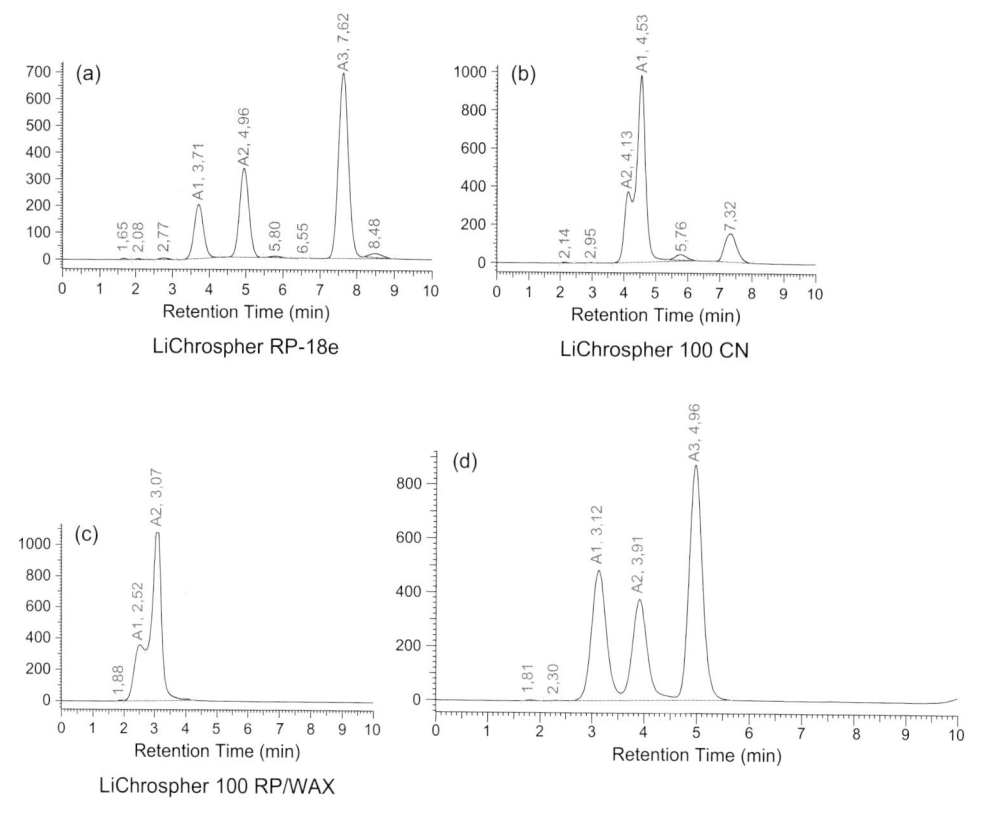

Figure 3.67 (a–c) Chromatograms on the single sorbents. (d) Chromatogram on a combined sorbent of LiChrospher RP18e/LiChrospher RP/WAX 2:3.

phases in general should be able to interact with the chosen mobile phase without problems. It has to be especially taken into account that problems might occur with hydrophobic RP-phases due to insufficient wetting of the stationary phase surface by the mobile phase or collapse of the surface modifications due to insufficient solubility in the mobile phase.

Figure 3.67a–c shows the single chromatograms of a three-component mixture on three different stationary phases: a highly unpolar RP-18e phase, a combined reversed phase/weak anion exchange phase (RP/WAX), and a cyano-modified phase. All three stationary phases have been prepared on the same base silica, a 12 μm, 100 Å LiChrospher®. While on the RP-18e-phase the mixture can be separated, the retention on the other two phases is shorter and insufficient for complete resolution. Through the combination of the different stationary phases the retention time of a single run with sufficient resolution can be reduced from 5 to 3 min which is equivalent to a capacity increase of the system of 40%. This can be achieved by a selection of RP-18e and RP/WAX-phases in a 2:3-combination (Figure 3.67d) (Horn, 2004).

The optimization of the stationary-phase combination can be approached by following the PRISMA-Model (Section 3.2.4.5). Instead of using different solvents, in this case the solvent is fixed and the different selectivities are obtained by using different surface modifications on the stationary phases.

For the combination of the stationary phases in a preparative system the stationary-phase particles can be either packed together in one column or used in segmented columns. In the first case, care has to be taken that no demixing of the particles occur and that the slurry solvent is equally suited for both types of packings. In the latter case, there should be a minimum dead volume between the two or more columns to avoid any backmixing. Sreedhar and Seidel-Morgenstern (2008) have shown that the serial coupling of single-column segments has benefits. In addition it has to be taken into account that the order of coupling is of importance for the separation if the whole system is operated in the nonlinear range of the adsorption isotherm. In that case the optimal length of the column segments is also different from the one used for a simple analytical separation.

The prejudices faced for preparative chromatography are often related to the mobile phase, for example, they are expensive, toxic, explosive, difficult to remove, yielding only low concentrations. By choosing an optimized mobile phase and later developing the appropriate stationary-phase gradient a lot of simplification in the phase system can be achieved. The concept of stationary phase gradients is thus a so far underutilized tool to improve preparative chromatography economics.

The following example illustrates the usefulness of the combination of stationary phases to solve a separation problem on production-scale level. An intermediate used in the last step of a drug synthesis contained small levels of three highly unwanted process critical impurities, resulting in side reaction products in the final product, which could not be removed using the generally applied purification methods. The relatively apolar product showed both on bare silica as well as on a polar modified silica a good retention using a hydrocarbon as the mobile phase. Therefore, preparative chromatographic experiments were performed on both types of stationary phases to study the behavior of the unwanted impurities. The experiments were immediately performed on a somewhat larger scale to provide the process research chemists with sufficient material to immediately evaluate the final reaction step using the purified material.

The results of these experiments are summarized in Table 3.35.

Chromatograms of a 50 g crude product injection on a column filled with 2000 g of cyano-modified silica and on a combination of two columns filled with 500 g of silica gel and 2000 g of cyano-modified silica gel respectively are depicted in Figures 3.68 and 3.69.

The experiments performed indicated that silica gel was highly effective in separating impurity number 1 from the main component, while the cyano-modified silica gel was more selective in the separation of impurities number 2 and 3 from the target product. After further optimization work with regard to the most efficient ratio of silica gel/cyano-modified silica gel an amount of 100 g of silica gel and 1250 g of cyano-modified silica gel was chosen to fill the 110 mm I.D. test columns.

Table 3.35 Summary of the purification results for an API and its impurities on different stationary phases.

		IMP-1	IMP-2	IMP-3
	Specification limit	0.35%	0.10%	0.10%
	Starting material	1.25%	0.17%	0.11%
	Injection amount (g)	% impurity in the purified product		
10 μm Si 60 Å	10	0.00	0.19	0.12
	30	0.00	0.17	0.11
	60	0.00	0.18	0.11
	100	0.50	0.16	0.10
60 Å – Cyano modified silica gel	20	0.86	0.00	0.00
	50	1.17	0.00	0.00

For the production-scale purification of about 600 kg of this intermediate, a 110 mm I.D DAC column filled with 350 g of 10 μm Kromasil 60 Å – silica gel was combined with a 200 mm I.D DAC column filled with 4150 g of 10 μm Kromasil 60 Å – cyano-modified silica. To speed-up the purification process, a 40 s rinsing plug of pure ethanol was introduced after detection of the impurities eluting in front of the main peak to fasten the elution of the impurities eluting behind the

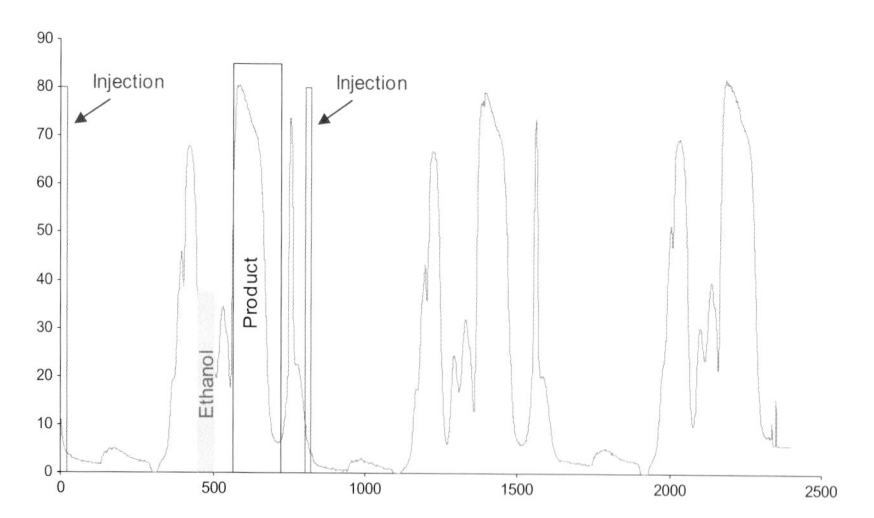

Figure 3.68 Injection of 50 g of product on a cyano-modified silica. Column: 110 mm I.D. DAC filled with 2000 g of 10 μm Kromasil 60 Å – cyano-modified silica, mobile phase: n-heptane, flow rate: 750 ml min^{-1}, temperature: eluent: 28 °C Column: 30 °C, sample amount: 50 g dissolved in 292 ml of n-heptane.

Figure 3.69 Injection of 50 g on a column filled with silica connected with a column filled with a cyano-modified silica. Pre-Column: 110 mm I.D. DAC filled with 500 g of 10 μm Kromasil 60 Å – Silica gel. Main Column: 110 mm I.D. DAC filled with 2000 g of 10 μm Kromasil 60 Å – cyano-modified silica, mobile phase: n-heptane, flow rate: 750 ml min^{-1}, temperature: eluent: 28 °C, column: 30 °C, sample amount: 50 g dissolved in 292 ml of n-heptane.

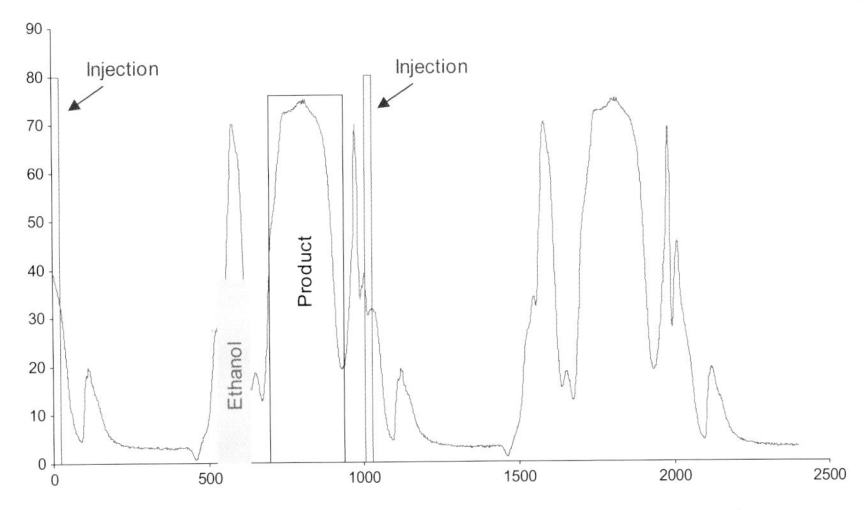

target component. Using this methodology, it was possible to very easily remove the unwanted impurities.

This example clearly illustrates that column coupling is not only a tool to be used in analytical chemistry but can certainly be very convenient and efficient to solve challenging separation problems.

References

Adams, B.A. and Holmers, E.L. (1935) Adsorbtive properties of synthetic resins. *J. Soc. Chem. Ind. (London)*, **54**, 1T.

Alexander, C., Andersson, H.S., Andersson, L.I., Ansell, R.J., Kirsch, N., Nicholls, I.A., O'Mahony, J., and Whitcombe, M.J. (2006) Molecular imprinting science and technology: a survey of the literature for the years up to and including 2003. *J. Mol. Recognit.*, **19**, 106–180.

Andersson, S., Allenmark, S., Moeller, P., Persson, B., and Sanchez, D. (1996) Chromatographic separation of enantiomers on N,N'-diallyl-L-tartardiamide-based network-polymeric chiral stationary phases. *J. Chromatogr. A*, **741**, 23–31.

Angerbauer, R., Grosser, R., Hinsken, W., and Rehse, J. Process for preparing sodium 3R,5S-(+)-erythro-(E)-7-7-[4-(4-fluorophenyl)-2,6-diisopropyl-5-methoxy-methyl-pyrid-3-yl]-3,5-dihydroxy-hept-6-enoate, DE 43 09 553,4, Mar 24, 1993.

Arangio, M. (1998) *Chromatographische Charakterisierung von Umkehrphasen*, Dissertation, Mathematisch-Naturwissenschaftliche Fakultät, Saarbrücken.

Armstrong, D.W., Tang, Y., Chen, S., Zhou, Y. Bagwill, C., and Chen, J.R.

(1994) Macrocylcic antibiotics as a new class of chiral selectors for liquid chromatography. *Anal. Chem.*, **66**, 1473–1484.

Arnell, R., Forssen, P., Fornstedt, T., Sardella, R., Lämmerhofer, M., Lindner, W. (2009) Adsorption behaviour of a quinidine carbamate-based chiral stationary phase: Role of the additive. *J. Chromatogr. A*, **1216**, 3480–3487.

Arnott, S.A., Fulmer, A., Scott, W.E., Dea, I.C.M., Moorehouse, R., and Rees, D.A. (1974) Agarose double helix and its function in agarose gel structure. *J. Mol. Biol.*, **90**, 269–284.

Ates, H., Mangelings, D., Vander-Heyden, Y. (2008) Chiral separations in polar organic solvent chromatography: Updating a screening strategy with new chlorine-containing polysaccharide selectors. *J. Chromatogr. B*, **875**, 57–64.

Bangs, L.B. (1987) Uniform latex particles. *Am. Biotechnol. Lab.*, **5** (3), **10**, 12–16.

Barnhart, W., Gahm, K., Hua, Z., Goetzinger, W. (2008) Supercritical fluid chromatography comparison of the poly(trans-1,2-cyclohexanediyl-bis-acrylamide) (P-CAP) column with several derivatized polysaccharide-based stationary phases. *J. Chromatogr. B*, **875**, 217–229.

Berthod, A., Liu, Y., Bagwill, C., and Armstrong, D.W. (1996) Facile liquid chromatographic enantioresolution of native amino acids and peptides using a teicoplanin chiral stationary phase. *J. Chromatogr. A*, **731**, 123–137.

Bluhm, L.H., Wang, Y., and Li, T. (2000) An alternative procedure to screen mixture combinatorial libraries for selectors for chiral chromatography. *Anal. Chem.*, **72**, 5201–5205.

Borman, P., Boughtflower, B., Cattanach, K., Crane, K., Freebairn, K., Jonas, G., Mutton, I., Patel, A., Sanders, M., and Thompson, D. (2003) Comparative performances of selected chiral HPLC, SFC, and CE systems with a chemically diverse sample set. *Chirality*, **15**, 1–12.

Brandt, A. and Kueppers, S. (2002) Practical aspects of preparative HPLC in pharmaceutical and development production. *LC GC Eur.*, **15** (3), 147–151.

Busini, V., Molani, D., Moscatelli, D., Zamolo, L., and Cavallotti, C. (2006) Investigation of the influence of spacer arm on the structural evolution of affinity ligands supported on agarose. *J. Phys. Chem. B*, **110**, 23564–23577.

Casey, J.L., Keep, P.A., Chester, K.A., Robson, L., Hawkins, R.E., and Begent, R.H. (1995) Purification of bacterially expressed single chain Fv antibodies for clinical applications using metal chelate chromatography. *J. Immunol Methods*, **179**, 105–116.

Cavalloti, C. (2008) *Univ Milano*, personal communication in the frame of AIMs.

Chang, C. and Lenhoff, A.M. (1998) Comparison of protein adsorption isotherms and uptake rates in preparative cation-exchange materials. *J. Chromatogr. A*, **827**, 281–293.

Cox, G. B. and Amoss, C. W. (2004) Extending the range of solvents for chiral analysis using a new immobilized polysaccharide chiral stationary phase CHIRALPAK (R) IA. *LC GC North America*.

Curling, J. and Gottschalk, U. (2007) Process chromatography: five decades of innovation. *BioPharm Int.*, **20**, 10–19.

Cutler, P. (2004) Size exclusion chromatography in protein purification protocols. *Methods Mol. Biol.*, Humana Press, **244**, 239–252.

Dallenbach-Tölke, K., Nyiredy, S., Meier, B., Sticher, O. (1986) Optimization of overpressured layer chromatography of polar, naturally occuring compounds by the PRISMA model, *J. Chromatogr. A.*, **365**, 63–72.

de Neuville, B., Thomas, H., Tarafder, A., Morbidelli, M. (2012) Simulation of porosity decrease due to protein adsorption using the distributed pore model. Submitted to publication

DePhillips, P. and Lenhoff, A.M. (2000) Pore size distributions of cation-exchange adsorbents determined by inverse size-exclusion chromatography. *J. Chromatogr. A*, **883**, 39–54.

Dingenen, J. (1994) Polysaccharide phases in enantioseparations, in *A Practical*

Approach to Chiral Separations by Liquid Chromatography (ed. G. Subramanian), Wiley-VCH, Weinheim.

Dungelova, J., Lehotay, J., Krupcik, J., Cizmarik, J., Welsch, T., and Armstrong, D.W. (2004) Selectivity tuning of serially coupled (*S,S*) Whelk-O 1 and (*R,R*) Whelk-O 1 columns in HPLC. *J. Chromatogr. Sci.*, **42**, 135–139.

Eckermann, C., Ebert, S., Rubenwolf, S., and Ambrosius, D. (2007) Patent WO2007144353.

Engelhardt, H., Grüner, R., and Scherer, M. (2001) The polar selectivities of non-polar reversed phases. *Chromatographia*, **53**, 154–161.

Eppert, G.J. and Heitmann, P. (2003) Selectivity optimization of stationary phases. *LCGC Europe*, **16** (10), 698–705.

Epton, R. (1978) Hydrophobic, ion exchange and affinity methods, in *Chromatography of Synthetic and Biological Polymers*, vol. 2, E. Horwood, Chichester, pp. 1–9.

Fassina, G., Ruvo, M., Palombo, G., Verdoliva, A., and Marino, M. (2001) Novel ligands for the affinity-chromatographic purification of antibodies. *J. Biochem. Biophys. Methods*, **49**, 481–490.

Franco, P., Senso, A., Oliveros, L., and Minguillón, C. (2001) Covalently bonded polysaccharide derivatives as chiral stationary phases in high-performance liquid chromatography. *J. Chromatogr. A*, **906**, 155–170.

Francotte, E. (1998) Achiral derivatization as a means of improving the chromatographic resolution of racemic alcohols on benzoylcellulose CSPs. *Chirality*, **10**, 492–498.

Francotte, E. and Wolf, R.M. (1991) Benzoyl cellulose beads in the pure polymeric form as a new powerfull sorbent for the chromatographic resolution of racemates. *Chirality*, **3**, 43–55.

Freyre, F.M., Vazquez, J.E., Ayala, M., Canaan-Haden, L., Bell, H., Rodriguez, I., Gonzalez, A., Cintando, A., and Gavilondo, J.V. (2000) Very high expression of an anti-carcinoembryonic antigen single chain Fv antibody fragment in the yeast *Pichia pastoris*. *J. Biotechnol.*, **76**, 157–163.

Füglistaller, P. (1989) Comparison of immunoglobulin binding capacities and ligand leakage using eight different Protein A affinity chromatography matrices. *J. Immunol. Meth.*, **124**, 171–177.

Gaberc-Porekar, V. and Menart, V. (2001) Review – perspectives of immobilized-metal affinity chromatography. *J. Biochem. Biophys. Methods*, **49**, 335–360.

Galushko, S.V. (1991) Calculation of retention and selectivity in reversed-phase liquid chromatography. *J. Chromatogr.*, **552**, 91–102.

Gant, J.R., Dolan, J.W., and Snyder, L.R. (1979) Systematic approach to optimizing resolution in reversed phase liquid chromatography, with emphasis on the role of temperature. *J. Chromatogr.*, **185**, 153–177.

Gedicke, K., Kaspereit, M., Beckmann, W., Budde, U., Lorenz, H., and Seidel-Morgenstern, A. (2007) Conceptual design and feasibility study of combining continuous chromatography and crystallisation for stereoisomer separations. *Chem. Eng. Res. Des.*, **85**, 928–936.

Geiss, F., Schlitt, H., and Klose, A. (1965) Reproducibility in thin-layer chromatography: Influence of humidity, chamber form, and chamber atmosphere. *Zeitschrift für Anal. Chem.*, **213**, 5.

Ghethie, V. and Schell, H.D. (1967) Electrophoresis and immunoelectrophoresis of proteins on DEAE-agarose gels. *Rev. Roum. Biochim.*, **4**, 179–184.

Hahn, R., Schlegel, R., and Jungbauer, A. (2003) Comparison of Protein A affinity sorbents. *J. Chromatogr. B*, **790**, 35–51.

Hansen, S.H., Helböe, P., and Thomasen, M. (1986) Agar derivates for chromatography, electrophoresis and gel-bound enzymes – VII. Influence of apparent surface pH of silica compared with the effects in straight-phase chromatography. *J. Chromatogr.*, **368**, 39–47.

Heitz, W. (ed.) (1970) Gel chromatography. *Angew. Chem. Int. Ed.*, **9**, 689–702.

Helmreich, M., Niesert, C.P., Schulte, M., Lindner, W., Lämmerhofer, M., and Hoffmann, C.WO WO/2010/108583 (30.9.2010) Process for separating enantiomers of 3,6-dihydro-1,3,5-triazine derivatives.

Hjelm, H., Hjelm, K., and Sjoquist, J. (1972) Protein A from *Staphylococcus aureus* its isolation by affinity chromatography and its use as an immunosorbent for isolation of immunoglobulins. *FEBS Lett.*, **28**, 73–76.

Hjerten, S. (1964) Preparation of agarose spheres for chromatography of molecules and particles. *Biochim. Biophys. Acta*, **79**, 393–398.

Hjerten, S. (1983) *Protides of Biological Fluids*, vol. **30** (ed. H. Peeters), Pergamon Press, Oxford, pp. 9–17.

Hochuli, E., Doebeli, H., and Schacher, A. (1987) New metal chelate adsorbent selective for proteins and peptides containing neighbouring histidine residues. *J. Chromatogr.*, **411**, 177–184.

Hochuli, E., Bannwarth, W., Doebeli, H., Gentz, R., and Stueber, D. (1988) Genetic approach to facilitate purification of recombinant proteins with a metal novel chelate adsorbent. *Bio-Technology*, **6**, 1321–1325.

Horn, F. (2004) Diploma thesis, Selektivitätsoptimierung von Festphasen in der HPLC; Fachhoch-schule Lippe und Höxter, Fachbereich Lebensmitteltechnologie.

ICH-Topic (Feb 2009) Q3C Impurities: Guidelines for residual solvents http://www.ema.europa.eu/docs/en_GB/document_library/Scientific_guideline/2009/09/WC500002674.pdf.

Ikai, T., Yamamoto, C., Kamigaito, M., Okamoto, Y. (2007) Preparation and chiral recognition ability of crosslinked beads of polysaccharide derivatives. *J. Sep. Sci.*, **30**, 971–978

Ikai, T., Yamamoto, C., Kamigaito, M., and Okamoto, Y. (2008) Immobilized-type chiral packing materials for HPLC based on polysaccharide derivatives. *J. Chromatogr. B*, **875**, 2–11.

Ilisz, I., Berkecz, R., and Peter, A. (2006) HPLC separation of amino acid enantiomers and small peptides on macrocyclic antibiotic-based chiral stationary phases: a review. *J. Sep. Sci.*, **29**, 1305–1321.

Ishikawa, A. and Shibata, T. (1993) Cellulosic chiral stationary phase under reversed-phase condition. *J. Liq. Chromatogr.*, **16**, 4, 61–68.

Jacob, L. and Frech, Ch. (2007) Ion-exchange chromatography in biopharmaceutical manufacturing, in *Bioseparation and Bioprocessing* (ed. G. Subramanian), Wiley-VCH, Weinheim, pp. 127–149.

Jandera, P. and Guiochon, G. (1991) Effect of the sample solvent on band profiles in preparative liquid chromatography using non-aqueous reversed-phase high-performance liquid chromatography. *J. Chromatogr.*, **588**, 1–14.

Janowski, F. and Heyer, W. (1982) *Poröse Gläser*, VEB Deutscher Verlag für Grundstoffindustrie, Leipzig.

Janson, J.-C. (1987) On the history of the development of Sephadex. *Chromatographia*, **23**, 361.

Jiang, C., Liu, J., Rubacha, M., Shukla, A. (2009) A mechanistic study of Protein A chromatography resin lifetime. *J. Chromatogr. A*, **1216**, 5849–5855

Johansson, B.I., Belew, M., Eriksson, S., Glad, G., Lind, O., Maloisel, J.L., and Norrmann, N. (2003a) Preparation and characterisation of prototypes for multi-modal separation aimed for capture of negatively charged biomolecules at high salt concentrations. *J. Chromatogr. A*, **1016**, 21–33.

Johansson, B.I., Belew, M., Eriksson, S., Glad, G., Lind, O., Maloisel, J.L., and Norrmann, N. (2003b) Preparation and characterisation of prototypes for multi-modal separation aimed for capture of positively charged biomolecules at high salt concentrations. *J. Chromatogr. A*, **1016**, 35–49.

Jungbauer, A. (2005) Chromatographic media for bioseparation. *J. Chromatogr. A*, **1065**, 3–12.

Kaiser, R.E. and Oelrich, E. (1981) *Optimisation in HPLC*, Dr. Alfred Hüthig Verlag, Heidelberg.

Kaslow, D., Shiloach, J. (1994) Production, Purification and Immunogenicity of a Malaria Transmission-Blocking Vaccine Candidate: TBV25H Expressed in Yeast and Purified using Nickel-NTA Agarose. *Biotechnology*, **12**, 494–499

Kato, Y., Nakamura, K., and Hashimoto, T. (1987) High-performance hydroxyapatite chromatography of proteins. *J. Chromatogr. A*, **398**, 340–346.

Kinkel, J.N. (1994) Optivally active polyacrylamide/silica composites and related packings and their applications in A practical approach to chiral separations by liquid chromatography (ed G. Subramanian), Wiley-VCH, Weinheim, pp. 217–.

Kopaciewicz, W., Fulton, S., and Lee, S.Y. (1987) Influence of pore and particle size on the frontal uptake of proteins: implications for preparative anion-exchange chromatography. *J. Chromatogr.*, **409**, 111–124.

Kraml, Ch., Zhou, D., Byrne, N., McConnell, O. (2005) Enhanced chromatographic resolution of amine enantiomers as carbobenzyloxy derivatives in high-performance liquid chromatography and supercritical fluid chromatography. *J. Chromatogr. A*, **1100**, 108–115.

Kronvall, G. (1973) Purification of staphyllococcus protein A using immunosorbents. *Scand. J. Immunol.*, **2**, 31–36.

Kummer, M. and Palme, H.J. (1996) Resolution of enantiomeric steroids by high-performance liquid chromatography on chiral stationary phases. *J. Chromatogr. A*, **749** (1–2), 61–68.

Kurganov, A., Trüdinger, U., Isaeva, T., and Unger, K.K. (1996) Native and modified alumina, titania and zirkonia in normal and reversed-phase high-performance liquid chromatography. *Chromatographia*, **42**, 217–222.

Laas, T. (1975) Agar derivates for chromatography, electrophoresis and gel-bound enzymes – IV. Benzylated dibromopropanol cross-linked sepharose as an amphophilic gel for hydrophobic salting-out chromatography of enzymes with special emphasis on denaturing risks. *J. Chromatogr.*, **111**, 373–387.

Lämmerhofer, M. (2010) Chiral recognition by enantioselective liquid chromatography: mechanisms and modern chiral stationary phases. Review article. *J. Chromatogr. A*, **1217**, 814–856.

Lämmerhofer, M. and Lindner, W. (1996) Quinine and quinidine derivatives as chiral selectors I. Brush type chiral stationary phases for high-performance liquid chromatography based on cinchonan carbamates and their application as chiral anion exchangers. *J. Chromatogr. A*, **741**, 33–48.

Lea, D.J. and Sehon, A.H. (1962) Preparation of synthetic gels for chromatography of macromolecules. *Can. J. Chem.*, **40**, 159–160.

Lowe, C., Lowe, A., Gupta, G. (2001) New developments in affinity chromatography with portential application in the production of biopharmaceuticals. *J. Biochem. Biophys. Methods*, **49**, 561–574.

Maa, Y.-F. and Horvath, C. (1988) Rapid analysis of proteins and peptides by reversed-phase chromatography with polymeric microcellular sorbent. *J. Chromatogr.*, **445**, 71–86.

Macaudiere, P. Tambute, A., Caude, M., Rosset, R. (1986) Resolution of enantiomeric amides on a Pirkle-type chiral stationary phase. A comparison of subcritical fluid and liquid chromatographic approaches. *J. Chromatogr.*, **371**, 177–193.

Macquarrie, D.J., Tavener, S.J., Gray, G.W., Heath, P.A., Rafelt, J.S., Saulzet, S.I., Hardy, J.J.E., Clark, J.H., Sutra, P., Brunel, D., di Renzo, F., and Fajula, F. (1999) The use of Reichardt's dye as an indicator of surface polarity. *New J. Chem.* **23**, 725–731.

Majors, R.E. (2003a) The cleaning and regeneration of reversed-phase HPLC columns. *LC-GC Eur.*, **16** (7), 404–409.

Matthijs, N., Maftouh, M., Vander-Heyden, Y. (2006) Screening approach for chiral separation of pharmaceuticals. Part IV:

Polar organic solvent chromatography. *J. Chromatogr. A*, **1111**, 48–61.

McCarthy, J.P. (1994) Direct enantiomeric separation of the 4 stereoisomers of nadolol using normal-phase and reversed-phase high performance liquid chromatography with Chiralpak AD. *J. Chromatogr. A*, **685** (2), 349–355.

Meyers, J.J. and Liapis, A.I. (1998) Network modeling of the intraparticle convection and diffusion of molecules in porous particles packed in a chromatographic column. *J. Chromatogr. A*, **827**, 197–213.

Meyers, J.J. and Liapis, A.I. (1999) Network modeling of the convective flow and diffusion of molecules adsorbing in monoliths and in porous particles packed in a chromatographic column. *J. Chromatogr. A*, **852**, 3–23.

Mikeš, O., štrop, P., Zbrožek, J., and Čoupek, J. (1976) Chromatography of biopolymers and their fragments on ion-exchange derivates of the hydrophilic macroporous synthetic gel spheron. *J. Chromatogr.*, **119**, 339–354.

Mikes, O. (1988) High-Performance Liquid Chromatography of Biopolymers and Biooligomers Parts A and B, J. Chromat Libr Vol 41A and B, Elsevier, Amsterdam.

Moiani, D., Salvalaglio, M., Cavallotti, C., Bujacz, A., Redzynia, I., Bujacz, G., Dinon, F., Pengo, P., and Fassina, G. (2009) Structural characterization of a Protein A mimetic peptide denrimer bound to human IgG. *J. Phys Chem B*, **113**, 16268–16275.

Moore, J.C. (1964) Gel permeation chromatography. I. A new method for molecular weight distribution of high polymers. *J. Polymer. Sci., Part A*, **2**, 835.

Murer, P., Lewandowski, K., Svec, F., Frechet, J.M. (1999) On-bead combinatorial approach to the design of chiral stationary phases. *Anal. Chem.*, **71**, 1278–1284.

Nawrocki, J., Rigney, M., Mc Cormick, P.A., and Carr, P.W. (1993) Chemistry of zirconia and its use in chromatography. *J. Chromatogr. A*, **657**, 229–282.

Neue, U.D., Mazza, C.B., Cavanaugh, J.Y., Lu, Z., and Wheat, T.E. (2003) At-column dilution for improved loading in preparative chromatography. *Chromatographia*, **57**, 121–127.

Ning, J.G. (1998) Direct chiral separation with Chiralpak AD converted into the reversed-phase mode. *J. Chromatogr. A*, **805** (1–2), 309–314.

Nyiredy, S. (2002) Planar chromatographic method development using the PRISMA optimization system and flow charts. *J. Chromatogr. Sci.*, **40** (10), 553–563.

Nyiredy, S. and Fater, Z. (1995) Automatic mobile phase optimization, using the "PRISMA" model for the TLC separation of apolar compounds. *Jpc-J. Planar Chromatography-Mod. TLC*, **8** (5), 341–345.

Nyiredy, S., Meier, B., Erdelmeier, C.A.J., and Sticher, O. (1985) PRISMA: a geometrical design for solvent optimization in HPLC. *J. High Res. Chrom. Chrom. Comm.*, **8**, 186–188.

Nyiredy, S., Szucs, Z., and Szepesy, L. (2006) Stationary phase optimized selectivity LC (SOS-LC) separation examples and practical aspects. *Chromatographia*, **63**, S3–S9.

Nyiredy, S., Szucs, Z., and Szepesy, L. (2007) Stationary phase optimized selectivity liquid chromatography: basic possibilities of serially connected columns using the "PRISMA" principle. *J. Chromatogr. A*, **1157** (1–2), 122–130.

Okamoto, Y., Kawashima, M., Hatada, K. (1984) Usefull Chiral Packing Materials for High-Performance Liquid Chromatographic Resolution of Enantiomers: Phenylcarbamates of Polysaccharides Coated on Silica Gel. *J. Am. Chem. Soc.*, **106**, 5357–5359.

Okamoto, M. and Nakazawa, H. (1991) Reversal of elution order during direct enantiomeric separation of pyriproxyfen on a cellulose-based chiral stationary phase. *J. Chromatogr.*, **588**, 177–180.

O'Leary, R.M., Feuerhelm, D., Peers, D., Xu, Y., Blank, G.S. (2001) Determining the useful lifetime of chromatography resins. *BioPharm.*, **14** (9), 10–17 O'Leary 2003.

Pandey, R.C. (1997) Isolation and purification of paclitaxel from organic matter containing paclitaxel, cephalomannine and other related taxanes, US patent.

Pandey, R.C., Yankov, L.K., Poulev, A., Nair, R., and Caccamese, S. (1998) Synthesis and separation of potential anticancer active dihalocephalomannine diastereomers from extracts of *Taxus yunnanensis*. *J. Nat. Prod.*, **61**, 1, 57–63.

Perrin, C., Vu, V.A., Matthijs, N., Maftouh, N., Massart, D.L., Vander Heyden, Y. (2002a) Screening approach for chiral separation of pharmaceuticals. Part I. Normal-phase liquid chromatography. *J. Chromatogr. A*, **947**, 69–83.

Perrin, C., Matthijs, N., Mangelings, D., Granier-Loyaux, C., Maftouh, N., Massart, D.L., Vander Heyden, Y. (2002b) Screening approach for chiral separation of pharmaceuticals. Part II. Reversed-phase liquid chromatography. *J. Chromatogr. A*, **966**, 119–134.

Pirkle, W.H. and Däppen, R. (1987) Reciprocity in chiral recognition – comparison of several chiral stationary phases. *J. Chromatogr.*, **404**, 107–115.

Pirkle, W.H., Welch, C.J., and Lamm, B. (1992) Design, synthesis and evaluation of an improved enantioselective naproxen selector. *J. Org. Chem*, **57**, 3854–3860.

Poole, C.F. (2003) *The Essence of Chromatography*, Elsevier, Amsterdam.

Porath, J. and Flodin, P. (1959) Gel filtration: a method for desalting and group separation. *Nature*, **183**, 1657.

Porath, J., Janson, J.C., and Laas, T. (1971) Agar derivates for chromatography, electrophoresis and gel-bound enzymes – I. Desulphatet and reduced cross-linked agar and agarose in spherical bead form. *J. Chromatogr.*, **60**, 167–177.

Porath, J., Laas, T., and Janson, J.-Ch. (1975) Agar derivates for chromatography, electrophoresis and gel-bound enzymes – III. Rigid agarose gels cross-linked with divinyl sulphone (DVS). *J. Chromatogr.*, **103**, 49–62.

Rabel, F. (1980) Use and maintenance of microparticle high performance liquid chromatography columns. *J. Chromatogr. Sci.*, **18**, 394.

Rathjen, T., Wenzel, D., Studts, J.M., and Ambrosius, D. (2009) High-Troughput Downstream Development, *Innov. Pharmaceut. Technol.*, 56–60, http://www.iptonline.com/articles/public/IPT_29_p56non%20print.pdf.

Regnier, F.E. (1987) HPLC of biological macromolecules: the first decade. *Chromatographia*, **24**, 241–251.

Reichardt, C. (2003) *Solvents and Solvent Effects in Organic Chemistry*, 3rd edn, VCH, Weinheim.

Rogl, H., Kosemund, K., Kühlbrandt, W., Collinson, I. (1998) Refolding of Escherichia coli produced membrane protein inclusion bodies immobilised by nickel chelating chromatography. *FEBS Letters*, **432**, 21–26.

Roussel, C., Piras, P., and Heitmann, I. (1997) An approach to discriminating 25 commercial chiral stationary phases from structural data sets extracted from a molecular database. *Biomed. Chromatogr.*, **11**, 311–316.

Ruthven, D.M. (ed.) (1997) *Encyclopedia of Separation Technology*, vol. **1**, Wiley, New York.

Schnabel, R. and Langer, P. (1991) Controlled-pore glass as a stationary phase in chromatography. *J. Chromatogr.*, **544**, 137–146.

Schulte, M., Ditz, R., Devant, R., Kinkel, J.N., Charton, F. (1997) Comparison of the specific productivity of different chiral stationary phases used for Simulated Moving Bed – Chromatography. *J. Chromatogr.*, **769**, 93–100.

Schulte, M. (2001) Chiral derivatization chromatography, in *Chiral Separation Techniques* (ed. G. Subramanian), Wiley-VCH, Weinheim.

Schulte, M., Devant, R., Grosser, R. (2002) Enantioseparation of Gantofiban precursors on chiral stationary phases of the poly-(N-acryloyl amino acid derivative)-type. *J. Pharm. Biomed. Anal.*, **27**, 627–637.

Schulze-Wierling, P., Bougmil, P., Knieps-Grünhagen, E., and Hubbuch, J. (2007) High-throughput screening of packed-bed chromatography coupled with SELDI-TOF MS analysis: monoclonal antibodies versus host cell protein. *Biotechnol. Bioeng.*, **98** (2), 440–450.

Scott, R., Kucera, P. (1975) Solute interactions with the mobile and stationary phases in liquid-solid chromatography. *J. Chromatogr.*, **112**, 425–442.

Seidel-Morgenstern, A. (1995) *Mathematische Modellierung der präparativen Flüssigchromatographie*, Deutscher Universitätsverlag, Wiesbaden.

Seidl, J., Malinsky, J., Dušek, D., and Heitz, W. (1967) Macroporöse styrol-divinylbenzol-copolymere und ihre Verwendung in der chromatographie und zur darstellung in ionenaustauschern. *Adv. Polymer Sci.*, **5**, 113–213.

Smith, G.P. (1985) Filamentous fusion phage: novel expression vectors that express cloned antigens on the virion surface. *Science*, **228**, 1315–1317.

Snyder, L.R. (1968) *Principles of Adsorption Chromatography*, Marcel Dekker, New York.

Snyder, L.R. and Dolan, J.W. (1998) The linear-solvent-strength model of gradient elution. *Adv. Chromatogr.*, **38**, 115–187.

Snyder, L.R. and Kirkland, J.J. (1979) *Introduction to Modern Liquid Chromatography*, Wiley, New York.

Snyder, L.R., Kirkland, J.J., and Glajch, J.L. (1997) *Practical HPLC Method Development*, Wiley, New York.

Soczewinski, E. (1969) Solvent composition effects in thin-layer chromatography systems of the type silica gel-electron donor solvent. *Anal. Chem.*, **41** (1), 179–182.

Sreedhar, B. and Seidel-Morgenstern, A. (2008) Preparative separation of multi-component mixtures using stationary phase gradients. *J. Chromatogr. A*, **1215**, 133–144.

Staby, A., Sand, M.-B., Hansen, R.G., Jacobsen, J.H., Andersen, L.A., Gerstenberg, M., Bruus, U.K., and Jensen, I.H. (2004) Comparison of chromatographic ion-exchange resins: III. Strong cation-exchange resins. *J. Chromatogr. A*, **1034** (1–2), 85–97.

Stanislawski, B., Schmit, E., and Ohser, J. (2010) Imaging of fluorophores in chromatographic beads, reconstruction of radial density distributions and characterisation of protein uptaking processes. *Image Anal. Stereol.*, **29**, 181–189.

Strube, J., Arlt, W., Schulte, M. (2006) Technische Chromatographie in *Fluidverfahrenstechnik: Grundlagen, Methodik, Technik, Praxis* (ed. R. Gödecke), Wiley-VCH, Weinheim, pp. 381–495.

Svec, F., Wulff, D., and Fréchet, J.M. (2001) Combinatorial approaches to recognition of chirality: preparation and use of materials for the separation of enantiomers, in *Chiral Separation Techniques* (ed. G. Subramanian), Wiley-VCH, Weinheim.

Szanya, T., Argyelan, J., Kovats, S., and Hanak, L. (2001) Separation of steroid compounds by overloaded preparative chromatography with precipitation in the fluid phase. *J. Chromatogr. A*, **908** (1–2), 265–272.

Tanaka, N., Hashizume, K., and Araki, M. (1987) Comparison of polymer-based stationary phases with silica-based stationary phases in reversed-phase liquid chromatography. *J. Chromatogr.*, **400**, 33–45.

Thomas, H., Storti, G., Joehnck, M., Schulte, M., and Morbidelli, M. (2012) *Role of tentacles and protein loading on pore accessibility and mass transfer of proteins in cation exchange materials*, submitted to publication.

Trapp, O. and Schurig, V. (2002) Novel direct access to enantiomerization barriers from peak profiles in enantioselective dynamic chromatography: enantiomerization of dialkyl-1,3-allene- dicarboxylates. *Chirality*, **14**, 465–470.

Tweeten, K.A. and Tweeten, T.N. (1986) Reversed-phase chromatography of proteins on resin-based wide-pore packings. *J. Chromatogr.*, **359**, 111.

Ugelstad, J. (July 10 (1984)) Monodisperse polymer particles and dispersions thereof, US 4,459,378.

Ugelstad, J., Mørk, P.C., Kaggernd, K.H., Ellingsen, T., and Berge, A. (1980) Swelling of oligomer–polymer particles. New methods of preparation of emulsions and polymer dispersions. *Adv. Colloid Interface Sci.*, **13**, 101–140.

Unger, K.K. (1990) *Packings and Stationary Phases in Chromatographic Techniques*, Marcel Dekker, New York.

Unger, K.K. and Weber, E. (1999) *A Guide to Practical HPLC*, GIT Verlag.

Unger, K.K., Janzen, R., and Jilge, G. (1987) Packings and stationary phases for biopolymer separations by HPLC. *Chromatographia*, **24**, 144–154.

Unger, K.K., Bidlingmaier, B., du Fresne von Hohenesche, C., and Lubda, D. (2002) Evaluation and comparison of the pore structure and related properties of particulate and monolithic silicas for liquid phase separation processes, COPS VI Proceedings. *Stud. Surf. Sci. Catal.*, **144**, 115–122.

Urmann, M., Graalfs, H., Joehnck, M., Jacob, L.R., and Frech F Ch. (2010) Cation-exchange chromatography of monoclonal antibodies – characterization of a novel stationary phase designed for production-scale purification. *mAbs*, **2**, 395–404.

Walters, R.R. (1982) High-performance affinity chromatography. Pore size effects. *J. Chromatogr.*, **249**, 19–28.

Welch, C., Protopopova, M.N., Bhat, G. (1998) Microscale synthesis and screening of chiral stationary phases. *Enantiomer* **3**, 471–476.

Welsch, T., Dornberger, U., and Lerche, D. (1993) Selectivity tuning of serially coupled columns in high-performance liquid-chromatography. *HRC-J. High Res. Chrom.*, **16** (1), 18–26.

Wiendahl, M., Schulze-Wierling, P., Nielsen, J., Fomsgaard Christensen, D., Karup, J., Staby, A., and Hubbuch, J. (2008) High throughput screening for the design and optimization of chromatographic processes – miniaturization, automation and parallelization of breakthrough and elution studies. *Chem. Eng. Technol.*, **31**, 893–903.

Wu, Y., Wang, Y., Yang, A., and Li, T. (1999) Screening of mixture combinatorial libraries for chiral selectors: a reciprocal chromatographic approach using enantiomeric libraries. *Anal. Chem.*, **71**, 1688–1691.

Yamamoto, S. and Kita, A. (2005) Theoretical background of short chromatographic layers. Optimization of gradient elution in short columns. *J. Chromatogr. A*, **1065**, 45–50.

Yamamoto, S., Nakanishi, K., and Matsuno, R. (1988) *Ion-Exchange Chromatography of Proteins*, Marcel Dekker, New York.

Yang, R.T (2003) *Adsorbents: Fundamentals and Applications*, John Wiley & Sons, New York.

Yau, W.W., Kirkland, J.J., and Bly, D.D. (1979) *Modern Size-Exclusion Chromatography*, Wiley, New York.

Yau, W.W., Kirkland, J.J., and Bly, D.D. (1979) *Modern Size-Exclusion Chromatography*, Wiley, New York.

Zamolo, L., Busini, V., Moiani, D., Moscatelli, D., and Cavallotti, C. (2008) Molecular dynamic investigations of the interaction of supported affinity ligands with monoclonal antibodies. *Biotechnol. Progr.*, **24**, 527–539.

Zhang, T., Schaeffer, M., and Franco, P. (2005a) Optimization of the chiral separation of a Ca-sensitizing drug on an immobilized polysaccharide-based chiral stationary phase. Case study with a preparative perspective. *J. Chromatogr. A*, **1083**, 96–101.

Zhang, T., Kientzy, Ch., Franco, P., Ohnishi, A., Kagamihara, Y., Kurosawa, H. (2005b) Solvent versatility of immobilized 3,5-dimethylphenyl-carbamate of amylose in enantiomeric separations by HPLC. *J. Chromatogr.*, **1075**, 65–75

Zhang, T., Nguyen, D., Franco, P., Isobe, Y., Michishita, T., Murakami, T. (2008) Cellulose tris-(3,5-dichlorphenyl-carbamate) immobilised on silica: A novel chiral stationary phase for resolution of enantiomers. *J. Pharm. Biomed. Anal.*, **46**, 882–891.

Zhang, T., Nguyen, D., Franco, P., Murakami, T., Ohnishi, A., Kurosawa,

H. (2006) Cellulose 3,5-dimethylphenyl-carbamate immobilised on silica. A new chiral stationary phase for the analysis of enantiomers. *Anal. Chim. Acta*, **557**, 221–228.

Zhang, T., Nguyen, D., Franco, P., Isobe, Y., Michishita, T., Murakami, T. (2008) Cellulose tris(3,5-dichlorophenyl-carbamate) immobilised on silica: A novel chiral stationary phase for the resolution of enantiomers. *J. Pharm. Biomed. Analysis*, **46**, 882–891.

Zhong, Q., Han, X., He, L., Beesley, Th., Trahanovsky, W., Armstrong, D.W, (2005) Chromatographic evaluation of poly (trans 1,2-cyclohexanediyl-bis-acrylamide) as a chiral stationary phase for HPLC. *J. Chromatogr. A*, **1066**, 55–70.

Zhou, L., Welsh, C., Lee, C., Gong, X., Antonucci, V., Ge, Z. (2009) Development of LC chiral methods for neutral pharmaceutical related compounds using reversed phase and normal phase liquid chromatography with different types of polysaccharide stationary phases. *J. Pharm. Biomed. Anal.*, **49**, 964–969.

4
Chromatography Equipment: Engineering and Operation

Abdelaziz Toumi, Jules Dingenen, Joel Genolet, Olivier Ludemann-Hombourger, Andre Kiesewetter, Martin Krahe, Michele Morelli, Henner Schmidt-Traub, Andreas Stein, and Eric Valery

4.1
Introduction

Among the stages in a purification process, liquid chromatography still remains the "work horse." It is now a mature market with countless applications in a diverse range of industries and has emerged in critical sectors such as life sciences, environmental engineering, pharmaceuticals, R&D, food and beverage, and industrial chemicals, whether used alone or in tandem with other high performance technologies. The total US Market for protein separation systems alone (including hardware and chromatography resins) is estimated to be valued at almost $3.0 billion in the base year of 2006. This total market is expected to grow with an average annual growth rate of 11.1% and will cross the $5.0 billion by 2011. The liquid chromatography market was $944 million in 2004 and will rise to more than $1.1 billion in the year 2006. This market is expected to reach $2.1 billion by 2011, an annual growth rate of 13.0% (BBC-Research, 2007).

Preparative liquid chromatography is conducted at moderate (often ambient) temperatures that are beneficial for products sensitive to heat degradation like proteins or heat-sensitive pharmaceuticals. The selectivity is often higher compared with other separation technologies (distillation, crystallization) and today's chromatography systems are often characterized by a higher flow velocity and binding capacity. One essential drawback still remains the large usage of solvents, a drawback that is counter-balanced by the trends toward using continuous processes and integrated recycling technologies.

In fermentation processes, chromatography systems have to cope with the rapid increase in concentration of target molecules. According to DePalma (2005, 2007), protein titers have increased a 1000-fold from 1980 to today. In the year 1980, the titer was 5–50 mg l^{-1}. Nowadays, chromatography steps have to deal with titers around 5 g l^{-1} with a target of 10–20 g l^{-1} for the next three to five years. Higher titers mean it is necessary to scale up chromatography steps and find more robust and effective separation technologies in a continuous mode of operation. You can

---

Preparative Chromatography, Second Edition. Edited by H. Schmidt-Traub, M. Schulte, and A. Seidel-Morgenstern.
© 2012 Wiley-VCH Verlag GmbH & Co. KGaA. Published 2012 by Wiley-VCH Verlag GmbH & Co. KGaA.

often scale fermentation bioreactors without changing the size of the equipment (footprint); this is just not possible with chromatography columns and related equipment.

Purification experts and engineers responded with numerous improvements including

- process integration and intensification;
- the use of compound filters during clarification (due to the fact that more biomass is produced along with a higher titer);
- more efficient resin utilization (including reuse) and higher product recovery;
- alternative downstream strategies that employ membrane absorbers, liquid–liquid separations, and of course disposables.

Economics is a driving force in general chromatography separations to minimize waste and maximize output. For many years it could be debated as to whether upstream or downstream operations were rate-limiting step within a bioprocess. Now, with fermentation titers or product concentrations rising inexorably, purification is likely to remain rate limiting, for a given facility layout, for a very long time.

Today half of the purification costs within pharmaceutical or biotechnological separations for instance lie in the chromatography steps, approximately 30% of the total cost, and this increases with the further concentration of the product. In other sectors, this factor is not so accentuated but still the operation costs of chromatography steps compared with other separation processes are relatively high.

Chromatography costs are dependent on three major aspects:

- **Hardware costs:** The cost of the equipment depends upon its installed cost amortized over its lifetime, which will generally range from 3 to 10 years.
- **Stationary phase costs:** With the increase of molecules' titers in fermentation processes and the increasing complexity of small pharmaceutical molecules the suppliers of the stationary phase always work to increase the performances of their phases.

  Protein A sorbents used in antibody purification processes are typically the most expensive but have also seen the most improvement with time. To decrease costs, the lifetime of media has to be increased while keeping the adsorption characteristics.

  The lifetime of a stationary phase in small molecule purification strongly depends of course on the type of crude material that has to be purified. In general, the current spherical phases on the market (bare silica, C8, C18, cyano, diol, amino ...) based on purified metal-free silica have a very good mechanical as well as a long-term chemical stability if they are used under the appropriate conditions. For example, on production scale, a cyano-modified silica stationary phase could be kept in the same column for about 4 years allowing to purify multiton quantities of product.

  Chiral stationary phases, chemically grafted on a silica matrix, can also be used for quite a long period of time. The same holds true for the very versatile physically coated cellulose and amylose phases if they are treated with care.

- **Mobile phase costs:** Solvent, salts, and also water (especially highly purified water (HPW) and water for injection (WFI) typically in bioprocesses) could have a big impact on the purification costs. Similarly in small molecule purification, solvent cost is a very important cost factor and requires an intensive investigation regarding solvent recovery and reuse.

Also a limited production capacity of some specific organic solvents, for example, acetonitrile, can strongly influence the cost of a chromatography process. In large-scale preparative reversed phase chromatography, an alcohol is preferentially used as the organic modifier in the mobile phase, although in some cases the use of acetonitrile is required to obtain the desired solubility or selectivity. Trends are going toward the usage of semicontinuous process steps like the simulated moving bed technology as well a recycling technology.

4.2
Engineering and Operational Challenges

Purification processes in general, and chromatography steps in particular, need to be scaled-up to match the improved reactor productivity. For example, in some biotechnology applications such as mammalian cell culture technology multikilogram production of drugs are possible in a single batch (Aldington and Bonnerjea, 2007).

Fine chemicals are generally produced in large-scale batch reactors. However, in the past few years, a trend toward process intensification can be observed and more and more continuous production processes appear. Besides the advantage of a better reaction selectivity, these types of manufacturing processes also offer a lot of opportunities and challenges for engineers to combine continuous synthesis with a continuous purification method.

Processes are typically developed in the laboratory with milligram quantities of product, and this is the case for the majority of drug substances. Purely chemically synthesized products could start at a larger scale. However, in any case, if the product is a clinical or commercial success, a cGMP (for regulated applications) at pilot-scale manufacturing process will be needed to produce many tens of kilograms or even tons per batch before commercial launch. A design of a production facility for pharmaceutical applications is often built before the drug achieves full commercial approval, so this will occur in parallel to the second and third phases of clinical investigations. This is mainly due to the fact that qualification and validation processes, especially those involving the automation systems, require substantial time to complete.

The starting point for process design is often a small-scale example or a laboratory recipe of the worst-case scenario. All available information has to be collected and used as basis for the first step of process design, which defines the different process sequences and specifies all boundary conditions, like binding capacity of the resin, filtrate throughput of a given membrane, and reactor concentration (titer) in biotechnological applications. Next, a first attempt is made to size the equipment (reactors, product vessels, chromatography columns, solvent/buffer vessels).

Depending on the complexity of the installation, scheduling logistic software can be used to analyze the capacity of the plant and de-bottleneck while it is still in the design phase. In addition to the process side, requirements on auxiliary systems like utilities, drains, and waste water treatment systems have to be reliably estimated.

The scale-up of manufacturing processes can be divided into a number of stages (Aldington and Bonnerjea, 2007). In some organizations, this is formalized in a kind of manufacturability reviews that correspond to the project moving from one department to the next, for example, from research to process development, from process development to pilot plant, and from pilot plant to large-scale manufacturing. The desired outcomes of final large-scale manufacturing have to be kept firmly in mind even for the early research phase. This is the reason why many manufacturers and associated equipment suppliers tend to standardize a manufacturing platform to cope with the whole cycle from early development until large-scale manufacturing, that is, lab- and pilot-scale plants represent already the different process steps of the intended large-scale production plant. It might be also that no one size fits all existing solutions and one has to distinguish between two different platforms like an early clinical platform and a large-scale platform. Both platforms do not necessarily have the same objectives. While in early clinical phase, the major objective is time, costs become more relevant for a large-scale facility. The main challenge here is how to cope with the different objectives and assure at the same time scalable and compatible processes when you move from one platform to the next.

Multipurpose flexibility is another major requirement for process engineering given the high costs associated with process equipment and modern plants. Nowadays, plants are not built just for one single product; they have to cope rather with a certain variety of products. Some of the major multipurpose considerations are straightforward, for example the use of chromatographic resins that are chemically and physically robust and easy to clean. There are other issues that are less obvious and have to be considered early in the research activities, like the use of full traceable cell lines and the use of endotoxin-free components supplied with certificates of analysis (Aldington and Bonnerjea, 2007).

For the large-scale manufacturing, production costs have to be minimized. The use of large volumes of buffer at a slow flow rate to wash a chromatography column or the use of chemicals that cannot be routed to the waste water treatment plant has to be avoided. Thus process optimization is essential and often goes hand-in-hand with process scale-up. We will demonstrate all these different aspects by taking a look at the scale-up of a large-scale biotechnology facility.

Additionally, from the engineering and maintenance point of view ATEX (explosion proof) requirements heavily influence the investment and operational cost. Preparative chromatography is an established and mature technology and a large number of suppliers do exist, offering standard equipment as well as customized packages. Depending on the sector, whether pharmaceutical or chemical, further hygienic requirements and regulatory exigencies have to be considered. In the pharmaceutical and biotechnology industry sterile connections and drainable/steam able components have to be considered whereas in the fine-chemical and chemical sectors attention has to be paid to material resistance, explosion proof issues, and corrosion aspects. The operational nominal pressure is also one of the key elements and a driver for the size and costs of the installation.

During the conceptual design phase, the design of a chromatography system has to consider the following main criteria:

- space and layout requirements;
- required auxiliary equipment like recycling units or PAT (process analytical technology) instrumentation;
- packing and unpacking units for chromatography columns;
- the level of automation integration required;
- batch versus continuous process.

The maintenance of chromatography equipment might also dictate the addition of cranes and support tools to be able to exchange the different soft parts of a chromatography column. Easy maintainability and operability are very important, but in the design phase they are often overlooked. For instance care has to be taken for

- easy access to different valves, instruments, and components like pumps and pump heads;
- easy and ergonomic exchange of valves or inline filters;
- oil replacement of the pumps;
- calibration of measuring devices in place (e.g., temperature and pressure sensors, detectors);
- In case of pharmaceutical equipment all soft parts (e.g., gaskets) have to be regularly exchanged.

The following case study will demonstrate that engineering of chromatography equipment is not a stand-alone exercise and often needs to consider the whole plant design, as dependences between the different process steps and bottlenecks in utilities system might affect the size of the equipment.

Case Study: Large scale biotechnology project: The objective of a large scale biotech (LSB) project was to extend (double) the production capacity at an existing manufacturing site of Merck Serono in Vevey, Switzerland. The plant will initially be dedicated to the parallel production of two different molecules, a monoclonal antibody (mAb) and a fusion protein. Additional mAbs and related molecules from the Merck Serono pipeline are expected to be manufactured in the same facility in the future. The limited space available for the construction of the new facility made the design very challenging and the project highly complex. A computerized process model built in SuperPro Designer has been developed in a very early stage of the basic design phase of the

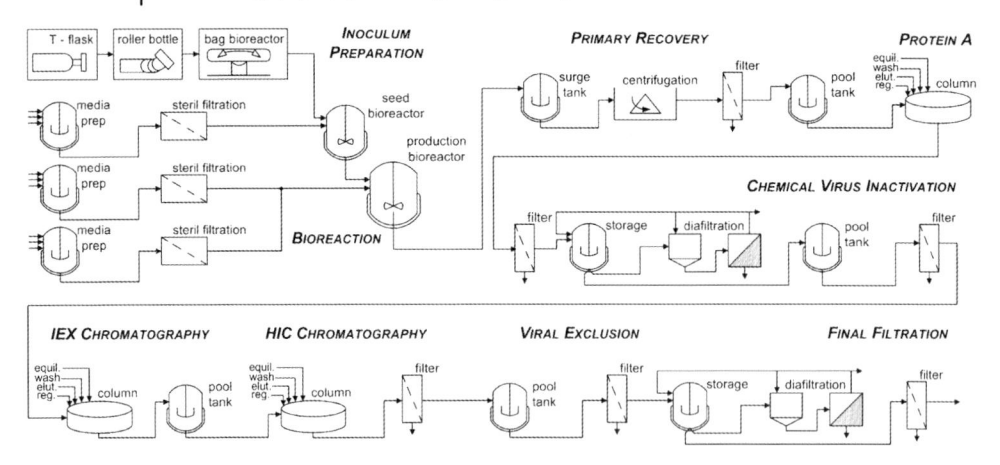

Figure 4.1 Principle of monoclonal antibody production (reproduced from Toumi et al., 2010).

project to support all design activities and facilitate scenario analysis and evaluation (Toumi et al., 2010).

Three chromatography steps are used: a Protein A binding step, flow-through step, and a hydrophobic interaction step (Figure 4.1). Such processes utilize a large number of buffer and cleaning solutions (usually 20–30) that must be prepared on time and be ready for delivery when required by the main process. The preparation and storage of such buffers involve a large number of tanks. Most of the tanks are used for the preparation and storage of multiple solutions and require cleaning and validation of cleanliness after each use. For every chromatography step, a list of operational steps are defined according to Table 4.1 and are introduced into a mathematical model.

The results of process scheduling and cycle time analysis are typically visualized with Gantt charts that display equipment occupancy as a function of time. These types of charts enable engineers to resolve scheduling conflicts and reduce the cycle time of a process.

The models were mainly used to size shared resources (e.g., chromatography columns, utilities, and media/buffer preparation tanks) and evaluate various capacity scenarios. The impact of different shift patterns on equipment demand for buffer preparation was also evaluated. Using such tools makes it easier to quantify the trade-off between labor cost and capital investment. The model was able to generate different options based on different work organizations. For example, two options were explored:

1) buffers either are prepared 24 h a day/7 days a week;
2) buffers prepared only during the day shifts.

Option 1 involves lower labor cost but higher capital investment. However, it also provides flexibility for future changes in increased plant operation and capacity challenges. More specifically, if product titers increase in the future and there is a need for reduced purification cycle times, the plant may switch to a three-shift

Table 4.1 Modeling of a chromatographic step (Toumi et al., 2010).

	Chromatographic Sequences				Capture				
Step	Description	No.	Buffer	Column volume	Total volume per cycle (l)	Cycle (s-start e-end)	Number of cycles required	Flow rate (l min^{-1})	Linear flow rate (cm h^{-1})	Processing time (min)
1	Rinse I	W1	HPW	2.0	420	S	1	50.9	270	8.3
2	Equilibrium			0.0	0	S	1	0.2	1	0.0
3	Regeneration I			0.0	0	S	1	0.2	1	0.0
4	Equilibration	E1	PPAEB	5.0	1050	S-E	2	71.6	380	14.7
5	Load	Pr	Product	n.a.	8098	S-E	2	71.6	380	113.0
6	Wash I	E1	PPAEB	5.0	1050	S-E	2	71.6	380	14.7
7	Wash II	E1	PPAEB	5.0	1050	S-E	2	71.6	380	14.7
8	Wash III			00	0	S-E	2	0.2	1	0.0
9	Elution	E7	PPAEB	4.0	840	S-E	2	50.9	270	16.5
10	Regeneration	E6	Acetic acid	2.0	420	S-E	2	50.9	270	8.3
11	Regeneration 1	E11	0.1 M NaOH	2.0	420	E	1	12.3	65	34.3
12	Regeneration 2	E6	Acetic acid	2.0	420	E	1	12.3	65	34.3
13	Regeneration 3	W1	HPW	1.0	210	E	1	33.0	175	6.4
14	Storage	E10	PPAEB	2.0	420	E	1	33.0	175	12.7

PPAEB: Poros Protien A equilibrium buffer, HPW: highly purified water.

operation for buffer preparation and accommodate the increased demands of the purification trains.

The tools also were used to analyze the impact of buffer expiration times, shift patterns, equipment sizes, and number of equipment items. Approximately, 35 different scenarios were evaluated during the project and most of the scenarios included major model updates. As the project evolved, the team's understanding for the processes, the facility, the underlying links, and constraints improved and the knowledge gain was used to improve the models. Figure 4.2 shows the evolution of the models up to scenario no. 15.

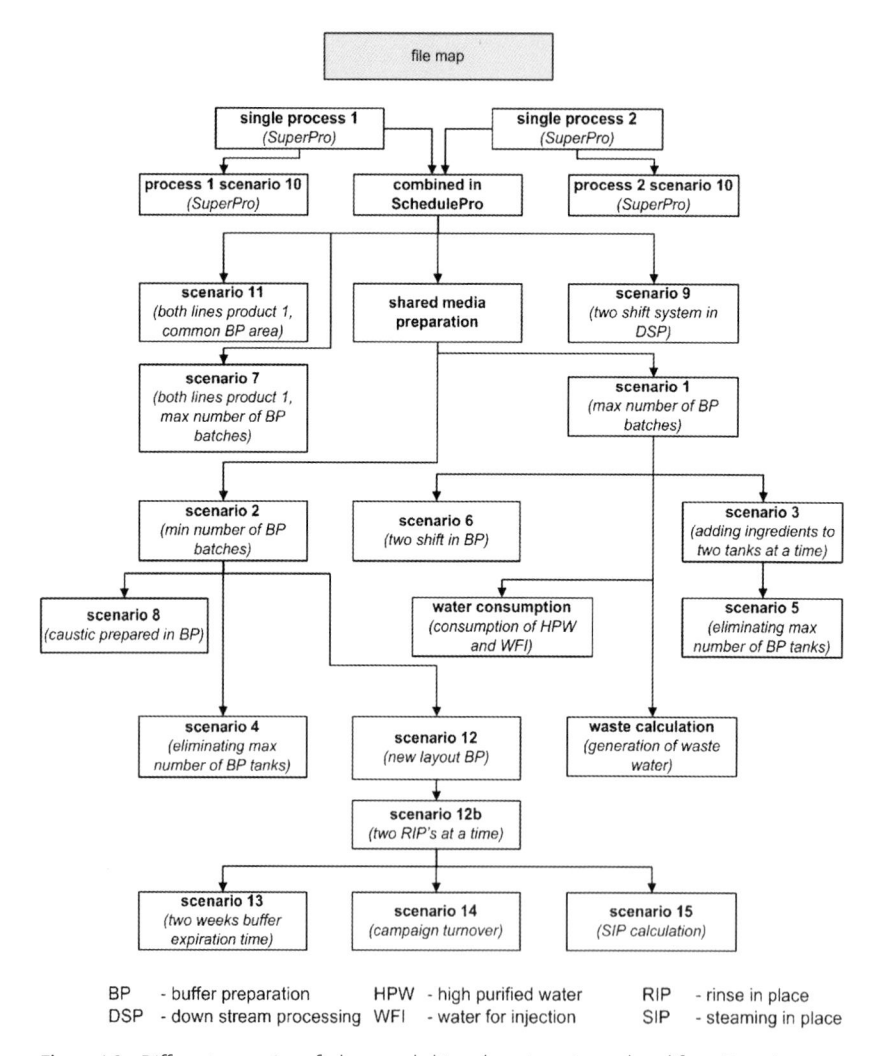

Figure 4.2 Different scenarios of a large-scale biotech project (reproduced from Toumi et al., 2010).

BP – buffer preparation
DSP – down stream processing
HPW – high purified water
WFI – water for injection
RIP – rinse in place
SIP – steaming in place

It can be summarized that when applied early, simulation tools can support plant design and technology transfer and can facilitate the communication between the engineering and operations teams. It is an important engineering tool to have the ability to design the proper size of the chromatography steps (column diameter and piping skid size). The insight that modeling provided for the design of the support areas, such as buffer preparation and holding, utilities, and equipment cleaning requirements, was of particular importance.

4.3 Chromatography Columns Market

4.3.1 Generalities – The Suppliers

When discussing columns, several items should be addressed: the column technology, the column design, the packing material, and the packing procedure. Particular emphasis is put on the column because this is a most critical part of the equipment. However, the column is just one critical aspect of chromatography purification. A good column associated with a poorly designed pumping and auxiliary unit is often worth nothing.

The design of the columns has to be optimized for the chromatographic technique to be used. For instance, the pressure and protocols to successfully and robustly pack middle or low pressure media are quite different. In the end, the operating condition of the column determines the design of the chromatographic system, for example, if it is a high- or low-pressure system. Here important differences due to size of the solute molecules, thermodynamics, and the mass transfer resistance have to be realized. For liquid chromatography, the challenges for separating fine chemicals and other low-weight molecules or biopolymers with molecular weights ranging from several thousand to several million Daltons are quite different. Stationary phases and appropriate chromatographic systems for these different separation tasks are explained in Chapter 3. One dominating factor for column design is the diffusivity that is by a factor of 100 or more smaller for proteins than for small molecules. Consequently, biopolymers need much more time to diffuse into and out of the pores and the linear velocity of the eluent in the column has to be about 100 times lower in order to achieve a sufficient separation (Unger *et al.*, 2010). Comparing high performance liquid chromatography (HPLC) chromatography with low pressure liquid chromatography (LPLC) can be summarized considering kinetics of diffusions of molecules on the stationary phase: on HPLC, particle size is smaller and diffusion is elevated, therefore, elevated mobile phase flow rates are applied while the separation is still very efficient (HP means high performance) but generates pressure (explaining that many consider that HP means high pressure). Column length is fine-tuned on both kind of chromatographies regarding pressure constraints and kinetic effects, 10 and 20 cm bed length columns can be found at industrial scale for HPLC and LPLC applications.

In the following discussion, it is considered that a preparative column is a column with an internal diameter exceeding 50 mm. This is obviously an arbitrary limit but it is more or less the maximum diameter of commercially available pre-packed columns.

Equipment used for preparative and large size biochromatography is very different compared to the equipment used for fine chemicals and pharmaceutically active small molecules. Biological separations are mostly using high concentration of salts and there are some specific aspects of the equipment for such separations. Today different technologies have been developed by companies like General Electric, Pall, Novasep, and Merck Millipore. These columns are often used at high flow rates (200–1000 cm h^{-1}) in order to maximize productivity and or minimize purifications costs.

The purpose of this chapter is to discuss equipment for preparative chromatography in general and to outline additionally aspects that are typical for the separation of fine chemicals and low-molecular-weight products as well as the separation of biopolymers like proteins.

4.3.2
General Design

What is the main purpose of a chromatography column? This question can be answered in several ways:

- contain the media;
- provide uniform plug flow;
- minimize band broadening;
- withstand pressure;
- compatible materials (solvent/chemical);
- scalable performance;
- facilitate packing/unpacking;
- meet regulatory compliance needs;
- allow cleaning in place (CIP).

Today we can find different established column designs. Concurrent technologies exist to lower and upper the piston like hydraulic pumps, manual moving, or an electrical motor. The different designs have different drawbacks and advantages, but have to obey the following criteria:

- **Reproducibility:** complex standard working procedure may impact reproducibility and operators may also require specific training.
- **Operator interaction and safety:** column manipulation, contact with media, and GMP.
- **Plant constraint:** size-floor space, compatibility with solvent/chemicals, pressure vessels, ex-proof for explosive solvents.

Chromatography columns have three basic components:

- **top flow cell assembly:** distributes flow evenly across the bed;
- **column tube unit:** houses the media;
- **bottom flow cell assembly:** retains media, collects effluent flows.

Manually moved piston: Generally used on column packed with low-pressure media, this technology could be applied on columns up to 600 mm. One of the advantages of this system is the reduction of column cost due to the fact that the use of automated technology is limited. This kind of system is often used in the biotechnology field for column packed with low-pressure bed which does not require high pressure for packing.

Electrically or hydraulically moved piston: In terms of column technology, there is a consensus today that the best (and probably only) feasible approach at any size is dynamic axial compression (DAC). The piston can be driven by a hydraulic jacket, or pushed by a liquid, by a spring, or by an electrical motor. Radial compression is an alternative but it does not seem to be available at (very) large column diameter sizes. It also requires the use of prepacked cartridges, thus reducing the operator's freedom compared to DAC. Under appropriate conditions, the DAC technology ensures bed stability and reproducible performance. DAC can be scaled-up to very large size with remarkable reproducibility in performance. This has been verified with columns up to 1600 mm internal diameter.

When fragile and/or soft packing media are used, special care must be taken to ensure that the compression energy is compatible with the mechanical properties of the particles. Also, the shrinking–swelling phenomena observed with some organic materials under certain mobile phase conditions should not result in over-compressed beds. To address this issue, columns with an "intelligent" piston are available. With such columns, a given bed compression can be programmed and the hydraulic system is able to adjust the piston position so that a constant compression is maintained.

In terms of column technology, another special situation is encountered in SFC (supercritical fluid chromatography). Here, rather elevated pressures must be used in order to attain the supercritical state. The fluid pressure in the column varies a lot, depending on whether the column is in use or not. The effective pressure exerted by the piston on the bed is the difference between the compression pressure and the fluid pressure under the piston. The compression pressure should then be adjusted depending on the internal pressure of the column, since too high a piston pressure may crush the stationary phase, whereas a too low pressure may prevent the piston from effectively compressing the bed. It seems that only DAC columns are suitable for large-scale SFC. These columns are, for example, equipped with "smart" compression units that maintain adequate bed compression without allowing the packing material to be overpressurized when the mobile phase pressure is released. Alternatively, it is also possible to use for SFC a specifically designed

Table 4.2 Typical dimensions and operation conditions for stainless steel columns.

Type of column	Column dimensions $L_c \times d_c$ (mm)	Flow-rate range (ml min^{-1})	Max. pressure drop (bar)
Preparative	100 × 20	20–40	300
	300 × 50	100–200	200
Large scale	300 × 100	500–1000	150
	200 × 300	8,000–12,000	100
	100 × 600	32,000–48,000	70
	100 × 1000	80,000–13,0000	50

DAC column equipped with high-pressure valves in the two lines of the hydraulic oil circuit, which are closed during normal use of the column. After the process is finished and the gas pressure has been released from the column, the valves in the oil circuit can be opened and the oil pump activated to compensate for an eventually created void volume during the process. Thereafter, both valves are closed again and the column is ready for further use.

4.3.3
High- and Low-Pressure Columns

Stainless steel, glass, and plastic materials (cross-linked organic polymers) are employed as tube materials. Usually stainless steel is used for high-pressure columns. The stainless steel tubes have a mirror finish inside with a surface roughness smaller than 1 µm. The column material should be mechanically stable toward high flow rates and high pressure and chemically resistant toward aggressive eluents. Table 4.2 gives for certain columns dimensions an overview of typical flow rates and maximum allowable pressure drops. If stainless steel is used for biopharmaceuticals' separations it is recommended that only high quality stainless steel is used to avoid any problems related due to corrosion of the metal from the use of salts in biotechnology processing. Usually stainless steel 316L (stainless steel with low ferrite ratio) is used and the inside of the column has a mirror finished (electro-polishing) with a surface roughness of approximately 0.6 µm. In any case, the construction materials for biopharmaceuticals' separations have to conform to relevant regulations (Section 4.4.3).

For biopharmaceuticals' separation, we can also find glass or plastic material (cross-linked polymers like acrylic). Table 4.3 gives typical dimensions and operating pressures of glass and large-scale acrylic columns for low-pressure bioseparations or intermediate purifications that are characterized by a height to diameter ratio < 1.

Construction materials for columns can be chosen according to several attributes; Table 4.4 shows an example of selecting criteria for the most common materials:

Table 4.3 Typical dimensions and operating conditions for acrylic and glass columns (Chromaflow^{TP}, BGP^{TM}, GE Health Care, 2011).

Material	Diameter (mm)	Height (mm)	Max. operating pressure (bar)
Acryl	400, 600, ... , 2000	100–300	3
Glass	100	50–95	8
Glass	140	50–95	6
Glass	200	50–95	6
Glass	296	50–95	4
Glass	446	50–100	2.5

4.3.3.1 Chemical Compatibility

Table 4.5 illustrates the chemical resistance of a series of columns materials across a range of organic solvents and chemical solutions. This list is not exhaustive and any detailed information regarding the column's compatibility to a particular solution should be provided by the column supplier.

4.3.3.2 Frits Design

Band broadening and thus the quality of chromatographic separations depends on axial dispersion within the column. Apart from the fluid flow through the packing axial dispersion is mainly influenced by the inlet and outlet of the column where the liquid should enter or leave, respectively, the column as a plug flow. Quite obviously, the bigger the column diameter the more severe the problems of equal distribution and collection are, for example, with a flow coming from a 2 mm tube diameter into a column with a diameter of more than 1000 mm and collecting the effluent again in a 2 mm tube (Figure 4.3). To optimize the fluid flow in a chromatographic column, several attempts have been made with regard to the fluid distributor and collector systems and the shape of the column tube itself.

Table 4.4 Construction materials and selecting attribute for chromatography columns.

Material	Visibility	Solvent/salt resistance	Pressure resistance	A-specific binding
Acrylic	Transparent	High salt	High pressure	Some binding
Glass	Transparent	Solvent and salt resistant	Pressure/size limits	Some binding
TPX, Polymethylpentene	Transparent	Solvent and salt resistant	High pressure	Low protein binding
Stainless steel	Not transparent	Solvent resistant, salt resistant limited low pH	High pressure	Low protein binding

4 Chromatography Equipment: Engineering and Operation

Table 4.5 Construction materials and selecting attribute for chromatography columns (courtesy of Merck-Millipore, Lit No. BCQ1080 Rev C).

Solvent	Acrylic	Glass	TPX	Stainless Steel 904L (1.4509)	Stainless Steel 316L (1.4404)	20% Glass filled poly-propylene	Poly-propylen	EPDM*	Santoprene™	PTFE	PVC
1,2-dichloroethane	−	+		+	+	+	+	L	−	+	+
Acetic acid 89%	−	+	−	+	+	+	+	+	+	+	−
Acetic acid 25%	+	+	+	+	+	+	+	+	+	+	−
Acetone 2%	+	+	+	+	+	+	+	+	+	+	−
Acetone 5%	+	+	+	+	+	+	+	+	+	+	−
Acetone 10%	−	+	L	+	+	+	+	+	−	+	−
Acetone 100%	−	+	L	+	+	+	+	+	−	+	−
Acetonitrile	−	+	−	+	+	+	+	−	L	+	−
Acrylonitrile	+	+	+	+	+	+	+	+	+	+	+
Ammonia aqueous <25%	+	+	+	+	+	+	+	+	+	+	+
Ammonium sulfate (10–40%)	−	+	L	+	+	+	+	+	L	+	+
Amyl alcohol (1-pentanol)	−	+	−	+	+	+	+	+	+	+	−
Aniline	−	+	−	+	+	+	+	+	L	+	+
Beer	+	+	+	+	+	+	+	+	L	+	−
Benzaldehyde	−	+	−	+	+	+	+	−	+	+	+
Benzene	−	+	−	+	+	+	+	−	−	+	−
Benzoic acid	+	+	+	+	+	+	+	+	+	+	+
Benzyl alcohol 1%		+	+	+	+	+	+	+	N/A	+	L

L: Low, N/A: not available.

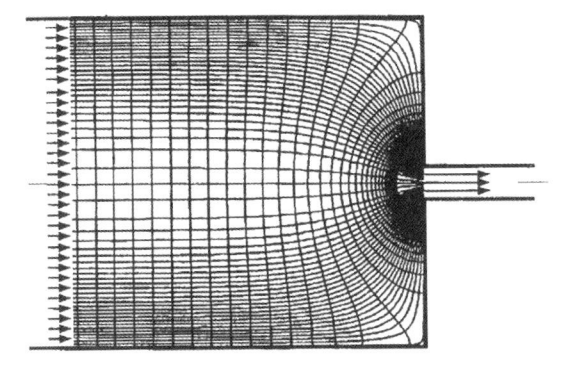

Figure 4.3 Isobars and streamlines for a cylindrical column outlet.

The easiest means of fluid distribution at the column inlet is by using a high-pressure drop of the packed bed, which forces the fluid inside the inlet frit into the radial dimension (Figure 4.4).

The layer design of the frits and the quality of the radial flow characteristics of the frit are of great importance to obtain a good distribution. Distribution by using the column pressure drop has nevertheless a severe drawback: every HPLC plant should be designed to show the lowest possible system pressure drop (i.e., to allow the maximum linear velocity with a given maximum pressure drop). Therefore, efficient pressure-less distribution systems had to be designed. One approach is the integration of distribution plates into the column inlet. Nowadays, their design is optimized by computational fluid dynamics (CFD), as shown by Boysen *et al.* (2002). The most advanced approach uses

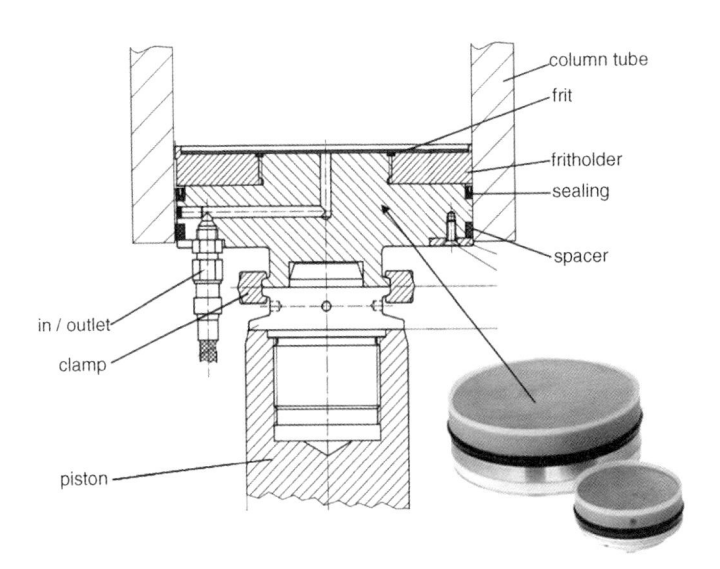

Figure 4.4 Cross-section of a column inlet with frit system.

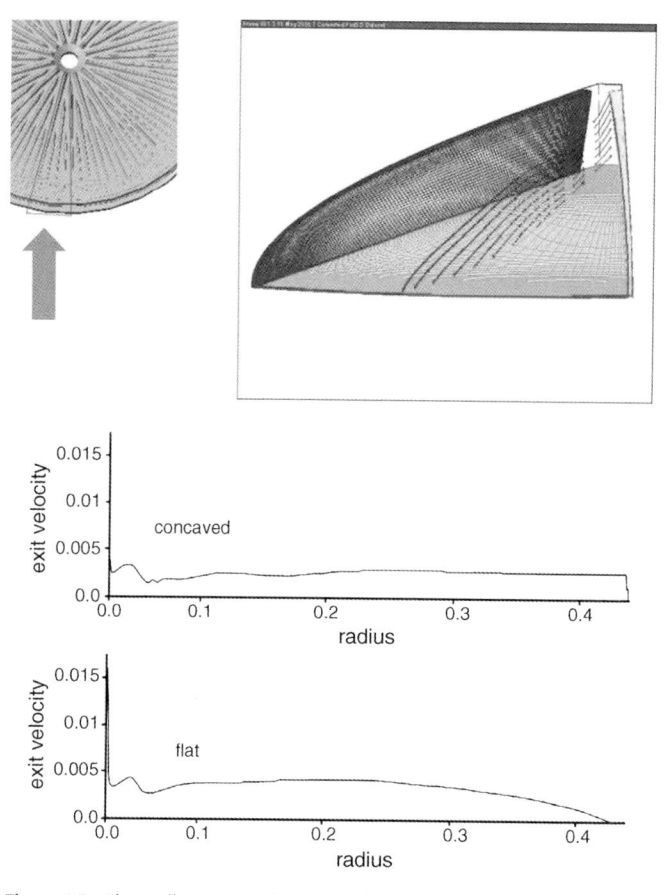

Figure 4.5 Flow-cell geometry design and impact on exit velocity.

distribution channels with engineered fractal geometries as an alternative to random-free turbulence (Kearney, 1999, 2000).

The homogeneity of flow distribution for LPCL column depends of the flow cell design; the objective is to have the fluid reaching the top of the resin at the same time. Flow cell geometry modeling has shown that concaved flow cell design provides constant velocity along the all flow cell radius (Figure 4.5).

Standard and engineered columns feature removable frit plates for easy replacement, in order to cleaning purpose or product change. Frit plates are available within a range of material and frit pore sizes from 2 to 75 μm. The materials used could be plastic (polymeric) or metal.

Stainless steel bed supports are used extensively in standard and engineered chromatography columns. In contrast with hydrophobic polyethylene sinters, they are easily "wetted". The multilayer woven mesh construction offers high porosity and mechanical strength, and may in many cases be cleaned of media particles or by ultrasonic means. However, all stainless steel products are susceptible to corrosion when in prolonged contact with chloride-containing solutions and especially at low pH.

In general, properly passivated mesh of DIN grade 1.4404 (approximately equivalent to type 316L) is suitable for prolonged immersion in 2 M sodium chloride solution at neutral pH and at ambient temperature (approximately 20 °C).

Typical cleaning procedures are:

- NaOH 0.5–2 M soaking solution contact time 24 h;
- ultrasonic bath (refer to http://www.southern-metal.com);
- passivation (refer to http://www.poligrat.de).

4.3.3.3 Special Aspects of Bioseparation

Biomolecular separation requires different techniques than chemical separations. Biological molecules are sensible to heat and solvents, which can destabilize the three-dimensional structure of proteins resulting in loss of the molecule's function (Carta and Jungbauer, 2010).

4.3.3.3.1 Protein Binding
Most materials used in the manufacture of chromatography columns exhibit some degree of nonspecific protein binding. In the purification of biological products it is of paramount importance for regulatory compliance that equipment suppliers provide experimental investigation and data into such binding and show that the levels involved are very low and the bound protein can be efficiently removed by a clean-in-place wash with a sodium hydroxide solution.

Table 4.6 gives some basic information on protein binding for common column construction materials.

4.3.3.3.2 Sanitary Design and Sanitization
Sanitary design is necessary to ensure hygienic performance and thorough cleaning; standard column attributes to minimize bacterial contamination are listed below:

- fully flushed flow path and adjuster seal for cleaning-in-place (CIP);
- minimum dead space cell seal arrangement;
- reduced risk of corrosion;
- leachable free tube material;
- minimize unswept areas in flow path.

Table 4.6 Protein binding for construction materials.

Material	Protein (mg cm^{-2})
Borosilicate glass	0.030
Polypropylene	0.027
TPX (polymethylpentene)	0.009
Acetal	0.030
Stainless steel	0.001
Acrylic (perspex)	0.012
Santoprene	0.023
EPDM	0.032

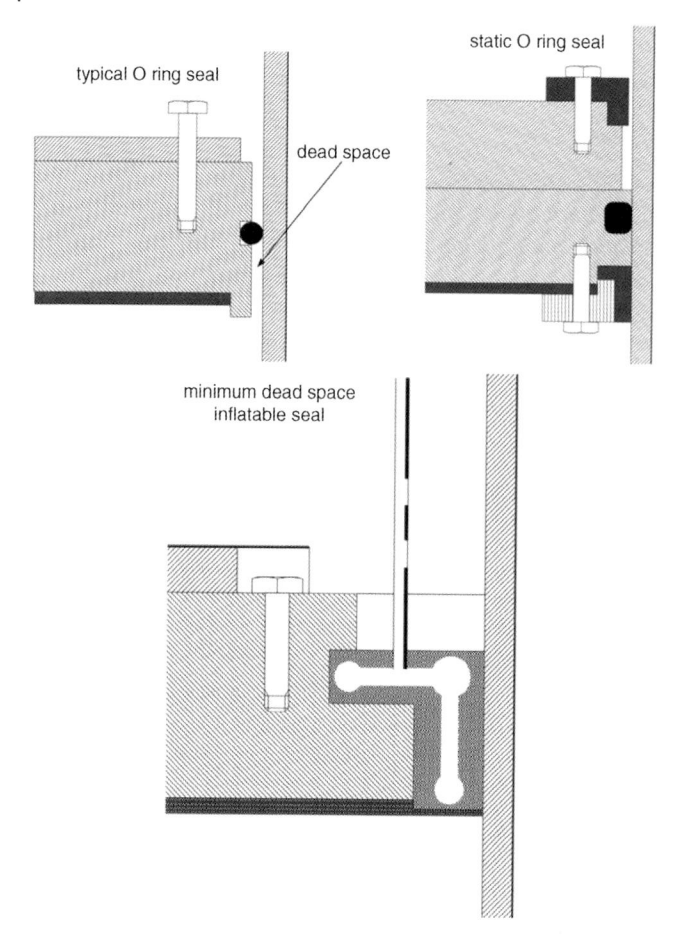

Figure 4.6 Hygienic criteria for pharmaceutical applications.

Figure 4.6 depicts design principles for equipment seals with minimum dead space.

For proof of cleanability column suppliers need to provide data of how the equipment performs under bacterial challenge. An example of bacterial challenge is expressed in the next paragraph.

The chromatography column can be challenged with E-coli, to test cleanability under standard operating conditions. Sodium hydroxide (1 Molar solution) can be used as a sanitizing agent and loaded with a minimum concentration of 105 cfu (colony-forming units) ml^{-1}. The column should be left to incubate overnight at room temperature and washed with two column volumes of 1 M NaOH recirculated for 1 h. After neutralization, the column should be sampled for residual microbial contamination. Sample should be taken from the critical points as shown in Figure 4.7.

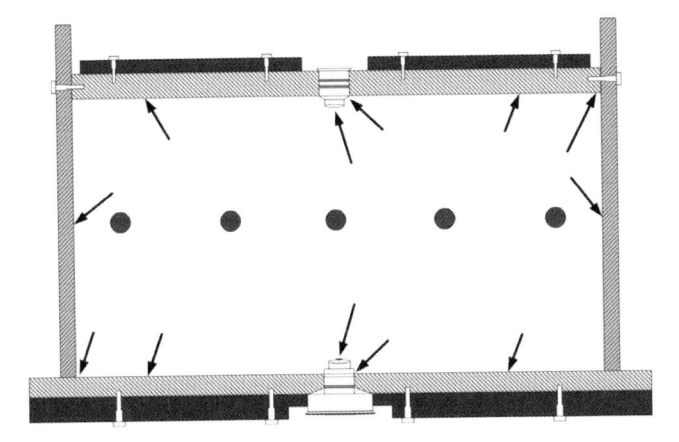

Figure 4.7 Sampling for cleanability test.

A result of no viable target organisms is found in the process liquid flow path, and no viable cells are detected on the clean sides, indicates that the column design provides an effective flow path for chromatography with no unswept areas, and that the cleanability is maintained.

4.4 Chromatography Systems Market

4.4.1 Generalities – The Suppliers

The market can be divided into different main segments:

- analytical and preparative equipments;
- high pressure/performance systems;
- low pressure systems.

Analytical systems do equip almost all laboratories. Many different kinds of particle sizes are used: from 20 to 2 µm (from HPLC to UPLC), using liquid or supercritical fluid mobile phases. An intermediate market is to be considered using 10 mm ID columns for small-scale purifications. The term preparative chromatography is dedicated to purification units using column diameters from 50 up to 1600 mm ID.

4.4.2 General Design Aspects – High Performance and Low-Pressure Systems

The chromatographic column is responsible for performing the purification, and the chromatography system is responsible for optimizing the separation.

A preparative system has to:

- minimize loss in efficiency of the chromatographic separation;
- allow the use of one or many mobile phases, with the ability to utilize isocratic conditions, step gradients, or sloped gradients;
- carry the injection of feed mixture;
- monitor and control chromatography separation;
- collect the different fractions;
- protect the column against misuses.

The system when connected to the column has to avoid any kind of mixing before or after the column. Therefore, the system needs to be designed to minimize the dispersion within its different elements. For a targeted flow rate, the different elements have to be minimized to find a good compromise between internal volume and induced pressure drop.

A chromatographic system is always a combination of compromises between chromatographic and technical constraints.

For HPLC systems, the stationary phases often use small particles from 10 to 50 µm, the number of plates generated by the chromatographic column can reach up to 50 000 plates per meter. The constraint of avoiding any kind of dispersion is then a major priority. An important action to minimize the dispersion from the moment the feed is introduced to the moment where the fractions are collected is to reduce volumes and to design smooth connections between the different parts of the plant. Small piping diameters are also a requirement because the operating pressure introduced by the column can go up to 50 bars or even higher. HPLC systems are mainly used for the separation of fine chemicals and small pharmaceutical molecules.

LPLC systems are generally used for the purification of biomolecules like proteins. In contrast to HPLC, these systems are mainly operated with step gradients and use stationary phases with large particle size (50–200 µm). Their column technologies are not designed to resist to high pressure. The volumes of the different equipments are larger than on HPLC systems and the corresponding pressures are of course significantly lower (<5 bar). Furthermore, the operating conditions have to be compatible for biological molecules, as particularly proteins are sensible to heat and solvents that can destabilize their three-dimensional structure resulting in loss of the molecule's properties (Carta and Jungbauer, 2010).

In order to satisfy these requirements chromatography systems need to have the following basic components:

- inlets for buffers, mobile phase, product, and sample addition;
- pumps and/or switching valves (for gradient mixing);
- specific protection devices like filters and bubble trap to avoid loss of efficiency due to potential introduction of particles or air;
- sensors (pre- and postcolumn);
- valves to control flow direction and collect fractions;

- automation system to control all electronic and pneumatic devices and run properly recipes;
- data collection and reporting.

Different preparative chromatographic equipment should be considered for the following applications:

- to carry out batch or continuous purifications;
- to carry out different kind of chromatographies: from chiral purifications to the binding of mAbs on protein A resins.

Different layouts will be presented in this chapter, considering:

- batch HPLC equipment;
- batch LPLC equipment;
- continuous HPLC equipment;
- continuous LPLC equipment.

4.4.3
Material

The construction material of the piping could be made of stainless steel or plastic according the application and the size of the equipment. Stainless steel is available in different grades that are resistant against high salt concentration. High surface finish is required for effective cleaning. Plastic is available in different materials that may have limited resistance to pressure than stainless steel.

In case of biochromatography, all materials used in the manufacture of columns for biopharmaceuticals' separations, and which come into contact with the process stream should be selected so as to either conform to relevant sections of the FDA Code of Federal Regulations, Volume 21, parts 170–199, or have passed USP class VI tests for *in vivo* toxicity.

Stainless steel can be supplied with a batch-specific material certificate that details the exact composition of the alloy. Plastics and rubber components cannot be so easily defined. For this reason, certification of particular polymers is rarely provided on a batch-by-batch basis, but by "generic" assurance that relates to a particular grade of material from a particular supplier.

Two types of conformance criteria are used, and a specific material may conform to one or both of these. The first is that material suppliers provide written confirmation that specific materials conform to the relevant section(s) of the FDA Code of Federal Regulations, Volume 21, parts 170–199.

The second is that representative samples of the material are independently tested according to a method described in USP 23. In the Biological Test for Plastics Class VI (70 °C), the test sample is extracted at 70 °C in a range of solutions, and both these extracts and the test material itself are checked for direct biological reactivity, leading to a pass/fail result.

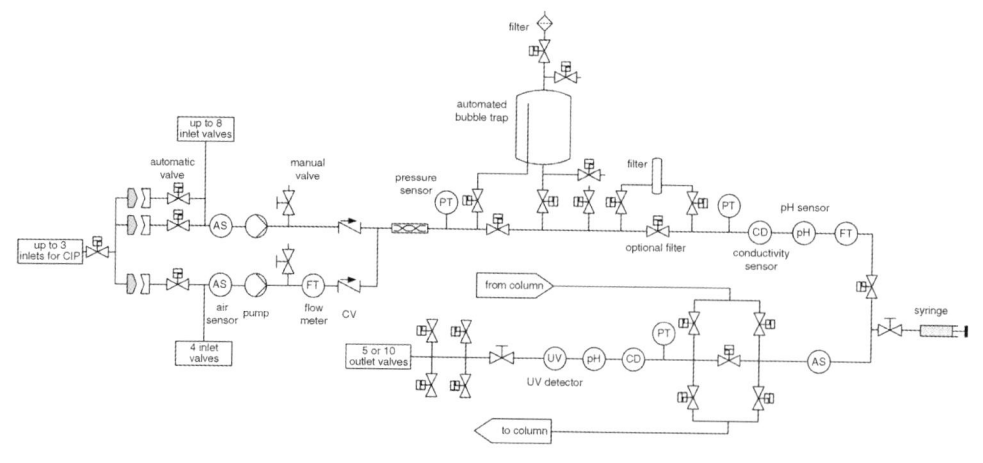

Figure 4.8 P&ID of a valve gradient based LPLC system (reproduced from HiperSep Bio, Novasep).

4.4.4
Batch Low-Pressure Liquid Chromatography (LPLC) Systems

Figure 4.8 shows a typical PID for a batch LPLC system.

4.4.4.1 Inlets

The system inlets' number may vary according to user requirements, 3–5 inlets are common in preparative chromatography where the process is in general simple and repetitive. Development or pilot system can be featured with larger number of inlets to allow more flexibility in the development or optimization of the chromatographic separation.

In case of bioseparations it is mandatory to be able to process effectively cleaning chemicals, that is, cleaning in place (CIP). CIP is executed with the use of chemicals with the purpose of minimizing bio-burden. In most applications sterility as in fermentation is not claimed. In large manufacturing installation, the chromatography system can be connected directly by stainless steel piping to the tanks containing product that needs to be purified and the buffers necessary for the chromatography process. These tanks as well as the transfer lines need to tolerate steam cleaning (SIP) and the chromatography systems connected to it should be designed to allow this sterilization process. An additional possibility needs to be foreseen to protect the column and any other sensitive material during this process step. This can be performed using valves equipped with a steam trap or even a switching valve and embedded into the system design.

4.4.4.2 Valves to Control Flow Direction

Valves are integral components in piping systems; they are the primary method of controlling the flow, pressure, and direction of the fluid. Valves may be required to

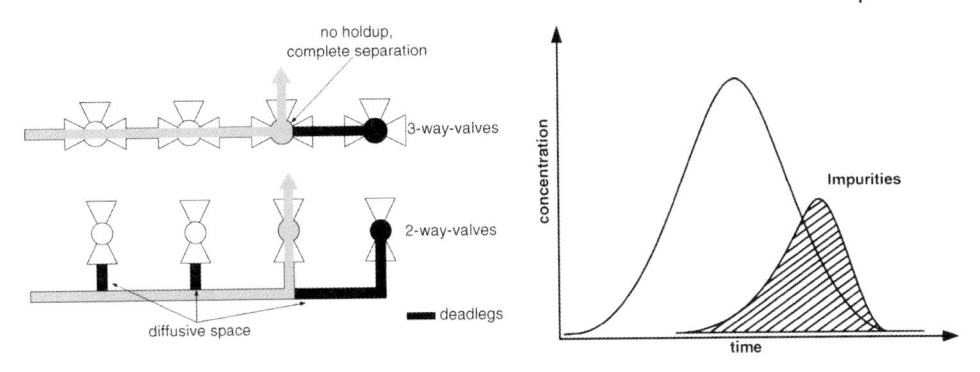

Figure 4.9 Influence of holdup volume and diffusive space in a chromatogram.

operate continuously for example, control valves, or they may be operated intermittently for example, isolation valves, or they may be installed to operate rarely if ever for example, safety valves. A valve can be an extremely simple, low cost item or it may be and extremely complicated, expensive item. In piping design the valves probably require more engineering effort than any other piping component.

The properties of the fluid to be controlled have a major impact on the design and materials of construction of the valve. The piping industry, over the years, had developed a wide range of valve designs and materials to handle virtually all of the fluids being handled. The selection of the valve should take into account fluid viscosity, temperature, density, and flow rate. The valve must be suitable to withstand resulting corrosion and erosion and if necessary the valve may have to be designed for no internal hold up of fluids. Figure 4.9 illustrates how impurities because of dead volumes influence the quality of a chromatographic separation.

In addition to the low or no dead volume feature, valves should also be fully drainable and of fast actuation especially if they are used for the gradient formation. In the past, there were Gemü, Saunders, and ITT; nowadays new vendors offer designs that feature improved chromatographic performance that is, NOVASEPTIC, Furon, Robolux (Burkert), and Swagelok.

4.4.4.3 Pumps

Pumps are major components of all chromatography systems as they have to induce the driving force for the elution and separation process. In any case, process pumps should be able to maintain their set-point flow rate in a smooth manner. Positive-displacement pumps, such as diaphragm pumps, are often used in preparative and large-scale chromatography systems.

Some key requirements for pumps used in chromatography systems are as follows:

- The set-point flow rate should be maintained irrespective of any back pressure.
- The pulsation should be minimized, to avoid disturbance of the chromatographic bed. Thus the usage of multihead pumps is favorable.

Table 4.7 Examples of available pumps features and limitations.

Pump type	Features	Limitations
Diaphragm pump	Precise flow control	Only true positive-displacement pump
		For low flow applications
		Must manage pulsation
Lobe pump	Low pulsations	High flow applications
Peristaltic pump	No shearing	Unreliable, tube breakage
		Not suitable for process chromatography

- The handling of any biological solution should be gentle, so as not to shear or denature any biological product.

None of the pumps currently on today's market can meet all those requirements, except smooth flow diaphragm pumps. The major disadvantages of peristaltic pumps are the likelihood of breakage of the flexible tubing, pressure limits with no more than 30 psi, potential contamination of product/matrices with leachables, and the difficulty of making truly sanitary and strong connections between flexible tubing and steel piping. Table 4.7 describes some features and limitations of different pump types.

With respect to the gear, rotary lobe or cog-wheel-type pumps they are not recommended for sensitive proteins, since high shear forces generated in the pump head, possibly in combinations with other factors such as air entrapment, can lead to protein aggregation and de-naturation.

4.4.4.4 Pump(s) Valves and Gradient Formation

In case of large retention factors of the last eluting component, which happens very often in bioseparations, chromatographic processes require the means to change the composition of the mobile phase in order to control the elution rate from the column. The rate at which the mobile phase composition changes concurs precisely with the elution time for each peak for a given process (Figure 5.26).

Two factors affect the viability of the elution profile:

- accuracy;
- reproducibility.

Each characteristic is important; however the most critical factor for process scale pharmaceutical purification applications is reproducibility. Processes that are required to be validated must yield consistent results between operations. For a chromatography process to be reliable, the gradient profile must be reproduced consistently during each operation. Several approaches are used to control the formation of a gradient profile. As the most important principle valve- or pump-based mixing systems can be identified.

4.4 Chromatography Systems Market

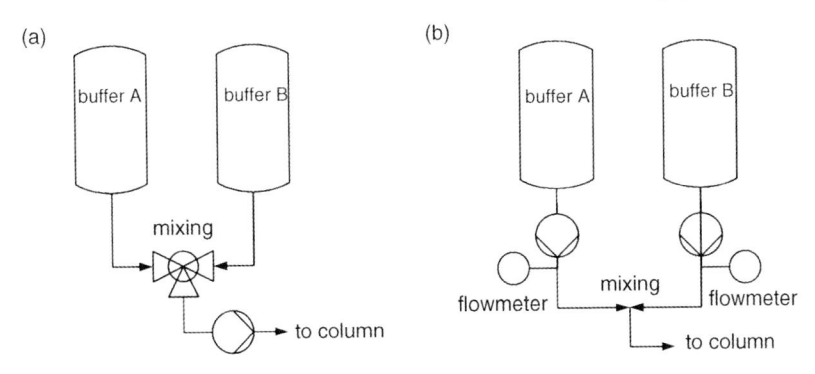

Figure 4.10 Valve-based and pump-based mixing.

For many applications, the system forms the gradient mix by cycling the valve at the ratio required by the process (Figure 4.10a). Simply cycling the valve will result in the proper gradient mix being formed if the mix frequency is properly tuned prior to use. In order to provide a steady flow gradient valves need to have the same inlet pressure; this means that inlet pressure should be controlled by a regulation valve at the tanks outlet.

Pump-based gradient (Figure 4.10b) works very well for bench scale equipment and it can be very accurate in a limited flow interval. Unfortunately, the individual reciprocating pumps may show inaccurate functioning at very low flow rates (Figure 4.11). In addition, these pumps require extensive calibration procedure and expensive maintenance. Compared to pump-based gradients, valve systems are in general cheaper and can achieve 1–2% accuracy in a broad operating interval.

Figure 4.11 depicts an example for the acceptance range for the mixing of two solvents A and B. The diagram makes clear that in the low range of flow rate from

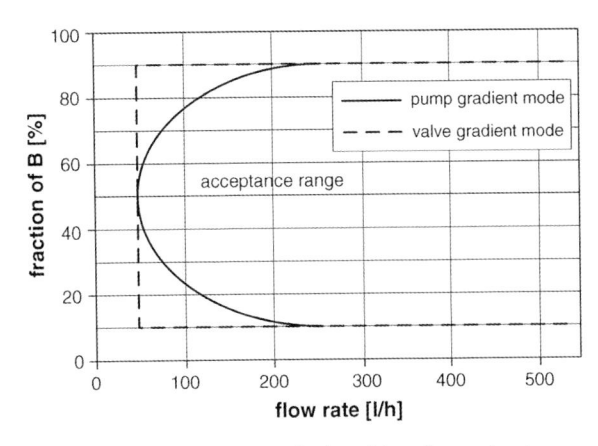

Figure 4.11 Acceptance range for the mixing of two solvents.

0 to $200 \, \text{l h}^{-1}$ valve-based gradients can be mixed effectively within the ratio of 10–90%, while pump-based mixing assures acceptable accuracies only between 40 and 60%. For higher flow rates, both methods offer effective gradient mixing. These differences between valve and pump mixing are valid independently from the flow rate or pump capacity.

Some applications require gradients to be formed to a prescribed set-point value. The most common parameter used to identify the set-point value is the ionic strength or conductivity of the buffer solution. Another possibility for alcohol or acetonitrile – water mixtures is the use of a near infrared sensor.

A very reliable, but more expensive gradient mixing unit is based on the use of regulation valves combined with mass flow meters.

4.4.5
Batch High-Pressure Liquid Chromatography (HPLC) Systems

4.4.5.1 General Layout

By design HPLC separations use small stationary phase particles to increase the column efficiency and improve the resolution between peaks, enhancing product purity and recovery yield. At a given linear eluent velocity, small particles cause higher pressures than large particles. Therefore, HPLC systems are operated at high pressures in order to increase the eluent flow rate, hence the productivity of the separation. HPLC systems represent different constraints compared to LPLC systems. In a large majority of the cases:

- To scale up the high efficiencies obtained at small scale, the preparative HPLC systems have to be designed in order to minimize all the dead volumes. High means that the time scales can be short.
- The mobile phases use organic solvents. It is rare to use only pure water, most of the time in reversed phase chromatography, water is combined with acetonitrile, or with an alcohol.

4.4.5.2 Inlets and Outlets

4.4.5.2.1 Mobile Phases
Figure 4.12 shows a P&ID of a typical gradient mode HPLC system that allows a binary analog gradient working from 5 to 95% in the dynamic range of the system flow rate, with an absolute accuracy of $\pm 0.5\%$ or better. The different gradient modes have already been described on LPLC systems.

4.4.5.2.2 Feed Injection
The number of plates of the HPLC column may be very high (e.g., 10 000 plates), often resulting in narrow peaks. The ideal injection is obtained when a perfect step of feed concentration "as thin as possible" is performed: in HPLC separations, the injected volume is generally small ($<1\%$) compared to the volume of the column.

The injection of the feed mixture can be performed through many different ways; three of them are described as follows:

4.4 Chromatography Systems Market

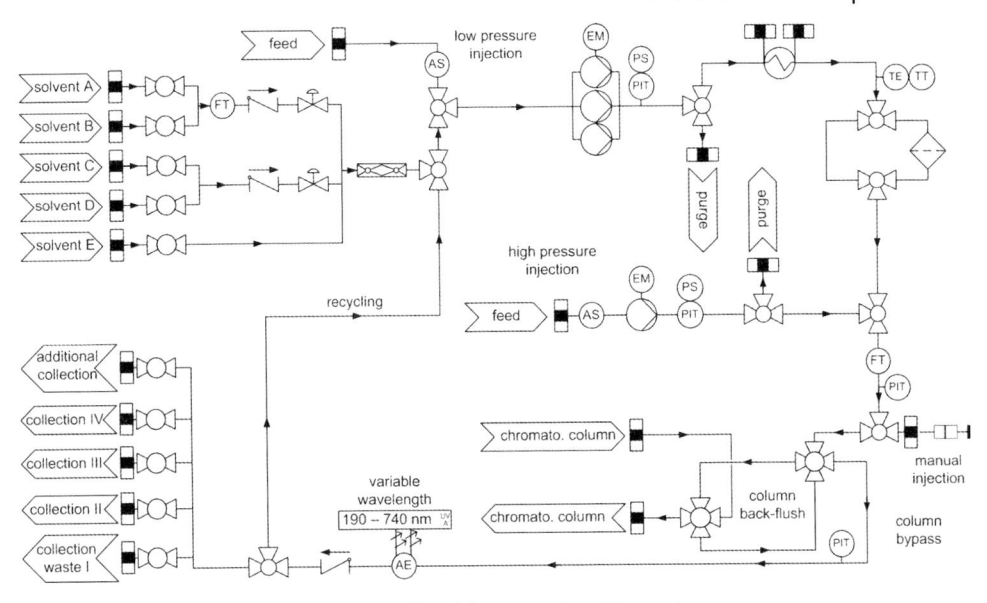

Figure 4.12 P&ID of an HPLC system (reproduced from HiperSep, Novasep).

- The feed injection is performed as a "step gradient", through the main eluent pump (called "low-pressure injection"). The advantage of this method is the simplicity and the speed: a valve is switched to the feed line without changes in the flow of the main pump. The injected volume is regulated by the opening time of the feed valve. The main drawback of this method is that for small injected volumes and in strongly nonlinear conditions, the reproducibility of the chromatograms may be poor. Additionally, the "pulse" of feed is sent through the main pump and may be potentially diluted, eventually leading to a loss of efficiency. However, this method is appropriate for gradient chromatography as well as for isocratic elution chromatography. Another possibility is the use of the combination of a tree-way valve combined with of a mass flow meter to improve the reproducibility of injection for smaller volumes.
- The feed injection is performed by an additional pump (called high-pressure injection). The advantage of this method is that the feed pump is designed and sized in order to inject with a high reproducibility a large range of volumes, avoiding loss of efficiency. One drawback of this method is the use of an additional pump, often with a reduced flow compared to the main pump, leading to an increase of the cycle time, and of the equipment cost.
- The feed injection is performed by an injection loop. Advantage of this method is that the loop can be filled during the elution of the column, so the accuracy of the injection can be combined with a single flow rate purification method. This method is the only way to perform the injection of crude material in a one-column steady-state recycling system. It is also the most suitable injection method in super critical fluid chromatography. The main drawback of this method is that

the geometry of the loop has to be carefully designed to avoid dilution or tailing in case of partially filled loops.

4.4.5.3 Pumps

The flow rate of the eluent and the sample is delivered to the column by a metering pump. The demand of HPLC pumps is very high, because they must guarantee a reproducible flow rate even against high back pressure generated by the column. To prevent mechanical stress to the fixed bed in the column the solvent flow should have low pulsations. In the start-up period of pumping the adjusted flow rate should be reached gently (avoiding too high flow rate) so as to prevent damage to the packed bed. The pumps have to be feedback controlled by flow-meters (in general mass flow meters).

Based on the system size:

- The main pump (eluent pump) is a 2- or 3-head membrane (PTFE) pump, able to operate at up to 100 bar, equipped with membrane failure indicators.
- The high pressure injection pump is a 1- to 3-head membrane (PTFE) pump, able to operate at 100 bar, equipped with membrane failure indicators.

Piston pumps with more than one pump head (in most cases two or three) or a pressure regulator downstream of the pump are not so frequently used in industrial units.

The working range of the pump, concerning the maximum pressure difference and flow rate, must be adjusted to the column. An inadequate system causes lower productivities of the separation and thus wastes time and money. Table 4.8 gives typical flow rates and tube diameters for different column geometries.

Piston pumps are used for analytical and small preparative HPLC systems (below 1000 ml min^{-1}). The major operating problem of piston pumps is dissolved air and the formation of bubbles in the eluent. Bubbles in the pump heads cause pulsations of volume flow and pressure. Bubble formation and cavitation problems are promoted at the inlet check valve because the minimum pressure in the system is

Table 4.8 Typical flow rates and tube diameters for different column geometries.

d_c (mm)	4	25	50	100	200	450	600	1000
$\dot{V}_{min}$ (ml min^{-1})	0.5	20	80	300	1200	6000	10×10^3	30×10^3
$\dot{V}_{max}$ (ml min^{-1})	2	80	300	1200	5000	25×10^3	45×10^3	120×10^3
Outer tube diameter (mm)	1.56	1.56–3.12	3.12	3–6	6–8	8	16	25
(inch)	1/16	1/16–1/8	1/8	1/8–1/4	1/4–3/8	3/8	5/8	1
Inner tube diameter (mm)	0.25	0.75–1.6	1.6	1.6–4.0	4.0–5.0	5.0	13.0	21.0
Sample loop volume (ml)	0.01–0.2	1–2	2–5	10–20	Feed introduction via feed pump			
Max. unit dead volume (ml)	0.5	3–5	5–10					

reached here. In analytical units degassing can be done online by a membrane degasser, by pearling helium offline through the eluent, or by using an ultrasonic bath. In Prep units, where water and an organic solvent are mixed, a bubble trap is used to solve this problem. Important to mention is also that degassing in the detector can also create very annoying noise problems, which eventually can be solved by creating a small back pressure in the collection line.

Cavitation is a general operation problem, which can be solved by using tubings with a higher internal diameter or by increasing the pressure at the suction side of the pump. The easiest way to increase the pressure is to lift the eluent reservoir to a higher level than the pump inlet. If this is not sufficient the pressure in the solvent reservoir must be increased. For this reason the solvent reservoir should be airtight and isolated from the environment. Isolation of the eluent has the positive side effect that the composition and water content can be held constant during operation. Especially in case of normal phase chromatography, it is important to control the water content in the range of a few ppm, because of the hygroscopic character of the adsorbent.

4.4.5.4 Valves and Pipes

As presented on the P&ID (Figure 4.12), different valves can be used:

- For the solvent module, 2-way ball valves are used in combinations with 2 way analog control valves to adjust the composition of the gradient.
- On the flow path: two pneumatic 3-way ball valves used to bypass the filter. Pneumatic 4-way (or an arrangement of 3-way) ball valves are used to bypass the chromatographic column and to revert the elution direction into the chromatographic column.
- On the collection module, 2–way ball valves are used, mainly being normally closed (to perform the collection) and one being normally opened (typically defined as the waste).

Volumes respectively diameters of the pipes are preassigned: large pipe diameters are used on the entrance (suction side) of the pumps, and thin pipes are used on the chromatographic flow path (pressure side). Basically, internal pipe diameters are designed to ensure a turbulent flow with fluid speeds around two meters per second. On industrial systems, all pipes have to be connected to the mass in order to prevent the build-up of electrostatic charges, which is especially a danger in case of non conducting solvents and high fluid velocities.

In large systems, pipes connect the different parts of the chromatographic plant. The tubes are made of stainless steel, Teflon or polyetheretherketone (PEEK). However, care has to be taken when pure tetrahydrofurane is used together with peek tubing at higher pressures, because after prolonged usage, the tube wall can become permeable for this type of solvent. Selection of the right material depends on the mobile phase used for the separation. The choice of the internal diameter of these tubes should be a compromise between small dead volume and low back pressure. Small diameters result in high pressure drops in the system and higher diameters lead to higher dead volume, causing back mixing and, thereby, peak

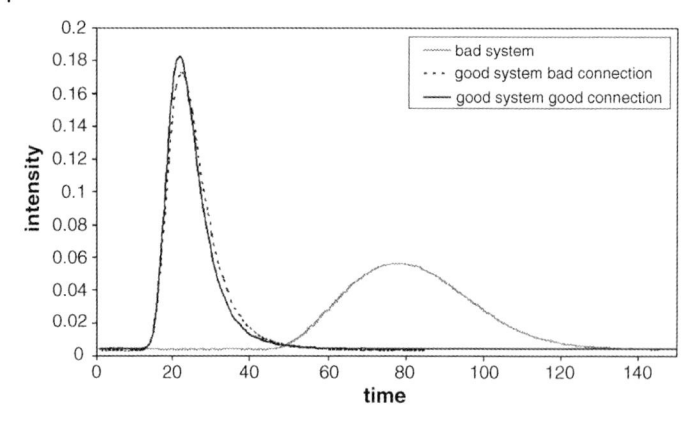

Figure 4.13 Influence of system design on dead time and peak distortion.

distortion. More important than the absolute tube length is the smooth connection of tubes via connectors, reducers, and ferrules to different system parts. Abrupt changes of cross-sections induce turbulence and vortices which increase axial dispersion and consequently band broadening. Another aspect of pipe design is cross-contamination. All pipe connections must avoid dead spots that cannot be sufficiently cleaned before a product change.

Figure 4.13 shows the influence of dead volume and connection design on the peak distortion and dead time of the analytical HPLC unit. Three runs were performed with the same flow rate and sample size (injection time: 2 s). The straight black line describes the elution profile of a tracer in an optimal system with small dead volume and optimal connections. In the next example one of the tube connections was insufficient because, within the connector, the tube endings did not touch and a small dead volume is formed. The dashed line shows the new elution profile. A poor connection results in higher back mixing and, thus, peak distortion. This effect should be kept in mind for systems with many connections and small tube diameters where bad connections can have a larger impact.

The third experiment, represented by the gray line, shows the elution profile of the same tracer in the same unit, but in this case one tube with a small internal diameter was replaced by a bigger one. Dead volume and peak distortion are dramatically increased. This increase in dead volume results in a loss of baseline separation. For preparative separations the higher back mixing in the unit leads to decreased productivity and economy of the process.

4.4.6
Batch SFC Systems

4.4.6.1 General Layout
Supercritical fluid chromatography could also be called very high performance chromatography.

- As the mobile phase is mainly made of supercritical or subcritical carbon dioxide, with a low viscosity and elevated diffusivity, there is almost no mass transfer resistance. As a consequence, the peaks are narrow, and the efficiency does not decrease with the speed of the mobile phase. For production purposes, it makes sense to have systems with elevated flow rates.
- The chromatographic system must be designed to handle the different states of carbon dioxide based on its phase diagram. The recycling of the carbon dioxide makes SFC a relatively green technology, particularly at large scale. In general to increase polarity of the mobile phase, as well as to enhance the solubility of the products to be separated, a certain amount of polar solvent (called modifier) is added in the mobile phase. As carbon dioxide can be solubilized in the modifier inside the section of the system dedicated to carbon dioxide recycling, the use of a modifier introduces some losses – approximately using 10% of modifier introduces losses of 5% of carbon dioxide.

Preparative SFC systems with different sizes are available:

- from lab scale: columns from 3 to 8 cm I.D. with CO_2 flow rates from 100 to 3000 g min^{-1}.
- to large scale: columns up to 20 cm I.D. with CO_2 flow rate up to 14 kg min^{-1}.

Figure 4.14 characterizes the different thermodynamic states of an SFC process. At the inlet of the main pump, CO_2 is in liquid state (40 bars and $-5\,°C$), and at the outlet of the pump, CO_2 is at the operating pressure. After heating, CO_2 is in supercritical state, ready to be used for chromatography. At the outlet of the column, a decompression turns the mobile phase to a biphasic system: gaseous CO_2 and droplets of solute. Separators allow the recovery not only of the solute, but also of gaseous CO_2 that can be recycled after cooling. At the end of this process, the CO_2 makes a complete turn around its critical point.

At the outlet of the separators, the solute can be recovered with an extremely high yield of 95–98% if the cyclones are designed correctly.

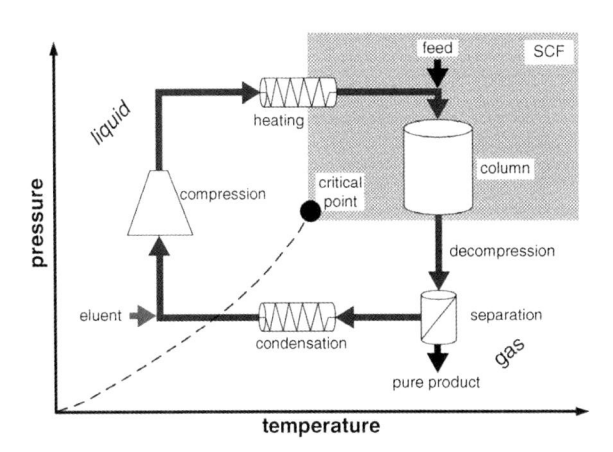

Figure 4.14 Principles of SFC chromatography with CO_2 recycling.

4.4.6.2 Inlets

4.4.6.2.1 Mobile Phases An organic solvent (called co-solvent or modifier) is very often used in SFC to increase the solubility of the crude mixture in the eluent and fine-tune the eluent strength. Mobile phases on SFC use from 0 to 40% of modifier. Fresh modifier is pumped in the system with a high pressure pump. The modifier is collected at the liquid outlet of the separators.

Carbon dioxide is the typical mobile phase on SFC. CO_2 can be supplied in liquid or gas state, but

- On SFC systems with recycling, liquid carbon dioxide is introduced as an eluent and gaseous carbon dioxide is recycled after separation from the product (Figure 4.14).
- On SFC systems that do not have a recycling needs liquid CO_2 is preferred.

As a substantial amount of the carbon dioxide can be recycled, it has to be considered that an important part of the CO_2 supply comes from the recycling.

ecycling devices have then to be efficient, to remove all the liquid droplets from gas carbon dioxide to avoid pollutions. The recycled gas is then cleaned from traces of products, and only contains vapors of modifier. Thanks to gas liquid equilibrium, the amount of vapors of the modifier in CO_2 is reduced and constant; this enables to scale up any method developed at analytical scale on production systems with an efficient CO_2 recycling and cleaning.

4.4.6.2.2 Feed Injection As in HPLC, an ideal injection is obtained when a perfect step of feed concentration "as thin as possible" is obtained. However, in SFC, the feed mixture has to be introduced at high pressure, into carbon dioxide that may be a weak solvent in terms of solubility. The feed mixture can be injected in heptane that has a solvent strength similar to carbon dioxide (but solubility of the crude material is often very limited), or can be introduced in the organic modifier. Different methods have been reported:

- **The feed injection is performed through the modifier pump:** Advantage of this method is that the composition of the mobile phase is constant even during the injection. Drawback is that the injection of large volumes involves the use of elevated ratios of modifier which can be a significant limitation.
- **The feed injection is performed by an additional pump:** Ideally, the injection has to be performed at the same flow rate as the mobile phase, which may request a large pump. If this is not suitable for the injections of small volumes, then there is an optimum to be found between the size of the pump and the volume of product to be injected.
- **The feed injection is performed by an injection loop:** Advantage of this method is that the loop can be filled during the elution of the column and the injected volume is accurate. Drawback is that the volume of the loop may have to be changed in order to avoid dispersion inside the loop itself.

4.4.6.3 Pumps, Valves, and Pipes

As in SFC time ranges are shorter than on any other preparative chromatography technologies (some purifications can be handled with cycle times shorter than a minute), flow control is then a key point to obtain reproducible chromatograms.

The pumps have to deliver a constant flow despite the high pressure that can be generated. The modifier pump pumps a liquid and its flow rate is easy to control.

As CO_2 is a compressible gas, a good control of the CO_2 flow rate requires:

- mass flow meter;
- excellent regulation of the pressure of the system; this regulation of pressure being made at the outlet of the chromatographic column;
- control of the pressure and temperature of the inlet of the CO_2 pump to ensure the liquid state.

To summarize, SFC uses high-pressure liquid pumps, to introduce CO_2, modifier, and the feed. These pumps are piston or membrane pumps.

Compared to HPLC, SFC provides much higher efficiencies, so the piping has to be accordingly designed. From the introduction of the feed to the collection of the different fractions, the pipes are designed to minimize the volume, to avoid peak tailing and pollutions.

At that point, we could summarize that pipes have decreasing internal diameters, from LPLC to HPLC and finally to SFC.

4.4.7
Continuous Systems – Simulated Moving Bed

4.4.7.1 General Layout

Principles and interests of continuous chromatography are presented in Chapter 5.

The SMB process was patented in 1961 by UOP (Broughton and Gerhold, 1961). The technical realization of this process is presented in Figure 4.15a.

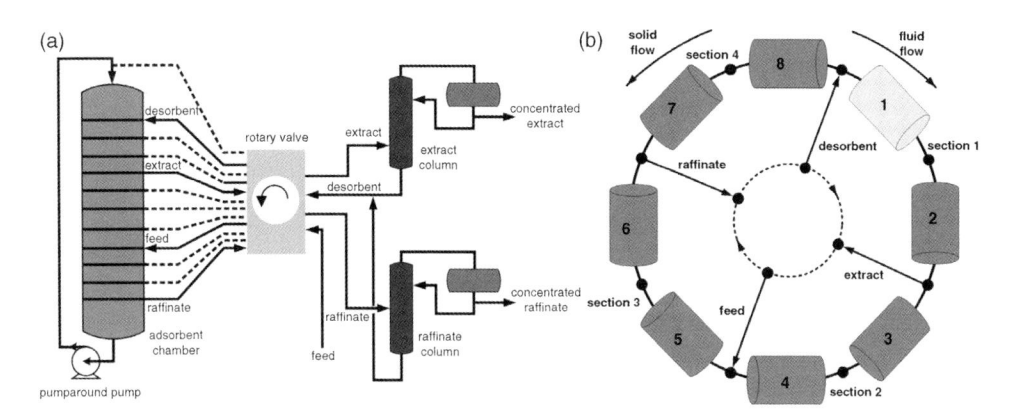

Figure 4.15 (a) UOP PAREX process with rotary valve, (b) SMB process with valves between the columns and port switching.

4 Chromatography Equipment: Engineering and Operation

It suits well the petroleum industry applications like the PAREX process where 24 columns were used in a system with a small pressure drop. This design is characterized by a recycling pump and in particular a rotary valve that periodically switches the lines so that a simulated counter current flow of the solid and liquid phase is achieved. The alternative design of SMB processes uses valves between all columns instead of one rotary valve (Figure 4.15b and 4.16, see also Figure 6.35). Here the counter current flow is realized by switching the ports in the direction of the liquid flow. This design has the advantage that for process optimization the ports can be switched individually as it is for instance applied by the VariCol process.

This chapter presents industrial solutions developed for preparative continuous chromatography of pharmaceuticals and fine chemicals. The technical choices will be presented for HPLC applications.

4.4.7.2 A Key Choice: The Recycling Strategy

As shown in Figure 5.16 fixed pump can be placed in different ways between two cells or columns.

In configurations Figure 5.16b as well as 5.16c and d the fluid loop is broken at the eluent line. Therefore the highest pressure is located at the entrance of the column connected to the pressure side of the pump and the lowest pressure at the outlet of the column connected to the suction side respectively the process outlet. In these cases, the pump operates at constant flow rate. When the eluent line is switched from one column to the next the points of high and low pressures are switched by one column as well. In configuration 5.16a, the pressure all along the system is more or less constant, irrespective of the position of the eluent, extract, feed, and raffinate lines. However, the pump has to change its flow rate if it enters a different zone of the SMB process.

In HPLC applications, the pressure drop along the columns of the system is around 30 bars, and the order of magnitude of the switching time is the

Figure 4.16 Novasep multicolumn (1 meter i.D.) continuous process.

minute. Therefore industrial applications promoted over the years configuration a with a fixed recycling pump and control of the flow rates as this leads to a smoother operation and lower stress on the stationary phase, hence increasing its lifetime.

4.4.7.3 Pumps, Inlets, and Outlets

Considering the recycling strategy with a fixed pump, five flow rates have to be controlled:

- **Eluent and feed inlets:** constant flow rates that can be achieved with 2- or 3- headed HPLC membrane pumps.
- **Extract and raffinate outlets:** constant flow rates that can be controlled using pumps, or by analog control valves.
- **Recycle stream:** high pressure liquid pump, mainly 3-headed HPLC membrane pumps are used, or high-pressure centrifuge pumps in large industrial systems. If because of the line switching the recycling pump is located in a different zone of the process, the flow rate has to be controlled accordingly.

It has to be noticed that as two inlets and two outlets are applied to a closed loop, even if the sum of the inlet flow rates are equal to the sum of the outlet flow rates, a small difference will strongly increase or decrease the internal pressure of the process. Thus, the outlet flow rates have to be regulated based on the internal pressure of the liquid loop.

4.4.7.4 Valves and Piping

Considering Figure 5.17a, it appears that two inlets and two outlets have to be installed between the columns of the system. Thus, four valves are necessary between all the columns.

The use of many valves instead of one large rotary valve offers many advantages: For maintenance small valves are easier and much cheaper to replace. However, the main advantage is that all ports can be shifted independently from each other. SMB plants are very often designed with a suction line of the recycling pump, which is much longer than the other lines between the columns. In the case of asynchronous shifting, it is necessary to compensate the unequal distribution of the dead volume. In the case of the Varicol process, an asynchronous multicolumn continuous system derived from SMB (Section 5.2.5.1) individual switching of the inlet and outlet lines is obligatory.

The internal diameters of the pipes are designed in the same manner as for batch HPLC systems.

4.4.8
Auxiliary Systems

Auxiliary systems surround the chromatography columns and their attached equipments. They are particularly required to pack the columns (Section 4.6), to maintain or qualify the system and to test the integrity of filters.

4.4.8.1 Slurry Preparation Tank

Slurry preparation tanks are used for several operations:

- pool new resins being delivered in 1 to 25 l bottles;
- resuspend already used resins in buffer;
- exchange storage solution of resins with packing buffer (sedimentation/decantation);
- wash used resins and remove fines (particle sedimentation/decantation);
- pack and unpack columns.

If a stainless steel tank is used, it must be able to suspend the resins (and fines) but avoid high shear stress, which would damage the resin. Axial mixers introduced from the top are appropriate. For decantation a height adjustable dip tube is required. High flow rates at the entrance of the dip tube must be avoided. To control the suspension/sedimentation, a viewing glass and a separate lamp are required. The bottom of the tank is often discussed; either dished bottom or conical bottoms are used. Dished bottoms are preferred to provide homogeneous conditions, conical bottoms have better sedimentation/decantation characteristics and are suitable to recover all resin materials. In both cases, a bottom valve, sealing with the vessel wall shall be used. Slurry preparation tanks need to be cleaned but a steam sanitization for bioburden reduction is typically not required.

An alternative to stainless steel tanks is open cylindrical storage tanks. Plastic tanks or single-use liner can be used as well. Closed single-use bags with a top mounted paddle mixer and a valve closing with the wall can also be taken into consideration.

Slurry tanks should be carefully designed to allow a homogeneous mixing with minimal impact on the resin. To avoid particles' degradation that could reduce resin lifetime shearing should be very low. Therefore, it is recommended to use a tank size that is four times of the packed bed volume, to use the following ratios for tank design $D/T \sim 0.35$ and $C/D \sim 0.9$, to locate the mixer centrally and to select mixers with a rotor tip velocity that does not exceed $3.3\,\mathrm{m\,s^{-1}}$ (Figure 4.17).

Chromatography resins for manufacturing scale are delivered in standard containers of 10, 20, or 25 l. Transferring sedimented resin from shipping containers (drums or jerry cans) to the mixing tank can be performed with specific equipment like drum rollers.

4.4.8.2 Slurry Pumps and Packing Stations

A pump is necessary to pack, unpack, and provide buffer from the slurry preparation tank to the column. The ideal pump provides constant pressure and flow rate at low shear stress. Because of shear stress reasons rotating parts have to be avoided, therefore diaphragm pumps are preferred. To dampen the pulsing characteristic of a diaphragm pump, typically 2 or 3 "pistons" are used.

Suppliers typically provide movable skids integrating the pump and a valve arrangement. To avoid turbulence and shear stress true bore valves such as ball and butterfly valves shall be used.

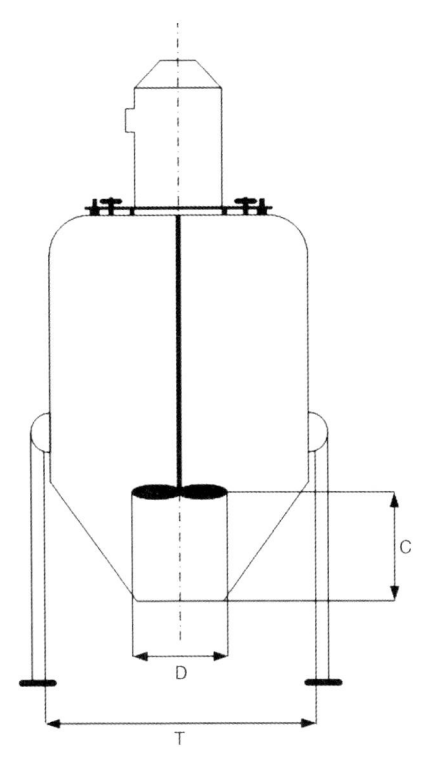

Figure 4.17 Slurry tank and stirrers for low shear stress.

4.4.8.3 Cranes and Transport Units

For maintenance the columns and/or top plates must be lifted. For this purpose hydraulic/pneumatic lifting pistons integrated in the column or other methods are applied. If several columns are in use a crane suitable for clean rooms can be used.

To move heavy columns electrically operated "Master Mover" can be used. Important is to check in advance the appropriate combination of the wheels (diameter and width) with the floor resistance (maximal peak pressure on floor). For fragile floors transport systems with an air cushion (hoover system for example from DELU-GmbH/Germany) can be used, but such systems require smooth floors without any cracks or gaps and a compressed air supply on the transport way.

4.4.8.4 Filter Integrity Test

In chromatography skids, liquid absolute filter, depth filter, and/or gas filter might be integrated. Absolute liquid and gas filter are typically integrity tested before and/or after use off-line. An *in situ* integrity testing, requiring additional connection points for the compressed air and water, is seldom designed for chromatography skids.

Hydrophobic gas filter can be tested with the water intrusion test (WIT). Hydrophilic liquid filter are usually tested with water, the often used test method is the pressure decay test (PDT) a variation of the forward flow test (FFT).

4.5
Process Control

As discussed in Chapters 6 and 7, the use of detailed process models in the design and in the choice of the operating parameters of chromatographic separation processes may lead to considerable improvements. However, optimal settings of the operating parameters do not guarantee an optimal operation of the real plant. This is due to nonperfect behavior of the column packing and the peripherals, the effect of external disturbances, changes of plant behavior over time, and to the inevitable discrepancy between the model and the real system.

Any real process is therefore operated using some sort of feedback control. Feedback control means that some variables (degree of freedom) of the plant are modified during the operation based upon observation of some measurable variables. These measurements may be available quasi-continuously or with a more or less large sampling period (results of a laboratory analysis), and accordingly the operation parameters (termed inputs in feedback control terminology) are modified in either a quasi-continuous fashion or intermittently. In chromatographic separations, feedback control is most often realized by operators, who change the operating parameters until the specifications are met.

Automatic feedback control is the continuous or repetitive modification of some operating parameters based on measurement data. Due to the dynamic nature of the process, the control algorithm is usually also dynamic and has to be designed carefully to avoid instability. We have to distinguish between two types of process control: standard and advanced one. Standard controllers have a static algorithm like PID-controller type or cascade mode for the temperature control. Advanced control schemes use a more sophisticated algorithm that could be based on short-cut models or dynamic process models.

4.5.1
Standard Process Control

Standard feedback laws are proportional (P), integral (I), and differentiating (I) control and combinations thereof, for example, PI control where the controlled input depends on the instantaneous control error and on its integral, guaranteeing steady-state accuracy and a more or less quick response to sudden disturbances.

Different standard PID controllers are used within chromatographic processes. Controllers are used for instance to adjust the flow rate, to temper properly the process, to control the air pressure, the sterilization temperature, the pressure before the column, and the level of the tank. In the latter case simpler schemes like a two-point controller can be adopted, which mainly close/open the tank bottom-valve based on two level switches placed at the top and bottom positions within the vessel.

Fine-tuning schemes exist to assist the process and control engineers in optimizing the PID parameters. Rules like the one proposed by Ziegler–Nichols (1942 and Microstar Laboratories Inc., 2009) can be used. They are based on some measurements of the dynamic system response. The Ziegler–Nichols rules allow furthermore the calculation of the PID parameter for different target criteria (classic rule, integral rule, some overshoot, no overshoot, etc.). Additionally, the majority of decentralized automation system vendors provide some sort of automatic fine-tuning software. Clear recommendations cannot be given as the authors have not collected reliable positive experience with such tools so far.

If the process dictates tighter process requirements, for instance following a precise gradient, then sophisticated and advanced process control schemes are required. Such schemes integrate in addition to classical PID-controller static and dynamic calculation and logic blocks. The so-called feed-forward controller integrates the dynamic trajectory of the manipulated variable to follow the target set point. Dynamic elements might be added to filter noise from measured variables or to add delays or ramp values to avoid abrupt changes in valve positions or pump flow rates. Figure 4.18 shows the advantage of an advanced process control scheme using the example of the control of an elution gradient. Figure 4.18a represents the change of conductivity controlled by a PID controller while in Figure 4.18b the conductivity is controlled by an advanced process control system using feed-forward controllers.

4.5.2
Advanced Process Control

Advanced process control schemes can be either smart combination of simple controllers, or nonlinear controllers using short-cut or even dynamic models. In both cases the design of such feedback loops is nontrivial when several actuated

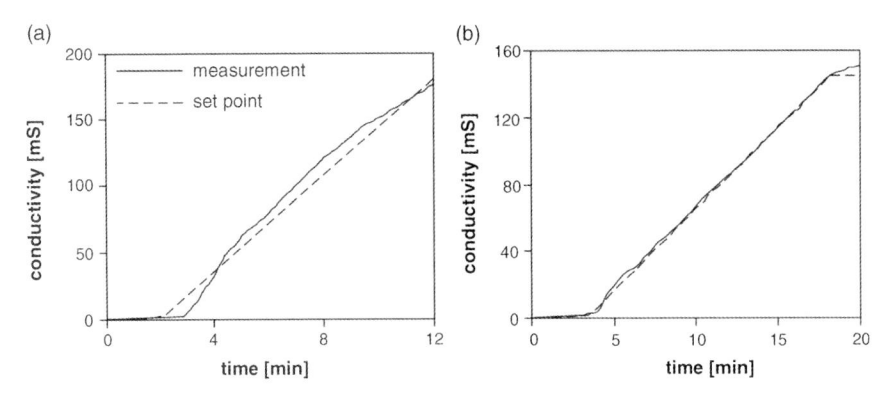

Figure 4.18 Linear gradient with and without feed-forward terms.

variables affect the parameters of interest in interacting fashion, and when the plant exhibits nonlinear behavior and/or complex dynamics. In the following, the concept of an advanced control system is explained using the example of an inline dilution system that has been integrated into a tangential flow filtration and chromatography systems of a large-scale biotech plant (Section 4.2 and Toumi et al., 2010). For further information on the principles of advanced process control see Section 7.8.

The large volume of buffers needed often for biotechnological and pharmaceutical applications led to the development of inline dilution systems. Such systems intend to produce the just in time solvent/buffers by mixing water with a concentrated solution, instead of storing the final very diluted solutions in tanks that require obviously a high storage capacity.

Storage costs can be so minimized as well as inventory and capital costs. By adjusting the amount of purified water to the concentrated buffer a large range of diluted buffer concentrations can be achieved. A better scalability and flexibility to process requirements is the result. Given the trend toward multiproduct facilities such systems become state-of-the art within modern biotechnology processing.

Another advantage pointed out by Matthews et al., 2009 is the opportunity to widely use disposable bags for smaller volumes (<2500 l) in place of traditional stainless steel or higher alloy tanks, further decreasing capital expenses. The use of disposables could also eliminate the need of cleaning and sterilization between two consecutive runs, reducing as well production costs. Disposable bags are also more corrosion-resistant to chloride-containing buffers.

The inline dilution system shall allow the preparation of buffers with an inline dilution factor ranging from 1 to 10. This factor determines the ratio of water to concentrated buffer. Larger ratios are of course feasible, but there is a trade-off between the process accuracy and flexibility to be taken into account. The systems shall cope as well with process disturbances and shall be robust again changing process conditions (such as pressure drop in a chromatography column or a sudden change in pressure or flow rate in the primary water supply). The robustness of such systems is crucial since pharmaceutical processes are validated to run in certain predefined limits. Any deviations to such limits will lead to additional investigation costs that could have severe consequences, in the extreme situation the withdrawal of the produced product.

Special attention has to be though paid to the robustness and overall performance of such systems. A common inline dilution system has been developed for the filtration and chromatography system. Additionally specific measures have been added to the chromatography systems that are more sensitive toward disturbances in the water loop.

Figure 4.19 presents an advanced control strategy for the inline dilution as implemented within the large-scale chromatography systems. The inline dilution controller takes into account the actual set point for conductivity SP_COND and maximal allowed pressure SP_PRESS of the chromatography system. The conductivity controller (1) corrects the inline dilution factor, that is, the ratio between the flow rates

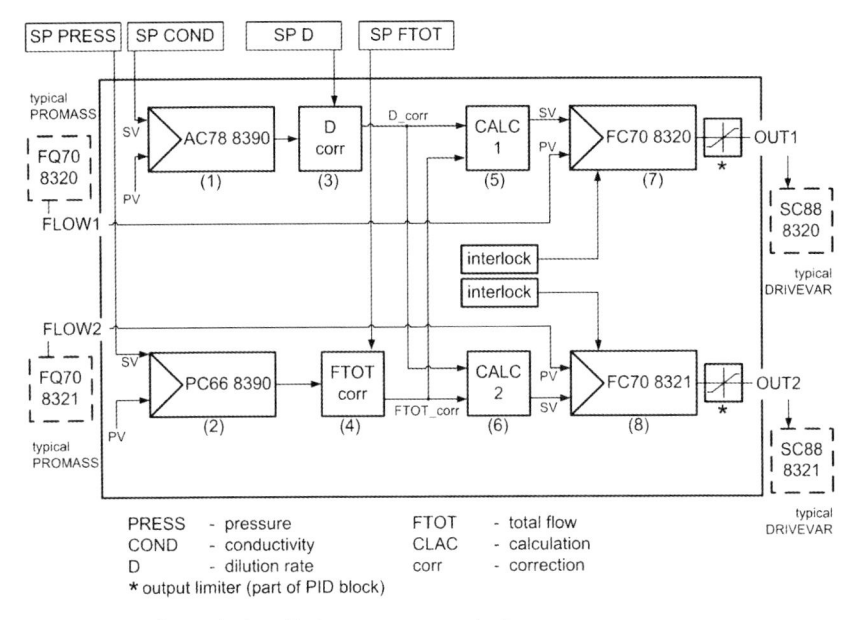

Figure 4.19 Advanced inline dilution process control scheme.

of both pumps, while the pressure controller (2) corrects the total flow rates of both pumps. The latter pressure controller is only active once if a certain threshold is reached and adapt the total flow rate incrementally according to the actual pressure. Based on the total flow rate and the ratio between both flow rates (buffer and feed) the individual flow rates of each can be calculated in the calculation blocks (5) and (6). Finally the two controllers (7) and (8) adapt the flow rates by manipulating the speed of the pumps.

To demonstrate the efficiency of this control scheme, different experimental scenarios for the adjustment of the elution gradient in a large-scale chromatography system have been investigated. The system was connected to a 1000 L HPW-tank (highly purified water) and a 200 l salt buffer tank (NaCl). The salt concentrations vary between 0.5 and 1 mol l^{-1}. This allowed experiments of 15–20 min each running at 60 l h^{-1} total flow and an initial dilution rate of 6.

To simulate the back pressure of a chromatography column, the column outlets were connected with a tube and a membrane valve in-between. The final buffer conductivity (after inline dilution) was measured precolumn using an Optek CF60 measurement cell.

This test was executed to show that the system is capable of keeping the final buffer conductivity constant at different total flow rates and system pressures (Figure 4.20). The system was run at a constant dilution rate of $D = 6$. The resulting conductivity was 28 mS cm^{-1}. Starting from 48 l min^{-1} the total flow rate was increased to 54 and 60 l min^{-1}. At 60 l min^{-1} the system pressure was increased to 1 bar using a membrane valve simulating the pressure drop of a chromatography column. Finally, the total flow was increased to 66 l min^{-1} and then reduced back to the starting value of 48 l min^{-1}.

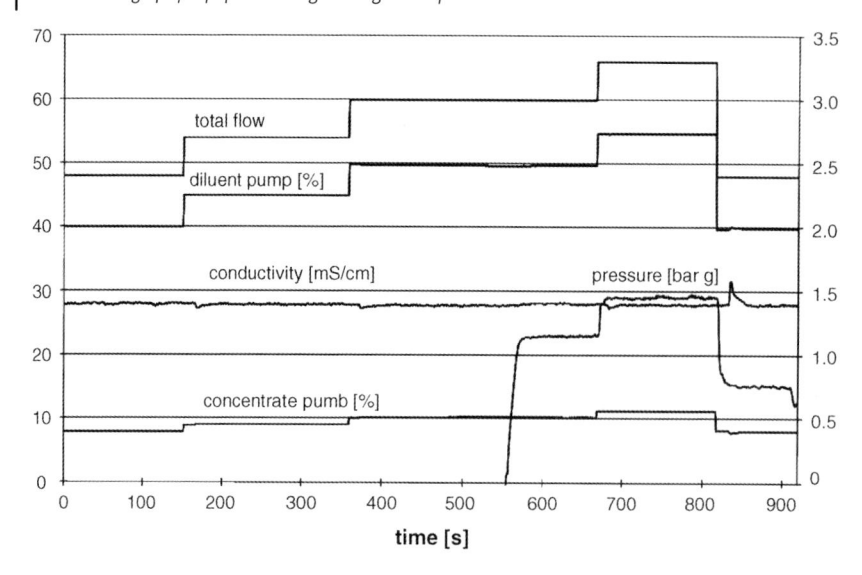

Figure 4.20 Stable conductivity during multiple changes of flow rate and system pressure.

In all cases the final buffer conductivity remained stable around 28 mS cm^{-1}. Only the abrupt flow reduction in the end of the experiment resulted in a short conductivity spike of approximately 4 mS cm^{-1}.

4.5.3
Detectors

For a process control system the accurate detection of the target components is mandatory. The detectors have to be selected and combined carefully in order to get signals as accurate as possible. The assembly of the different detectors is always very similar. The product stream flows continuously through the detector cell, where a physical or chemical property is measured online. The detected value is transformed into an electric signal and transferred to a PC or integrator to record the chromatogram.

The signal should be linear to the concentration of the target component over a wide range – as indicated in Figure 4.21 – because this simplifies the calibration of the detector and the signal represents the real shape of the peaks. The pure eluent signal is recorded as the baseline of the chromatogram. Fluctuations of the baseline are called the noise of the detection system. To qualitatively identify peaks, the signal-to-noise ratio (Figure 4.22) of the chromatogram should be higher than 3, and for quantitative analysis this ratio should be at least 10 (Meyer, 1999). The detection limit is defined as amount of a certain substance that gives a signal two times higher than the noise. A detector's dynamic range is the solute concentration range over which the detector will provide a concentration-dependent response. The minimum is bordered by the detection limit and the maximum is given at the

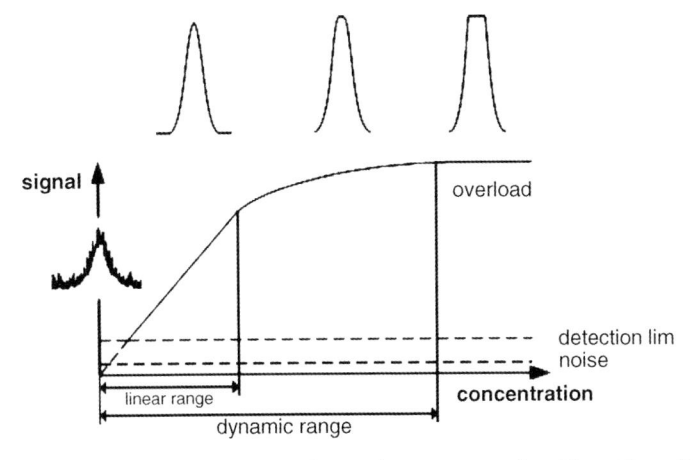

Figure 4.21 Detection regimes of HPLC detectors (reproduced from Meyer, 1999).

saturation concentration where the detector output fails to increase with an increase in concentration. The dynamic range is usually quoted in orders of magnitude (Scott, 1986).

An ideal detector for preparative chromatography has the following properties:

- good stability and reproducibility, even in the production environment;
- high reliability and ease of use in routine operation;
- sufficient sensitivity (need not to be too sensitive, as no trace detection is the goal in preparative chromatography, but the sensitivity should be high enough to keep the flow cell volume, which contributes to the axial dispersion, small);
- short reaction time, independent of flow rate;
- high linear and dynamic range.

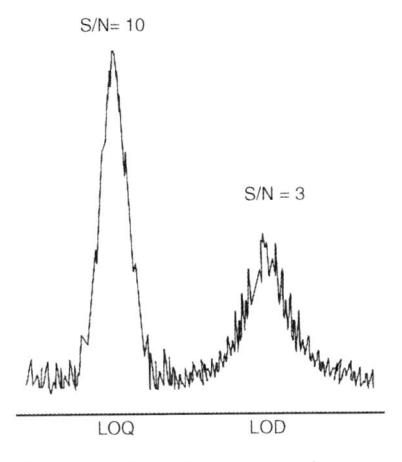

Figure 4.22 Signal-to-noise ratios for two peaks.

Table 4.9 Chromatographic detection systems and their detection principles.

Detection principle	Detection sensitivity (g)	Stability (sensitive to)	Dynamic range	Full flow or split flow (max flow rate in ml min^{-1})	Detection sensibility
Solvent sensitive					
Refractive index	10^{-6}	Temp shift, solvent composition	10^4	Full (100)	All analytes
Density	10^{-6}	Solvent composition	10^3	Full	All analytes
Conductivity	10^{-9}		10^6	Full	All ionic analytes
Analyte sensitive					
Light scattering ELSD	10^{-7}	Volatile analytes	10^3	Split (4.0)	All analytes
Electrochemical	10^{-12}	Temperature shift, flow rate, and mobile phase purity	10^8	Split	All analytes
UV/VIS (DAD)	10^{-9}	UV adsorption of solvents	10^5	Full (10 000)	Compound selective (DAD)
Fluorescence	10^{-12}	Fluorescent substances in eluent	10^3	Split	Compound selective
Mass spectrometer	10^{-10}	Inorganic salt in eluent	Wide	Split (0.1)	Compound selective
Polarimetry	10^{-9}	Temperature shift	10^3	Full (5000)	Compound selective (chirals)
Element selective (nitrogen, sulfur)	10^{-8}		10^2	Split (0.4)	Compound selective

Table 4.9 lists the different detectors and their operation principles that are used in (analytical as well as some of them in preparative) chromatography. The optimal detection system only detects the target components and all impurities but is not affected by the solvent. Detectors can be divided into two groups: (1) those that detect any change in composition of the solvent and can, therefore, obviously only be used under isocratic conditions; (2) analyte specific detectors – this group can be subdivided into detectors with a similar sensibility to all analytes and those that can detect certain components selectively. All the detection principles listed in Table 4.9 may be applied in preparative chromatography, but some of the detectors have limitations with regard to their applicable flow rate range.

If the maximum allowable flow rate of the detector is too low compared with the flow rate of the chosen column diameter, the total solvent stream has to be split into a detector stream and the eluent stream. The easiest splitting principle is the integration of a T-connector with a certain diameter for the two different streams. Integration of a connector into the eluent stream increases the system pressure drop and the back mixing of the system. This is a severe drawback of all passive splitting systems. In addition, care has to be taken that the time delay between detection and fractionation is carefully determined and taken into account for exact fractionation. The pressure drop problem is solved by active split principles. These are motor valves where, in short intervals, a portion of the eluent stream is captured and injected via an injection pump into the detector. The need for a second pump is one of the drawbacks of active splitting systems. In addition, the valve seal has to be carefully maintained and changed at regular intervals.

For further information see for instance Snyder, Kirkland, and Dolan (2010).

4.6
Packing Methods

4.6.1
Column and Packing Methodology Selection

The column design and the packing procedure need to be chosen in function of the characteristics of the media that need to be packed. The objective when packing a column is to obtain a homogeneous bed with a specific compression rate (or consolidation in case of rigid matrix as these are not compressible). The packing technology should be robust enough to reproduce consistently the same compression (consolidation) factor as this defines the amount of media for a given bed height required for the chromatographic separation. Semirigid media, for example crosslinked Agarose should be packed taking into consideration the pressure drop generated and the bed stability (Figure 4.23).

In case of capital investment in a multipurpose facility the option that offers a broader panel of packing methodologies should be considered.

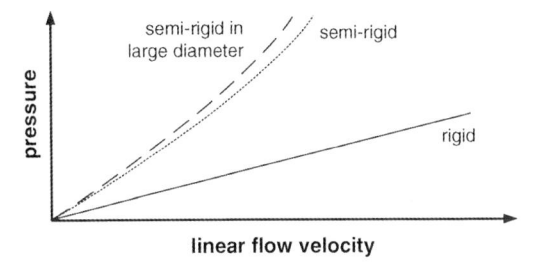

Figure 4.23 Pressure/flow curve according to chromatography media rigidity.

4.6.2
Slurry Preparation

The preparation of the slurry is a critical step to ensure a good packing. The critical parameters to be considered are as follows:

- **The quantity to be prepared**: depending on the packing method selected, an extra quantity has to be provided to take into account the dead volume of the packing system (hoses, packing skid) to avoid introducing air into the column. The other important parameter to consider is the compression factor.
- **Compression factor**: most of the resins are flexible and have to be compressed during packing; the ratio is provided by the supplier and has to be considered when calculating the volume of resin to prepare.

Settled chromatography resin volume = Packed bed volume × compression factor.

Table 4.10 gives some general remarks on the amount of packing material per ml respectively liter of column volume.

- **Media exchange**: the resin is delivered in a storage solution (e.g., 20% ethanol) that has to be replaced by the packing buffer. Several techniques can be used to change the media.
    - The exchange can be performed directly in the original container by removing the supernatant and replacing it by an equivalent volume of packing buffer. This operation has to be repeated several times (with a homogenization in between), until the resin is completely in the buffer wanted. All slurry containers can be then pooled in the slurry tank.
    - The second solution consists of pooling all the slurry contents needed from the different containers in the slurry tank with the storage solution. After the resin has settled, the supernatant is removed and replaced by the packing buffer. That operation has to be repeated several times (with a homogenization in between), until the resin is completely in desired buffer solution.
    - Dilution ratio (= slurry concentration): Slurry concentration refers to the ratio of settled bed volume to total slurry volume. To ensure a good packing performance, a defined appropriate dilution ratio between the slurry and the mobile phase has to be adjusted during the resin preparation. The information is usually provided by the supplier, and may range from 30 to 50% (v/v) depending on the packing technique or even up to 70% if hardware or construction restrictions exist.

Table 4.10 Amount of packing material per liter of bed volume.

Type of adsorbents	Adsorbent amount per liter bed volume
Silica, normal, and reversed phase	Approximately 0.4–0.6 kg L^{-1}
Polymeric adsorbents	Approximately 0.3–0.4 kg L^{-1}

Table 4.11 Suggestions for slurry solvents.

Normal phase	Methanol (vacuum), toluene–isopropanol (DAC), isopropanol (circle suspension flow – packing (CSF))
Reversed phase	Acetone, ethanolIsopropanol (CSF packing)
Polymeric adsorbents	Ethanol, isopropanol
Chiral stationary phases	Ethanol–isopropanol–n-heptane–isopropanol mixtures

The process of decanting the resin and removing the supernatant should also be taken into consideration especially for used resin. Chromatography media that have gone through the cleaning process of the API purification cycle and potentially through several packs can be partially modified in their particles' size distribution due to chemical and mechanical stress. This small debris generated is called "fines," which could partially block the frit and affect the chromatographic performances. A "defining" performed as described above for the buffer replacement is recommended prior to the use of the resin for purification processes. The only difference is the time required for settling that should be set according to the sedimentation speed of the resin, "fines" should not settle and will be removed siphoning the liquid on the top of the not fully sedimented slurry.

The cloudiness of the supernatant after (i.e., the concentration of fines) can be monitored measuring its UV absorbance after a referenced settling time. An absorbance value can be set as an acceptance criterion and crosschecked with a calibration curve to validate the resin status.

Table 4.11 gives some general suggestions for solvent buffers in preparative chromatography. Nevertheless, other solvents might be used that are either the eluent of the initial step or at least chemically close to that eluent to avoid other solvents and the resulting tedious conditioning steps.

For any column packing it is critical to know exactly the quantity of resin to be introduced in the column (corresponding to the final bed height). Table 4.12 lists the ratio of total slurry volume to packed bed volume for the different packing technologies. Slurry volumes might have to be adjusted to the given column geometry.

Once the slurry is ready at the right dilution ratio and homogenous, the resin quantity must be determined following procedure (i) or (ii):

Table 4.12 Slurry volumes for different packing technologies.

Packing technology	Volume slurry: volume packed bed
Dynamic axial compression	3:1
Vacuum packing	2:1
Circle suspension flow packing	9:1

i) remove a 1 liter sample in a graduated cylinder and allow the resin to settle overnight for percent slurry determination or;
ii) dispense the approximate amount of resin into the column. Re-suspend the resin to homogeneity and let settle overnight. Measure the settled bed height.

For transparent columns that do not have packing valves, it is easiest to let the resin settled overnight in the column. For all others, calculate the amount of settled resin using the total volume and percent slurry from the 1 liter graduated cylinder. If necessary, remove or add resin to achieve the desired packed bed height. Do not rely on the calibration marks on large containers as they are generally inaccurate.

Homogenize and wet the adsorbent intensively. For small slurry volumes and stable adsorbents this might be done by ultrasonification for 5 min. For larger slurry volumes this might be carried out in a mixing device or the suspension has to be stirred for at least 15 min with a paddle stirrer. Never use a magnetic stirrer in any slurry preparation and make sure that with a paddle stirrer no sediment is built up underneath the stirrer, as this might damage the adsorbent particles.

During ultrasonification, the adsorbent tends to form a sediment on the ground of the container, which has to be carefully rehomogenized. After this final homogenization the slurry should be immediately transferred to the prepared column and the packing procedure started.

4.6.3
Column Preparation

1) Clean the frits carefully and inspect them for obvious damage.
2) Mount, clean, and examine the column for proper sealing according to the manufacturer's operation manual.
3) Prior to packing, check and adjust the vertical orientation of the column using a spirit level. Move the piston to its maximum bed length position.
4) Flush the empty column with the slurry solvent prior to use to ensure that the column, column frits, and plumbing are clean and wetted. Empty the column before introducing the slurry suspension.

4.6.4
Flow Packing

The principle of flow packing is to introduce the resin slurry into the column and then to quickly apply a flow to pack the column. To be able to perform a good flow packing, the column tube should be large enough to receive all the slurry. The way to process flow packing may differ slightly because of different equipment. The main steps are as follows:

- Circulate buffer through the system (buffer vessel → pumping module → buffer vessel) in order to remove air.
- Adjust the position of the piston at the maximum height.

- Wet the bottom frit: with the packing module introduce buffer in upflow through the process valve (bottom) keeping the process top valve open.
- Introduce the slurry slowly down the inside of the column and rinse any resin particles from the inner wall of the column into the slurry and add fresh solvent to fill up the column. Insert the top adapter in the top position. Prevent air entrapment. Let the resin settle just a few (2–5) centimeters and lower the top frit (process top valve open) into the buffer (Figures 4.25a and 4.26a). Make sure that the column tube and entire length of the inlet tubing is full of liquid and free of air.
- An alternative to flow packing is circle suspension flow packing (Figure 4.24). Here the slurry suspension is introduced by lateral openings in the upper part of the otherwise closed column. The slurry suspension is pumped from a stirred tank into the column, which is filled with slurry solvent by means of a slurry pump and the column bed is built up over the outlet frit. The slurry solvent is recovered in the slurry suspension tank and circulated for a certain time. After introducing all adsorbents continue following the normal flow packing modes.
- Another good alternative is to start with the piston completely lowered and introducing the resin through the bottom packing valves. The piston is pressure free and is raised by the introduction of the resin. As the complete amount of resin has to be transferred, additional buffer has to be introduced in the slurry tank just before the end of the transfer, to avoid introducing air in the system. Let the resin settle just a few centimeters and lower the top frit (process top valve open) into the buffer, taking care to remove all the air.
- Place the end of the suction tube of the pump in a large container of packing solvent and the outlet tube to waste. Simultaneously open the column outlet (bottom process valve) turn on the pump, and pack the column at a starting flow rate of approximately $1/3$ of the final packing velocity (the packing flow rate has to be

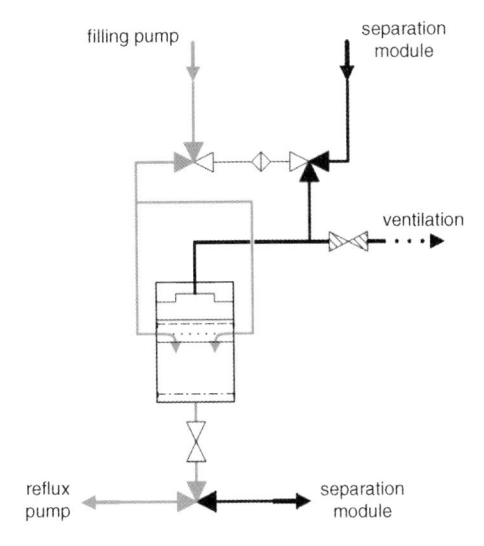

Figure 4.24 Circle suspension flow packing.

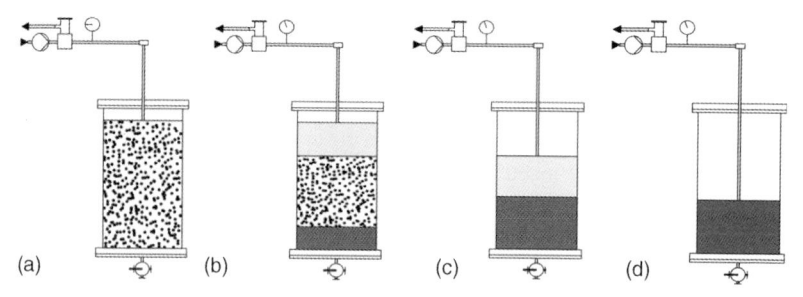

Figure 4.25 The principle of flow packing.

bigger than the process flow rate). Ramp up the speed until the packing flow rate is reached. This prevents hydraulic shock to the forming bed and therefore prevents uneven packing of the column. This operation should be performed with the chromatography skid. Some resins require starting with the packing flow rate, not ramping up the speed gradually. After flow has been started, the resin bed starts to form at the bottom, and the mobile phase begins to clear from the top (Figure 4.25b).
- Once the resin bed height is stable, stop the pump and quickly close the bottom process valve (Figures 4.25c and 4.26b).
- Open the process top valve and lower the top frit until the resin bed, then close the valve.
- Open the column outlet again and restart the pump to compress the bed further.
- Once the resin bed height is stable, stop the pump and, leaving the bottom valve open, lower the top frit to the desired (final) packed bed height (Figures 4.25d and 4.26c).

Figure 4.26 Flow packing of chromatography resins (46 cm i.d. Eastern Rivers column)

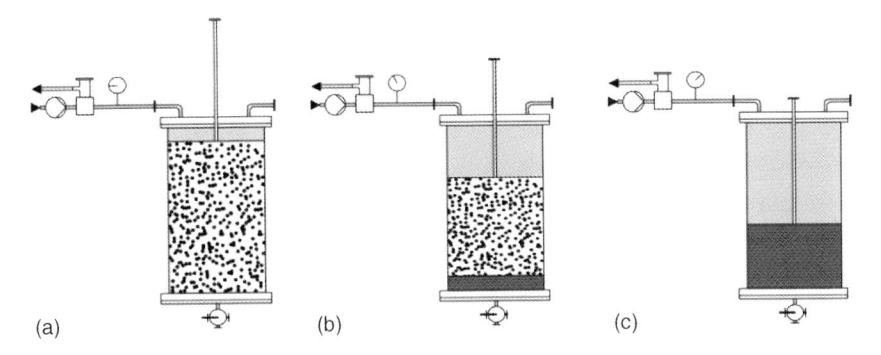

Figure 4.27 The principle of DAC packing.

4.6.5
Dynamic Axial Compression (DAC) Packing

The principle of the DAC packing is to transfer the slurry into the column and then pack using a hydraulic piston. To be able to perform a good DAC packing, the column volume should be sufficient to receive all the slurry or an extension tube should be attached to the column body (Figure 4.27).

- Transfer the slurry into the column as already described for flow packing (Figures 4.24 and 4.25a).
- After the slurry has been transferred into the column the piston is slowly lowered with the top valve open to remove air. When the expelled liquid is free of air bubbles, the top valve is closed (Figures 4.27a and 4.28a).
- Open the process bottom valve (column outlet) and lower quickly the piston (Figures 4.27b and 4.28b) until the final bed height (Figures 4.27c and 4.28c). Block then the piston or keep it in "down" position with the required pressure.

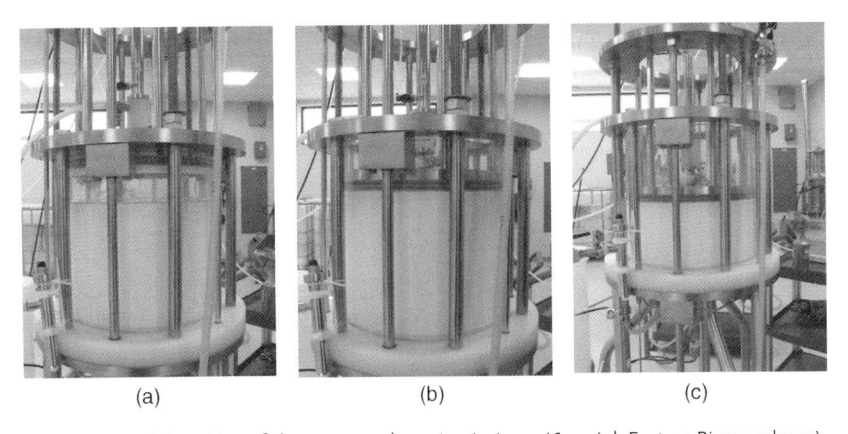

Figure 4.28 DAC packing of chromatography resins (using a 46 cm i.d. Eastern Rivers column).

4.6.6
Stall Packing

The principle of the stall packing method is to fix the final bed height of the column in advance, before introducing the media.

- Circulate buffer through the system (buffer vessel → pumping module → buffer vessel) to remove air.
- Wet the bottom frit: with the packing module introduce buffer upflow through the process valve (bottom) keeping the process top valve open, until the column is filled with a few centimeters.
- The column may or may not be filled with packing buffer before introducing the resin slurry. It is usually advantageous to have the column filled with buffer, as this will result in more homogenous slurry distribution. As shown in Figure 4.29 the bottom valve will be used for packing (introduction of the slurry) and is therefore switched to the "Pack" position. Alternatively, the top valve may be used for packing.
- Introduce the resin into the column by the bottom packing valves with the requested flow rate (top process valve open) controlling the pressure (Figure 4.29a); After the column is filled with slurry (Figure 4.29b), slurry is continued to be introduced into the column, with the bed starting to build up from the top (Figure 4.29c).
- Once the packing pressure is reached, decrease the flow rate in order to keep the packing pressure.
- Once no more resin can be introduced into the column, the pump must be stopped and the column's valves set to process mode ("RUN" position) (Figure 4.29d). Finally flow is started and the packed bed conditioned with buffer in downflow mode (Figure 4.29e).

4.6.7
Combined Method (Stall + DAC)

The principle of this method is to fix first the piston height around 5 cm above the final bed height, pack against the piston (stall), and finally fix the final bed with the DAC.

- Circulate buffer through the system (buffer vessel → pumping module → buffer vessel) to remove air.
- Adjust the position of the piston around 5 cm higher than the final bed height. Fill the column with buffer: packing buffer is sent upflow into the column in order to remove air of it. Excess of buffer is sent back to buffer vessel. This operation is performed through the process valves of the column.
- Pump packing buffer with the packing skid through the slurry valves to remove all the air.
- Column filling with slurry: open the bottom piston process valve and inject the slurry upflow. This direction is preferred if the hardware allows it otherwise,

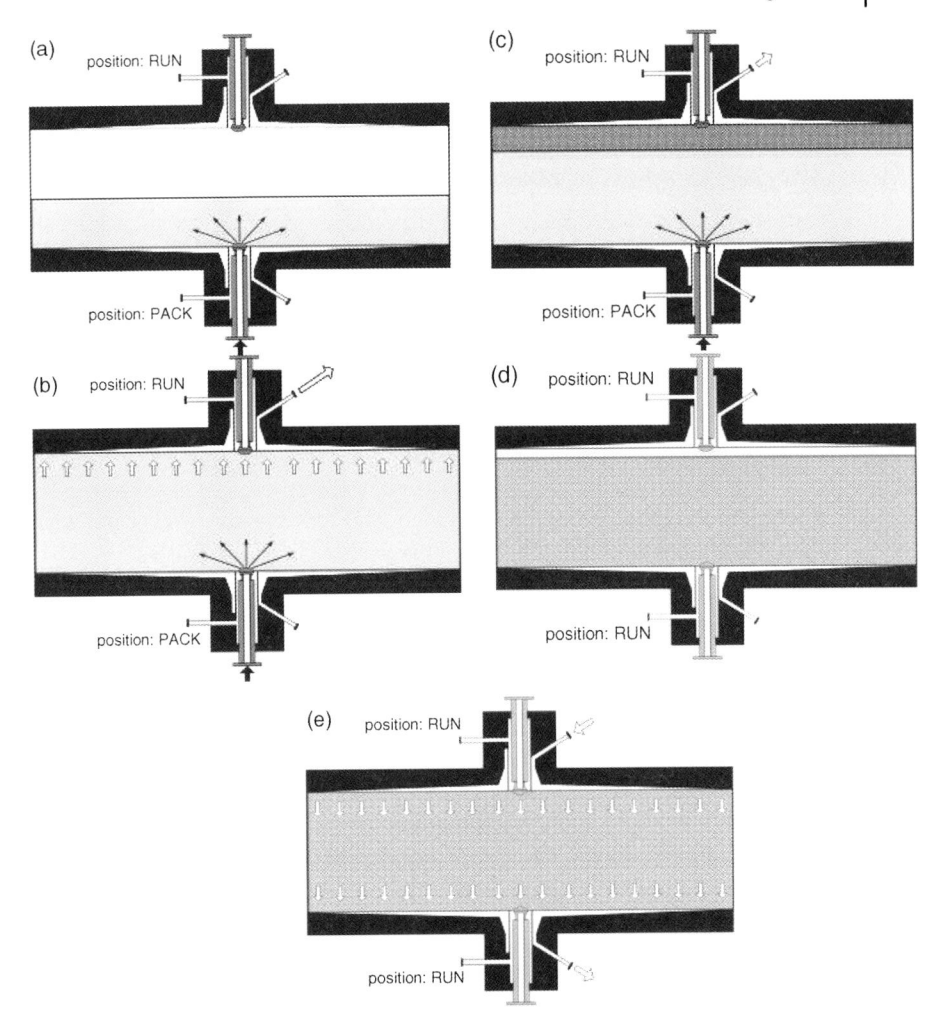

Figure 4.29 The principle of stall packing.

slurry can be introduced through the top valve in downflow (Figure 4.30a). Buffer is sent back to buffer tank. The flow of slurry is kept at the desired flow rate until the packing pressure is reached. Thereafter, the flow rate is reduced to avoid that the maximum packing pressure is exceeded.
- Once slurry introduction is almost finished (Figure 4.30b) and the slurry vessel is almost empty, the buffer is sent back to the slurry tank instead of the buffer tank. It is very important not to introduce air in the system at that step!
- Keep that operation running a few minutes in order to introduce all the resin in the column and stabilize the bed (Figure 4.30c).
- Inject buffer upflow through the packing valves in order to flush them. Buffer is sent back to the buffer tank.

Figure 4.30 Stall packing phase of the combined method (a 46 cm i.d. Eastern Rivers column).

- Bed height adjustment: close the packing valves but keep the process top valve open. Then move down the piston until the final bed height is reached.
- Close the top process valve and block the piston.

4.6.8
Vacuum Packing

The workflow for vacuum packing is as follows (Figure 4.31):

- Transfer the slurry to the column, equipped with the packing reservoir. *After* introducing the slurry apply the vacuum by connecting the column inlet line with the vacuum source through a safety bottle. The vacuum should be

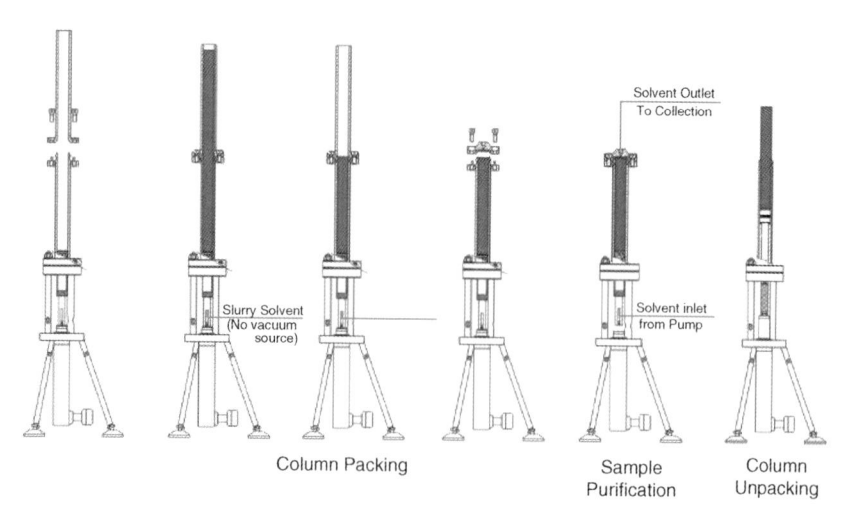

Figure 4.31 Vacuum packing.

approximately 100 mbar. Check the quality of the vacuum before introducing the slurry by closing the top of the packing reservoir.
- Maintain the vacuum constant over the suction time (5–45 min, depending on the particle size and the adsorbent quality). Immediately after the top of the column bed falls to dryness disconnect the vacuum line.
- Remove the packing reservoir, close the column with the top flange, and connect the outlet line to a solvent container. Compress the column with the hydraulic pump to the appropriate compression and secure the piston from moving back by means of the safety stop. The compression varies from 10 bar for polymeric materials up to 80 bar for very rigid silica adsorbents.
- Place the column in the system; do not connect it to any subsequent unit yet to avoid damaging it by particles that may come out of the column.
- Start your preparative system with a mobile phase of known viscosity, for example, methanol–water (80:20%) for RP systems, or isopropanol for normal phase systems. Use a linear flow rate at least as high as the flow rate later used in operation of the column. The pressure drop should be slightly higher than the operational pressure drop to be used. Let the system run for a short time.
- Stop the pump, recompress the column to the compression pressure, and readjust the safety stop.
- Repeat steps 5 and 6 up to three to four times. During the last repetition there should be no further movement of the safety stop when compressed.

4.6.9
Vibration Packing

Chromatography media for LPLC applications characterized by particles with rigid and irregular shape, like irregular silica or controlled pore glass (CPG), require a different packing approach. If a force is exerted on such a bed the incompressible particles form stable bridges that hinder the particles to settle uniformly and to a form stable packing.

At laboratory scale stable beds can be achieved easily just by filling the column and tap packing the column tube with a hard object. This will allow the particles to settle in a uniform bed. However, at large scale the most common packing methods suitable for soft or semirigid resins cannot be used for rigid and irregular matrixes as they may cause nonuniform compression, bed deformation, channeling, void formation, and crushing.

Vibration packing is the method developed specifically for this type of matrixes. A vibration device applied externally at the column tube generates the right amount of energy to break the bridges (similar to the tap packing at lab scale), allowing particles to settle properly and occupy the minimum bed volume in the column. Alternating vibration sequences and downflow through the bed is also applied to consolidate rigid particles and obtain homogeneous and stable beds without damaging the particles (Figure 4.32).

Figure 4.32 Vibration packing of rigid matrixes in a process column; the vibration device is clamped externally on the column flange.

4.6.10
Column Equilibration

Once the packing is finished, it is recommended to stabilize the bed. Monitoring the pressure/flow curve during the following steps allows to confirm the good packing.

- Connect the column to the chromatography skid (or other type of pumping module).
- Rinse the column with 10 BV of packing buffer downflow.
- Rinse the column with 10 BV of packing buffer upflow (caution: for some resins, reverse flow could damage the bed; in that case, skip that step).
- The column is ready for testing.

4.6.11
Column Testing and Storage

4.6.11.1 Test Systems

For analytical scale columns a simple test mixture consisting of a t_0-marker and different phthalic acid esters can be used to check the performance and some basic selectivities of packed columns. The advantage of phthalic acid esters as test substances is their liquid nature and good miscibility with different mobile phases. In addition they

Table 4.13 Test systems for normal phase and reversed phase columns.

Normal phase silica	Reversed phase silica
Mixture of 1 ml dimethyl phthalate, 1 ml diethyl phthalate, and 1 ml dibutyl phthalate	
Filled up with n-heptane–dioxane 90:10 to 100 ml	Filled up with MeOH–H_2O (80:20) to 100 ml
Addition of 50 µl toluene	Addition of 10 mg uracil
Mobile phase: n-heptane–dioxane (90:10)	Mobile phase: MeOH–H_2O (80:20)
Injection amount: 0.1 µl per ml V_c (40 µl for a 50 mm × 200 mm column)	

Note: If dimethylphtalate is substituted for dipropylphtalate, the dioxane (usage avoided in industry)-based eluent can be replaced by a mixture of n-heptane–ethyl acetate in the same 90–10 volume ratio.

can be used in straight phase as well as reversed phase systems. Table 4.13 gives the compositions of test mixtures for normal and reversed phase columns.

Figure 4.33 shows a typical chromatogram of the test mixture described in Table 4.13 on a $d_c = 50$ mm column packed with LiChrospher Si 60, 12 µm.

The diethyl phthalate peak (third peak of the test mixture in normal phase systems as well as reversed phase ones) can be used to calculate the number of theoretical plates (N) per meter of bed height according to Equation 2.27.

Production scale columns can be easily tested using a small quantity of reference standards of the products to be separated using the standard elution conditions. Also a good strategy (especially for very challenging separations) is to provide the supplier of the used stationary phase with these standards to evaluate a new batch of their product prior to shipment, to always assure a product fulfilling the requested purification task.

For LPLC production with large-scale packing the efficiency is usually checked by introducing a simple marker that is detected after elution with an appropriate detector. Most common markers are:

- 1 M NaCl detected by conductivity;
- acetone (around 2%) detected by UV (280 nm).

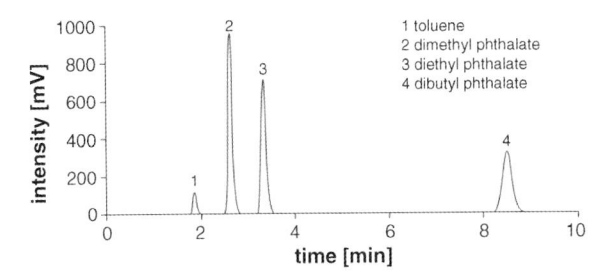

Figure 4.33 Test chromatogram of a t_0-marker and three phthalic acid esters.

The volume of the marker to be introduced on the column is usually 2.5% of the bed volume.

The chromatogram is then analyzed and the acceptance criteria are typically based on

- peak profile (a unique peak without shoulders);
- HETP (height equivalent to the theoretical plate);
- asymmetry of the peak.

The values expected for HETP or asymmetry may vary depending on the resin and the equipment used. Nevertheless it is important to reduce the dead volume between the inlet port of the tracer and the column itself to avoid dilution of the tracer.

4.6.11.2 Hydrodynamic Properties and Column Efficiency

For complete characterization of the column performance the efficiency has to be tested over the whole operation range of linear velocities. The injection of a test mixture at different flow rates can be easily automated with modern HPLC equipment and yields the pressure drop versus flow curve of the adsorbent as well as its HETP curve. By using marker substances with different molecular weights the influence of the mass transfer resistance (C term) can be investigated. Figure 4.34 shows the HETP curve for two similar adsorbents, which exhibit large differences at high flow rates.

Two test compounds with molecular weights of 110 and 11 000, respectively, were used to obtain the data. The minimum plate height at the lowest velocity was set to 100% and the percentage increase in plate height is plotted against the linear velocity. For both adsorbents, the plate height for the large molecule clearly shows a steeper increase than for the small molecule. At a linear velocity of approximately $8 \, \text{mm s}^{-1}$ the plate height for the large molecule increases dramatically for one of

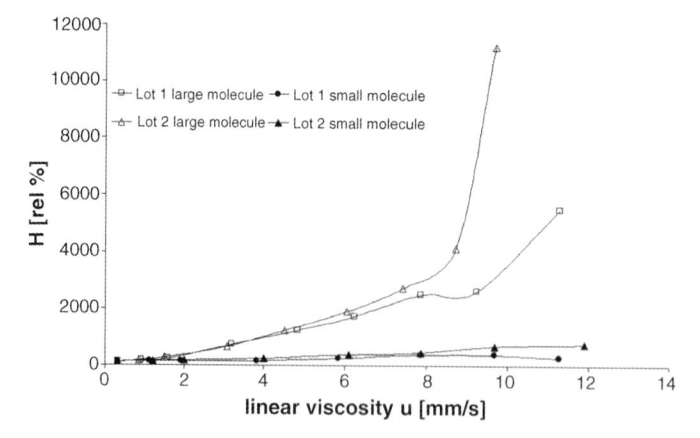

Figure 4.34 Plate height – linear velocity curve of two different adsorbents obtained with a low molecular weight (MW = 110) and a high molecular weight (MW = 11 000) marker.

Table 4.14 Recommended solvents for storage and flushing.

Adsorbent type	Flushing solvent	Storage solvent
Normal phase	Methanol or other polar solvent	Heptane + 10% polar compound (isopropanol, ethyl acetate)
Reversed phase	Methanol or acetonitrile, THF, or dichloromethane for oily samples	>50% organic in water
IEX (silica based)	Methanol	>50% organic in water

the adsorbents. This gives a hint of some obstacles within the pore system of that adsorbent, which are of course more severe for larger molecules. Even if the preparative process is not operated at such high linear velocities, the adsorbent properties should be carefully examined beforehand, to avoid any unnecessary extra contribution to the axial dispersion.

4.6.11.3 Column and Adsorbent Storage

If packed columns or adsorbents removed from columns are to be stored for a period of time the adsorbent should be carefully washed and then equilibrated with a storage solvent. Compounds strongly adsorbed to the adsorbent and, especially, any source of microbial growth has to be carefully removed before the adsorbent is stored. In addition, any buffer substances or other additives, which might precipitate during storage, have to be washed out. For flushing and storage different solvents are recommend (Table 4.14).

4.7
Process Troubleshooting

A chromatography equipment is quite complex and many technical failures may happen when operating a preparative device. Most of the time, a technical problem is directly visible or even detected by the system if this one is automated.

However, several technical problems are not trivial to identify. It is impossible to give an exhaustive list of the problems that may occur, but the attempt of this section is to give advice to troubleshoot the most frequent problems observed on a preparative chromatographic process (for additional information and discussion on troubleshooting see monthly columns of Dolan *et al.* in LC-GC Chromatographyonline.com or Dolan and Snyder, 1989).

This section is divided into three parts: the first one will describe the main technical failures that occur in the process, give the way to diagnose the failure, and the way to troubleshoot it. In a second part, we will examine the loss of performances that can be observed on a system, list the most probable root causes, and the way to avoid or correct it. In the third part, we will deal with the stability of the column and the way to monitor it.

4.7.1
Technical Failures

Table 4.15 describes technical failures that may happen in chromatographic systems.

Table 4.15 Possible technical failures in a chromatography system.

Equipment part	Problem	Diagnosis	Corrective action
Gradient valve	Leakage	Wrong gradient composition– variation retention	Check tightness and repair/replace valve
Injection pump	Wrong flow rate	Wrong injected volume– modification peak shape and height	Check flow rate, Clean/replace check valve Check membrane integrity Check volume hydraulic oil
Eluent pump	Wrong flowrate	When flow rate is controlled by a flow meter, an increase of the pump frequency is observed	Check frequency compared to expected value
	Pressure fluctuation		Clean/replace check valve Check membrane integrity Check volume hydraulic oil
Column	Pressure increase Loss of efficiency Variation retention Loss of purity/yield	See next section for diagnosis and corrective action	
Detector	No signal Smaller signal Noise increase	Check UV light intensity Check solvent degassing (noise)	Change UV light Check bubble trap/degas solvent
Collecting valve	Leakage of one valve	Loss of purity/yield	Check tightness and repair/replace valve

4.7.2
Loss of Performance

During the operation of a chromatographic process, a loss of performance may occur. It is important to track the process drifts that may announce a more serious issue. Sometimes, the problem can be quickly corrected in order to prevent the loss of the purified products.

The most common issues encountered when operating a chromatographic device are:

- an increase in the pressure measured after the eluent pump;
- a progressive loss of peak resolution (due to loss of efficiency);
- a variation of the elution profile (change of retention);
- a loss of purity/yield of the purified products without any significant change in the preparative chromatogram.

The most common root causes for these problems are examined to define the way to troubleshoot it.

The lifetime of a chromatographic column strongly depends on the application and on basic precautions to protect it. Some applications require to repack the column with fresh packing material after each injection (e.g., purification of specific natural substances) whereas several large-scale HPLC equipments are used continuously during few years without any significant loss of performances (e.g., chiral separations).

4.7.2.1 Pressure Increase

Most of the time, the pressure is measured after the eluent pump. On well-designed HPLC equipment, the operating pressure is mainly due to the pressure drop in the column.

A pressure increase can also be observed when inline filtration is implemented. Appropriate positioning of the pressure sensors is required to measure the pressure drop on the filters. When the pressure drop is too large, the filters need to be replaced or cleaned. If the frequency is too high, prefiltration of the crude can be implemented or should be revised.

When the pressure increase is linked to the column, several root causes are possible:

- **Insolubles in the crude or in the eluent:** a chromatographic column contains a bed of particles, which constitutes an excellent filter media. Most of the micron-size particles pumped through the column will be systematically retained on an HPLC system packed with 10–20 µm particles. It is therefore strongly recommended to filter the inlet streams of the process. Prefiltration and/or online filtration of the crude solution are strongly recommended. It is also recommended to filter the eluent used for the separation. Even if the insoluble content is low, the solvent volume pumped through the column is large and could rapidly lead to an accumulation of these particles in the column and to a pressure increase. Most of the time, this kind of problem is difficult to detect; the pressure increase starts slowly but the evolution of the pressure in a contaminated column is exponential. A column backflush can potentially remove the particles accumulated on the inlet frit.

- **Strongly retained compounds – gelation at the inlet of the column:** an increase in the column pressure drop can also be linked to the retention of strongly retained compounds. When the injected crude mixture contains impurities with a strong affinity to the stationary phase, a progressive accumulation of these compounds happens at the inlet of the column and could lead to a gelation or precipitation of these accumulated products.

This will progressively limit the flow and increase the pressure in the column.
In order to avoid it, several solutions exist:

1) **Washing:** Column wash with the appropriate eluent in order to desorb the strongly retained impurities. This can be performed between the injections or at an appropriate periodic interval.
2) **Online precolumn:** this is a good way to preserve the stationary phase of the column. This precolumn can be washed and/or repacked, keeping the integrity of the column. However, the type of the precolumn can also impact the efficiency and therefore reduce the achieved resolution.
3) **Pretreatment of the crude:** an alternative to the preceding solution is to prepurify the crude solution on another column filled with a similar material or a simple filtration aid as, for example, diatomeous earth or activated carbon to remove strongly retained impurities. If eventually chromatographic conditions for this pretreatment should be developed to remove the critical impurities, most of the time, a two-step approach is more favorable than a single step. This could also help to work under isocratic conditions and to get favorable process conditions to recycle solvent.

- **Apolar solvents and polar impurities:** When a very apolar solvent is used as the eluent, a column can get completely blocked within a very short period of time by the introduction of small traces of a viscous polar substance (e.g., glycerol, dipropylene glycol), which are not soluble in the used solvent system. This might happen if for instance the filling equipment used to fill the drums has been insufficiently cleaned by the solvent supplier. Therefore, a thorough analysis of the incoming solvent batch is absolutely necessary.
- **Fines in the stationary phase:** According to Darcy's law (Section 2.3.5) describing the pressure drop through a bed of monodispersed particles, the pressure is inversely proportional to the square of the particle size.

The situation changes dramatically when the stationary phase is polydispersed and contains small particles. These fines will fill the empty space between the beads and will therefore significantly increase the pressure drop through the particle bed. The smallest particles will also migrate through the bed and get accumulated on the outlet frit, which will be rapidly clogged.

Most of the modern stationary phases offer a good control of the particle size dispersity, but fines can appear under stressed conditions: frictions of particles, inappropriate packing conditions, too high hydraulic pressure on a DAC column, repetitive packing/unpacking of the same stationary phase, and so on.

Irregular stationary phases are generally more sensitive to mechanical attrition than spherical material.

The following advice should be followed to avoid an unexpected increase of the pressure drop during operation:

1) **Visual check of the stationary phase before packing:** A small sample can be used for a quick lab test. A sedimentation trial can be performed in the lab with less than a gram of stationary phase. The solvent should be selected so that the particles can sediment in a few minutes. Methanol is often an appropriate candidate for this test.

    The stationary phase should sediment and the turbidity of the supernatant clearly shows if fines are present in the sample. A reference trial can be made with a new batch of stationary phase to evaluate the aging of a reused batch.

    When the supernatant is turbid, fines are present that will potentially lead to a too high pressure drop on the column. It is advised either to change the stationary phase or to regenerate it. A mechanical sedimentation process can be applied to remove fines (repetition of 5–10 successive sedimentations and removal of the supernatant to eliminate fines).

2) **Check of the packing procedure:** The packing procedure could potentially create a mechanical stress to the stationary phase and break the silica beads, leading to fines in the column. This mechanical stress could be linked to the slurry preparation (inappropriate mixing device, transfer pump), to the hydraulic pressure, and so on. . . . Most of the packing suppliers assist their customers with detailed protocols for an efficient column packing.

3) **Check of the hydraulic system:** A too low hydraulic pressure compared to the column operating pressure could potentially damage the stationary phase. The hydraulic pressure should always stay higher than the operating pressure to ensure the best stability of the stationary phase. When the hydraulic pressure gets closer to the operating pressure, the piston could have small displacements and stress the particles.

4) **Check pressure pulsation of the elution module:** Depending on pumps' technologies, pulsations of a single or multiple head pump could be quite important and some industrial systems are equipped with a pulse dampener to smooth the flow variations. A lack of dampening could lead to a mechanical stress on the phase particles (small displacement of piston, mechanical stress on fragile packings). It has to be empathized that with multiple heads pumps pulse dampener is not necessary on HPLC units with traditional preparative packings.

5) **Check the hydraulic pressure:** Each packing has its own mechanical resistance. For example, a 100 A silica material resists to a mechanical pressure of

more than 100 bars, whereas wide pore silica (for example 3000 A) has a more limited mechanical resistance and could potentially break with a hydraulic pressure of 30–40 bars. The hydraulic pressure of a DAC column must be selected according to the stationary phase specifications.

A column back flush could temporary reduce the pressure drop by removing the fines accumulated in the outlet frit. However, a back flush will not solve the problem and the pressure will rapidly increase again. The only way to come back to an acceptable pressure is to unpack the column and to change the stationary phase or to remove the fines.

- **Cleaning of the frits:** The column frits are often reused when repacking a large-scale column. However, small particles (fines or insoluble particles) are probably present in the frit. De-packing the column also generates fines that would stay trapped in the frits. This would potentially lead to an increased pressure drop on the column after repacking. It is therefore strongly advised to systematically clean the frits or replace the frits when repacking a chromatographic column.

  Two main ways can be used to clean the frits:
  1) **Chemical cleaning:** Concentrated caustic soda is often used to dissolve silica.
  2) **Mechanical cleaning:** Mechanical cleaning used to clean the frit. A high pressure water system is often the most efficient way to clean blocked frits. This can be combined with a chemical cleaning to get rid of all residues in the frits (Section 4.3.3.2).

4.7.2.2 Loss of Column Efficiency

The column efficiency characterizes the dispersion of the product in the chromatographic column. A loss of efficiency will directly impact on the peak resolution and therefore on the achieved purity and/or yield for the purified product(s).

Several physical causes could disturb the elution stream through the column, leading to a loss of efficiency.

- **Clogged frits:** Accumulated insoluble particles from the crude or fines from the stationary phase can progressively block the frits and disturb the flow distribution in the column. This phenomenon is often associated with an increased pressure drop on the column, as explained in Section 4.7.2.1. A lack of efficiency can also be observed on a newly packed column when the frits are not appropriately cleaned before used.
- **Fine particles in the stationary phase:** Fines will progressively migrate through the bed and block the outlet frit, leading to a disturbance of the flow within the column. The column efficiency can be good right after the packing, but could progressively become worse and the pressure drop could increase due to the migration of the fines. A column backflush could improve the situation, but for a very limited duration. The only way to come back to a stable and efficiency column is to unpack it and to remove fines.
- **Heterogeneous packing:** Packing a preparative HPLC column is a tactful operation. Inappropriate conditions (e.g., slurry concentration, slurry solvent, speed) could lead to a heterogeneous packing with unordered particles, leading to

preferential paths through the bed. This will disturb the flow and will strongly impact the column efficiency. A backflush could sometimes help, but most of the time, the column needs to be repacked.

- **Temperature:** This criterion is often underconsidered in the operation of a preparative chromatographic process. Temperature has a direct impact on the adsorption isotherms and modifies the retention of the separated compounds. A significant loss of efficiency can be observed with a radial heterogeneity of the temperature, which will modify the retention of the products. This heterogeneity is linked to a difference between the temperature of the eluent and the temperature of the column body. This is particularly important to consider on small-scale preparative devices (5–10 cm ID columns), the wall effect is less significant for larger columns.

    A favorable effect on the column efficiency can be achieved when the column body is slightly warmer than the eluent (1–2 °C) (Dapremont et al., 1998) to correct the bed heterogeneity near the column wall.

- **Viscous Fingering effect:** This phenomenon is observed when the viscosity of the injected sample is significantly larger than the eluent viscosity. A hydrodynamic instability appears within the column, leading to a fingering elution of the viscous solution in the bed. This disturbance of the flow will strongly impact the obtained efficiency. A reduction of the injected concentration is efficient to correct this effect, which can appear during a process scale-up.

4.7.2.3 Variation of Elution Profile

This section mainly focuses on the parameters that could lead to an evolution of the elution profile during a purification campaign. Several root causes are listed, which could impact the elution profile and modify the retention or disturb the peak shape.

- **Variation of temperature:** Temperature has a major impact on the retention and could potentially impact the column efficiency, as described in the preceding section.

    This should be taken into account to ensure the reproducibility of the chromatographic process and a temperature control with an online heat exchanger is preferred to get a reproducible process.

    A variation of the elution profile could potentially occur when the temperature is not controlled and when a fresh solvent is loaded in the eluent tank used to feed the system. The stored solvent could potentially be cooler or warmer, leading to a disturbance of the elution profile.

- **Solvent composition of the eluent:** Most of the chromatographic separation uses a binary solvent mixture for the separation and sometimes a buffer solution. A wrong solvent preparation could lead to a wrong composition and therefore modify the retention of the products. A technical failure of the gradient module could also disturb retention.

- **Solvent contamination:** Contamination of the solvent used for the eluent preparation could disturb the separation. This could for example happen with a recycled solvent: this is particularly critical when the crude contains volatile

impurities, which could be evaporated together with the solvent and progressively concentrated in the recycled eluent.
- **Solvent grade:** A modification of the solvent grade could potentially impact the obtained separation. For example, the variation of the water content of the organic solvent used under normal phase conditions could modify the activity of silica. It is therefore suggested to keep the same solvent grade between the development and the production stage.
- **Solvent composition to dissolve crude:** The crude product needs to be dissolved before injection. The solvent composition used to dissolve the crude should theoretically be identical to the eluent composition used after injection. Sometimes, another solvent composition is used to get a better solubility of the crude. This solvent could potentially have a higher eluent strength and could therefore disturb the retention of the products.
- **Remaining impurities/solvent in the crude:** The crude could potentially contain remaining traces of solvent or impurities from the preceding process step, which could disturb the chromatographic separation. A variation of the solvent content in the crude could lead to a variability of the elution profile, which could be strongly disturbed.
- **Loss of capacity of the stationary phase:** Impurities could be strongly retained on the stationary phase, leading to a progressive loss of loading capacity. This will progressively lead to a loss of retention and of resolution. Regeneration of the stationary phase with a strong desorbent could potentially help to recover the initial properties.
- **Problem pumping module:** A pumping issue could potentially disturb the obtained elution profile. Stability of the obtained flow rate should be controlled if an online flow meter is available.
- **Valve leakage:** A leak of one valve, for example in the gradient module, could disturb the elution profile.

4.7.2.4 Loss of Purity/Yield

The last (and probably most difficult) situation described is when a loss of purity and/or yield occurs without any modification of the preparative chromatogram.

- **Modification impurity profile crude:** The impurity content may change, leading to a more difficult separation. The injected crude mixture may also contain new impurities, which might be coeluted under the applied preparative conditions. An LC-MS analysis using analytical method based on the preparative conditions (same stationary phase, same eluent, but small injection amount) could be used to identify new coeluted impurities.
- **Valve leakage:** A leak of one valve of the collecting module could potentially lead to a cross-contamination between the collected fractions. This could be checked by changing the collecting ports.
- **Design of the collector:** An inappropriate design of the collecting module can lead to a cross-contamination between the collected fractions due to the dead volumes of the common pipe. Changing the order of the collecting port could potentially reduce the impact of the cross-contamination.

4.7.3
Column Stability

Any resin, once packed, has a "shelf life" for a determined process. The number of cycles a packed chromatography column can be used has to be validated in order to guarantee specified yields and product quality.

Nevertheless, some unexpected events may occur during process, influencing the resin integrity or the stability of the chromatography bed:

- introduction of air into the column;
- overpressure inside the column;
- use of an inadequate solution;
- use of a correct sanitizing solution but during a too long period;
- defect of the inflatable gasket of the column;
- move of the piston that is not stabilized or fixed correctly.

Only referring to the validated number of cycles is of course not enough to guarantee a reproducible performance of the column. Additional measures can be foreseen in order to identify quickly any problem:

- regular HETP and asymmetry test (for example after each production batch or any other period);
- comparison of each elution peak with a reference chromatogram;
- calculation of the column performance by integrating the chromatogram profile.

Different symptoms allow to suspect a problem with the resin integrity or the bed stability:

- decrease of the yield from batch to batch;
- increase of impurity rate from batch to batch;
- several peaks or shoulder during elution (comparison to reference peak);
- increase of the back pressure of the column;
- tailing or fronting increasing from batch to batch;
- shifting of the retention time.

Note that these symptoms can also have other causes, like defaults on the batch loaded, and are not necessarily linked to the column itself.

4.8
Disposable Technology for Bioseparations

4.8.1
Market Trend

Pharmaceutical companies are continually striving to improve and optimize the process of manufacturing biotherapeutics in order to increase productivity, reduce capital investments, reduce operational downtime, and minimize risk of process failure due to cross-contamination. A lot of money, time, and resources are spent on equipment cleaning and sterilization – steps that require large volumes of chemicals. Most bioprocess chromatography equipment is made of stainless steel and requires cleaning-in-place (CIP) or steaming-in-place (SIP) operations, all of which result in manufacturing downtime.

Prepacked column and membrane chromatography are nowadays available to solve most of these issues and are rapidly gaining market adoption; disposable and single-use technologies are the fastest growing segment of the chromatography market with a CAGR of >15%. In the next session, we provide an overview of these new formats/trends regarding purification tools.

4.8.2
Prepacked Columns

Prepacked columns for pilot and large-scale application are the scale-up version of the small scouting bench-scale devices (Figure 4.36). These larger scale columns, packing with between 10 and 30 l of resin, are designed for easy installation and for use in any clinical or commercial manufacturing setting. These columns are packed, tested, and released with specifications for HETP, asymmetry, and pressure drop, all materials in the fluid flow are made from USP Class VI compliant raw materials and have used no animal-derived components in their processing. Columns are first packed under GMP conditions at the vendor manufacturing site and after quality control release each column is tested for specific criteria such as HETP and asymmetry. These columns are manufactured from suppliers presanitized in a similar manner as bulk media so the column is immediately useable for all applications. These columns are ideal in any application where reproducible column packing is a critical success parameter and where reduced process time and frequent batch switch are of paramount importance (Table 4.16).

There are nevertheless some limitations for the implementation of prepacked column as the limited range of diameters and bed heights that may limit the scouting options. This may require multiple cycles on the prepacked column resulting in greater buffer and time consumption than a traditional column set-up. These are typically offset by the fact that customers do not need to pack and validate the column packing.

Clinical batches are considered to be an ideal application for disposable columns since the batch sizes required at this stage of development are particularly suited to the volumes of resins provided in disposable columns. Because disposable columns can be reused and stored between campaigns the columns require shelf life data to confirm stability and no leachability from the storage buffers. Additionally, the customers will need to validate and determine what their policies are regarding columns that fail qualification tests such as HETP and asymmetry after they have been used and stored in between campaigns.

Another important aspect is the GMP requirement of quality checking the resin prior to any clinical or manufacturing batch. Resin needs to analyzed by QA; hence

Table 4.16 Advantages and limitations of using prepacked columns.

Cost savings
Consumables for packing, testing, and qualifying columns are no longer required
Time savings
Column packing and testing are no longer necessary
Space savings
Simple, robust design enables "on-a-shelf" storage of used columns
Reproducibility
Improved column-to-column reproducibility
Greater range of resins available for process development
Reduces need to learn multiple protocols for packing different base matrixes
Limitations
Not customizable
Restricted to the sizes offered for the prepacked columns
Bed stability/usage over time
This could increase costs if low cylces are used unless the column can be repacked

a sample of the resin that was from the same lot as the disposable column should be supplied by the manufacturer and be available for testing.

4.8.3
Membrane Chromatography

An alternative disposable format for chromatography that is seeing increased adoption is single-use or disposable membranes. Chromatography membranes are typically much more efficient for chromatography and flow rates than typical chromatography resins. Most membranes can in flow through mode process several kilograms of mAb per liter of membrane as compared to only <0.1 kg of mAb per liter of traditional resins.

The advantages of membranes for the biopharmaceutical manufacturer are savings in cost and volumes of WFI, buffers and chemicals costs since single use membranes do not require cleaning to, regenerate the packed columns between cycles/campaigns and the validation and testing of the cleaning process every time. Additional advantages of membranes are time saving since membranes do not require dedicated resources to pack/unpack resins and qualify traditional column chromatography. This greatly reduces operating time for single use membranes. Finally, due to the high throughput of membranes large diameter stainless steel columns weighing several tons can be replaced by a capsule of few liters, which means reduced plant space utilization and increased manufacturing process flexibility.

Due to these early advantages, chromatography membranes were launched in the early 1990s. Unfortunately due to their limited capacity versus chromatography media as well as their lack of capacity at high salt they did not achieve significant market penetration versus traditional chromatography media (<10%).

Figure 4.35 Buffer tanks and plant space requirements in a typical biotech facility.

Moreover, the reduced salt tolerance of these membranes compared to typical AIEX and CIEX agarose resins requires at a minimum twofold dilution doubling the volume to be processed. This puts an even greater burden at a critical bottleneck for drug manufacturers since additional buffer tanks are required in the limited floor space of the plant negating the advantage of membrane technology. As discussed earlier, a major component of inflexibility in plant design is efficient utilization of floor space. Membrane technology substituted column footprint but at the expense of increasing the buffer tank footprint required to operate them. (Figure 4.35).

Other major limitations are related to scaling issues due to distorted or poor inlet flow distribution; nonidentical membrane pore size distribution, and uneven membrane thickness (Zhou and Tressel, 2006).

Recent market forces have pressured biotherapeutic manufacturers to increase their capacity and flexibility; in essence, to generate larger batches of drugs and using the same facility for different campaigns of products. Plants that were already at close to maximum capacity are now being tasked with operating under these conditions and upstream manufacturing has uncovered bottlenecks and limitations with traditional chromatography resins. Increasing adoption of personalized medicine will require manufacturers to increase the frequency of batch switching and campaigns. These problems are driving the manufacturers to seek more flexible solutions in the manufacturing process. These market reasons have caused suppliers to invest additional capital in the research of more performing and scalable single-use technology and a new generation of membrane chromatography has been launched during the past several years (Figure 4.36).

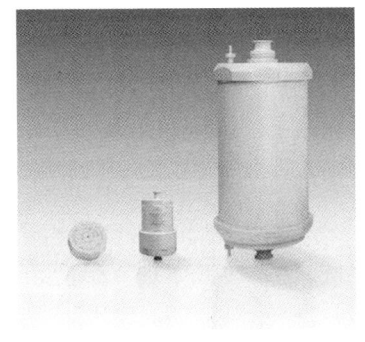

Figure 4.36 Examples of most common disposable membrane devices available on the market – Suppliers from left to right: Merck-Millipore, Satorius, Pall.

These new membranes have increased affinity toward impurities and novel membrane chemistry has increased the capacity for impurities under high salt conditions. This new membrane technology offers manufacturers enhanced flexibility, reduced processing time, allows smaller and more flexible production facilities, and reduces operating costs. New device formats have improved scalability with the membranes designed as multiple stacked flat sheets versus the traditional pleated or wrapped membranes.

4.8.4
Membrane Technology

Membranes are designed with same base concept of resins that is, a support structure made of casted polymer (cellulose, hydrophilic polyethersulfone polyethylene, etc.) in a form of sheet rather than particles. A critical difference between the two technologies is related to the flow characteristics; chromatography beads have diffusive flow with mass transfer limited by pore size and polymer structure.

Membranes have more open geometry allowing convective flow and faster diffusion. Membranes behave exactly as the internal surface area of beads accessed by pores. The internal surfaces of membranes are coated with affinity chemistries, "Q" groups (trimethyl ammonium quaternary salts or primary amine for anionic exchange), and "S" groups (sulfopropyl salts for ionic exchange) or alkyl group in case of HIC. These coatings create a high surface area for binding, and as liquid flows through these channels the active affinity molecules are immediately available for reaction. This phenomenon is enhanced especially in more modern membrane structured as a hydrogel formed over the supporting matrix acting as a suitable support (www.natrixseparation.com). This becomes an advantage for membranes versus beads when purifying large biomolecules like HCP, DNA, and viruses. Binding of these larger particles is restricted because in many instances they are too large to diffuse into the pore structure of traditional resins. Therefore, binding is limited to those active chemistry groups that are on the outer surface of the beads. The more open structure and immediate availability of all active chemistry groups on the surface of a membrane allow significantly greater binding capacities of even very large particles such as plasmids, DNA, and viruses.

The capability of treating large volumes with a small surface/volume is facilitated by using anion exchange membranes in flow-through (FT) systems as a polishing step. Membrane chromatography systems are currently available in two different device formats. Membranes are pleated or wrapped into cartridges; alternatively they are flat sheets stacked into disposable capsules. Both formats allow high flow rates, good adsorption capacities, and good resolving capabilities in an easy-to-use disposable format with the second providing superior scalability from small to product scales. The superior resolution provided by membranes allows immediate inclusion into an existing process chain for applications such as, DNA viral clearance and removal of endotoxin, and so on. An alternative device format consists of up to 80 membranes stacked together in a stainless steel housing. This format is designed to withstand higher pressure and is reusable although this brings back the threat of operation downtime due to stacks installation, regeneration cleaning, and also validation paperwork negating the true advantages of single use membranes.

References

Aldington, S. and Bonnerjea, J. (2007) Scale-up of monoclonal antibody purification processes. *J. Chromatogr. B*, **848**, 64–78.

BBC-Research (2007) http://www.marketresearch.com/BCC-Research-v374/Separation-Systems-Commercial-Biotechnology-1477810/, viewed Sept 2011.

Boysen, H., Wozny, G., Laiblin, T., and Arlt, W. (2002) CDF simulation of preparative chromatography columns considering adsorption isotherms. *Chem. Ing. Tech.*, **74**, 294–298.

Broughton, D.B. and Gerhold, C.G. (1961) Continuous sorption process employing fixed bed of sorbent and moving inlets and outlets, US Patent No. 2,985,589.

Carta, G. and Jungbauer, A. (2010) *Protein Chromatography: Process Development and Scale-Up*, Wiley-VCH Verlag GmbH, Weinheim.

Dapremont, O., Cox, G., Martin, M., Hilaireau, P., and Colin, H. (1998) Effect of radial gradient of temperature on the performance of large-diameter high-performance chromatography columns, I. Analytical conditions. *J. Chromatogr. A*, **796**, 81–99.

DePalma, A. (2005) Cost-driven Chromatography, PharmaManufacturing.com, http://www.pharmamanufacturing.com/articles/2005/401.html.

DePalma, A. (2007) Optimized facility design and utilization. *Genet. Eng. Biotechnol. News*, **27** (9), http://www.genengnews.com/gen-articles/optimized-facility-design-and-utilization/2114/.

Dolan, J.W. and Snyder, J.J. (1989) *Troubleshooting LC Systems: A Comprehensive Approach to Troubleshooting LC Equipment and Separations*, Humana Press Totowa, New Jersey.

GE Health Care (2011) Ion Exchange Chromatography & Chromatofocusing, Principles and Methods, http://www.gelifesciences.com/aptrix/upp00919.nsf/content/LD_273053445-R350 viewed Oct. 2011.

Kearney, M. (1999) Control of fluid dynamics with engineered fractals –

Adsorption process applications. *Chem. Eng. Comm.*, **173**, 43–52.

Kearney, M. (2000) Engineered fractals enhance process applications. *Chem. Eng. Prog.*, **96** (12), 61–68.

Matthews, T., Bean, B., Mulherkar, P., and Wolk, B. (2009) An integrated approach to buffer dilution and storage. *Pharma. Manufacturing, Pharmaceutical Technology*, http://www.pharmaceutical-technology.com/features/feature62165/, viewed 10.Nov.2009.

Meyer, V.R. (1999) *Fallstricke und Fehlerquellen der HPLC in Bildern*, Wiley VCH, Heidelberg.

Microstar Laboratories Inc., http://www.mstarlabs.com/control/znrule.html viewed Sept. 2009.

Scott, R.P.W. (1986) *Liquid Chromatography Detectors*, Elsevier Science Publishers, Amsterdam.

Snyder, L.R., Kirkland, J.J., and Dolan, J.W. (2010) *Introduction to Modern Liquid Chromatography*, 3rd edn, John Wiley & Sons, Hoboken, New Jersey.

Toumi, A., Jürgens, C., Jungo, C., Maier, B.A., Papavasileiou, V., and Petrides, D.P. (2010) Design and optimization of a large scale biopharmaceutical facility using process simulation and scheduling tools. *Pharm. Eng.*, **30**, 1–9.

Unger, K., Ditz, R., Machtejevas, E., and Skudas, R. (2010) Liquid chromatography – its development and key role in life science applications. *Ang. Chemie*, **49**, 2300–2312.

Zhou, J.X. and Tressel, T. (2006) Basic concepts in Q membrane chromatography for large-scale antibody production. *Biotechnol. Prog.*, **22**, 341–349.

Ziegler, J.G. and Nichols, N.B. (1942) Optimum settings for automatic controllers. *Trans. ASME*, **64**, 759–768.

5
Process Concepts
Malte Kaspereit, Michael Schulte*, Klaus Wekenborg*, and Wolfgang Wewers*

The basis for every preparative chromatographic separation is the proper choice of the chromatographic system. The most important aspects in this context are selectivity, capacity, and solubility, which are influenced and can be optimized by the deliberate selection of stationary and mobile phases, as is discussed extensively in Chapter 3.

Besides the selection of the chromatographic system its implementation in a preparative process concept plays an important role in serving the different needs in terms of, for example, system flexibility and production amount. Depending on the operating mode, several features distinguish chromatographic process concepts:

- batchwise or continuous feed introduction;
- operation in a single- or multi-column mode;
- elution under isocratic or gradient conditions;
- co-, cross-, or counter-current flow of mobile and stationary phases;
- withdrawal of two or a multitude of fractions.

Starting with the description of the main components of a chromatographic unit, this chapter gives an overview of process concepts available for preparative chromatography. The introduction is followed by some guidelines to enable a reasonable choice of the concept, depending on the production rate, separation complexity, and other aspects.

5.1
Discontinuous Processes

5.1.1
Isocratic Operation

The easiest chromatographic set-up consists of one solvent reservoir, one pump that can deliver the necessary flow rate against the pressure drop of the packed

*These authors have contributed to the first edition.

Preparative Chromatography, Second Edition. Edited by H. Schmidt-Traub, M. Schulte, and A. Seidel-Morgenstern.
© 2012 Wiley-VCH Verlag GmbH & Co. KGaA. Published 2012 by Wiley-VCH Verlag GmbH & Co. KGaA.

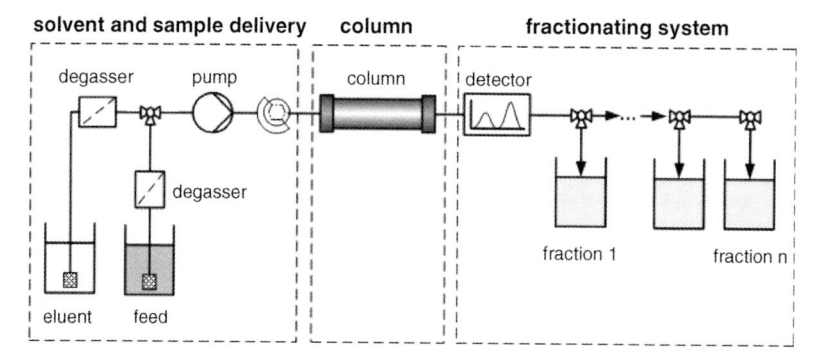

Figure 5.1 General design of a chromatographic plant.

column, a column, and a valve for the collection of pure fractions (Figure 5.1). If the feed mixture contains no early-eluting impurities or components with high affinity to the adsorbent, the stationary phase can be used very efficiently. With a proper choice of the operating conditions, such as injected amount of substances, flow rate, and time between two injections, a "stacked injections" situation can be achieved (Figure 5.2). This means that the eluting peaks of two consecutive injections do not overlap, but just "touch" each other (e.g., B_1 and A_2). Complete product recovery requires baseline separation, where the component peaks of the same injection do not overlap when leaving the column (A_1 and B_1). If these two peaks "touch", this is denoted as "touching band" situation. To produce fractions of high purity, baseline separation is not necessarily required, but waste fractions can be implemented between the product fractions or between the elution profiles of consecutive injections. Section 7.2 gives a detailed description of design and optimization strategies for such chromatographic processes in order to maximize productivity, minimize eluent consumption or total separation costs.

Processes where the composition of the solvent is not altered during elution, as is the case for the one described above, are called *isocratic*. They exhibit the following benefits:

- simplicity of set-up – the number of pumps and reservoirs is reduced to the absolute minimum;

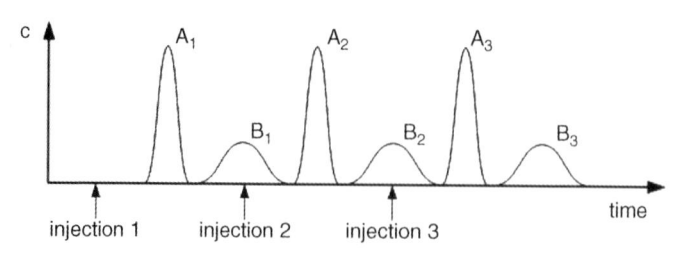

Figure 5.2 Separation of two components under touching band and stacked injections conditions for three consecutive injections.

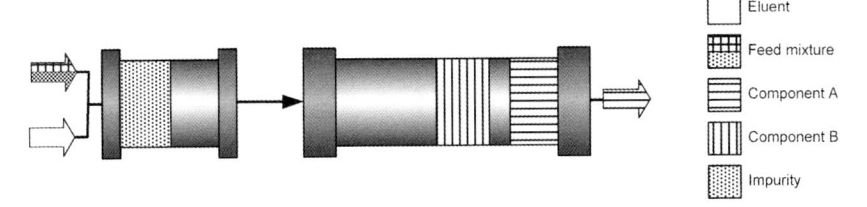

Figure 5.3 Precolumn for the adsorption of late-eluting impurities.

- cycle time can be reduced, because no reconditioning of the column is necessary;
- work-up and reuse of the solvent is easier due to its constant composition.

Whenever possible the development of a chromatographic separation should start with an isocratic elution mode. Its economic feasibility has to be checked afterwards.

If the feed mixture is contaminated with late-eluting impurities and their adsorption in the main column has to be avoided, the implementation of a pre- or guard column might be helpful (Figure 5.3).

This column should be of short bed length and can be packed with a more coarse material. Its function is only the adsorption of late-eluting impurities. As soon as a breakthrough of those components can be observed the separation has to be stopped and the precolumn cleaned or emptied and repacked. Repacking with a cheap, coarse adsorbent is often more advisable than cleaning the precolumn adsorbent. If the separation should not be stopped for precolumn cleaning, a second precolumn and a switching valve are installed, so that the contaminated precolumn can be cleaned or repacked offline.

Further options to treat mixtures with late-eluting components using a single-column set-up are, for example, the flip-flop concept or the use of gradients as discussed in Sections 5.1.2 and 5.1.5, respectively.

5.1.2
Flip-Flop Chromatography

A different concept dealing with mixtures containing highly adsorptive as well as fast-eluting components is the so-called flip-flop chromatography developed by Martin et al. (1979) and further elaborated by Colin, Hilaireau, and Martin (1991) (Figure 5.4). The feed mixture is injected at the one end of the column and the early-eluting components are pushed through the column (Figure 5.4a) and collected at the other end (Figure 5.4b). When the fast-eluting components have been withdrawn from the column, the flow direction is reversed while the late-eluting component is still retained within the column (Figure 5.4c). After a predetermined time a new portion of feed mixture is injected and the elution is performed in the reversed flow direction (Figure 5.4d). At the end of the column the late-eluting impurity from injection 1 is now eluted first (Figure 5.4e), followed by the early-eluting components

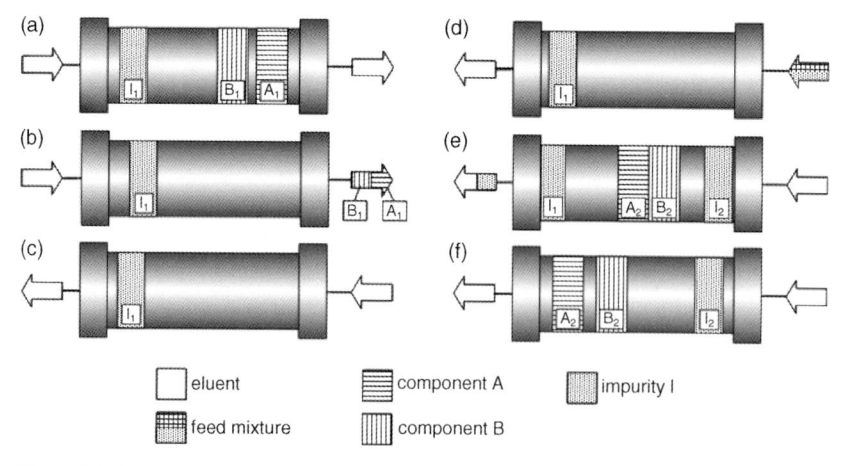

Figure 5.4 Flip-flop concept.

from injection 2. When the latter have been withdrawn from the column, the flow direction is reversed again and a new injection is performed.

The flip-flop concept is an elegant way of dealing with feed mixtures consisting of components with very different adsorption behavior. However, it is not very often used due to the complexity of its operation and the system set-up. If the feed mixture contains components of very different elution behavior it is more advisable to use the above-mentioned concepts of precolumns or to divide the whole separation process into two steps: a simple adsorption of the late-eluting impurities (see above) and the main separation.

5.1.3
Closed-Loop Recycling Chromatography

One of the constraints in preparative chromatography is the inherent interdependency of efficiency and pressure drop. If the plate number of a column has to be increased to achieve the desired separation, the pressure drop will increase because a longer column or a stationary phase with smaller particles has to be chosen. One possible way of overcoming this situation is to increase the number of plates in a column *virtually*. When the column outlet is connected to the column inlet and the partially separated mixture is again injected into the column without "destroying" the elution profile, the components of interest can be separated during the second or following passages through the column. This process concept is called closed-loop recycling chromatography (CLRC).

The critical point in CLRC is the increase in axial dispersion due to the additional holdup volume generated by the pump head and additional piping. This can be overcome by the "alternate pumping" approach that employs two columns (Duvdevani, Biesenberger, and Tan, 1971). In most cases CLRC is combined with "peak shaving" to remove the front and tail ends of the chromatogram (Figure 5.5).

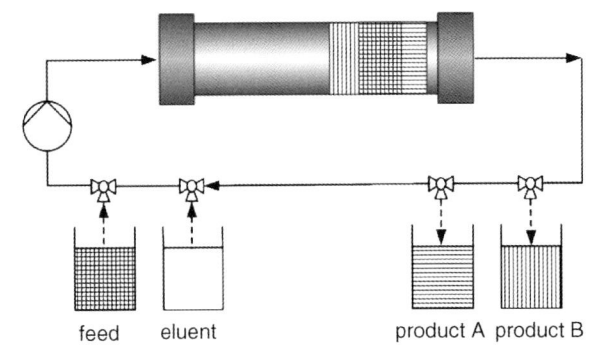

Figure 5.5 Closed-loop recycling chromatography with peak shaving. A corresponding chromatogram is shown in Figure 5.6.

The main advantage of CLRC is the possibility of isolating even components of low selectivity with good purity and yield, as no fractions with insufficient purities are withdrawn from the system. A second interesting aspect is that the consumption of mobile phase is not increased by this concept. During the closed-loop operation no eluent is consumed as the complete mobile phase is circulated within the system. Fresh eluent has to be introduced only during the intervals when the original feed is injected and the fractions are collected.

One drawback of CLRC with peak shaving is the decreasing amount of solutes in the column (Figure 5.6). After the third or fourth cycle only a small amount of the original feed mixture is left in the column. Therefore, the chromatographic productivity, which indicates the time–space yield, decreases rapidly and reduces process economics, although the latter might in part be compensated for by the high purities of the target components.

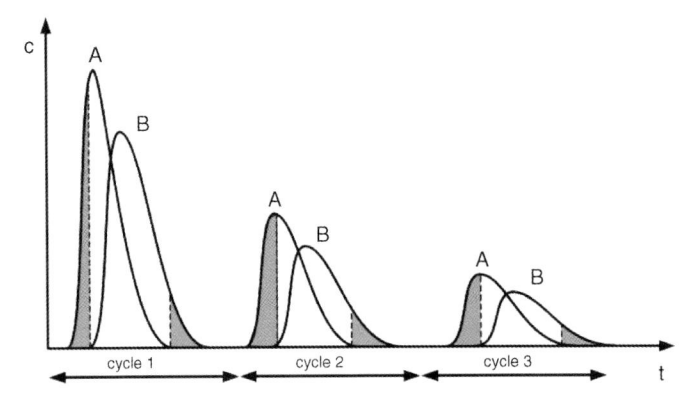

Figure 5.6 Development of concentrations during closed-loop recycling chromatography operation with peak shaving.

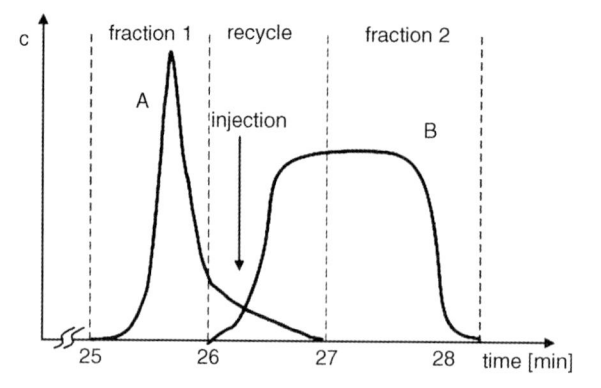

Figure 5.7 Steady-state recycling chromatography (SSRC).

5.1.4
Steady-State Recycling Chromatography

An interesting option to overcome this situation is to perform CLRC with a periodic injection of fresh feed mixture. The basic set-up of this process is similar to the CLRC concept shown in Figure 5.5. It is also known as steady-state recycling chromatography (SSRC) (Bailly and Tondeur, 1982; Grill and Miller, 1998; Grill, Miller, and Yan, 2004) and marketed as "CycloJet" concept. Figure 5.7 shows a chromatogram at the column outlet after one cycle. When the first component A elutes from the column ($t = 25$ min), the first fraction of pure product can be collected. When the region of nonseparated mixture elutes from the column ($t = 26$ min) a new portion of feed is injected into the system for a given time interval. After a new injection the chromatogram is again recycled in a closed loop. The desorption front is then withdrawn from the system to separate pure B ($t = 27$ min). When designing this process, the amount and the exact time point for the injection of fresh feed have to be optimized. Furthermore, the column length should be optimal to prevent unnecessary periods in which only mobile phase elutes from the column. A possible solution could also be to provide an additional column-external volume causing the holdup time required to achieve the optimal time point for reinjection of the recycle fraction (Scherpian and Schembecker, 2009).

Another option of operating SSRC is to include a reservoir in which the recycle fraction is *mixed* with the fresh feed before reinjection. While the achieved partial separation of the recycle fraction is given up by this strategy, it allows for a very simple adjustment of the optimal injection time point. It has been shown that this process option always allows for a reduced solvent consumption and – for systems with significant dispersion and/or difficult separations – also higher productivity than batch chromatography. Finally, in contrast to the closed-loop SSRC, there exist simple shortcut methods to design the mixed-recycle variant for any purity requirement (Bailly and Tondeur, 1982; Sainio and Kaspereit, 2009; Kaspereit and Sainio, 2011).

Concerning its performance, steady-state recycling chromatography can be seen as an intermediate between simple batchwise or CLRC operation and continuous simulated moving bed (SMB) processes (Section 5.2). It combines lower complexity and equipment requirements than an SMB system with higher productivities and lower eluent consumption than obtained in batch chromatography.

5.1.5
Gradient Chromatography

Under isocratic conditions, the strongest retained component might take a very long time to elute. Such late-eluting components are typically strongly diluted, which causes higher work-up costs. In extreme cases some components do not even elute under the chosen conditions. These drawbacks can be avoided by implementing gradients in order to manipulate the elution strength (Section 2.6.3) during the elution process.

Process conditions that might be altered during washing and/or elution steps are

- mobile phase composition (in terms of organic modifier, buffer amount or strength, salt concentration, or pH value);
- mobile phase flow rate;
- temperature;
- pressure (in supercritical fluid chromatography, SFC).

Figure 5.8 illustrates various elution modes based on different compositions of the mobile phase. In the diagrams, A represents the composition of the eluent while B indicates the modifier concentration versus time. The simplest gradient elution is the integration of a washing step (Figure 5.8b). It is mostly used if the component of interest can be adsorbed onto the stationary phase under noneluting conditions. After washing off all the impurities, the target component is eluted at increased modifier concentration.

If the feed mixture contains components with high affinity to the adsorbent so that they are not eluted with the current mobile phase composition, a second elution step with a different composition is necessary. Here, it is also advisable to

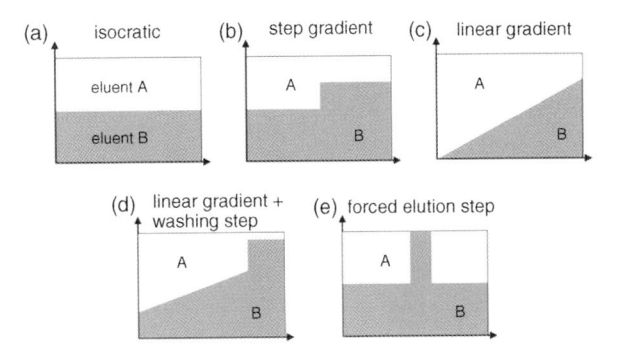

Figure 5.8 Elution modes with different mobile phase compositions. The picture shows eluent composition vs. time.

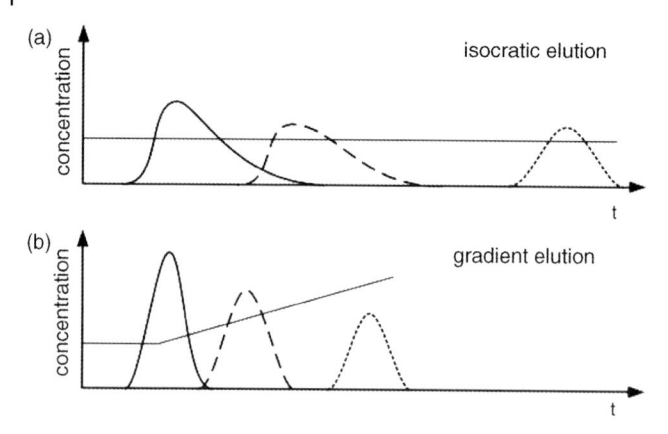

Figure 5.9 Comparison of isocratic and linear gradient elution.

install a bypass line that allows the flow direction to be reversed. By this approach components that "stick" close to the column's inlet can be withdrawn by flushing them back to the inlet rather than pushing them through the whole column. Such flow reversal can make a washing step much more efficient in terms of time and solvent consumption.

A continuous increase in elution strength of the eluent (Figure 5.8c) offers the best separation performance for components with very different retention times. If the retention time of the first and the last component differs by more than 30% (Unger, 1994) gradient elution should be applied or the whole separation should be divided into several process steps. Linear gradients (Figure 5.9) give good resolution and small peak width, allowing the isolation of pure fractions at higher product concentrations within a shorter cycle time.

Besides the two gradient types, combinations of linear and step gradients might be used. One option is to combine a flat gradient for the part of the chromatogram where the highest separation power is needed, with a step gradient for the quick removal of all unwanted components that elute after the component of interest (Figure 5.8d).

Gradient elution reduces the overall retention time of the mixed components. However, the cycle time can be longer than for isocratic elution because the mobile phase has to be readjusted to its starting composition in an additional process step.

A more efficient use of a step gradient can be the injection of a plug of mobile phase with high elution strength into the part of the chromatogram where the component of interest elutes (Figure 5.8e). This eluent plug desorbs the component of interest from the column. It, thus, increases the fraction concentration but not the cycle time due to solvent readjustment (Section 3.2.3). A similar approach is often applied if the sample has a poor solubility in the mobile phase. Here process performance can be improved significantly by injecting a more concentrated mixture dissolved in a stronger solvent (Gedicke, Antos, and Seidel-Morgenstern, 2007).

Recently, also the use of gradient of the *stationary* phase was suggested for preparative chromatography (Sreedhar and Seidel-Morgenstern, 2008). In this concept, column segments packed with different stationary phases are arranged in series. Type, order, and length of these segments can be optimized to enhance performance.

5.1.6
Chromatographic Batch Reactors

In conventional chemical productions a reaction unit is followed by one or more separation steps. This enables a straightforward design of each unit operation and offers a high degree of freedom for process design. However, for equilibrium-limited reactions the yield is limited and can only be enhanced by separating unreacted educts from the obtained mixtures, and reintroducing them into the reaction steps by additional recycle streams. This decreases the specific productivity and can necessitate increasing equipment dimensions and investment costs, especially for low conversion rates.

Another possibility for enhancing reaction yield and separation efficiency is to integrate both the reaction and the separation into a single apparatus. Examples of such integrated processes are reactive distillation, reactive extraction, reactive absorption, reactive membrane separation, reactive crystallization, and reactive adsorption (Kulprathipanja, 2002). While process integration can be attractive with respect to achievable conversion and lowered investment costs, it simultaneously reduces the degrees of freedom. Operating parameters of both the reaction and the separation must match, and they also have to fulfill restrictions due to equipment design. Therefore, a reduced window of possible operating points needs to be considered in the design process.

A chromatographic reactor combines a chemical or biochemical reaction with a chromatographic separation. Within the integrated process, homogeneous or heterogeneous reactions can occur. The latter require a solid catalyst or an immobilized enzyme, which are mixed with the chromatographic packing material. In special cases such as esterifications the adsorbent can act as catalyst as well (Mazzotti *et al.*, 1996). When different materials for catalyst and adsorbent are needed, establishing a reliable packing is a major problem. Due to different particle sizes and densities, a heterogeneous distribution of catalyst within the bed can occur. One possible way of overcoming these difficulties is to partition the bed with alternating layers of adsorbent and catalyst (Meurer *et al.*, 1997). For a homogeneously catalyzed reaction, the catalyst has to be either recovered or separated after the chromatographic reactor.

Since the late 1950s, analytical chromatographic reactors have been applied to studies of reactions (Coca *et al.*, 1993), determination of reaction kinetics (Bassett and Habgood, 1960), characterization of stationary phases (Jeng and Langer, 1989), or examination of interactions between mobile and stationary phases (Coca *et al.*, 1989). The main focus of recent developments has been on preparative-scale applications.

A reaction of type $A \rightleftarrows B + C$ is used in Figure 5.10 to explain the principles of chromatographic batch reactors. In the given example, the reactant A has

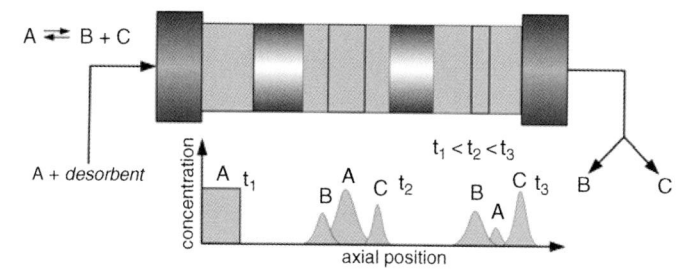

Figure 5.10 Chromatographic batch reactor.

intermediate adsorption behavior and products B and C are the more strongly and weakly adsorbed components. Reactant A is injected as a sharp pulse into the column. During its propagation along the column, A reacts to give B and C. Owing to their different adsorption behavior all components propagate with different velocities and are separated. Therefore, the restriction of an equilibrium-limited reaction can be overcome to convert the reactant completely. Additionally, no separation unit following the chromatographic reactor is necessary to gain the high-purity products. Due to the batchwise operation, only a limited part of the bed is used for reaction and separation. Therefore, the operation requires large amounts of desorbent and results in a high product dilution. Furthermore, the productivity of the process is usually rather low.

To separate the products and the reactant the chromatographic system has to provide reasonable selectivity for all components. For instance, for the reaction $A \rightleftharpoons B + C$ total conversion and total separation is possible for any order of adsorptivities. It is, however, favorable if the products have the highest and lowest affinity while the retention of the reactant is intermediate, as assumed in Figure 5.10. However, in case of the reaction type $2A \rightleftharpoons B + C$ reactant A must have an intermediate adsorptivity in order to achieve total conversion and total separation (Vu et al., 2005). For multiple reactants, a chromatographic system should be chosen that has a separation factor close to unity for the reactants. In all other cases the reactants separate and complete conversion is not possible, so that special operation modes have to be applied.

Chromatographic batch reactors are employed to prepare instable reagents on the laboratory scale (Coca et al., 1993) and for the production of fine chemicals. These applications include the racemic resolution of amino acid esters (Kalbé, Höcker, and Berndt, 1989), acid-catalyzed sucrose inversion (Lauer, 1980) and production of dextran (Zafar and Barker, 1988). Sardin, Schweich, and Villermaux (1993) employed batch chromatographic reactors for different esterification reactions such as the esterification of acetic acid with ethanol and the transesterification of methylacetate. Falk and Seidel-Morgenstern (2002) have investigated the hydrolysis of methyl formate. Ströhlein et al. (2006) measured the esterification of acrylic acid with methanol and validated the transport dispersive model for process simulation.

5.2
Continuous Processes

In the separation processes mentioned so far the introduction of the feed mixture is realized in a batchwise manner onto one single column only. Especially for separations in a preparative scale, modes for continuous operation have to be considered to increase productivity, product concentration, and to save fresh eluent.

5.2.1
Column Switching Chromatography

The easiest way of transforming a batch separation into a (pseudo-)continuous one is the column switching approach, which can be applied for relatively simple adsorption–desorption processes. The feed is pumped through a column until a breakthrough of the target component is observed. At that moment the injection is switched to a second column, while the first one is desorbed by introducing eluent by a second pump. After desorption is finished the initial eluent conditions have to be readjusted before the first column can again be used for adsorption. Obviously, this set-up can be used only for relatively simple separations, where the component of interest can be separated from the impurities by adsorbing it to the stationary phase. This implies that the target component is either the last-eluting component having a significantly higher affinity to the stationary phase than the other components. Or, alternatively, all other impurities stick to the stationary phase and only the component of interest shows no or little adsorption affinity.

5.2.2
Annular Chromatography

For more complex feed mixtures other approaches for continuous separation have to be considered. One example is annular chromatography with a rotating stationary phase as anticipated already by Martin (1949). This concept was developed in the 1950s as a continuous method for paper chromatography by Solms (1955). In annular chromatography, the stationary phase is packed between two concentric cylinders and rotates around a central axis (Figure 5.11).

At the top of the column the feed injection is performed at a fixed position. Over the whole remaining circumference of the packed bed the eluent is introduced into the column. The fluid phase moves toward the bottom while the annular column rotates slowly. This operation corresponds to a crosscurrent movement of packing material and mobile phase. The components of the feed mixture migrate in helical bands downwards through the bed and are separated into several product streams eluting at different angles at the bottom of the bed. From a design point of view, annular chromatography is a continuous alternative to batch chromatography. Its performance is comparable to multiple batch columns arranged in a circle. Therefore, productivity is similar to batch systems under the same operational conditions. In fact, a batch chromatogram can be transferred to an annular chromatographic

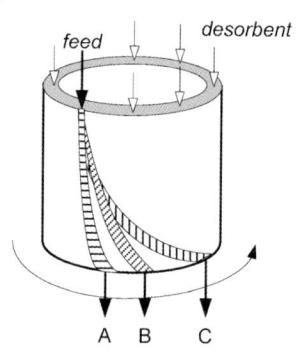

Figure 5.11 Annular chromatographic separation.

system by a coordinate transform that replaces the retention time by corresponding angular positions.

As in batch chromatography the solvent strength can be altered along the circumference by introducing further solvents, resulting in gradients for improved process performance. Also reactive variants were suggested (Sarmidi and Barker, 1993a, 1993b).

One drawback of annular chromatography is the difficulty of packing efficient columns. Because of the angular rotation, axial and tangential dispersions affect the peak broadening. In order to decrease tangential dispersion Brozio and Bart (2004) recommend to build apparatuses with a thicker diameter and a smaller annular width. They also discuss criteria for a simplified scale-up from batch chromatography to annular chromatography. To operate an annular chromatograph only rigid adsorbents should be used to obtain an improved distribution of the target components over the different withdrawal ports (Schmidt et al., 2003). The allowable pressure drop of the stationary phase is limited by the quality of the sealings. Wolfgang and Prior (2002) have given a more detailed overview of the basic principle, technical aspects, and industrial applications.

5.2.3
Multiport Switching Valve Chromatography (ISEP/CSEP)

As mentioned above, the concept of an annular chromatographic system can be realized by a multicolumn set-up where several single columns are mounted to a rotating "column carrousel." The distribution of the different liquid streams as well as the collection of all outlet streams is realized by one central multiport switching valve (Figure 5.12). The inlet and outlet lines of the unit are connected to the stationary part of the valve, while the columns are connected to the rotating part. In contrast to annular chromatography, where the bed rotates with a constant velocity, the switching valve performs this movement in discrete steps, from one position to next at given switching intervals.

The introduction of this valve and the fact that the stationary phase is packed into fixed beds offers a great variety of possible column interconnections. Besides

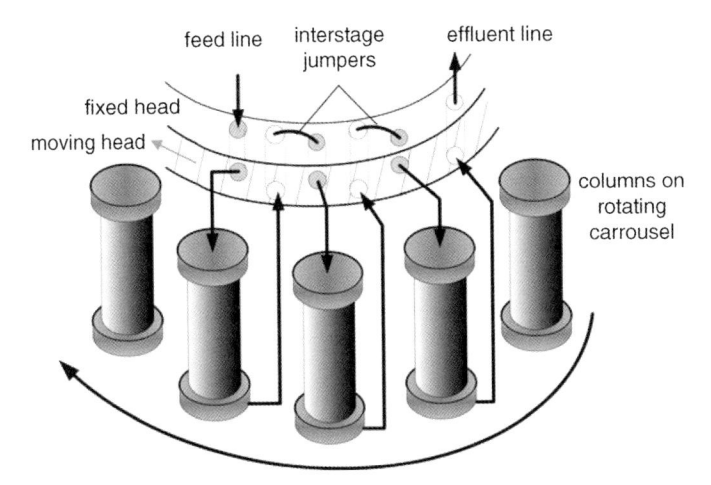

Figure 5.12 ISEP principle; the rotating valve.

performing the mentioned cross-current flow, columns can be connected very flexibly in series or parallel, allowing a multitude of different process set-ups (Figure 5.13). Different sections can be realized to fulfill special tasks within the process. Common tasks for such sections are

- introduction and adsorption of feed mixture;
- washing of the columns and removal of impurities;
- desorption of target component by a different mobile phase;
- reconditioning of columns with the mobile phase of the adsorption step.

The number of columns in certain sections and their interconnection can be chosen according to the process needs. To ensure continuous operation of the system the number of columns has to be adjusted to the time needed for the operation of each individual step. Many systems are known, for example, for the purification of amino acids and sugars on ion exchange resins.

As mentioned before, the complete parallel interconnection of the columns is similar to annular chromatography. The second border case is to connect all beds

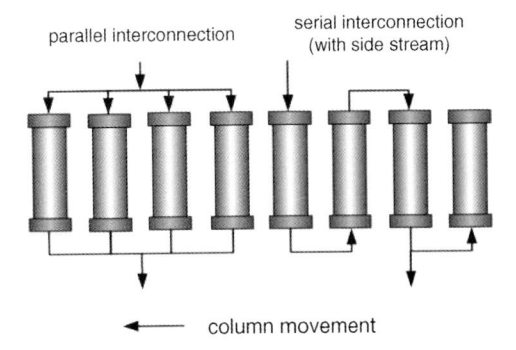

Figure 5.13 ISEP principle involving parallel and serial column interconnection.

in series. With an appropriate choice of inlet and outlet streams as well as internal flow rates, this set-up is better known as simulated moving bed (SMB) chromatography, which will be discussed below.

5.2.4
Isocratic Simulated Moving Bed (SMB) Chromatography

In general, a counter-current operation can achieve a more efficient separation than the cross-current operation considered above. In chromatography this is realized by the SMB technology. Chromatographic counter-current operation requires the mobile and stationary phases to move in opposite directions. Figure 5.14 illustrates the more or less hypothetical concept of an ideal counter-current system, denoted as true moving bed (TMB). This concept utilizes an actual circulation of the solid phase with a constant flow rate.

For the chromatographic separation, four external streams are present: a feed inlet containing the components to be separated, a desorbent inlet, the extract outlet containing the more retained component B, and the raffinate outlet enriched with the less retained component A. These streams divide the system into four sections or zones with different flow rates of the liquid phase. Each of these sections plays a specific role in the process. Separation is performed in sections II and III, where the more retained component B has to be adsorbed and carried toward the extract port by the moving solid, while the less retained component A has to be desorbed and carried by the mobile phase toward the raffinate port. In section I, the solid phase is regenerated by a fresh eluent stream that desorbs the strongly adsorbed component. Finally, in section IV the liquid is regenerated by adsorbing

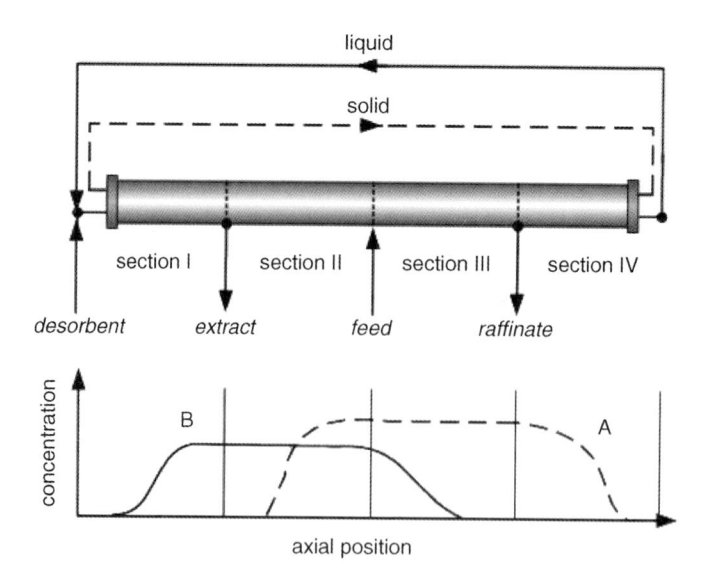

Figure 5.14 True moving bed (TMB) chromatography with internal concentration profiles.

the amount of less retained component not collected in the raffinate. In this way both the solid and the liquid phase can be recycled to sections IV and I, respectively. In summary, the four sections have to fulfill the following tasks:

- **Section I:** Desorption of the strongly adsorptive component;
- **Section II:** Desorption of the less adsorptive component;
- **Section III:** Adsorption of the strongly adsorptive component;
- **Section IV:** Adsorption of the less adsorptive component.

With a proper choice of all individual internal fluid flow rates in sections I–IV and the velocity of the stationary phase, the feed mixture can be completely separated into two pure products. Complete separation corresponds to axial concentration profiles at steady state as displayed in Figure 5.14.

Unfortunately, an efficient, dispersion-free movement of a stationary phase, which in most cases consists of porous particles in the micrometer range, is technically not possible. Therefore, other technical solutions had to be developed. The breakthrough was achieved with SMB systems, which were developed by Universal Oil Products (UOP) for the petrochemical industry in the 1960s (Broughton and Gerhold, 1961). Here the stationary phase is packed into single, discrete columns, which are connected to each other in a circle.

Figure 5.15 illustrates the principle of SMB processes. The mobile phase passes the fixed bed columns in one direction. Counter-current flow of both phases is achieved by switching the columns periodically upstream in the direction opposite to the liquid flow. Alternatively, not the columns are shifted but all ports are moved

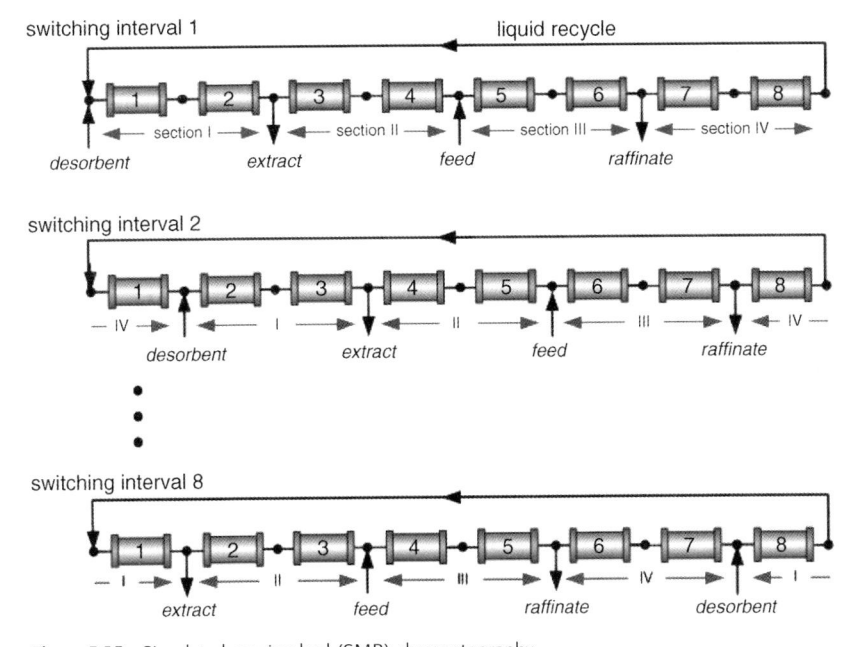

Figure 5.15 Simulated moving bed (SMB) chromatography.

in the direction of the liquid flow by means of valves. The counter-current character of the process becomes more obvious when the relative movement of the packed beds to the inlet and outlet streams during several switching intervals is observed. After a number of switching events or shifting intervalsequal to the number of columns in the system, one so-called cycle is completed and the initial positions for all external streams are reestablished. The analogy to the TMB concept above becomes even clearer when considering an SMB process with a large number of very short columns shifted in very short intervals.

The total number of columns and their distribution over the different sections are not fixed to eight with two columns per zone (2/2/2/2) as assumed in Figure 5.15. Another quite common set-up is 1/2/2/1 with only one column in sections I and IV. Naturally, this reduction in total number of columns can reduce investment costs. Conversely, the costs for fresh solvent will increase since regeneration of the solid phase has to take place in a shorter time and, therefore, requires a higher flow rate in section I. For a further decrease of column number, other continuous process concepts like VariCol or ISMB should be implemented (Section 5.2.5).

The SMB concept is, in general, realized for pharmaceutical and fine chemical separation purposes in two different ways. In the first alternative, one central rotating valve, as introduced in Section 5.2.3, is used to distribute and collect all inlet and outlet streams. The second design concept switches the ports by means of two-way valves between all columns. While the first approach corresponds to a less complex set-up, the advantage of the second is that higher pressures can be realized and the switching of the individual ports can be handled very flexibly, allowing for implementations like the VariCol process (Section 5.2.5).

Another characteristic of the SMB set-up is the implementation of a so-called recycle pump to ensure the liquid flow in one direction. In general, four cases can be distinguished (Figure 5.16).

In the first case (Figure 5.16a), the recycle pump is at a fixed position between two columns. Since all columns are moving upstream according to the SMB principle, the recycle pump performs the same migration. This means that the flow rate of this pump has to be adjusted depending on which section it is currently located in. This design results in a locally increased dead volume because of the recycle line and the pump itself. Such an unequal distribution of the dead volume can be compensated by asynchronous shifting of the external ports, as introduced by Hotier and Nicoud (1995).

The second approach (Figure 5.16b) is characterized by a "moving" recycle pump that is always located near the desorbent line. This has the advantage that the recycle pump never comes into contact with the products to be separated and is operated at a constant flow rate. However, this design requires additional valves.

Besides the first alternatives, more approaches exist that do not require an additional recycle pump. In the third case (Figure 5.16c), the outlet of section IV is not directly recycled to section I but introduced to the desorbent tank instead. Figure 5.16d shows another possibility where no recycling of the solvent takes place (Ruthven and Ching, 1989). This method is applied when the regeneration of the

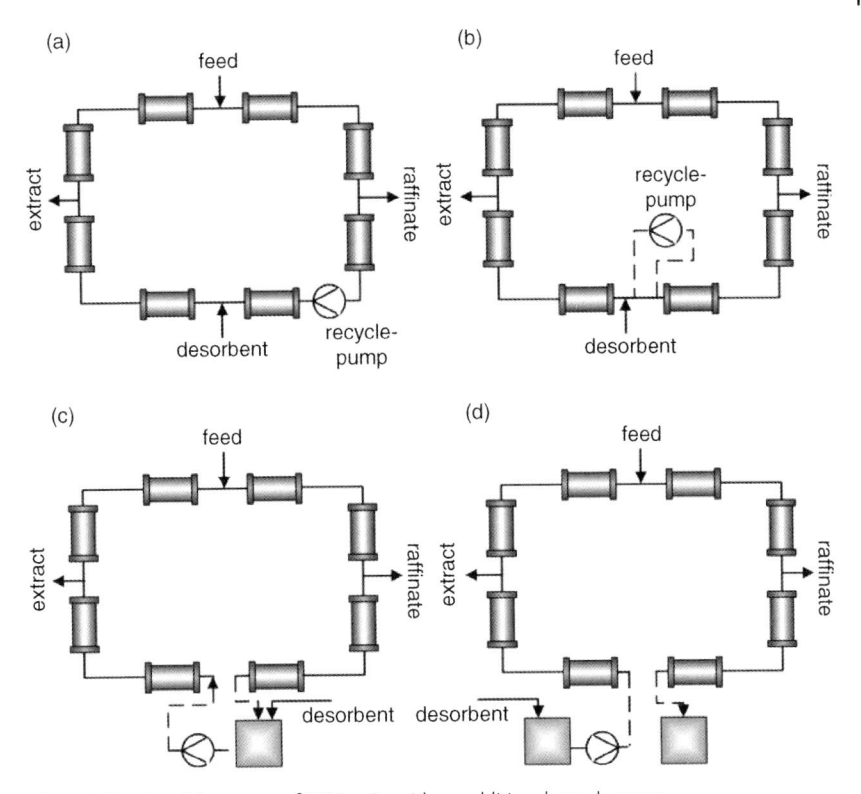

Figure 5.16 Possible set-ups of SMB units with an additional recycle pump.

desorbent turns out to be very difficult and fresh solvent is rather inexpensive. In such case, the regeneration of the fluid is no longer necessary and, consequently, section IV is no longer required. This leads to a three-section SMB concept without solvent recycle, as depicted in Figure 5.17. Finally, also two recycle pumps may be used that are positioned in zones II and IV, respectively.

In addition to three- and four-section SMBs, a system with five or more sections can be used if a third fraction is required, see section 5.2.7. These allow for a third product stream in addition to the extract and raffinate. For very strongly retained components a second solid regeneration section can be implemented by introducing a second desorbent stream with higher solvent strength.

The technology of SMB chromatography has been widely used in the petrochemical (xylene isomer separation) and food industries (glucose–fructose separation) in a multiton scale. Since the 1990s, with the advent of stable chiral stationary

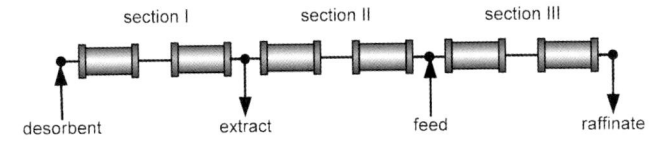

Figure 5.17 Three-section SMB concept.

phases for enantioseparations, the SMB technology is applied successfully by the pharmaceutical industry. Today it is possible to build SMB systems with column sizes from 4 to 1000 mm inner diameter and to produce pharmaceuticals at small scale as well as up to a 100 ton scale.

Further advanced SMB processes are developed utilizing more complex set-ups and operating strategies. This makes an empirical design quite difficult. Therefore, modeling and simulation are necessary for process design and optimization. Modeling approaches are presented in Section 6.6. Aspects regarding model-based design and optimization are introduced in Section 7.4.

5.2.5
SMB Chromatography with Variable Process Conditions

Continuous counter-current separation by SMB chromatography can improve process economics. Nevertheless, more advanced processes have been developed recently. They are all based on the standard SMB technology but operated under variable process conditions to reduce the costs of column hardware and stationary phase as well as for fresh eluent and eluent work-up. New trends and suggestions to improve the operation of SMB processes have been reviewed, for example, by Seidel-Morgenstern, Kessler, and Kaspereit (2008) and Kaspereit (2009).

5.2.5.1 VariCol
Classical SMB systems are characterized by the synchronous and constant downstream shift of all inlet and outlet lines after a defined switching period. In that case four defined sections can be distinguished over the complete cycle time (Figure 5.18a). Within these sections the columns are distributed equally (e.g., 2/2/2/2) or in any other given configuration. The number of columns per section is constant and an integer number (e.g., 1/2/2/1). At periodic steady state, the same process conditions are always reached after one switching interval (t_{shift}).

The VariCol approach, as introduced by Ludemann-Hombourger, Bailly, and Nicoud (2000), increases the flexibility of the continuous separation system by an asynchronous movement of the injection and withdrawal ports. Within a complete process cycle, this leads to mean numbers of columns per section that are typically non-integer. As the minimum number of columns per section in an SMB system is 1, it is possible in VariCol systems to reduce the mean number to virtually any value less than 1. Owing to the asynchronous shift of the inlet and outlet lines in a VariCol process it may even happen that, during a certain interval, there are no columns in a section. In this case the inlet and outlet lines of this section coincide in one valve block. By placing the outlet lines upstream from the inlet lines a pollution of the product lines is avoided.

As a simple example, the initial column configuration of a VariCol process at time t might be 1/2/2/1. After a predetermined time interval (e.g., $t + 0.5$ t_{shift}) only the feed line is shifted to the next downstream position (Figure 5.18b). Now section II contains three columns for the rest of the time

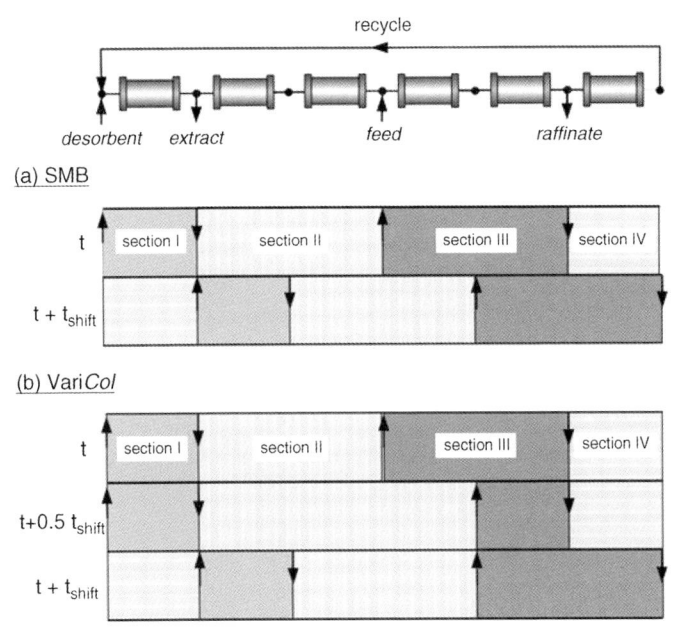

Figure 5.18 Switching strategies for SMB and a simple variant of VariCol.

interval while only one column remains in section III. Thus the new configuration is 1/3/1/1. At the end of the switching time ($t + t_{shift}$) all other ports are shifted and the initial set-up is reestablished. The section configuration can now be calculated according to the number of columns and their residence time within a section. In our example, the overall distribution of columns within one process cycle is 1/2.5/1.5/1. These figures demonstrate that the VariCol process offers high flexibility, especially for systems with a low number of columns (five or less columns). The main goal of the VariCol process is, therefore, to decrease the amount of stationary phase and the number of columns, overall decreasing investment costs for column hardware and stationary phase.

5.2.5.2 PowerFeed

Another multicolumn process based on the SMB principle is the so-called PowerFeed approach proposed by Kloppenburg and Gilles (1999) and Zhang, Mazzotti, and Morbidelli (2003). In contrast to the VariCol process, the switching is the same as in a classical SMB plant. The main idea of the PowerFeed approach is the variation of the internal ($\dot{V}_I - \dot{V}_{IV}$) and external flow rates ($\dot{V}_{feed}$, $\dot{V}_{des}$, $\dot{V}_{ext}$, $\dot{V}_{raf}$) within one switching interval t_{shift}. In the example illustrated in Figure 5.19 the flow rate in section IV $\dot{V}_{IV}$ is lowered during the second part of the switching interval for improved adsorption of the less retained component. However, as a consequence, also other flow rates may change and might have to be manipulated to maintain, for example, a constant flow rate in section I. Designing the various flow rates during the sub-intervals is an optimization problem.

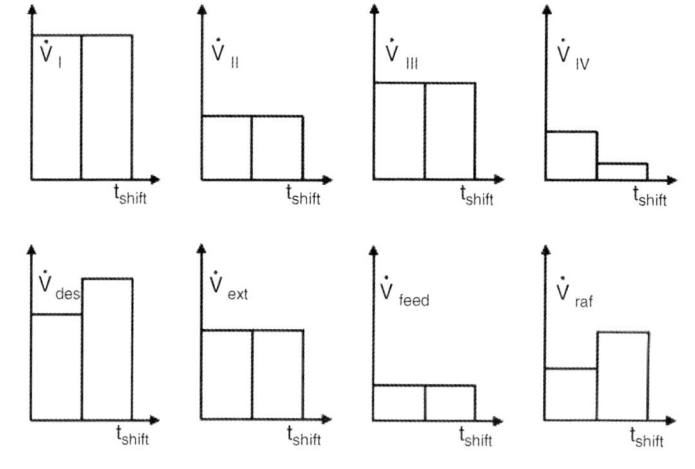

Figure 5.19 Flow rates during PowerFeed operation (illustrative example).

As reported by Zhang, Mazzotti, and Morbidelli (2003), the performance of the PowerFeed process is similar to VariCol and significantly improved with respect to classical SMB processes. Furthermore, the extent of improvement is particularly increasing for more difficult separations. However, this improved performance is paid for by the increased complexity of operation and design; for more details see Section 7.5.

5.2.5.3 Partial-Feed, Partial-Discard, and Fractionation-Feedback Concepts

Similar to the PowerFeed concept explained above, the partial-feed approach introduced by Zang and Wankat (2002a, 2002b) exploits the feed duration and feed time as degrees of freedom. Figure 5.20a compares the feed flow of classical SMB and partial-feed processes. As a consequence of this operation strategy, the raffinate flow rate changes during each switching interval according to the change in the feed flow (Figure 5.20b). In comparison to a conventional SMB process, the feed interval is shorter, but its flow rate can be higher. By this procedure process performance can be improved for the same throughput.

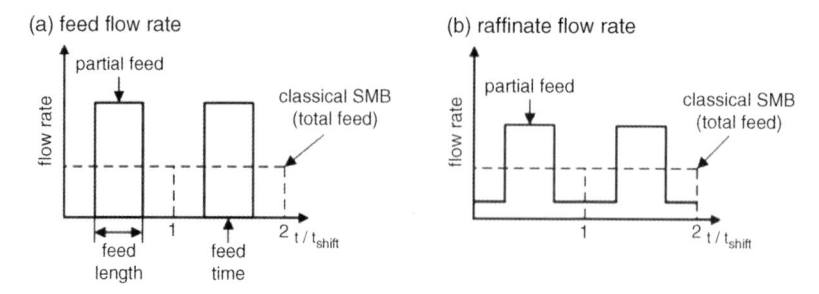

Figure 5.20 Flow rates during partial-feed operation (according to Zang and Wankat (2002a, 2002b).

The partial-feed concept seems promising for applications where the separation is made difficult by significant band broadening effects. Another option for such problem is the partial-discard strategy proposed by Bae and Lee (2006). Here, purity is improved by fractionating off impurities that elute at the outlet ports in the beginning or toward the end of a switching interval. This allows choosing more productive operating points. The resulting reduction of the recovery yield can be acceptable in certain cases.

An extension of the above is the fractionation-feedback approach (FF-SMB, see section 5.2.5.6) suggested by Keßler and Seidel-Morgenstern (2008), wherein collected fractions with insufficient purity are collected, recycled, and refed into the SMB unit, which overcomes yield limitations.

Besides the concepts discussed above there are ongoing activities in the development of further refined operating modes for SMB systems. A complete review is out of scope here.

5.2.5.4 Improved/Intermittent SMB (iSMB)

Partial-feed processes switch the feed stream on and off within one time interval while the other inlet and outlet lines are active all the time. The iSMB (improved or intermittent SMB) process (Tanimura, Tamusa, and Teshima, 1995) partitions the switching interval in a different way. In the first part of the period (Figure 5.21a), the unit is operated basically as a conventional SMB process; hence all external streams (desorbent and feed inlets as well as extract and raffinate outlets) are in operation. However, in contrast to a classical SMB unit, the outlet of section IV is not recycled during this part of the switching interval and, consequently, the flow rate in section IV is zero. This first "injection period" is followed by a recirculation period in the second part of the switching interval (Figure 5.21b). During this time all external ports are closed and recirculation is performed with a constant flow rate in all sections of the plant. This aims at repositioning the internal concentration profiles in an optimal manner. This operating procedure allows a separation to be carried out with a rather small number of columns, which of course has a positive impact on investment costs. Furthermore, the concept has been demonstrated to achieve a superior performance (Katsuo and Mazzotti, 2010a, 2010b). The main application area of the iSMB concept is so far the sugar industry.

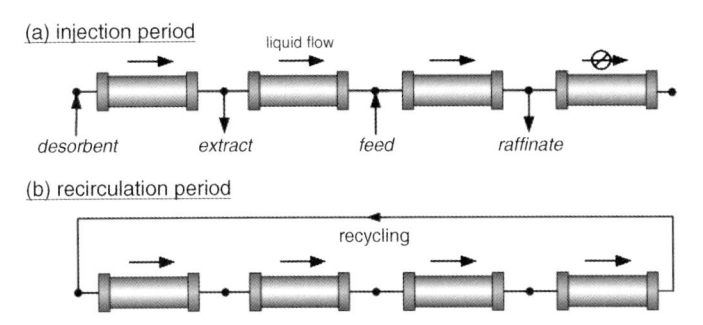

Figure 5.21 Flow rates during iSMB operation.

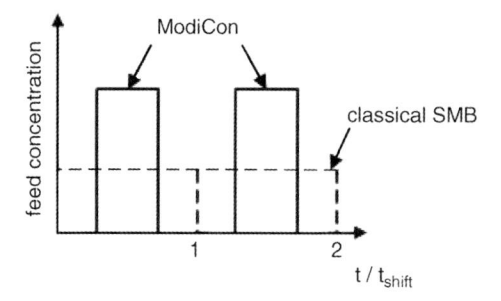

Figure 5.22 Feed concentration during ModiCon operation.

5.2.5.5 ModiCon

Another modification of the classical SMB process is the so-called ModiCon approach by Schramm *et al.* (2003). In contrast to the partial-feed process, where the feed flow rate is varied during one switching interval, in ModiCon the feed concentration is altered instead of the feed flow rate. Figure 5.22 illustrates this variation of the feed concentration during the switching interval t_{shift}. In the example shown, which is suitable for systems with favorable isotherms, at the beginning only pure solvent is fed to the plant while in the second part feed at a rather high concentration is used. The modulation of the feed concentration allows for improved process performance for systems with nonlinear adsorption isotherms.

5.2.5.6 FF-SMB

Another modification of SMB technology, referred to as fractionation and feedback SMB (FF-SMB), is based on fractionating of one or both outlet streams and feeding the off-spec fractions back into the unit alternatively with the original feed mixture. Based on results of an extensive optimization study Li *et al.* (2010a, 2010b) point out that the simultaneous fractionation of both outlet streams is the most efficient operating scheme in terms of maximum throughput while already the single outlet fractionation mode can be significantly superior to the classical SMB process.

5.2.6
SMB Chromatography with Variable Solvent Conditions

One restriction of SMB chromatography as described so far is the operation under isocratic conditions. When selecting the stationary and the fluid phases, it is therefore one of the main goals to find a separation method where the first component is fairly well adsorbed on the stationary phase while the second one is still eluting under those conditions. This is quite often a tedious task in which a number of compromises have to be made. In many cases, isocratic elution conditions lead to a severe tailing of the disperse front of the second component. However, similar to batch chromatography, a gradient can improve the SMB separation if the selectivity of the components is very large or a separation under isocratic conditions is impossible.

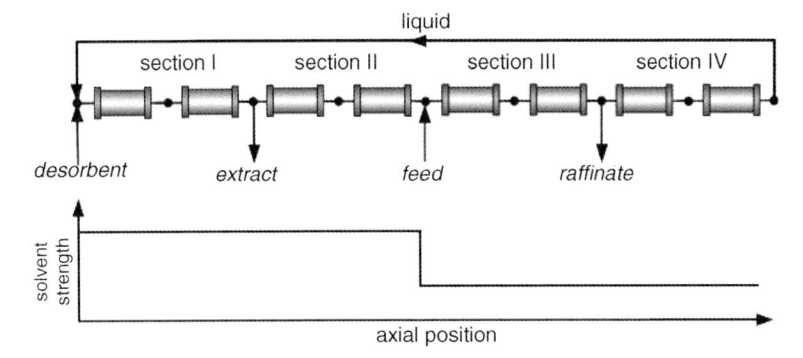

Figure 5.23 Solvent-gradient operation of a SMB unit.

5.2.6.1 Gradient SMB Chromatography

The easiest and most adopted way to adjust different elution strengths within an SMB plant is a two-step gradient (Figure 5.23). This is achieved, for example, by using a desorbent with high elution strength while the feed stream contains a "weaker" eluent. Such solvent-gradient SMB operation leads to the formation of a step gradient with a regime of high desorption power in sections I and II and a region of improved adsorption in sections III and IV. According to the tasks of the different sections (Section 5.2.4), it becomes obvious that such gradient in an SMB process can improve its performance with respect to productivity, eluent consumption, and product concentration.

Improved performance is achieved with a more complex operation and layout of the process, which requires precise process design and control – especially when a recycling of the section IV outlet is applied. An open-loop operation without eluent recycling between sections I and IV is more robust and allows for more pronounced gradients, which is particularly attractive when regeneration of the adsorbent is difficult.

Some prerequisites for the choice of solvent and stationary phase should be considered when developing the chromatographic system. The solvents used for the gradient should exhibit low viscosities, high diffusivity, and thus good miscibility. In addition, heat of mixing as well as volume changes should be low. Several research projects have investigated the application of gradient SMB processes (Jensen et al., 2000; Antos and Seidel-Morgenstern, 2001; Abel, Mazzotti, and Morbidelli, 2002; Beltscheva, Hugo, and Seidel-Morgenstern, 2003; Abel, Mazzotti, and Morbidelli, 2004). The concept is of special interest in the separation of bioproducts, such as, for example, the separation of proteins by ion exchange (Houwing, Billiet, and van der Wielen, 2002; Wekenborg, Susanto, and Schmidt-Traub, 2004) or based on hydrophobic interaction (Palani et al., 2011; Gueorguieva et al., 2011).

Manipulating adsorption strength is most readily achieved by variation of solvent composition (or column pressure in the case of SFC-SMB, see Section 5.2.6.2). However, there exist further options for implementing gradients. In particular, the use of salt and pH gradients is attractive in the context of bioseparations (see some of the references above). Another, obvious option is to introduce temperature gradients (e.g., Eagle and Rudy, 1950; Ching and Ruthven, 1986; Ching, Ho, and Ruthven, 1986; Kim et al., 2005). For example, a dedicated heating of section I of the SMB process enhances desorption and solid regeneration (Migliorini et al., 2001). A drawback is here the heat capacity of the system and the slow change of temperature – also of the connecting valves and pipes. This is particularly critical since the columns eventually enter another section where its temperature is to be lowered again. A useful alternative is to manipulate the temperature of the mobile phase (Jin and Wankat, 2007).

5.2.6.2 Supercritical Fluid SMB Chromatography

An additional degree of freedom is introduced in supercritical fluid chromatography (SFC) where a pressure gradient can be adjusted across the sections of the unit, which influences the adsorption strength of the solutes. SFC systems are operated above the critical pressure and temperature of the mobile phase systems. In most cases, the main component of the mobile phase is carbon dioxide, for which the critical point is reached at 31 °C and 74 bar. In the supercritical region the density and, therefore, the solvating power of the fluid is highly dependent on pressure and temperature, and so is the affinity of a given solute for the supercritical fluid phase itself. With a higher operating pressure, and thus a density increase, the elution strength is improved and smaller retention times can be realized.

Supercritical fluids can be seen as intermediates between gases and liquids with liquid-like solvating powers in combination with gas-like diffusion coefficients and viscosities. This results in the following main benefits: low mobile phase viscosity, high diffusivity, easy recovery of substances from the eluent stream by simple solvent expansion as well as the low cost and nontoxicity of the mobile phase ("green solvent"). On the other hand, such system is more complex and the solubility of most pharmaceutical components in pure CO_2 is rather limited, because drug substances often exhibit a certain hydrophilicity to achieve solubility and resorption in the gastro-intestinal tract. This drawback is often overcome by introducing an additional modifier, in most cases an alcohol or ether.

As already pointed out, the adsorption properties of a given separation system can be influenced by adjusting the column pressure. Different adsorption strengths within the four sections of an SMB system are achieved by pressure variation (Figure 5.24). In section I desorption of the more adsorptive component has to take place; therefore, the elution capability should be the highest and thus the system pressure is also the highest. Through sections II and III toward IV the pressure can be constantly lowered due to adsorption and desorption requirements in the single section. More detailed information about this process concept can be found in Clavier, Nicoud, and Perrut (1996), Denet et al. (2001), Depta et al. (1999) and Giovanni et al. (2001).

5.2.7
Multicomponent Separations

The classical SMB process has two outlet ports and is, thus, limited to binary separations. A number of approaches have been suggested to perform separations of

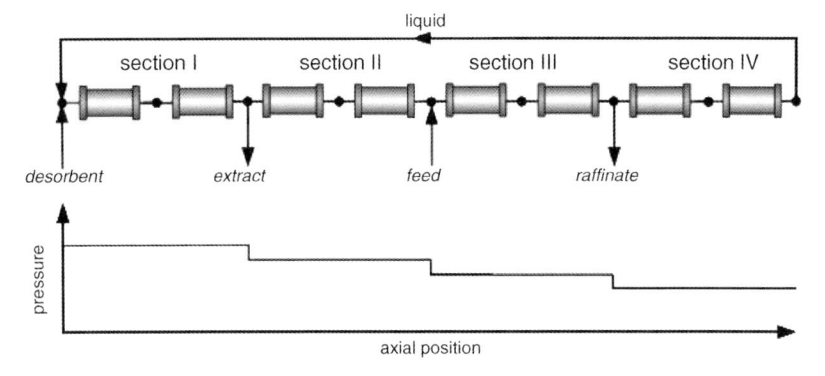

Figure 5.24 Supercritical fluid operation of a SMB unit.

multicomponent mixtures based on SMB technology. A straightforward attempt is to implement several SMB units in series as shown in Figure 5.25 (left). A number of corresponding options were investigated by Wankat (2001). Another way is to increase the number of zones of the SMB unit and to perform an internal recycling of partially separated streams as is the case, for example, in the nine-zone system proposed by Wooley, Ma, and Wang (1998) and the eight-zone set-up by Keßler and Seidel-Morgenstern (2006), see Figure 5.25 (middle and right). An overview on such and further possible multicomponent set-ups is given in, for example, Chin and Wang (2004).

The mentioned concepts exhibit a higher degree of integration making necessary some specific design considerations. Interconnected SMB processes might require a partial solvent removal between the units to balance the dilution of the product delivered by the upstream SMB unit. In systems with internal recycles usually purge streams or solvent removal steps are necessary, as indicated in Figure 5.25 (middle and right). Furthermore, a restrictive design aspect of such multizone processes is the identical switching time applied to all zones.

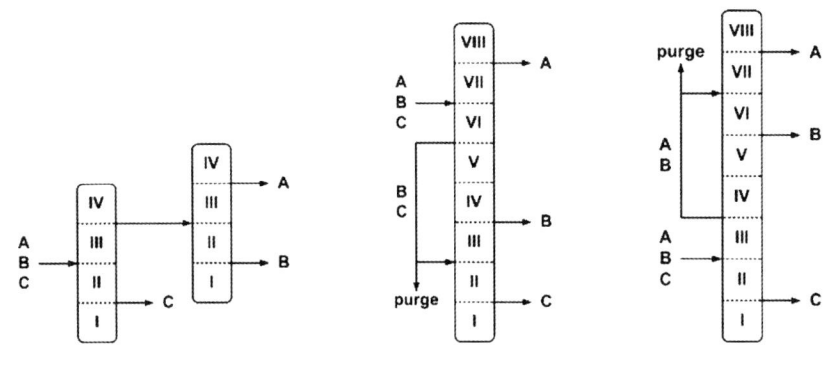

Figure 5.25 Examples for possible approaches for multicomponent separations by SMB chromatography for the case of separating a three-component mixture of A, B and C. Left – interconnected SMB units. Middle and right – different eight-zone set-ups with internal recycle streams (reproduced from Kaspereit, 2009).

5.2.8
Multicolumn Systems for Bioseparations

Above sections already describe plant concepts where chromatographic columns can be combined very flexibly according to process needs and optimization. Process schemes for bioseparations vary as well because in comparison to fine chemicals the separation of bioproducts is characterized by a wide range of the adsorption strength of large molecules and hence the necessity to change the eluent compositions, that is, the solvent strength within the process (Carta and Jungbauer, 2010).

Sometimes bioseparation reduces to a sequence of capture steps instead of real chromatographic separation. In order to obtain the product molecules, a capture step includes besides the loading of the stationary phase additional process steps like elution, washing, and regeneration. As shown in Figure 5.26 the column is first equilibrated before the sample is injected. Unbounded molecules pass through the column within the total dead time. Next the solvent strength is increased by a linear gradient. This sharpens the peaks and reduces the retention time. Finally, tightly bound impurities are eluted by a step gradient and are washed out. Before starting the batch cycle again the column has to be reequilibrated.

5.2.8.1 Sequential Multicolumn Chromatography (SMCC)

A standard task in biochromatography is to separate the product molecules, for instance a protein, from not or weakly retained impurities and very strongly adsorbing impurities. In case of batch chromatography this implies the following process steps with increasing solvent strength: loading the column with feed, washing out weakly adsorbed impurities, eluting the product, and regenerating the column from strongly adsorbed products. Finally, the solid phase is equilibrated and the cycle starts again. Besides high solvent consumption the low productivity is the

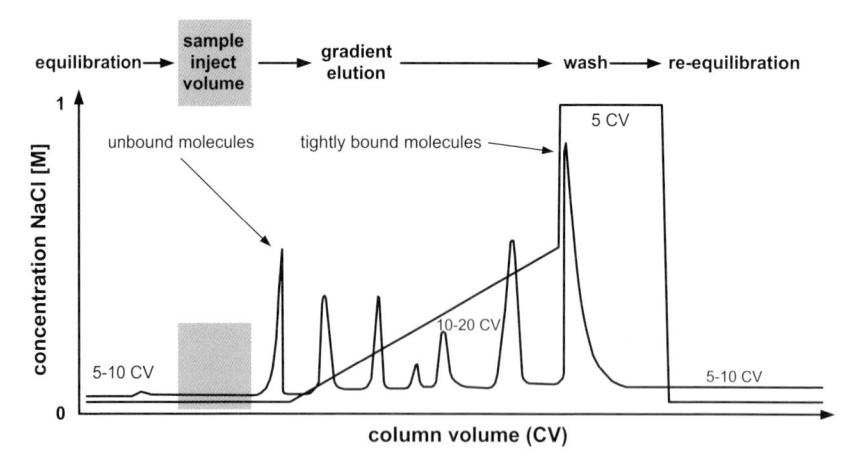

Figure 5.26 Typical ion exchange separation using linear gradient elution (reproduced from GE Healthcare Handbook 11-0004-21, 2011).

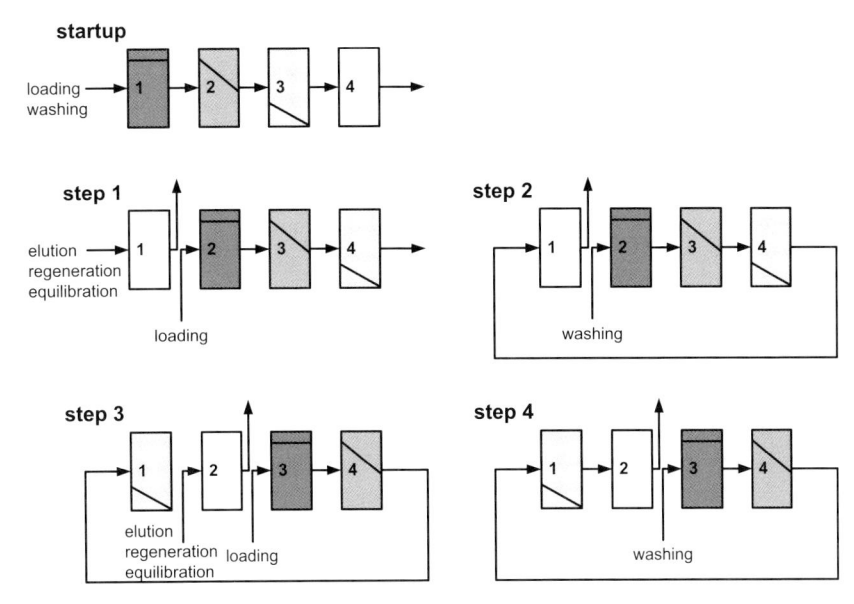

Figure 5.27 Sequential multicolumn chromatography.

major disadvantage of such process. This drawback is eliminated if instead of a unique batch column the different process steps are carried out in a sequence of shorter columns. Figure 5.27 illustrates the sequential multicolumn chromatography (SMCC) for using four columns (Holzer, Osuna-Sanchez, and David, 2008).

Initially the sequence of columns 1–3 is loaded like a batch column and afterwards columns 1–4 are washed until all product molecules within the liquid phase of column 1 are transferred into the subsequent columns. Then the sequential operation starts.

- Step 1: Column1 is eluted, regenerated, and finally equilibrated, while columns 2–4 are loaded without a breakthrough of the product.
- Step 2: Column 2 is washed until all product molecules within the liquid phase of column 2 are transferred into the subsequent columns.
- Step 3: As shown in Figure 5.27, the process scheme is changed again. Now column 2 is eluted, regenerated, and equilibrated while columns 3, 4, and 1 are loaded.
- Step 4: Now column 3 is washed until all product molecules within the liquid phase of column 3 are transferred into the subsequent columns.
- Steps 5 and 6 as well as further steps: The process scheme is changed analog to steps 3 and 4 and afterwards the process cycle starts again.

5.2.8.2 Multicolumn Countercurrent Solvent Gradient Purification (MCSGP)

The multicolumn countercurrent solvent gradient purification (MCSGP) process (Ströhlein et al., 2006) aims at the purification of biomolecules from earlier and later

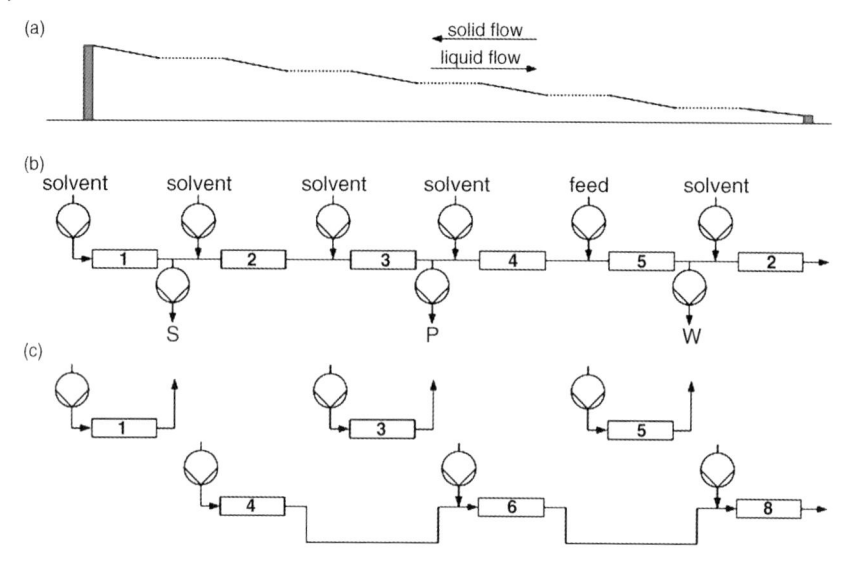

Figure 5.28 Continuous multicolumn countercurrent solvent gradient purification (MCSGP) process.

eluting impurities. It takes advantage of the reduction of the overall retention time by nonisocratic elution strength and the efficiency of countercurrent flow separation (Aumann, Ströhlein, and Morbidelli, 2007a, 2007b). The concept has been applied successfully to, for example, the purification of antibodies (Müller-Spath et al., 2010). Figure 5.28b describes the principles of the MCSGP process. Like in SMB processes, the counter-current flow of the stationary and the liquid phases is achieved by switching the column periodically upstream, that is, in the opposite direction of the liquid flow. The solvent gradient is adjusted by solvent pumps in front of each column. The solvent strength is the highest at the inlet of column 1 and decreases until the outlet of column 6. For the example shown in Figure 5.28a, the solvent concentration at the adjoining inlets and outlets of the columns is equal. However, this is not mandatory, depending on thermodynamics stepwise or other forms of the overall gradient may be chosen. The strongly adsorbing component S elutes on the left while the product P is gained in the middle and the weakly adsorbed component W elutes on the right.

Aumann, Ströhlein, and Morbidelli (2007a) describe the process operation as follows: "Only weak impurities W and product P are present in column 6, where they strongly adsorb. In column 5, the feed that also contains strongly adsorbing impurities S is added to the unit. P and S strongly adsorb in column 5, so that they never leave column 5 with the liquid flow. But the conditions in column 5 are designed, so that W leaves the system after several switches through the outlet of column 5. In column 4, all W has to leave the column before it switches to the position 3. The mobile phase conditions are designed so that P and S adsorb, but W desorbs. After W has left completely, column 4 is switched to position 3 and pure product can be collected. No W is present in column 3 and S adsorbs strongly. Before S starts to contaminate the product outlet flow $Q_{3,out}$, column 3 is switched to position 2 from where the side fraction containing P is washed for recycling into column 4. When column 2 does not contain product anymore, it is switched to position 1 to clean out S and to perform some cleaning in place if required."

Figure 5.28c shows the real scheme of the MCSGP process. Columns 1, 3, and 5 operate in batch mode while columns 2, 4, and 6 are interconnected. It is not necessary to operate all columns at the same time, therefore the operation time of the batch columns as well as the interconnected columns can be chosen individually, which offers additional parameters for process optimization. Aumann and Morbidelli (2008) also point out that in order to reduce investment costs the number of columns can be halved. In this case, columns 1, 3, and 5 are operated in batch mode. Afterwards, the column switching takes place so that the former columns 1, 3, and 5 are now the interconnected columns 2, 4, and 6.

5.2.9
Countercurrent Chromatographic Reactors

5.2.9.1 SMB Reactor

The simulated moving bed reactor (SMBR) based on the SMB process is a practical alternative for implementing counter-current continuous reactors. Counter-current movement of the phases is simulated by sequentially switching the inlet and outlet ports located between the columns in direction of the liquid flow (Figure 5.29). As with the SMB process, two different concepts are known to realize the counter-current flow. One is based on switching the ports and the other on the movement of columns. Owing to the periodic switching the process reaches, after start up, a cyclic steady state. Depending on the stoichiometry of the reaction and the adsorptivity of the components, at most two pure products can be withdrawn. The concept

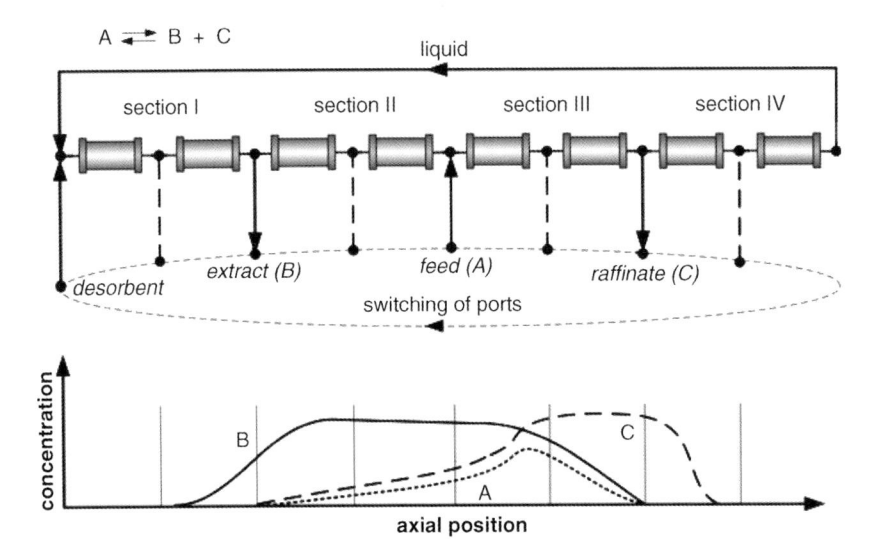

Figure 5.29 Chromatographic simulated moving bed reactor.

is usually proposed in the context of increasing conversion for equilibrium-limited reactions.

The design of SMBR processes has to take into account the requirements of different types of reactions. Therefore, different types of flowsheets and operating modes can be chosen. As with a semicontinuous operation, a process without section IV (regeneration of the solvent), a five-section process or a four-section process without recycle of the eluent may be advantageous.

The chromatographic SMB reactor has been examined for various reaction stoichiometries, with the main focus on reactions of the type $A + B \rightleftarrows C + D$. Examples are esterifications of acetic acid with methanol (Lode et al., 2003; Ströhlein, Mazzotti, and Morbidelli, 2005), ethanol (Mazzotti et al., 1996), and β-phenethyl alcohol (Kawase et al., 1996) as well as the production of bisphenol A (Kawase et al., 1999). The same reaction type can also be found for various hydrocarbons, such as the transfer reaction of sucrose with lactose to lactosucrose (Kawase et al., 2001) and the hydrolysis of lactose (Shieh and Barker, 1996). Barker et al. (1992), Kurup et al. (2004) and Ströhlein, Mazzotti, and Morbidelli (2005) focused on reactions of the type $A \rightleftarrows B + C$, such as enzyme-catalyzed sucrose inversion and the production of dextran.

Reactions of the type $A \rightleftarrows B$ have been investigated, for example, the isomerization of glucose to fructose, by Fricke (2005) as well as Toumi and Engell (2004). However, for isomerizations conventional SMB reactors are suitable only if the purity requirements are moderate. In other cases, a spatial distribution of the reaction and separation is necessary as discussed below.

Based on the triangle theory for TMB reactors Lode et al. (2002, 2003) as well as Fricke et al. (2003, 2005) developed short cut methods for the preliminary design of SMB reactors. They derived analytical solutions for different types of reactions by taking into account the conditions for different subdivisions of the separation region.

5.2.9.2 Processes with Distributed Functionalities

Conventional chromatographic SMB reactors as described in Section 5.2.9.1 are not well suited for reactions of the type $A \rightleftarrows B$. Due to the reaction proceeding everywhere throughout the whole unit it is only possible to slightly overcome the chemical equilibrium and total conversion is not possible. However, specially designed processes can overcome this limitation.

Hashimoto et al. (1983) proposed a reactive counter-current chromatographic process for the production of higher fructose syrup. This process, also known as the three-section Hashimoto process, can be applied for these reactions. It is based on a partial de-integration of reaction and separation. Columns and reactors are alternately arranged in one or several sections of the SMB process (Figure 5.30a). Characteristic for this process is that the reactors remain stationary in the assigned section. For the example shown in Figure 5.30a, after shifting the nodes the reactors will be located in front of the columns numbered 6, 7, 8, and 1. Movement of the reactors is also responsible for the characteristic internal axial concentration profiles and process dynamics of the Hashimoto process. By shifting the reactors, their liquid content is moved from the inlet to the outlet of the column. At the reactor outlet, a much

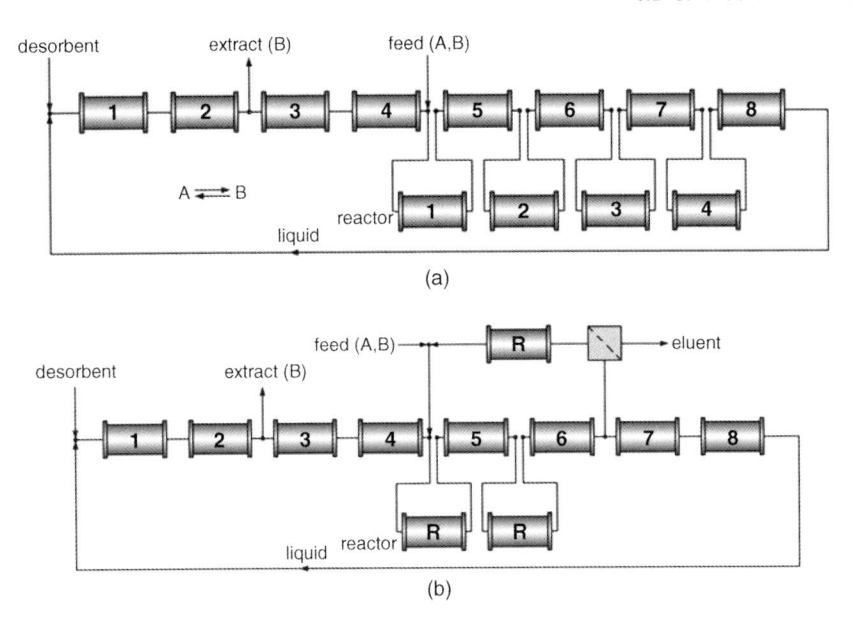

Figure 5.30 (a) Three-section Hashimoto process, (b) four-section process variant with raffinate recycle.

higher concentration level results than at the inlet of the following column. Therefore, a discontinuity in the axial concentration profile arises, which is balanced over the switching interval by adsorption and hydrodynamic effects. To obtain high product purity, a considerable number of reactors and columns are required in the "reactive" zone; in particular if the weaker adsorbing component should be produced. Michel et al. (2003), (2007) proposed electrochemical side reactors for producing arabinose. Several modifications of the concept were proposed, including four-zone processes aiming at the removal of side products or improved performance. Figure 5.30b depicts such process variant where the raffinate stream is recycled via an outside reactor to the feed stream.

The strength of the Hashimoto process stems from the fact that it separates the functionalities separation and reaction as well as its flexibility. Various types of (bio)-chemical reactions can be carried out in the side reactors. The concept offers the possibility of exchanging adsorbent and catalyst (e.g., immobilized enzymes) separately after different operating times. It also allows for different operating conditions (e.g., temperatures) for reaction as well as separation. An interesting application of this kind is the thermal racemization of enantiomers (Borren, 2007) if the stronger adsorbed enantiomer should be produced. Michel et al. (2003), (2007) proposed electrochemical side reactors for producing arabinose. Several authors investigated the isomerization of glucose to fructose (Hashimoto et al., 1983, Borren and Schmidt-Traub, 2004, Borges da Silva et al. (2006), Zhang, Hidajat, Ray, 2007). A possible economic benefit is also expected for p-xylene isomerization using four-zone schemes similar to Figure 5.30b (with a product outlet instead of the recycle) by Minceva et al. (2008) and Bergeot et al. (2010).

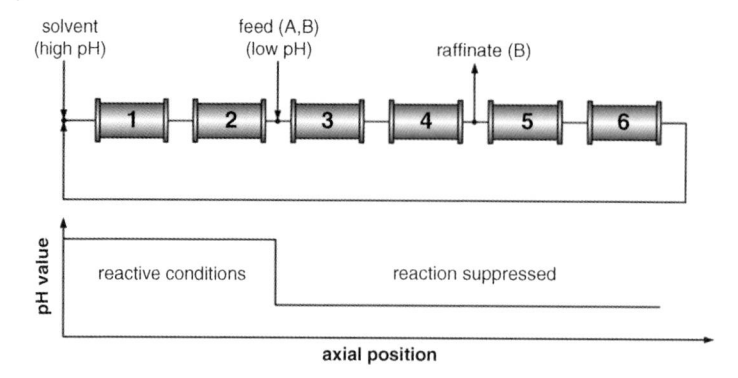

Figure 5.31 Reactive SMB process with a pH gradient for the production of the weaker adsorbing enantiomer B from a racemic mixture at 100% purity and yield.

An application for reactive systems with more than two components is also possible if special catalysts or operating conditions are required. However, an even more complex design, as well as control of the process, has to be taken into account.

The Hashimoto concept is particularly suited for separations with low to moderate purity requirements, but has limited performance for very high product purities (García Palacios, Kaspereit, and Kienle, 2009). As regards isomers, it is not capable of producing the weaker adsorbing species at high purity and 100% yield. A possible work-around for such systems is to replace the side reactors by chromatographic reactors in the corresponding zone. This concept suggested by García Palacios, Kaspereit, and Kienle (2009, 2011a) can be realized by a step gradient of, for example, the pH value, the temperature, or a homogeneous catalyst. Goal is to accelerate the reaction in one zone, while it should be suppressed in the other zones, that is to spatially distribute reaction and separation functionalities. Figure 5.31 explains the approach, which can be adapted to produce either the weaker or stronger adsorbing isomer. Such set-up was successfully applied to produce the weaker adsorbing enantiomer from a racemic mixture at 100% yield (García Palacios, Kaspereit, and Kienle, 2011a; García Palacios *et al.*, 2011b). In practical applications, the major challenge related to this concept is certainly to find a suitable variable to manipulate and a gradient thereof that provides for a sufficient change of the reaction rate between the zones.

5.3
Choice of Process Concepts

Every new separation is an individual task that needs the skills of experts to find the optimal chromatographic system and to scale-up the process according to proven routes for successful process development. Developing a new separation based on routines can ensure a trouble-free development but might also risk not finding the optimum process economy and thus endanger the whole project. Therefore, every separation should be developed as open-mindedly as possible.

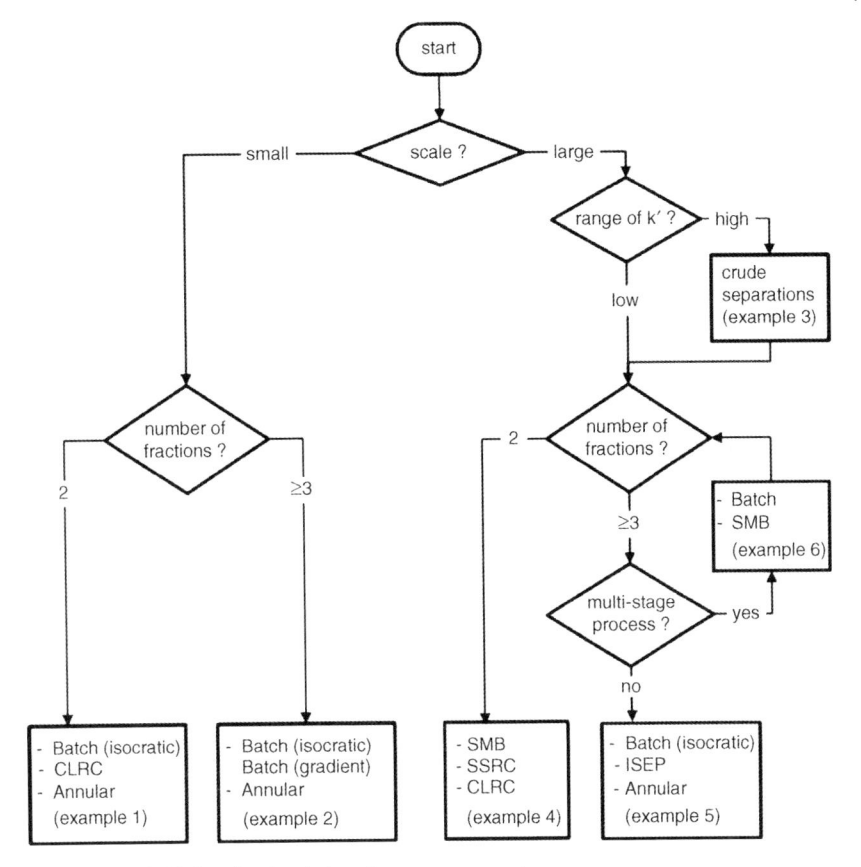

Figure 5.32 Guidelines for the choice of a process concept.

This section provides some guidelines to assist the decision for an appropriate process concept. The main decision criteria of the approach, as shown in Figure 5.32, are "scale," "range of k'," and "number of fractions." Processes resulting from this decision tree are explained by examples consisting of an analytical chromatogram representing the early stage of process development and the chromatogram for the final realization of the preparative process.

5.3.1
Scale

The main criterion to be considered is the "scale" of the project, which distinguishes between large or production scale (kg a^{-1} to t a^{-1}) and small or laboratory scale (mg a^{-1} to g a^{-1}). This implies the question of whether the separation justifies a time-consuming method development and process design to improve process performance in terms of productivity, eluent consumption, and yield.

Influencing aspects in this context are for instance:

- total amount of feed mixture to be separated;
- duration and frequency of the project;
- equipment available in the laboratory or at the production site;
- experience and knowledge about certain process concepts.

5.3.2
Range of k'

Chromatographic processes show their best performance when the adsorption behavior of the components to be separated is not too different. Target components of the feed mixture should elute within a certain window. Under optimized conditions, k' is in the range of 2–8. Within that window the selectivity between the target component and the impurities should be optimized (Chapter 3).

If early- or late-eluting components are present in the feed, the front or rear end components can be removed by simple prepurification steps. These steps need not be chromatographic separations. Alternatives are extraction techniques, as the components normally differ substantially in their polarity and, thus, solubility behavior.

5.3.3
Number of Fractions

To find a suitable process concept, the final differentiation is given by the number of fractions to be collected. Notably, one fraction can contain either one single component (e.g., one target product) or a group of many similar components (e.g., several impurities). Some process concepts, such as SMB and SSR chromatography, can only separate a feed mixture into two fractions. In contrast, batch elution chromatography, annular chromatography, ISEP, and so on separate feed streams into three or more fractions.

Multistage processes should be considered for production-scale processes with three or more fractions. An intermediate step by SMB or batch separation reduces the separation problem, finally, to a two-fraction problem that can be performed by applying one of the above-mentioned concepts.

5.3.4
Example 1: Lab Scale; Two Fractions

A racemic mixture has to be separated to produce several grams of a pure enantiomer for further reactions. By optimizing the batch conditions, a good selectivity but no baseline separation is obtained. Details are given in Table 5.1 and Figure 5.33.

Because only a small amount of the target component was available the yield of the chromatographic separation should be as high as possible. Therefore, CLRC (Section 5.1.3) with peak shaving was chosen to avoid fractions with insufficient purity, which otherwise had to be reworked. Figure 5.34 depicts the resulting

Table 5.1 Example 1.

	Analytical scale	Production scale
Sample	Chiral sulfoxide	
Adsorbent	Chiralcel® OD, 20 µm	
Mobile phase	Heptane–isopropanol (85:15) isocratic	
Column	250 × 4.6 mm, stainless steel	200 × 50 mm, stainless steel

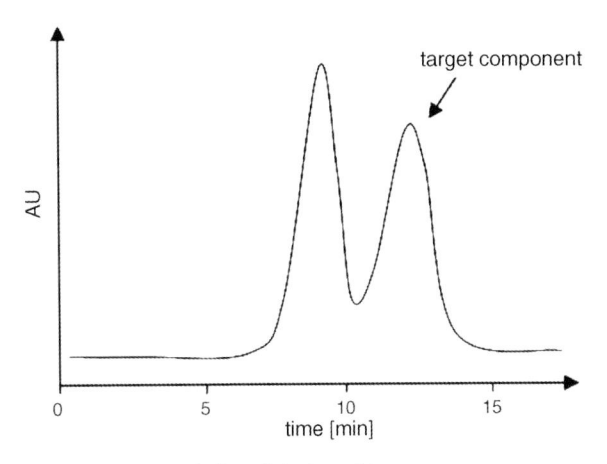

Figure 5.33 Example 1: analytical conditions.

chromatogram of the preparative process with four cycles. Both enantiomers were collected directly during the first cycle. In cycle two, the main part of the second eluting component was withdrawn from the system so that in cycle three nearly no second enantiomer is left.

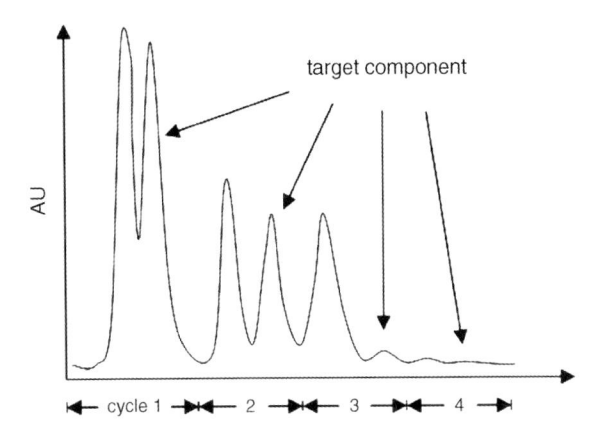

Figure 5.34 Example 1: preparative conditions.

Table 5.2 Example 2.

	Analytical scale	Production scale
Sample	Drug metabolites from urine	
Adsorbent	LiChrospher® 100 RP-18, 5 µm	Superspher® 100 RP-18, 4 µm
Mobile phase	Acetonitrile–water; gradient	Acetonitrile–water; gradient
Column	250 × 4 mm, stainless steel	200 × 50 mm, stainless steel

5.3.5
Example 2: Lab Scale; Three or More Fractions

Only a few milligram of a drug metabolite had to be separated from a patient's urine for structure elucidation purposes. Table 5.2 summarizes the conditions of orienting experiments.

In the resulting chromatogram, the target component is surrounded by other components (Figure 5.35). In addition, all components elute within a wide range of k'. However, since the total amount to be separated is very small, no additional preseparation steps were implemented. For the same reason no further effort was made to improve the chromatographic system. The elution conditions were linearly transferred to a larger column (Table 5.2) and the loading factor increased until the target component eluted immediately behind the earlier eluting impurity (Figure 5.36). In this case, a smaller adsorbent particle diameter was chosen for the preparative column to achieve the necessary plate number in the column.

As only milligrams or grams of product were to be purified in the above examples, relatively little effort was made to optimize the chromatographic system and

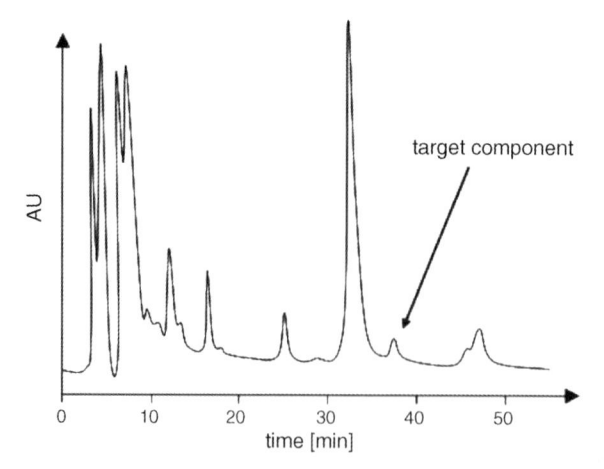

Figure 5.35 Example 2: analytical conditions.

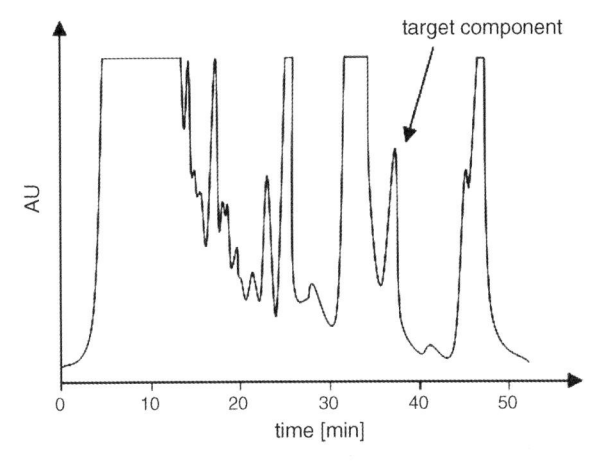

Figure 5.36 Example 2: preparative conditions.

quite simple concepts were applied. This is completely different to processes with production rates of 100 kg a^{-1} or higher.

5.3.6
Example 3: Production Scale – Wide Range of k'

Several flavonoids have to be separated from a plant extract, where the desired production rate is in the range of several tons per year. At the very beginning of the project the complete feed stock has been analyzed under the conditions listed in Table 5.3.

The resulting chromatogram (Figure 5.37) reveals a rather wide range of retention times for the multiple components in the mixture, even though a gradient has been applied. This indicates that the components have very different physical properties, and a crude separation at the beginning is sufficient to remove the main impurities.

Adsorption under isocratic conditions was chosen as a crude separation step. Chromatographic parameters were adjusted to a simple water–alcohol solvent on a coarse RP-18 silica (40–63 µm). Figure 5.38 shows the corresponding chromatogram. Using this prepurification step, the components of interest are obtained at

Table 5.3 Example 3.

	Analytical scale	Production scale
Sample	Flavonoids from plant extract	
Adsorbent	LiChrospher® 100 RP-18, 5 µm	LiChroprep® 100 RP-18, 40–63 µm
Mobile phase	Acetonitrile–water; gradient	Water–ethanol (90:10), isocratic
Column	250 × 4 mm, stainless steel	125 × 25 mm, stainless steel

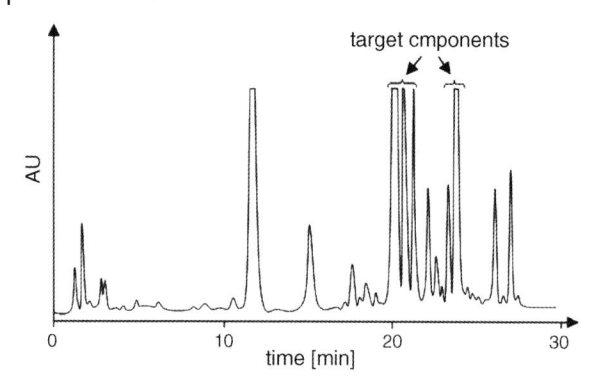

Figure 5.37 Example 3: analytical conditions.

higher concentrations. In subsequent polishing steps single flavonoids can be further purified.

5.3.7
Example 4: Production Scale; Two Main Fractions

A pharmaceutical component had to be isolated from a multicomponent mixture at a production rate of $>100\,\text{kg\,a}^{-1}$. After adjusting conditions for the preparative separations (Table 5.4), a chromatogram was obtained on a small column (Figure 5.39).

This analytical chromatogram indicates that some early-eluting impurities are present in the feed mixture as well as two main impurities eluting after the target component. No baseline separation could be obtained between the component of

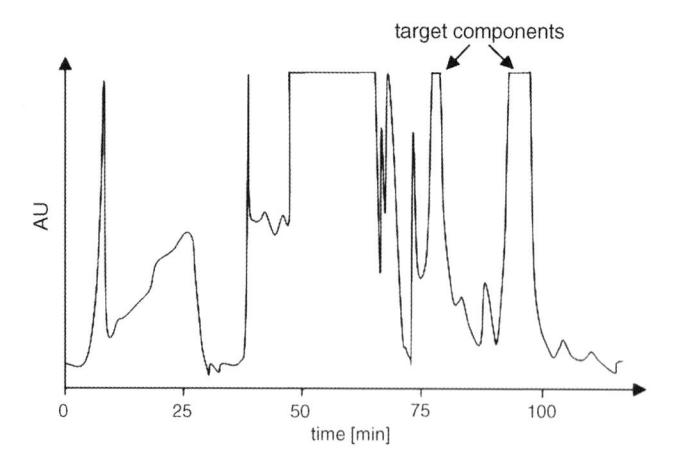

Figure 5.38 Example 3: preparative conditions.

Table 5.4 Example 4.

	Analytical scale	Production scale
Sample	Pharmaceutical component	
Adsorbent	LiChrospher® Si 60, 15 µm	
Mobile phase	n-Heptane/ethylacetate, isocratic	
Column	250 × 4 mm, stainless steel	250 × 25 mm, stainless steel

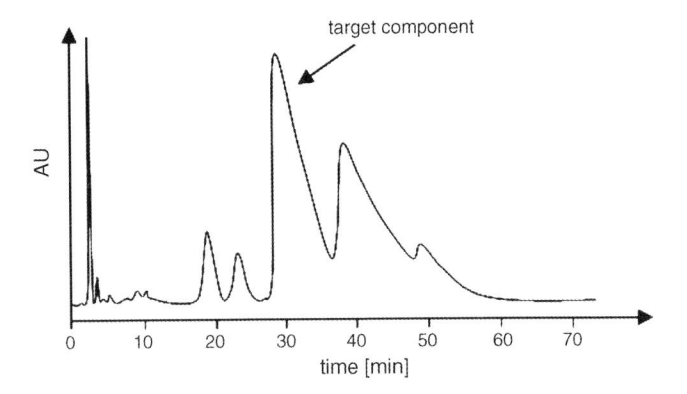

Figure 5.39 Example 4: analytical conditions.

interest and the late-eluting impurities. To avoid intermediate fractions with insufficient purities, which would have to be stored and reworked afterwards CLRC was applied. Here the early-eluting impurities are sent to waste during the first cycle, reducing the separation problem to a two-component separation. By only one additional cycle the target component can be obtained in good purity and yield.

Complete conditions for the preparative separation and corresponding chromatogram are given in Table 5.4 and Figure 5.40.

5.3.8
Example 5: Production Scale; Three Fractions

For the production scale separation of a prostaglandine derivative (about 100 kg a^{-1}), the chromatographic system in terms of stationary and mobile phases was optimized in advance according to the considerations described in Chapter 3 (Table 5.5).

Figure 5.41 indicates that, besides the target product, early- and late-eluting impurities are also present. Following the considerations made in the beginning of this section the decision has to be made as to whether the separation should be performed with a process concept, allowing a multifraction withdrawal, or a

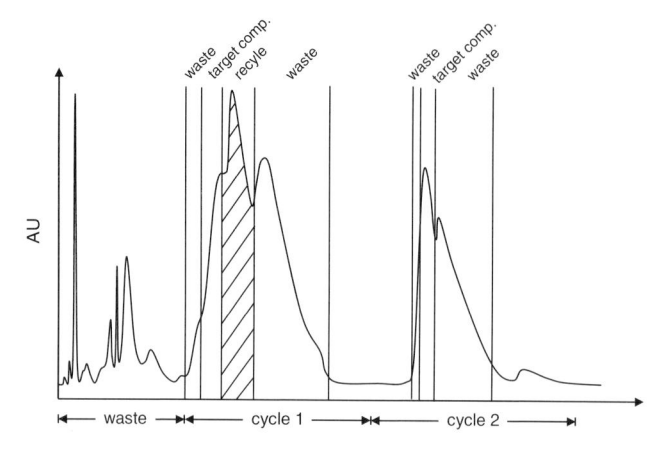

Figure 5.40 Example 4: preparative conditions.

multistage process. In the present case, isocratic batch chromatography on an existing process-scale HPLC system was used. The chromatographic system was kept constant with regard to stationary and mobile phase. Only the particle diameter of the silica sorbent was increased. As the stationary phases are manufactured by the same production process, the different particle sizes did not change the adsorbent's properties, especially in terms of selectivity. Figure 5.42 shows the chromatogram under preparative conditions.

Besides the chosen batch separation, preparative annular or ISEP chromatographic processes are alternatives if the required equipment is available.

5.3.9
Example 6: Production Scale; Multi-Stage Process

100 kg a^{-1} of a cyclic peptide had to be separated from a fermentation broth. Analytical experiments (Table 5.6) afforded an initial chromatogram (Figure 5.43).

Table 5.5 Example 5.

	Analytical scale	Production scale
Sample	Prostaglandine	
Adsorbent	LiChrosorb® Si 60, 10 μm	LiChroprep® Si 60, 25–40 μm
Mobile phase	n-Heptane–isopropanol–methanol–THF (96/2.4/1/0.6)	
Column	250 × 4 mm, stainless steel	600 × 200 mm, stainless steel

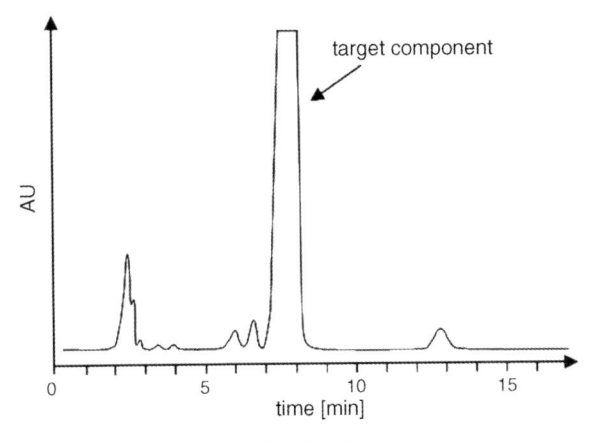

Figure 5.41 Example 5: analytical conditions.

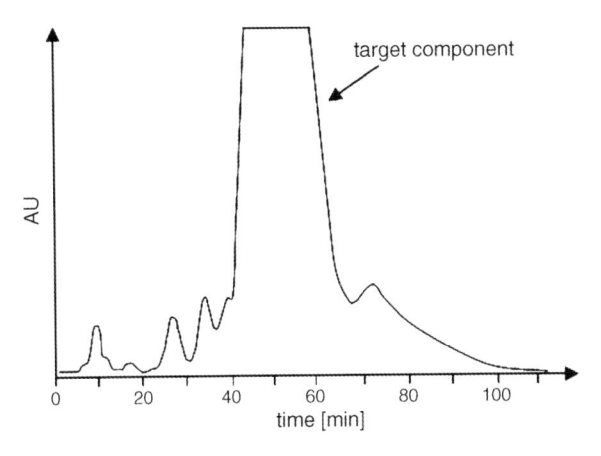

Figure 5.42 Example 5: preparative conditions.

Again the target component is surrounded by both early- and late-eluting impurities. However, in contrast to the foregoing example, here a multistage process consisting of a batch separation followed by two SMB separations was applied. The first batch unit divided the complete feed into two fractions. The

Table 5.6 Example 6.

	Analytical scale (step 1)	Analytical scale (step 2)
Sample	Cyclic peptide from fermentation broth	
Adsorbent	LiChrospher® 100 RP-18e. 5 µm	LiChrospher® 100 RP-18. 12 µm
Mobile phase	Acetonitrile–water; gradient	Acetonitrile–water, isocratic
Column	250 × 4 mm, stainless steel	250 × 4 mm, stainless steel

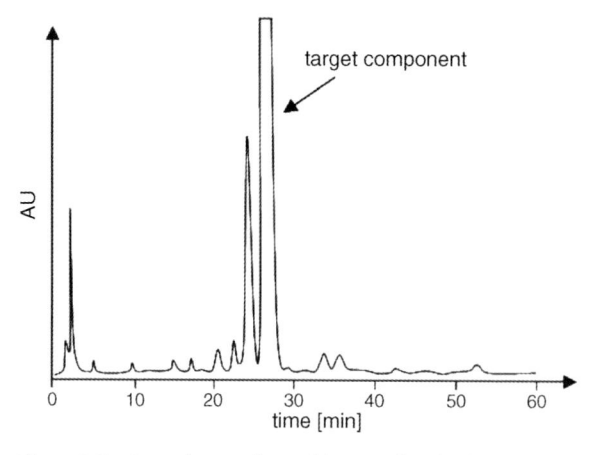

Figure 5.43 Example 6: analysis of the complete feed stock.

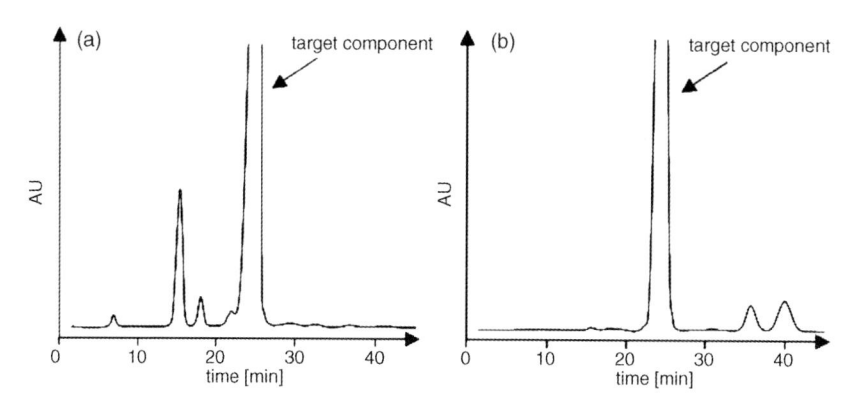

Figure 5.44 Example 6: analysis of the two fractions obtained by the initial batch separation.

first fraction contains early-eluting impurities and approximately half of the target component, while the rest of the main product and late-eluting impurities are collected in the second fraction. Figure 5.44 shows an analysis of these two fractions.

Both fractions obtained from the first separation can be processed in subsequent SMB units where, again, the respective portion is split into two fractions. In this case fraction (a) is fed to a SMB unit and the target product is collected as extract. An SMB separation of fraction (b) leads to a raffinate stream containing the main product and the extract line where all impurities are collected.

Even though three chromatographic steps are now involved instead of just one batch separation, the overall process economy is increased because a higher yield is achieved and no intermediate fractions with lower purities have to be collected and reworked (Voigt, Hemple, and Kinkel, 1997).

References

Abel, S., Mazzotti, M., and Morbidelli, M. (2002) Solvent gradient operation of simulated moving beds I: linear isotherms. *J. Chromatogr. A*, **944**, 23–39.

Abel, S., Mazzotti, M., and Morbidelli, M. (2004) Solvent gradient operation of simulated moving beds II: Langmuir isotherms. *J. Chromatogr. A*, **1026**, 47–55.

Antos, D. and Seidel-Morgenstern, A. (2001) Application of gradients in simulated moving bed processes, *Chem. Eng. Sci.*, **56**, 6667–6682.

Aumann, L. and Morbidelli, M. (2008) A semicontinuous 3-column countercurrent solvent gradient purification (MCSGP) process. *Biotech. Bioeng.*, **99**, 728–733.

Aumann, L., Ströhlein, G., and Morbidelli, M. (2007a) Parametric study of a 6-column countercurrent solvent gradient purification (MCSGP) unit. *Biotech. Bioeng.*, **98**, 1029–1042.

Aumann, L., Ströhlein, G., and Morbidelli, M. (2007b) A continuous multicolumn countercurrent solvent gradient purification (MCSGP) process. *Biotech. Bioeng.*, **98**, 1043–1055.

Bae, Y.-S. and Lee, C.-H. (2006) Partial-discard strategy for obtaining high purity products using simulated moving bed chromatography. *J. Chromatogr. A*, **1122**, 161–173.

Bailly, M. and Tondeur, D. (1982) Recycle optimization in non-linear productive chromatography – I Mixing recycle with fresh feed. *Chem. Eng. Sci.*, **37**, 1199–1212.

Barker, P.E., Ganetsos, G., Ajongwen, J., and Akintoye, A. (1992) Bioreaction-separation on continuous chromatographic systems. *Chem. Eng. J.*, **50**, B23–B28.

Bassett, D.W. and Habgood, H.W. (1960) A gas chromatographic study of the catalytic isomerisation of cyclopropane. *J. Phys. Chem.*, **64**, 769–773.

Beltscheva, D., Hugo, P., and Seidel-Morgenstern, A. (2003) Linear two-step gradient counter-current chromatography: analysis based on a recursive solution of an equilibrium stage model. *J. Chromatogr. A*, **989**, 31–45.

Bergeot, G., Le-Cocq, D.L., Wolff, L., Muhr, L., and Bailly, M. (2010) Intensification of paraxylene production using a simulated moving bed reactor. *Oil Gas Sci. Technol.*, **65**, 721–733.

Borges da Silva, E., de Souza, A., de Souza, S., and Rodrigues, A. (2006) Analysis of high-fructose syrup production using reactive SMB technology. *Chem. Eng. J.*, **118**, 167–181.

Borren, T. (2007) Untersuchungen zu chromatographischen Reaktoren mit verteilten Funktionalitäten, Fortschritt-Berichte VDI: Reihe 3 Nr.876, VDI Verlag GmbH, Düsseldorf.

Borren, T. and Schmidt-Traub, H. (2004) Vergleich chromatographischer Reaktorkonzepte. *Chem. Ing. Tech.*, **76** (6), 805–814.

Broughton, D.B. and Gerhold, C.G. (1961) Continuous sorption process employing fixed bed of sorbent and moving inlets and outlets, US Patent No. 2.985.589.

Brozio, J. and Bart, H.-J. (2004) A rigorous model for annular chromatography. *Chem. Eng. Technol.*, **27**, 962–970.

Carr, R.W. (1993) Continuous reaction chromatography, in *Preparative and Production Scale Chromatography* (eds G. Ganetsos and P.E. Barker), Marcel Dekker Inc., New York.

Carta, G. and Jungbauer, A. (2010) *Protein Chromatography: Process Development and Scale-Up*, Wiley-VCH Verlag, Weinheim.

Chin, C.Y. and Wang, N.-H.L. (2004) Simulated moving bed equipment designs. *Sep. Purif. Rev.*, **33**, 77–155.

Ching, C.B. and Ruthven, D.M. (1986) Experimental study of a simulated counter-current adsorption system – IV. Non-Isothermal operation. *Chem. Eng. Sci.*, **41**, 3063.

Ching, C.B., Ho, C., and Ruthven, D.M. (1986) An improved adsorption process for the production of high-fructose syrup. *AIChE J.*, **32**, 1876.

Clavier, J.-Y., Nicoud, R.M., and Perrut, M. (1996) A new efficient fractionation process: the simulated moving bed with supercritical eluent, in *High-Pressure*

Chemical Engineering (eds P.R. von Rohr and C. Trepp), Elsevier Science, London.

Coca, J., Bravo, M., Abascal, E., and Adrio, G. (1989) Dicyclopentadiene dissociation in a chromatographic reactor – effect of the liquid phase polarity on the reaction rate. Chromatographia, **28**, 300–302.

Coca, J., Adrio, G., Jeng, C.Y., and Langer, S.H. (1993) Gas and liquid chromatographic reactors, in Preparative and Production Scale Chromatography (eds G. Ganetsos and P.E. Barker), Marcel Dekker Inc., New York.

Colin, H., Hilaireau, P., and Martin, M. (1991) Flip-Flop elution concept in preparative liquid chromatography. J. Chromatogr. A, **557**, 137–153.

Denet, F., Hauck, W., Nicoud, R.M., Giovanni, O.D., Mazzotti, M., Jaubert, J.N., and Morbidelli, M. (2001) Enantioseparation through supercritical fluid simulated moving bed (SF-SMB) chromatography. Ind. Eng. Chem. Res., **40**, 4603–4609.

Depta, A., Giese, T., Johannsen, M., and Brunner, G. (1999) Separation of stereoisomers in a simulated moving bed-supercritical fluid chromatography plant. J. Chromatogr. A, **865**, 175–186.

Duvdevani, I., Biesenberger, J.A., and Tan, M. (1971) Recycle gel permeation chromatography. III. Design modifications and some results with polycarbonate. J. Polym. Sci., Part B: Polym. Lett., **9**, 429–434.

Eagle, S. and Rudy, C.E. Jr., (1950) Separation and desulfurization of cracked naphtha. Application of cyclic adsorption process. Ind. Eng. Chem., **42**, 1294–1299.

Falk, T. and Seidel-Morgenstern, A. (2002) Analysis of a discontinuously operated chromatographic reactor. Chem. Eng. Sci., **57**, 1599–1606.

Fricke, J. (2005) Entwicklung einer Auslegungsmethode für chromatographische SMB-Reaktoren, Fortschritt-Berichte VDI: Reihe 3 Nr. 844, VDI Verlag GmbH, Düsseldorf.

García Palacios, J., Kaspereit, M., and Kienle, A. (2009) Conceptual design of integrated chromatographic processes for the production of single (Stereo-)isomers. Chem. Eng. Technol., **32**, 1392–1402.

García Palacios, J., Kaspereit, M., and Kienle, A. (2011a) Integrated simulated moving bed processes for the production of single enantiomers. Chem. Eng. Technol., **34**, 688–698.

García Palacios, J., Kramer, B., Kienle, A., and Kaspereit, M. (2011b) Experimental validation of a new integrated simulated moving bed for the production of single enantiomers. J. Chromatogr. A, **1218**, 2232–2239.

GE Healthcare Handbook 11-0004-21: Ion Exchange Chromatography & Chromatofocusing, Principles and Methods, http://www.gelifesciences.com/aptrix/upp00919.nsf/Content/3E56FFCAFE43BCADC1257628001D0EC6/$file/11000421AB.pdf, viewed Oct. 2011.

Gedicke, K., Antos, D., and Seidel-Morgenstern, A. (2007) Effect on separation of injecting samples in a solvent different from the mobile phase. J. Chromatogr. A, **1162**, 62–73.

Giovanni, O.D., Mazzotti, M., Morbidelli, M., Denet, F., Hauck, W., and Nicoud, R.M. (2001) Supercritical fluid simulated moving bed chromatography. II Langmuir isotherm. J. Chromatogr. A, **919**, 1–12.

Grill, C.M. and Miller, L. (1998) Separation of a racemic pharmaceutical intermediate using closed-loop steady state recycling. J. Chromatogr. A, **827**, 359–371.

Grill, C.M., Miller, L., and Yan, T.Q. (2004) Resolution of a racemic pharmaceutical intermediate – A comparison of preparative HPLC, steady state recycling, and simulated moving bed. J. Chromatogr. A, **1026**, 101–108.

Gueorguieva, L., Palani, S., Rinas, U., Jayaraman, G., and Seidel-Morgenstern, A. (2011) Recombinant protein purification using gradient assisted simulated moving bed hydrophobic interaction chromatography. Part II: process design and experimental validation. J. Chromatogr. A, **1218**, 6402–6411.

Hashimoto, K., Adachi, S., Noujima, H., and Ueda, Y. (1983) A new process combining adsorption and enzyme reaction for producing higher-fructose syrup. *Biotechnol. Bioeng*, **25**, 2371–2393.

Herbsthofer, R., Bart, H.J., and Brozio, J. (2002) Einfluss der Reaktionskinetik auf die Funktion eines chromatographischen Reaktors. *Chem. Ing. Tech.*, **74**, 1006–1011.

Holzer, M., Osuna-Sanchez, H., and David, L. (2008) Multicolumn chromatography. *Bioprocess Int.*, **6** (8), 74–84.

Hotier, G. and Nicoud, R.M. (1995) Separation by simulated moving bed chromatography with dead volume correction by desynchronization of periods. European Patent EP 688589 A1.

Houwing, J., Billiet, H.A.H., and van der Wielen, L.A.M. (2002) Optimization of azeotropic protein separation in gradient and isocratic ion-exchange simulated moving bed chromatography. *J. Chromatogr. A*, **944**, 189–201.

Jeng, C.Y. and Langer, S.H. (1989) Hydroquinone oxidation for the detection of catalytic activity in liquid chromatographic columns. *J. Chromatogr. Sci.*, **27**, 549–552.

Jensen, T.B., Reijns, T.G., Billiet, H.A., and van der Wielen, L.A. (2000) Novel simulated moving-bed method for reduced solvent consumption. *J. Chromatogr A*, **873**, 149–162.

Jin, W. and Wankat, P.C. (2007) Thermal operation of four-zone simulated moving beds. *Ind. Eng. Chem. Res.*, **46**, 7208–7220.

Kalbé, J., Höcker, H., and Berndt, H. (1989) Design of enzyme reactors as chromatographic columns for racemic resolution of amino acid esters. *Chromatographia*, **28**, 193–196.

Kaspereit, M. (2009) Advanced operation concepts for simulated moving bed processes, in *Advanced Chromatography*, Vol. 47 (eds E. Grushka and N. Grinberg), CRC Press, Boca Raton, pp. 165–192.

Kaspereit, M. and Sainio, T. (2011) Simplified design of steady-state recycling chromatography under ideal and nonideal conditions. *Chem. Eng. Sci.*, **66**, 5428–5438.

Katsuo, S. and Mazzotti, M. (2010a) Intermittent simulated moving bed chromatography: 1. Design criteria and cyclic steady-state. *J. Chromatogr. A*, **1217**, 1354–1361.

Katsuo, S. and Mazzotti, M. (2010b) Intermittent simulated moving bed chromatography: 2. Separation of Tröger's base enantiomers. *J. Chromatogr. A*, **1217**, 3067–3075.

Kawase, M., Suzuki, T.B., Inoue, K., Yoshimoto, K., and Hashimoto, K. (1996) Increased esterification conversion by application of simulated moving-bed reactor. *Chem. Eng. Sci.*, **51** (11), 2971–2976.

Kawase, M., Inoue, Y., Araki, T., and Hashimoto, K. (1999) The simulated moving-bed reactor for production of bisphenol A. *Catal. Today*, **48**, 199–209.

Kawase, M., Pilgrim, A., Araki, T., and Hashimoto, K. (2001) Lactosucrose production using a simulated moving bed reactor. *Chem. Eng. Sci.*, **56**, 453–458.

Keßler, L.C. and Seidel-Morgenstern, A. (2006) Theoretical study of multicomponent continuous countercurrent chromatography based on connected 4-zone units. *J. Chromatogr. A*, **1126**, 323–337.

Keßler, L.C. and Seidel-Morgenstern, A. (2008) Improving performance of simulated moving bed chromatography by fractionation and feed-back of outlet streams. *J. Chromatogr. A*, **1207**, 55–71.

Kim, J.K., Abunasser, N., Wankat, P.C., Stawarz, A., and Koo, Y.-M. (2005) Thermally assisted simulated moving bed systems. *Adsorption*, **13**, 579.

Kloppenburg, E. and Gilles, E.D. (1999) A new concept for operating simulated moving-bed processes. *Chem. Eng. Technol.*, **22** (10), 813–817.

Kulprathipanja, S. (2002) *Reactive Separation Processes*, Taylor & Francis, New York.

Kurup, A., Subramani, H., Hidajat, K., and Ray, A. (2004) Optimal design and operation of SMB bioreactor for sucrose inversion. *Chem. Eng. J.*, **108**, 19–33.

Lanckriet, H. and Middelberg, A. (2004) Continuous chromatographic protein refolding. *J. Chromatogr. A*, **1022**, 103–113.

Lauer, K. (1980) Technische Herstellung von Fructose. *Starch/Stärke*, **32**, 11–13.

Li, S., Kawajiri, Y., Raisch, J., and Seidel-Morgenstern, A. (2010a) Optimization of simulated moving bed chromatography with fractionation and feedback: Part I. Fractionation of one outlet. *J. Chromatogr. A*, **1217**, 5337–5348.

Li, S., Kawajiri, Y., Raisch, J., and Seidel-Morgenstern, A. (2010b) Optimization of simulated moving bed chromatography with fractionation and feedback: Part II. Fractionation of both outlets. *J. Chromatogr. A*, **1217**, 5349–5357.

Lode, F., Francesconi, G., Mazzotti, M., and Morbidelli, M. (2003) Synthesis of methylacetate in a simulated moving bed reactor: experiments and modelling. *AIChE J.*, **49** (6), 1516–1524.

Ludemann-Hombourger, O., Bailly, M., and Nicoud, R.-M. (2000) The VARICOL-Process: a new multicolumn continuous chromatographic process. *Sep. Sci. Technol.*, **35**, 1829.

Machold, C., Schlegl, R., Buchinger, W., and Jungbauer, A. (2005) Continuous matrix assisted refolding of α-lactalbumin by exchange chromatography with recycling of aggregates combined with ultradiafiltration. *J. Chromatogr. A*, **1080**, 29–42.

Martin, A.J.P. (1949) Summarizing paper. *Discuss. Faraday Soc.*, **7**, 332.

Martin, A.J.P., Halász, I., Engelhardt, H., and Sewell, P. (1979) "Flip-flop" chromatography. *J. Chromatogr. A*, **186**, 15–24.

Mazzotti, M., Kruglov, A., Neri, B., Gelosa, D., and Morbidelli, M.A. (1996) A continuous chromatographic reactor: SMBR. *Chem. Eng. Sci.*, **51** (10), 1827–1836.

Mazzotti, M., Storti, G., and Morbidelli, M. (1996) Robust design of countercurrent adsorption separation processes: 3. Nonstoichiometric systems. *AIChE J.*, **42**, 2784–2796.

Meurer, M., Altenhöner, U., Strube, J., and Schmidt-Traub, H. (1997) Dynamic simulation of simulated moving bed chromatographic reactors. *J. Chromatogr. A*, **769**, 71–79.

Michel, M. (2007) Integration elektrochemischer Mikroreaktoren in chromatographische Trennverfahren, Fortschritt-Berichte VDI: Reihe 3 Nr.869, VDI Verlag GmbH, Düsseldorf.

Michel, M., Schmidt-Traub, H., Ditz, R., Schulte, M., Kinkel, J., Stark, W., Küpper, M., and Vorbrodt, M. (2003) Development of an integrated process for electrochemical reaction and chromatographic SMB-separation. *J. Appl. Electrochem.*, **33**, 939–949.

Migliorini, C., Wendlinger, M., Mazzotti, M., and Morbidelli, M. (2001) Temperature gradient operation of a simulated moving bed unit. *Ind. Eng. Chem. Res.*, **40**, 2606–2617.

Minceva, M., Gomes, P.S., Meshko, V., and Rodrigues, A.E. (2008) Simulated moving bed reactor for isomerization and separation of p-xylene. *Chem. Eng. J.*, **140**, 305–323.

Müller-Spath, T., Krattli, M., Aumann, L., Ströhlein, G., and Morbidelli, M. (2010) Increasing the activity of monoclonal antibody therapeutics by continuous chromatography (MCSGP). *Biotechnol. Bioeng.*, **107**, 652–662.

Palani, S., Gueorguieva, L., Rinas, U., Seidel-Morgenstern, A., and Jayaraman, G. (2011) Recombinant protein purification using gradient-assisted simulated moving bed hydrophobic interaction chromatography. Part I: selection of chromatographic system and estimation of adsorption isotherms. *J. Chromatogr. A*, **1218**, 6396–6401.

Ruthven, D.M. and Ching, C.B. (1989) Counter-current and simulated counter-current adsorption separation processes. *Chem. Eng. Sci.*, **44**, 1011–1038.

Sainio, T. and Kaspereit, M. (2009) Analysis of steady state recycling chromatography using equilibrium theory. *Sep. Purif. Technol.*, **66**, 9–18.

Sardin, M., Schweich, D., and Villermaux, J. (1993) Preparative fixed-bed chromatographic reactor, in *Preparative and Production Scale Chromatography* (eds G. Ganetsos and P.E. Barker), Marcel Dekker Inc., New York.

Sarmidi, M.R. and Barker, P.E. (1993a) Saccharification of modified starch to maltose in a continuous rotating annular chromatograph. *J. Chem. Tech. Biotechnol.*, **57**, 229–235.

Sarmidi, M.R. and Barker, P.E. (1993b) Simultaneous biochemical reaction and separation in a rotating annular chromatograph. *Chem. Eng. Sci.*, **48**, 2615–2623.

Scherpian, P. and Schembecker, G. (2009) Scaling-up recycling chromatography. *Chem. Eng. Sci.*, **64**, 4068–4080.

Schmidt, S., Wu, P., Konstantinov, K., Kaiser, K., Kauling, J., Henzler, H.-J., and Vogel, J.H. (2003) Kontinuierliche Isolierung von Pharmawirkstoffen mittels annularer Chromatographie. *Chem. Ing. Tech.*, **75** (3), 302–305.

Schramm, H., Kienle, A., Kaspereit, M., and Seidel-Morgenstern, A. (2003) Improved operation of simulated moving bed processes through cyclic modulation of feed flow and feed concentration. *Chem. Eng. Sci.*, **58**, 5217–5227.

Seidel-Morgenstern, A., Kessler, L., and Kaspereit, C.M. (2008) New developments in simulated moving bed chromatography. *Chem. Eng. Technol.*, **31**, 826–837.

Shieh, M.T. and Barker, P.E. (1996) Combined bioreaction and separation in a simulated counter-current chromatographic bioreactor-separator for the hydrolysis of lactose. *J. Chem. Tech. Biotechnol.*, **66**, 265–278.

Solms, J. (1955) Kontinuierliche Papierchromatographie. *Helv. Chim. Acta*, **38**, 1127–1133.

Sreedhar, B. and Seidel-Morgenstern, A. (2008) Preparative separation of multi-component mixtures using stationary phase gradients. *J. Chromatogr. A*, **1215**, 133–144.

Ströhlein, G., Mazzotti, M., and Morbidelli, M. (2005) Optimal operation of simulated-moving-bed reactors for nonlinear adsorption isotherms and equilibrium reactions. *Chem. Eng. Sci.*, **60**, 1525–1533.

Ströhlein, G., Assuncao, Y., Dube, N., Bardow, A., Mazzotti, M., and Morbidelli, M. (2006) Esterification of acrylic acid with methanol by reactive chromatography: experiments and simulation. *Chem. Eng. Sci.*, **61**, 5296–5306.

Ströhlein, G., Aumann, L., Mazzotti, M., and Morbidelli, M. (2006) A continuous, counter-current multi-column chromatographic process incorporating modifier gradients for ternary separations. *J. Chromatogr. A*, **1126**, 338–346.

Tanimura, M., Tamura, M., and Teshima, T. (1995) Japanese Patent JP-B-H07–046097.

Toumi, A. and Engell, S. (2004) Optimization-based control of a reactive simulated moving bed process for glucose isomerization. *Chem Eng. Sci.*, **59**, 3777–3792.

Unger, K.K. (ed.) (1994) *Handbuch der HPLC, Teil 2: Präparative Säulenflüssig-Chromatographie*, GIT Verlag, Darmstadt.

Voigt, U., Hemple, R., and Kinkel, J.N. (1997) Deutsches Patent DE 196 11 094 A1.

Vu, T., Seidel-Morgenstern, A., Grüner, S., and Kienle, A. (2005) Analysis of ester hydrolysis reaction in a chromatographic reactor using equilibrium theory and a rate model. *Ind. Eng. Chem. Res.*, **44**, 9565–9574.

Wankat, P.C. (2001) Simulated moving bed cascades for ternary separations. *Ind. Eng. Chem. Res.*, **40**, 6185–6193.

Wekenborg, K., Susanto, A., and Schmidt-Traub, H. (2004) Nicht-isokratische SMB-trennung von proteinen mittels ionenaustauschchromatographie. *Chem. Ing. Tech.*, **76** (6), 815–819.

Wolfgang, J. and Prior, A. (2002) Modern advances in chromatography. *Adv. Biochem. Eng./Biotech.*, **76**, 233–255.

Wooley, R., Ma, Z., and Wang, N.H.L. (1998) A nine-zone simulating moving bed for the recovery of glucose and xylose from biomass hydrolyzate. *Ind. Eng. Chem. Res.*, **37**, 3699–3709.

Zafar, I. and Barker, P.E. (1988) An experimental and computational study of a biochemical polymerisation reaction in a chromatographic reactor separator. *Chem. Eng. Sci.*, **43**, 2369–2375.

Zang, Y. and Wankat, P.C. (2002a) SMB operation strategy – partial feed. *Ind. Eng. Chem. Res.*, **41**, 2504–2511.

Zang, Y. and Wankat, P.C. (2002b) Three-zone simulated moving bed with partial feed and selective withdrawal. *Ind. Eng. Chem. Res.*, **41**, 5283–5289.

Zhang, Z., Mazzotti, M., and Morbidelli, M. (2003) PowerFeed operation of simulated moving bed units: changing flow-rates during the switching interval. *J. Chromatogr. A*, **1006**, 87–99.

Zhang, Y., Hidajat, K., and Ray, A. (2007) Modified reactive SMB for production of high concentrated fructose syrup by isomerisation of glucose to fructose. *Biochem. Eng. J.*, **35**, 341–351.

6
Modeling and Model Parameters

Andreas Seidel-Morgenstern, Henner Schmidt-Traub, Mirko Michel, Achim Epping*, and Andreas Jupke**

Essential steps in the design of production-scale chromatographic processes are the selection of the chromatographic system and the most suitable process concept as well as the scale-up from laboratory-scale experiments to economically relevant plant sizes. The complex nonlinear system behavior makes an empirical design difficult and time consuming. Predictions based on approximations and numerical simulations can considerably reduce material and time needed for process analysis and optimization. Validated process models can be used for optimal plant design and identification of suitable operating parameters. Other clear benefits offered by process simulation are an improved process understanding and the possibility for efficient training of operators.

This chapter starts with an introduction to modeling of chromatographic separation processes, focusing on different models capable to describe the dynamics of front propagation phenomena in the columns and the plant peripherals. A short introduction into numerical solution methods as well as an overview regarding methods for the consistent determination of the free model parameters, especially those of the thermodynamic submodels, is given. Methods of different complexity and experimental effort are presented. Finally, it will be illustrated that appropriate models can simulate experimental data with rather high accuracy. This validation is demonstrated both for standard batch elution and for a more complex multicolumn operation mode.

6.1
Introduction

Basic physical phenomena occurring during a chromatographic separation are described in Chapter 2. A quantitative description is possible using suitable mathematical models, which are typically based on material, energy, and momentum balances, in addition to equations that quantify the thermodynamic equilibria of the distribution of the solutes between the different phases. A good model has to be as

*These authors have contributed to the first edition.

Preparative Chromatography, Second Edition. Edited by H. Schmidt-Traub, M. Schulte, and A. Seidel-Morgenstern.
© 2012 Wiley-VCH Verlag GmbH & Co. KGaA. Published 2012 by Wiley-VCH Verlag GmbH & Co. KGaA.

detailed as necessary but also as simple as possible. For steady state, that is, time-independent processes, often encountered in the production of bulk chemicals, a corresponding steady-state model is sufficient. However, if the system variables change with time as during chromatographic processes, the process has to be described by means of dynamic models. Another classical way to distinguish different model types is based on the nature of the balance space. A macroscopic balance can be applied if all physical quantities are assumed to be constant throughout the whole volume of interest, whereas microscopic balances are necessary for processes with spatial changes of the state variables.

Since the evolution of chromatographic profiles is time and space dependent, dynamic and microscopic balances must be employed to describe the process behavior.

Chromatographic processes occur in several sub-steps that must be described by different connected models. Flow-sheeting systems could be used, which link all submodels by streams representing the material flow and solve the individual and overall heat and mass balances. The core of the overall separation processes is always, of course, the chromatographic column.

6.2
Models for Single Chromatographic Columns

6.2.1
Classes of Chromatographic Models

Different kinds of modeling approaches, including many analytical solutions, are comprehensively summarized in monographs by Ruthven (1984), Seidel-Morgenstern (1995), Guiochon and Lin (2003), and Guiochon et al. (2006), as well as in various review articles, for example, by Bellot and Condoret (1991) and Klatt (1999). In addition to forced convection, most of these models take into account one or more of the following effects:

- dispersion;
- mass transfer from the bulk phase into the boundary layer of the adsorbent particle;
- diffusion inside the pores of the particle (pore diffusion);
- diffusion along the surface of the solid phase ("surface diffusion");
- finite adsorption kinetics.

Figure 6.1 gives an example for a one-dimensional model. Mass balances for the fluid mobile phase as well as the stationary adsorbent phase are derived on the basis of differential volume elements. Section 6.2.2 gives further information regarding the derivation of the material balance.

Figure 6.2 shows the classification of the various models based on the number and type of effects considered.

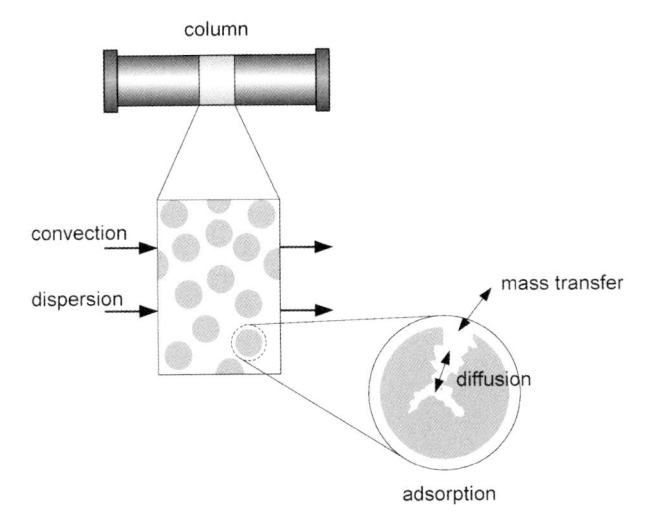

Figure 6.1 Principle of differential mass balances for a chromatographic column.

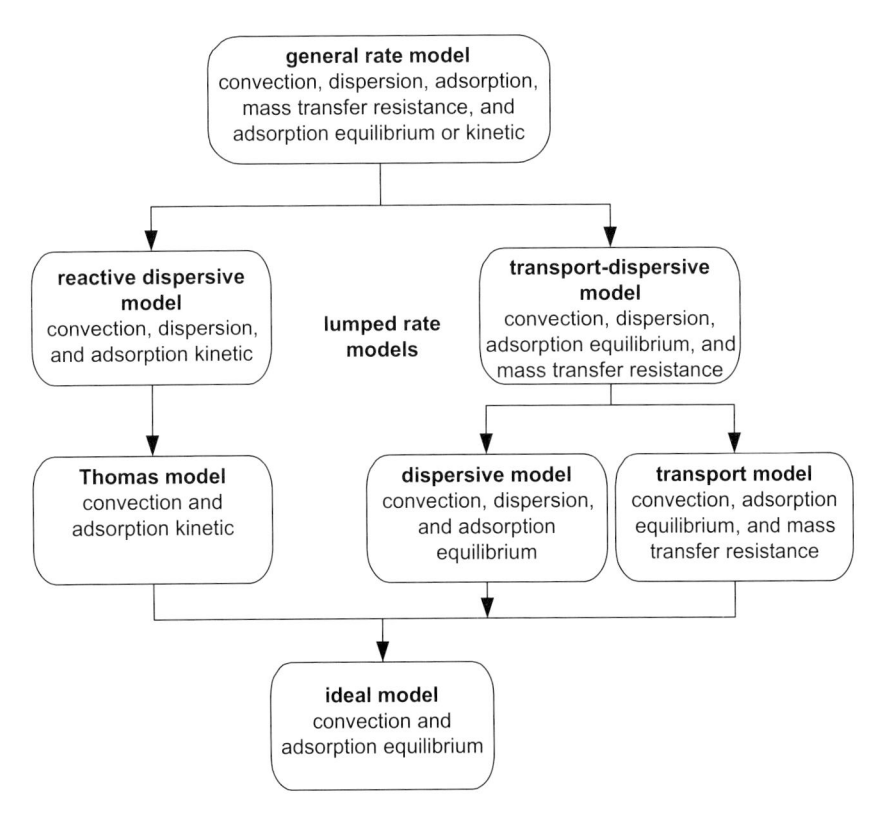

Figure 6.2 Classification of different model approaches for a chromatographic column (reproduced from Klatt, 1999; Guiochon et al., 2006).

In modeling liquid chromatographic processes, frequently the following assumptions are justified:

- The adsorbent bed is homogeneous and packed with spherical particles of constant diameter.
- Fluid density and viscosity are constant.
- Radial distributions are negligible.
- The process is isothermal.
- The eluent is inert. Its influence on the adsorption process is taken into account only implicitly by the parameters of the adsorption isotherm.
- There is no convection inside the particles.

Consequently, standard models describing chromatographic columns consist typically of one-dimensional mass balances. The pressure drop can be calculated by Equation 2.25.

6.2.2
Derivation of the Mass Balance Equations

Figure 6.3 visualizes differential volume elements and all ingoing and outgoing streams as well as source and sink terms for a quite general model. Elements include the mobile fluid and the stationary adsorbent phase, which have to be accounted for separately. Additionally, the adsorbent is split into the range of stagnant fluid inside the particle pores and the actual solid structure of the particles.

In the following are provided the mass balance and expressions quantifying the different possible mass transport mechanisms.

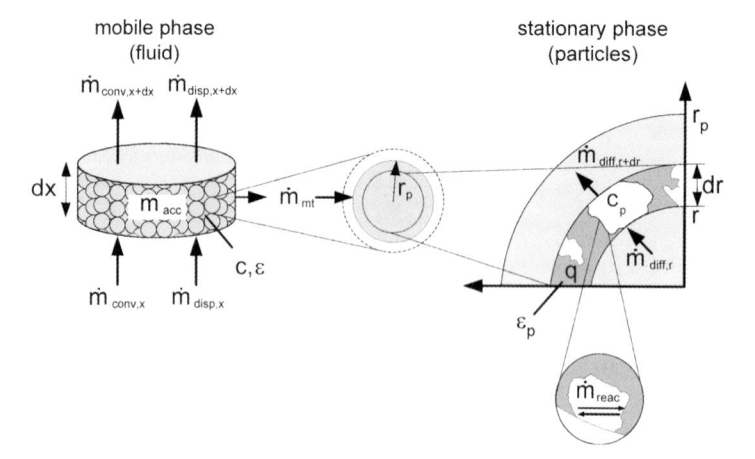

Figure 6.3 Differential model elements for a chromatographic column.

6.2.2.1 Mass Balance Equations

According to the assumptions in Section 6.2.1 the general mass balance for one component i in a differential volume dV_c of the mobile phase is (Figure 6.3)

$$\frac{\partial}{\partial t}(m_{acc,i}(x,t)) = \dot{m}^x_{conv,i}(x,t) - \dot{m}^{x+dx}_{conv,i}(x,t) + \dot{m}^x_{disp,i}(x,t) - \dot{m}^{x+dx}_{disp,i}(x,t) \\ - \dot{m}_{mt,i}(x,t) \quad (6.1)$$

Equation 6.1 includes mass accumulation in a storage term in the differential volume, the mass transport by ingoing and outgoing convection and dispersion, as well as the mass transfer into the particles. Using a first-order Taylor series approximation for the outgoing streams

$$\dot{m}^{x+dx} \approx \dot{m}^x + \frac{\partial \dot{m}^x}{\partial x} dx \quad (6.2)$$

Equation 6.1 can be written as

$$\frac{\partial}{\partial t}(m_{acc,i}) = -\frac{\partial(\dot{m}^x_{conv,i} + \dot{m}^x_{disp,i})}{\partial x} dx - \dot{m}_{mt,i} \quad (6.3)$$

The superscript x is dropped in the following.

Mass transfer into the particle is equal to the overall accumulation of component i in the adsorbent:

$$\frac{\partial}{\partial t}(\bar{m}_{acc,ads,i}) = \dot{m}_{mt,i} \quad (6.4)$$

Inside the adsorbent particles, mass transport is assumed to take place due to pore and/or surface diffusion. The resulting mass balances for the two reservoirs in the adsorbent including the adsorption kinetics by a quasi-reaction term (Figure 6.3) are

$$\frac{\partial}{\partial t}(m_{acc,pore,i}) = -\frac{\partial}{\partial r}(\dot{m}_{diff,pore,i}) dr - \dot{m}_{reac} \quad (6.5)$$

$$\frac{\partial}{\partial t}(m_{acc,solid,i}) = -\frac{\partial}{\partial r}(\dot{m}_{diff,solid,i}) dr + \dot{m}_{reac} \quad (6.6)$$

Alternatively, either Equation 6.5 or (6.6) can be replaced by their sum (Equation 6.7):

$$\frac{\partial}{\partial t}(m_{acc,pore,i} + m_{acc,solid,i}) = -\frac{\partial}{\partial r}(\dot{m}_{diff,pore,i} + \dot{m}_{diff,solid,i}) dr \quad (6.7)$$

If adsorption equilibrium is assumed, the reaction kinetics are infinitively fast and the term $\dot{m}_{reac}$ is no longer well defined (Section 6.2.2.7). This makes independent mass balances for the pore and solid phase impractical as they are coupled through an isotherm equation. In this case Equation 6.7 together with an isotherm equation (Section 6.2.2.7) must be used instead of adsorption kinetics (Section 6.2.2.6) and two separate equations – Equations 6.5 and 6.6.

To transfer these equations into mass balances based on concentrations, it is necessary to introduce characteristic volumes. The overall differential volume dV_c is the sum of the mobile phase volume dV_{int} and the stationary phase volume dV_{ads}. By means of the void fraction (Equation 2.6) and the cross-section A_c of the column, those volumes can be calculated as

$$dV_{int} = \varepsilon \cdot A_c \cdot dx \tag{6.8}$$
$$dV_{ads} = (1 - \varepsilon) \cdot A_c \cdot dx \tag{6.9}$$

Hereby, the following holds:

$$dV_c = A_c \cdot dx = dV_{int} + dV_{ads} \tag{6.10}$$

The same considerations apply for the differential pore phase volumes dV_{pore} and solid phase volumes dV_{solid} of the adsorbent, for which the phase distribution is given by the porosity (Equation 2.7) of the adsorbent:

$$dV_{pore} = \varepsilon_p \cdot dV_{ads} = \varepsilon_p \cdot (1 - \varepsilon) \cdot A_c \cdot dx \tag{6.11}$$
$$dV_{solid} = (1 - \varepsilon_p) \cdot dV_{ads} = (1 - \varepsilon_p) \cdot (1 - \varepsilon) \cdot A_c \cdot dx \tag{6.12}$$

Introducing the concentration in the liquid (bulk) phase c_i, the mean overall adsorbent loading $\bar{q}_i^*$, the mean pore concentration $\bar{c}_{p,i}$, and the mean solid loading $\bar{q}_i$, the mass balance contributions transform into

$$m_{acc,i} = c_i \cdot dV_{int} = c_i \cdot \varepsilon \cdot A_c \cdot dx \tag{6.13}$$
$$\bar{m}_{acc,ads,i} = \bar{q}_i^* \cdot dV_{ads} = \bar{q}_i^* \cdot (1 - \varepsilon) \cdot A_c \cdot dx \tag{6.14}$$
$$\bar{m}_{acc,pore,i} = \bar{c}_{p,i} \cdot dV_{pore} = \bar{c}_{p,i} \cdot \varepsilon_p \cdot (1 - \varepsilon) \cdot A_c \cdot dx \tag{6.15}$$
$$\bar{m}_{acc,solid,i} = \bar{q}_i \cdot dV_{solid} = \bar{q}_i \cdot (1 - \varepsilon_p) \cdot (1 - \varepsilon) \cdot A_c \cdot dx \tag{6.16}$$

The overall adsorbent loading is equal to the sum of the pore concentration and the solid loading (Equation 2.47):

$$\bar{q}_i^* = \varepsilon_p \cdot \bar{c}_{p,i} + (1 - \varepsilon_p) \cdot \bar{q}_i \tag{6.17}$$

The bar above the symbols in Equations 6.14–6.17 denotes average values. These average concentrations and the concentration in the liquid phase, c_i, are a function of the time t and the axial column coordinate, x. The overall balance (Equation 6.17) has to be distinguished from the balance that takes into account radial distributions within the particles:

$$q_i^*(r) = \varepsilon_p \cdot c_{p,i}(r) + (1 - \varepsilon_p) \cdot q_i(r) \tag{6.18}$$

Average concentrations and loadings within spherical particles of radius r_p can be calculated with the following integrals:

$$\bar{c}_{p,i} = \frac{1}{(4/3) \cdot \pi \cdot r_p^3} \int_0^{r_p} c_{p,i}(r) \cdot 4\pi \cdot r^2 \, dr = \frac{3}{r_p^3} \int_0^{r_p} r^2 \cdot c_{p,i}(r) dr$$

$$\bar{q}_i = \frac{3}{r_p^3} \int_0^{r_p} r^2 \cdot q_i(r) dr \tag{6.19}$$

In case of general rate models (Section 6.2.6) the local mass transfer inside the particles is considered additionally to the transport outside the particles. Therefore, the mass balances have to take into account the number of (spherical) particles per volume element (Figure 6.3):

$$N_p = \frac{dV_{ads}}{\text{particle volume}} = \frac{(1-\varepsilon) \cdot A_c \cdot dx}{(4/3) \cdot \pi \cdot r_p^3} \tag{6.20}$$

6.2.2.2 Convective Transport

After having defined above the accumulation terms of the general mass balances (Equations 6.1–6.6), the transport and source terms can be evaluated as follows. Mass transport in the mobile phase occurs due to convection with the interstitial velocity u_{int} (Equation 2.9)

$$\dot{m}_{conv,i} = \varepsilon \cdot A_c \cdot u_{int} \cdot c_i \tag{6.21}$$

6.2.2.3 Axial Dispersion

It is assumed that axial dispersion (Equation 6.22) in the liquid phase can be defined in analogy to Fick's first law of diffusion:

$$\dot{m}_{disp,i} = -\varepsilon \cdot A_c \cdot D_{ax} \cdot \frac{\partial c_i}{\partial x} \tag{6.22}$$

The axial dispersion coefficient D_{ax} depends essentially on the quality of the packing and captures deviations of the fluid dynamics from plug flow. In preparative chromatography contributions of molecular diffusion are generally negligible (Section 6.5.6.2).

6.2.2.4 Intraparticle Diffusion

Diffusion inside the adsorbent particles can also be quantified using Fick's law. As for the accumulation terms, size and number of particles per unit volume (Equation 6.20) are taken into account:

$$\dot{m}_{diff,pore,i}(r) = -N_p \cdot \varepsilon_p \cdot 4\pi \cdot r^2 \cdot D_{pore,i} \frac{\partial c_{p,i}(r)}{\partial r} \tag{6.23}$$

$$\dot{m}_{diff,solid,i}(r) = -N_p \cdot (1-\varepsilon_p) \cdot 4\pi \cdot r^2 \cdot D_{solid,i} \frac{\partial q_i(r)}{\partial r} \tag{6.24}$$

In Equation 6.23 the transport in the pore fluid is modeled as free diffusion in the macropores and mesopores, but the diffusion coefficient $D_{pore,i}$ is usually smaller than the molecular diffusivity characterizing transport in the liquid mobile phase due to the random orientations and variations in the diameter of the pores (tortuosity) (Section 6.5.8).

In Equation 6.24 transport is assumed to occur by micropore or surface diffusion, where the molecules are under the influence of a force fields of the inner adsorbent surface. The surface diffusion concept is applied to quantify the transport in the adsorbed phase.

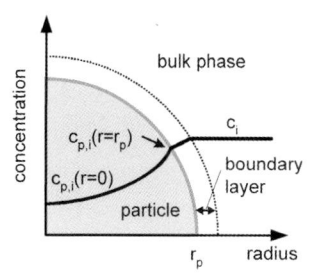

Figure 6.4 Concentration profile assumed in liquid film linear driving force models.

6.2.2.5 Mass Transfer

According to the simplifying assumptions made in Section 6.2.1, the liquid phase concentrations change only in axial direction and are constant in a cross-section. Therefore, mass transfer between liquid and solid phases is not defined by a local concentration gradient around the particles. Instead, a general mass transfer resistance is postulated. A common method describes the (external) mass transfer $\dot{m}_{mt,i}$ as a linear function of the concentration difference between the concentration in the bulk phase and that on the adsorbent surface, which are separated by a film of stagnant liquid (boundary layer). This so-called linear driving force model (LDF model) has proven to be often suitable to describe chromatographic processes (Ruthven, 1984; Bellot and Condoret, 1991; Guiochon et al., 2006). Resulting concentration profiles are depicted in Figure 6.4.

In the corresponding equation for the mass transfer rate of component i, calculated by Equation 6.25, the film transfer coefficient k_{film} is related to the overall surface area dA_s of the adsorbent of all particles in the finite volume element (Equations 6.20 and 6.26):

$$\dot{m}_{mt,i} = k_{film,i} \cdot (c_i - c_{p,i}(r = r_p)) \cdot dA_s \tag{6.25}$$

$$dA_s = N_p \cdot 4\pi \cdot r_p^2 = \frac{3}{r_p} \cdot (1 - \varepsilon) \cdot A_c \cdot dx \tag{6.26}$$

A useful characteristic value is the specific surface area a_s of the particles per unit volume, which can be derived from Equation 6.26:

$$a_s = \frac{dA_s}{dV} = \frac{3}{r_p} \cdot (1 - \varepsilon) \tag{6.27}$$

Combining Equations 6.25 and 6.26 yields Equation 6.28 for the mass transfer rate:

$$\dot{m}_{mt,i} = k_{film,i} \cdot (c_i - c_{p,i}(r = r_p)) \cdot \frac{3}{r_p} \cdot (1 - \varepsilon) \cdot A_c \cdot dx \tag{6.28}$$

Note that in some publications the definition of a mass transfer coefficient includes the factor $3/r_p$.

According to Equation 6.4 the mass flow $\dot{m}_{mt}$ is generally equal to the overall accumulation in the adsorbent. If concentrations and loadings change inside the particles, the overall accumulation (Equation 6.14) must be calculated by integration (Equation 6.19) after the radial concentration profiles inside the particles are obtained. In this case it is reasonable to replace Equation 6.4 with the continuity equation around the particles, as the mass flow through the outer liquid film is equal to the mass flow entering the adsorbent particle (Equations 6.23 and 6.24):

$$\dot{m}_{mt,i} = -(\dot{m}_{diff,pore,i}(r = r_p) + \dot{m}_{diff,solid,i}(r = r_p)) \tag{6.29}$$

Equation 6.29 represents a boundary condition of the general rate model, which is further discussed in Section 6.2.6.

6.2.2.6 Adsorption Kinetics

The final elements of the mass balances are the adsorption kinetics and equilibria. According to Equations 6.5 and 6.6, adsorption and desorption steps are modeled as reactions with finite rate. The reaction rate based on the solid volume of all particles (Equation 6.20) in the volume element is

$$\dot{m}_{reac} = (1 - \varepsilon_p) \cdot N_p \cdot \psi_{reac,i} \cdot 4\pi \cdot r^2 \cdot dr \tag{6.30}$$

Like the equilibrium isotherms, the net adsorption rates $\psi_{reac,i}$ can be defined in quite different ways. One example is given in Equation 6.31 where a first-order equilibrium reaction with two rate constants for the adsorption (k_{ads}) and desorption (k_{des}) steps (Ma, Whitley, and Wang, 1996) is specified:

$$\psi_{reac,i}(r) = k_{ads,i} \cdot q_{sat,i} \cdot \left(1 - \sum_{j=1}^{N_{comp}} \frac{q_j(r)}{q_{sat,j}}\right) \cdot c_{p,i}(r) - k_{des,i} \cdot q_i(r) \tag{6.31}$$

The terms $q_{sat,i}$ represent again the maximum possible loadings for each component. Interaction between the different components is considered by the summation over all N_{comp} components. Equation 6.31 is the nonequilibrium form of the multicomponent Langmuir isotherm (Equation 2.57).

The nonequilibrium form of the simpler linear isotherm (Equation 2.48) can be obtained by neglecting the sum in Equation 6.31, leading to

$$\psi_{reac,i}(r) = k_{ads,i} \cdot q_{sat,i} \cdot c_{p,i}(r) - k_{des,i} \cdot q_i(r) \tag{6.32}$$

with

$$H = \frac{q_{sat} \cdot k_{ads}}{k_{des}} \tag{6.33}$$

6.2.2.7 Adsorption Equilibrium

Notably, adsorption and desorption steps are usually very fast. In the limit of very large rate constants, $\psi_{reac,i}$, large the pore concentrations and the solid loadings are in equilibrium, and connected through an isotherm equation:

$$q_i(r) = f(c_{p,1}(r), c_{p,2}(r), \ldots, c_{p,N_{comp}}(r)) \tag{6.34}$$

or simply

$$q_i = f(c_{p,1}, \ldots, c_{p,N_{comp}}) \tag{6.35}$$

Equation 6.34 can be derived from Equation 6.31 by setting $\psi_{reac,i}$ equal to zero, which corresponds to equally fast adsorption and desorption steps (a "dynamic" chemical equilibrium).

To formulate the accumulation term of the balance equation, an isotherm equation must be used in Equation 6.7 avoiding the need for distinctly different terms as used in Equations 6.5 and 6.6.

All considered terms of the mass balance of a quite comprehensive model for liquid chromatographic column have now been specified in detail. Table 6.1 summarizes them.

In the following the relevant models for liquid chromatography are derived in a bottom-up procedure related to Figure 6.2. To illustrate the difference between these models their specific assumptions are discussed and the level of accuracy and their field of application are pointed out. In all cases the mass balances must be complemented by initial and boundary conditions (Section 6.2.7). For the so-called transport-dispersive model a dimensionless representation will also be presented below.

6.2.3
Equilibrium ("Ideal") Model

The simplest model takes into account only convective transport and thermodynamics. It assumes a permanently established local equilibrium between mobile and stationary phases. This model is frequently called the ideal or basic model of chromatography (Guiochon et al., 2006). It was described first by Wicke (1939) for the elution of a single component. Subsequently, De Vault (1943) derived in more detail the corresponding mass balance.

Table 6.1 Phenomena evaluated and quantified in mass balances of chromatographic fixed-bed models.

Accumulation
Convection
Dispersion
Mass transfer
Porosities and numbers of particles per volume element
Diffusion
Adsorption kinetics
Adsorption equilibria
Averaged intraparticle loadings and concentrations

6.2 Models for Single Chromatographic Columns

The ideal equilibrium model neglects the influence of axial dispersion and all mass transfer and kinetic effects, that is:

$$\begin{aligned} D_{ax} &= 0 \\ D_{pore,i} &= D_{solid,i} = \infty \\ k_{ads,i}, k_{des,i}, k_{film,i} &= \infty \end{aligned} \quad (6.36)$$

Consequently, the loadings and concentrations within the adsorbent are constant and not a function of the particle radius. Further, the concentration in the liquid phase are identical to that in the particle pores:

$$\left. \begin{aligned} c_{p,i} &= \bar{c}_{p,i} \\ q_i &= \bar{q} \\ c_i &= c_{p,i} \end{aligned} \right\} \neq f(r) \quad (6.37)$$

Therefore, Equations 6.3 and 6.4 reduce to the following balance:

$$\frac{\partial}{\partial t}(m_{acc,i}) = -\frac{\partial(\dot{m}_{conv,i})}{\partial x} dx - \frac{\partial}{\partial t}(\bar{m}_{acc,ads,i}) \quad (6.38)$$

Introducing the appropriate terms given by Equations 6.13–6.16 and 6.21 leads to

$$\frac{\partial c_i}{\partial t} + u_{int} \cdot \frac{\partial c_i}{\partial x} + \frac{1-\varepsilon}{\varepsilon} \cdot \left(\varepsilon_p \frac{\partial c_i}{\partial t} + (1-\varepsilon_p) \frac{\partial q_i}{\partial t} \right) = 0 \quad (6.39)$$

The only additional equation required for solution is the thermodynamic equilibrium:

$$q_i = f(c_1, \ldots, c_{N_{comp}}) \quad (6.40)$$

Further simplification can be obtained by introducing the total porosity:

$$\varepsilon_t = \varepsilon + \varepsilon_p(1-\varepsilon) \quad (6.41)$$

and an effective velocity u_m of a nonretained solute that enters the pore space:

$$u_m = \frac{L_c}{t_0} = \frac{\varepsilon}{\varepsilon_t} u_{int} = \frac{\varepsilon}{\varepsilon_t} \frac{L_c}{t_{0,int}} \quad (6.42)$$

Note that u_m is directly linked with the measurable dead time t_0 (Equation 2.10), while the larger velocity u_{int} is connected with the corresponding time $t_{0,int}$ of non-penetrating molecules (Equation 2.11).

After rearrangement and insertion of Equations 6.41 and 6.42 into Equation 6.39, the following results:

$$\frac{\partial c_i}{\partial t} + \frac{1-\varepsilon_t}{\varepsilon_t} \frac{\partial q_i(c_1, c_2, \ldots, c_{N_{comp}})}{\partial t} + u_m \frac{\partial c_i}{\partial x} = 0 \quad (6.43)$$

Today there is a well-established, comprehensive and rather complete theory available to solve the equations of the ideal model. This theory is called the equilibrium theory of chromatography. Major contributions were made by Helfferich and Klein (1970), Helfferich and Carr (1993), Helfferich and Whitley (1996) and

Helfferich (1997) and Rhee, Aris, and Amundson (1970, 1986, 1989), who derived instructive analytical solutions of the first-order system of nonlinear partial differential equations given by Equation 6.43. Most of the solutions are available for multicomponent Langmuir isotherms. Interested readers will find in the excellent treatment of Rhee *et al.* the basics of the method of characteristics applied to derive the solutions and to explain the wave phenomena that take place in chromatographic columns.

The equilibrium theory does not provide only a powerful tool to understand important phenomena that take place during the chromatographic process. One of the main advantages of equilibrium theory is the capability to explain some fundamental phenomena that occur in multicomponent chromatography such as the displacement effect and the tag-along effect (Section 2.6.2). An even more important application is the use of the equilibrium theory as a powerful shortcut method for preliminary process design. It is further an important starting point for several shortcut design methods. In particular, the understanding and design of the complex SMB process benefited in the last decades strongly from the equilibrium theory.

To illustrate the strength of the results that can be obtained applying the ideal model, we will consider below just the case of single-component elution of a component i. Using an established equilibrium the time derivative of the loading can be expressed by the isotherm slope and the corresponding liquid phase time derivative:

$$\frac{\partial q_i}{\partial t} = \frac{dq_i}{dc_i} \cdot \frac{\partial c_i}{\partial t} \tag{6.44}$$

Equation 6.43 can now be rearranged to

$$\frac{\partial c_i}{\partial t} + \frac{u_m}{1 + [(1 - \varepsilon_t)/\varepsilon_t] \cdot (dq_i/dc_i)} \cdot \frac{\partial c_i}{\partial x} = 0 \tag{6.45}$$

Equation 6.45 predicts the propagation velocity w of an arbitrary concentration c^+ inside the column depending on the isotherm slope:

$$w(c_i^+) = \frac{u_m}{1 + [(1 - \varepsilon_t)/\varepsilon_t] \cdot (dq_i/dc_i)|_{c_i^+}} \tag{6.46}$$

The velocity is connected with the observable retention time by

$$t_{R,i}(c_i^+) = \frac{L_c}{w(c_i^+)} \tag{6.47}$$

Combining Equations 6.42, 6.46 and 6.47 leads directly to the basic equation of chromatography:

$$t_{R,i}(c_i^+) = t_0 \cdot \left(1 + \frac{1 - \varepsilon_t}{\varepsilon_t} \cdot \frac{dq_i}{dc_i}\bigg|_{c_i^+}\right) \tag{6.48}$$

This fundamental equation was already introduced in Chapter 2 (Equation 2.17). In the special case of linear isotherms it reduces to

$$t_{R,\text{lin},i} = t_0 \cdot \left(1 + \frac{1-\varepsilon_t}{\varepsilon_t} \cdot H_i\right) \quad (6.49)$$

Thus, for linear isotherms the retention times do not depend on concentration (Lapidus and Amundson, 1952; Van Deemter, Zuiderweg, and Klinkenberg, 1956).

Already, through work by Glueckauf and Coates (1947) and Glueckauf (1949) considerable progress has been made in understanding the influences of the isotherm shapes on the shapes of elution profile for nonlinear isotherms.

For nonlinear systems the isotherm slopes decide whether smaller or larger concentrations propagate faster. In most cases the isotherms slopes decrease with increasing concentrations, as described, for example, by the Langmuir isotherm. This behavior causes a faster movement of high concentrations. Under these conditions, Equation 6.48 predicts that larger concentrations could overtake smaller concentrations. This would cause unrealistic discontinuities at the column outlet. In reality, shock fronts form instead, in which again all concentrations travel with the same velocity. To describe the movement of these shocks, the differentials in Equation 6.48 have to be replaced by difference quotients Δ, which evaluate a secant in the isotherm between the two equilibrium points framing the concentration shocks:

$$t_{R,i,\text{shock}} = t_0 \cdot \left(1 + \frac{1-\varepsilon_t}{\varepsilon_t} \cdot \frac{\Delta q_i}{\Delta c_i}\bigg|_{\text{shock}}\right) \quad (6.50)$$

Thus, Equation 6.48 is valid only for the disperse part of a peak (Section 2.2.3). Depending on the shape of the isotherm, this is the rear part ("Langmuir") or the front part ("anti-Langmuir") of the peak (Figure 2.6). The description of the velocity of the opposite shock front requires Equation 6.49. It should be mentioned that in case of inflection points in the course of an isotherm the situation is more complex and composite fronts form (Rhee, Aris, and Amundson, 1986; Zhang, Shan, and Seidel-Morgenstern, 2006).

For large sample sizes that cause complete concentration breakthroughs, Equation 6.50 can be directly used to calculate the position of the breakthrough curve or to determine from measured breakthrough times parameters of an isotherm model (see Section 6.5.7.5).

It should be noted that Equations 6.48 and 6.50 are only valid for retention times corrected by dead times of the plant (Sections 2.2.1 and 6.5.3.1). Further, the injection time t_{inj} of a real peak (cf. Figure 6.11) has to be considered when evaluating measured retention times belonging to rear parts of peaks:

$$t_{R,i}(c_i^+) = t_{\text{inj}} + t_0 \cdot \left(1 + \frac{1-\varepsilon_t}{\varepsilon_t} \cdot \frac{dq_i}{dc_i}\bigg|_{c_i^+}\right) \quad (6.51)$$

For multicomponent mixtures and competitive isotherms, total differentials need to be used and the analysis becomes more complicated.

For two components the following holds:

$$\frac{dq_1}{dc_1} = \frac{\partial q_1}{\partial c_1} + \frac{\partial q_1}{\partial c_2} \cdot \frac{dc_2}{dc_1}$$
$$\frac{dq_2}{dc_2} = \frac{\partial q_2}{\partial c_2} + \frac{\partial q_2}{\partial c_1} \cdot \frac{dc_1}{dc_2}$$
(6.52)

Thus, the elution behavior of the two components is coupled through the concentration dependence of both isotherm equations. To quantify the impact of the concentration of one component on the propagation velocity of other components the so-called coherence condition introduced by Helfferich and Klein (1970) needs to be applied, that is:

$$w_1(c_1^+, c_2^+) = w_2(c_1^+, c_2^+)$$
(6.53)

Regarding details of available analytical solutions for two and more components we refer the reader to the cited references by Helfferich et al., Rhee et al., and Guiochon et al. Here, a recently derived analytical solution for the prediction of the elution bands of two components in case of Langmuir isotherms and intermediate sample sizes should be mentioned. These explicit equations, which extend the status described by Guiochon et al. (2006), were used for efficient optimization of batch chromatography (Siitonen and Sainio, 2011).

6.2.4
Models with One Band Broadening Effect

Basically, models using only one effect to describe band spreading – lump all effects in only one model parameter. This approach is straightforward for linear isotherms (Section 6.5.3.1) but also commonly applied for describing chromatographic processes occurring in the nonlinear range. Among these models, the equilibrium-dispersive model plays a prominent role.

6.2.4.1 Dispersive Model

This model is very often used to design and optimize chromatographic systems (e.g., Guiochon et al., 2006). Compared with the ideal model (Equation 6.39) a term describing axial dispersion (Equation 6.22) is included in the mass balance of the mobile phase:

$$D_{\text{solid},i} = D_{\text{pore},i} = \infty$$
$$k_{\text{ads},i}, k_{\text{des},i}, k_{\text{film},i} = \infty$$
$$\left.\begin{array}{c} c_{p,i} = \bar{c}_{p,i} \\ q_i = \bar{q} \\ c_i - c_{p,i} \end{array}\right\} \neq f(r)$$
(6.54)

Using these simplifications, Equations 6.3 and 6.4 can be combined in the same way as Equation 6.38:

$$\frac{\partial}{\partial t}(m_{\text{acc},i}) = -\frac{\partial(\dot{m}_{\text{conv},i} + \dot{m}_{\text{disp},i})}{\partial x} dx - \frac{\partial}{\partial t}(\dot{m}_{\text{acc,ads},i}) \tag{6.55}$$

which can be evaluated using Equations 6.13–6.16, 6.21 and 6.22.

As already mentioned, the effects of several parameters are often lumped into a dispersion coefficient. The so-called apparent dispersion coefficient D_{app} is used here differing from the axial dispersion coefficient, D_{ax}, which is assumed to be independent of concentration and influenced only by the quality of the packing. The lumped parameter D_{app} includes peak broadening effects caused by the fluid dynamics of the packing (axial dispersion), as well as by all other mass transfer effects that might occur:

$$D_{\text{app}} = f(D_{\text{ax}}, D_{\text{solid}}, D_{\text{pore}}, k_{\text{film}}, c_i, \ldots c_{N_{\text{comp}}}, u_{\text{int}}) \tag{6.56}$$

Like most lumped parameters, the apparent dispersion coefficient is generally dependent on the interstitial velocity (Section 6.5.3.1). For nonlinear isotherms, it will generally also depend on the concentration.

Despite using only one parameter to describe mass transfer resistance and the fluid dynamics of the packing, experiments and simulation show good agreement for highly efficient columns ($N \gg 100$). In this region the difference between this model and the more detailed models almost vanishes (Golshan-Shirazi and Guiochon, 1992; Seidel-Morgenstern, 1995; Guiochon et al., 2006).

Using this new coefficient and term, Equation 6.55 becomes

$$\frac{\partial c_i}{\partial t} + u_{\text{int}} \cdot \frac{\partial c_i}{\partial x} + \frac{1-\varepsilon}{\varepsilon} \cdot \left(\varepsilon_{\text{p}} \frac{\partial c_i}{\partial t} + (1-\varepsilon_{\text{p}}) \frac{\partial q_i}{\partial t}\right) = D_{\text{app},i} \cdot \frac{\partial^2 c_i}{\partial x^2} \tag{6.57}$$

As the fluid concentration is in equilibrium with the solid loading, still no differential equation for the particle phase is needed and the adsorption equilibrium is given by the general function Equation 6.40.

Notably, Equation 6.57 can be defined in similar fashion to Equation 6.43 with the velocity u_{m} of Equation 6.42, while all constant parameters, such as volume fractions, are included in the apparent dispersion coefficient, which then differs from the one defined in Equation 6.57:

$$\frac{\partial c_i}{\partial t} + u_{\text{m}} \cdot \frac{\partial c_i}{\partial x} + \frac{1-\varepsilon_t}{\varepsilon_t} \cdot \frac{\partial q_i}{\partial t} = \tilde{D}_{\text{app},i} \cdot \frac{\partial^2 c_i}{\partial x^2} \tag{6.58}$$

$$\tilde{D}_{\text{app},i} = \frac{\varepsilon}{\varepsilon_t} D_{\text{app},i} = \frac{u_{\text{m}}}{u_{\text{int}}} D_{\text{app},i} \tag{6.59}$$

The equilibrium-dispersive model is widely applied in chromatography, owing to the equivalence with standard dispersion models well known in chemical engineering (Levenspiel and Bischoff, 1963; Danckwerts, 1953) and due to the availability of numerical solution techniques.

Instructive analytical solutions of Equation 6.57 for linear isotherms and different boundary and initial conditions have been reported (Levenspiel and Bischoff, 1963;

Guiochon et al., 2006; Guiochon and Lin, 2003). The resulting elution profiles are not symmetrical and show a certain tailing. This is understandable as the longer the "parts" of the profile are in the column, the more they are affected by band broadening. Thus, the late eluting parts of the peak are "broader" than the early eluting parts. As a consequence of this asymmetry, the position of the maximum is not identical to the first moment, which is a reason why first moments rather than the peak maxima should be preferably used when characterizing peaks (Section 6.5.3.1).

When D_{app} is not "very large" (number of stages N (see Section 6.2.9) >100; Guiochon, Golshan-Shirazi, and Katti, 1994), the solution of the equilibrium-dispersive model for small injections can be approximated by a symmetrical Gaussian distribution:

$$c_i(x,t) = \frac{m_{inj}}{\dot{V} \cdot t_{R,lin,i}} \cdot \frac{1}{\sqrt{4\pi \cdot [D_{app,i}/(u_{int} \cdot L_c)]}} \cdot \exp\left\{-\frac{[(t/t_{R,lin,i}) - (x/L_c)]^2}{(4 \cdot D_{app,i})/(u_{int} \cdot L_c)}\right\}$$

(6.60)

where m_{inj} is given by the mass balance at the column inlet:

$$m_{inj} = \dot{V} \cdot t_{inj} \cdot c_{feed} = \dot{V} \cdot \int_0^\infty c(L_c, t) dt$$

(6.61)

The first moment μ_t (Equation 2.38) and the second moment σ_t^2 (Equation 2.38) of the elution profile defined by Equation 6.60 are as follows (Section 6.5.3.1):

$$\mu_t(x = L_c) = t_{R,lin,i}$$

(6.62)

$$\sigma_t^2(x = L_c) = t_{R,lin,i}^2 \cdot \frac{2 \cdot D_{app,i}}{u_{int} \cdot L_c}$$

(6.63)

As this approximate profile is symmetrical, the first moment is identical to the position of the maximum.

Although Equation 6.60 is different from the exact analytical solution of the equilibrium-dispersive model (Levenspiel and Bischoff, 1963; Guiochon et al., 2006; Guiochon and Lin, 2003), the resulting moments derived from this analytical solution are equal to Equations 6.62 and 6.63.

When comparing the moments of Equations 6.62 and 6.63 with the definition of the number of stages given in Equation 2.37, a simple correlation between dispersion coefficient and the number of stages N or the HETP can be derived:

$$N_i = \frac{\sigma_t^2}{\mu_t^2} = \frac{L_c}{HETP_i} = \frac{L_c \cdot u_{int}}{D_{app,i} \cdot 2}$$

(6.64)

If HETP has been determined (Section 6.5.3.1), $D_{app,i}$ is easily estimated by Equation 6.64. Although Equation 6.64 is, strictly, valid only for linear isotherms, it is often used for estimations in case of nonlinear isotherms, too.

6.2.4.2 Transport Model

The so-called transport model has been used by some authors to simulate batch and SMB processes, when mass transfer resistance is assumed to be the limiting factor so that axial dispersion can be neglected (Hashimoto et al., 1983; Hashimoto, Adachi, and Shirai, 1993). This model considers adsorption equilibrium, convection, and mass transfer as the only rate-limiting and band broadening step. In analogy to the equilibrium-dispersive model, all mass transfer effects are lumped into an effective mass transfer coefficient, which is modeled in analogy to film transfer:

$$\tilde{k}_{\text{eff},i} = f(D_{\text{ax}}, D_{\text{pore}}, D_{\text{solid}}, k_{\text{film}}, c_i, \ldots c_{N_{\text{comp}}}, u_{\text{int}}) \tag{6.65}$$

The mass balance for the mobile phase is similar to Section 6.2.3, but the concentrations in the pores c_p are no longer equal to c:

$$\frac{\partial c_i}{\partial t} + u_{\text{int}} \cdot \frac{\partial c_i}{\partial x} + \frac{1-\varepsilon}{\varepsilon} \cdot \left(\varepsilon_p \frac{\partial c_{p,i}}{\partial t} + (1-\varepsilon_p) \frac{\partial q_i}{\partial t} \right) = 0 \tag{6.66}$$

An additional equation for the material balance for the particle phase is necessary (Equations 6.4, 6.14 and 6.28), where the accumulation in the stationary phase equals the mass transfer stream:

$$\varepsilon_p \frac{\partial c_{p,i}}{\partial t} + (1-\varepsilon_p) \frac{\partial q_i}{\partial t} = \tilde{k}_{\text{eff},i} \cdot \frac{3}{r_p} \cdot (c_i - c_{p,i}) \tag{6.67}$$

The isotherm equation is given by Equation 6.35.

6.2.4.3 Reaction Model

This model was first used by Thomas (1944) to simulate ion exchange processes. It postulates a rate-limiting adsorption kinetic as the only effect causing band broadening. Originally, homogeneous particles without pores ($\varepsilon_p = 0$) were assumed, but this can be modified to fit in the framework presented here. It need only be stated that the concentration in the bulk phase is the same as in the particle pores. Thus, the mobile phase balance is still the same as used for the derivation of the ideal model (Equation 6.39). Compared with the transport model, the right-hand side of the stationary phase balance equation (Equation 6.67) is replaced by an equation to quantify finite adsorption kinetics (expressed, for example, by Equation 6.31) and rewritten for constant concentrations inside the particles:

$$\varepsilon_p \frac{\partial c_i}{\partial t} + (1-\varepsilon_p) \frac{\partial q_i}{\partial t} = \tilde{k}_{\text{ads},i} \cdot q_{\text{sat},i} \cdot \left(1 - \sum_{j=1}^{N_{\text{comp}}} \frac{q_j}{q_{\text{sat},j}} \right) \cdot c_i - \tilde{k}_{\text{des},i} \cdot q_i \tag{6.68}$$

In Equation 6.68 $\tilde{k}_{\text{ads}}$ and $\tilde{k}_{\text{des}}$ are the lumped rate constants of the adsorption and desorption steps, respectively. Equation 6.31 is just one example how to express the kinetic equation – others have been published by Bellot and Condoret (1991) and Ma, Whitley, and Wang (1996).

Notably, for "standard" chromatography, the assumption that adsorption is the only rate-determining mechanism is unrealistic, because this step is generally much faster than other mass transfer–related effects (Guiochon et al., 2006; Ruthven, 1984). However, in special cases this simplification may be justified.

6.2.5
Lumped Rate Models

The next level of detail in the model hierarchy of Figure 6.2 is the so-called "lumped rate models." They are characterized by a second parameter describing rate limitations apart from axial dispersion. This second parameter subdivides the models into those where either mass transport or kinetic terms are rate limiting. No concentration distribution inside the particles is considered and, formally, the diffusion coefficients inside the adsorbent are assumed to be infinite:

$$D_{\text{solid},i} = D_{\text{pore},i} = \infty$$

$$\left.\begin{array}{c} c_{p,i} = \bar{c}_{p,i} \\ q_i = \bar{q} \end{array}\right\} \neq f(r) \tag{6.69}$$

The basic material balance of the mobile phase for all lumped rate models is based on Equations 6.3, 6.4 and 6.13–6.17 and can be derived in the same manner as the equilibrium-dispersive model (Equation 6.58):

$$\frac{\partial c_i}{\partial t} + u_{\text{int}} \cdot \frac{\partial c_i}{\partial x} + \frac{1-\varepsilon}{\varepsilon} \cdot \frac{\partial q_i^*}{\partial t} = D_{\text{ax}} \cdot \frac{\partial^2 c_i}{\partial x^2} \tag{6.70}$$

The lumped rate models are distinguished on the basis of different equations for the particle phase, considering either adsorption kinetics or mass transfer. In the former case the concentration inside the particle pores c_p is identical to the mobile phase concentration c, while they are different for the latter.

The dispersion coefficient D_{ax} (Section 6.5.6.2) is assumed to depend only on the packing properties and flow conditions (Equation 6.22) and, is therefore smaller than the apparent dispersion coefficient D_{app} defined in Section 6.2.4.1.

6.2.5.1 Transport-Dispersive Model

The transport-dispersive model (TDM) is an extension of the transport model and summarizes the internal and external mass transfer resistances in one lumped film (=effective) transfer coefficient, k_{eff} (compare Equation 6.28):

$$k_{\text{eff},i} = f(D_{\text{solid}}, D_{\text{pore}}, k_{\text{film}}, c_i, \ldots, c_{N_{\text{comp}}}, u_{\text{int}}) \tag{6.71}$$

In contrast to the transport model $k_{\text{eff},i}$ is theoretically assumed to be independent of axial dispersion and thereby of the packing quality (Sections 6.5.3.1 and 6.5.8). Compared, for example, with D_{ax}, k_{eff} is much less dependent on fluid velocity (Section 6.5.8).

The mass transfer term can be quantified by the linear driving force approach. Then, the transport-dispersive model consists of a balance equation for the mobile phase (Equation 6.70) written with the pore concentration

$$\frac{\partial c_i}{\partial t} + u_{\text{int}} \cdot \frac{\partial c_i}{\partial x} + \frac{1-\varepsilon}{\varepsilon} \cdot \left(\varepsilon_p \frac{\partial c_{p,i}}{\partial t} + (1-\varepsilon_p) \frac{\partial q_i}{\partial t} \right) = D_{\text{ax}} \cdot \frac{\partial^2 c_i}{\partial x^2} \tag{6.72}$$

as well as a balance for the stationary phase:

$$\varepsilon_p \frac{\partial c_{p,i}}{\partial t} + (1-\varepsilon_p) \frac{\partial q_i}{\partial t} = k_{\text{eff},i} \cdot \frac{3}{r_p} \cdot (c_i - c_{p,i}) \tag{6.73}$$

which is derived from Equations 6.4 and 6.14–6.17 as well as the assumptions of Equation 6.69. As discussed in Section 6.2.2, the driving force in Equation 6.73 is the difference between the concentration c_i in the bulk phase and the concentration $c_{p,i}$ on the particle surface, which in case of the lumped rate model is identical to the concentration inside the whole network of particle pores. The local adsorption equilibrium is again given by

$$q_i = f(c_{p,1}, \ldots, c_{N_{\text{comp}}}) \tag{6.35}$$

In Equation 6.73 the main resistance is modeled to lie within the liquid boundary layer surrounding the walls of the particle pores.

As already for the equilibrium-dispersive model (Section 6.2.4.1), the transport-dispersive model provides even under linear conditions asymmetrical peaks, and the asymmetry is enhanced by increasing D_{ax} as well as decreasing k_{eff} (Lapidus and Amundson, 1952).

Several modifications of this model can be found in the literature. One that is frequently used considers the mass transfer resistance in the solid phase to be dominant. As proposed by Glueckauf and Coates (1947), an analogue linear driving force approach for the mass transfer in the solid can then be applied and Equations 6.73 and 6.35 are replaced by Equations 6.74 and 6.75. Mathematically, this linear driving force is modeled as the difference between the overall solid loading of Equation 6.17 and an additional hypothetical loading q_{eq}, which is in equilibrium with the liquid phase concentration:

$$\frac{\partial q_i^*}{\partial t} = k_{\text{eff},s,i} \cdot \frac{3}{r_p} \cdot (q_{\text{eq},i}^* - q_i^*) \tag{6.74}$$

$$q_{\text{eq},i}^* = f(c_1, \ldots, c_{N_{\text{comp}}}) \tag{6.75}$$

For linear isotherms ($q_i = H_i c_{p,i}$), Equations 6.73 and 6.74 are equivalent, using the following relationships between the transport coefficients:

$$k_{\text{eff},i} = [\varepsilon_p + (1-\varepsilon_p) \cdot H_i] \cdot k_{\text{eff},s,i} \tag{6.76}$$

and

$$q_{\text{eq},i}^* = [\varepsilon_p + (1-\varepsilon_p) \cdot H_i] \cdot c_i \tag{6.77}$$

In this book, the term "transport-dispersive model" refers below to the liquid film linear driving force model (Equations 6.35, 6.72 and 6.73).

6.2.5.2 Reaction-Dispersive Model

Another subgroup of the lumped rate approach consists of the reaction-dispersive model where the adsorption kinetics are rate-limiting. It is an extension of the reaction model (Section 6.2.4.3). Like the mass transfer coefficient in the transport-dispersive model, the adsorption and desorption rate constants are considered as effective lumped parameters, $k_{\text{ads,eff}}$ and $k_{\text{des,eff}}$. Since no film transfer resistance is considered ($c_{p,i} = c_i$), the solid phase material balance can be described by Equation 6.78:

$$\varepsilon_p \frac{\partial c_i}{\partial t} + (1 - \varepsilon_p) \frac{\partial q_i}{\partial t} = k_{\text{ads,eff},i} \cdot q_{\text{sat},i} \cdot \left(1 - \sum_{j=1}^{N_{\text{comp}}} \frac{q_j}{q_{\text{sat},j}}\right) \cdot c_i - k_{\text{des,eff},i} \cdot q_i \tag{6.78}$$

and the material balance of the mobile phase (Equation 6.70) with $c_{p,i} = c_i$:

$$\frac{\partial c_i}{\partial t} + u_{\text{int}} \cdot \frac{\partial c_i}{\partial x} + \frac{1 - \varepsilon}{\varepsilon} \cdot \left(\varepsilon_p \frac{\partial c_i}{\partial t} + (1 - \varepsilon_p) \frac{\partial q_i}{\partial t}\right) = D_{\text{ax}} \cdot \frac{\partial^2 c_i}{\partial x^2} \tag{6.79}$$

6.2.6
General Rate Models

General rate models (GRM) are the most detailed continuous models considered in this book. In addition to axial dispersion, they incorporate a minimum of two other parameters describing mass transport effects. These two parameters may combine mass transfer in the liquid film and inside the pores as well as surface diffusion and adsorption kinetics in various kinds. Only a small representative selection of the abundance of different models suggested is given here in order to provide an overview. Alternatives not considered can be easily derived in a straightforward manner.

The underlying basic equations needed in a comprehensive approach using film transport, pore diffusion, surface diffusion, and adsorption kinetics (Berninger et al., 1991; Whitley, Van Cott, and Wang, 1993; Ma, Whitley, and Wang, 1996) were discussed in Section 6.2.2.

The column model equations are derived without further simplifications. Compared with models presented in the previous sections, radial mass transport inside the particle pores is here also taken into account, which results in concentration and loading distributions along the particle radius. Hence, averaged concentrations in Equations 6.14–6.16 have to be calculated using Equations 6.18 and 6.19.

The mass balance in the liquid phase (Equation 6.3) includes accumulation (Equation 6.13), convection (Equation 6.21), axial dispersion (Equation 6.22), and (external) mass transfer through the liquid film outside the particles (Equation 6.28):

$$\frac{\partial c_i}{\partial t} + u_{\text{int}} \cdot \frac{\partial c_i}{\partial x} + \frac{1 - \varepsilon}{\varepsilon} \cdot \frac{3}{r_p} \cdot k_{\text{film},i} \cdot (c_i - c_{p,i}(r = r_p)) = D_{\text{ax}} \cdot \frac{\partial^2 c_i}{\partial x^2} \tag{6.80}$$

The differential mass balances for the pores (Equation 6.5) and the solid (Equation 6.6) are formulated together with Equations 6.23 and 6.24 as well as Equations 6.30 and 6.31 as

$$\frac{\partial c_{p,i}(r)}{\partial t} = \frac{1}{r^2} \cdot \frac{\partial}{\partial r}\left(r^2 \cdot D_{\text{pore},i} \cdot \frac{\partial c_{p,i}(r)}{\partial r}\right) - \frac{1-\varepsilon_p}{\varepsilon_p} \cdot \psi_{\text{reac},i}(r) \tag{6.81}$$

$$\frac{\partial q_i(r)}{\partial t} = \frac{1}{r^2} \cdot \frac{\partial}{\partial r}\left(r^2 \cdot D_{\text{solid},i} \cdot \frac{\partial q_i(r)}{\partial r}\right) + \psi_{\text{reac},i}(r) \tag{6.82}$$

$$\psi_{\text{reac},i}(r) = k_{\text{ads},i} \cdot q_{\text{sat},i} \cdot \left(1 - \sum_{j=1}^{N_{\text{comp}}} \frac{q_j(r)}{q_{\text{sat},j}}\right) \cdot c_{p,i}(r) - k_{\text{des},i} \cdot q_i(r) \tag{6.31}$$

The internal mass transfer is modeled with Fick's diffusion inside the (macro) pores (Equation 6.81) as well as surface or micropore diffusion in the solid phase (Equation 6.82). Note that Equations 6.81 and 6.82 represent the balance in one particle.

Equation 6.31 is the net adsorption rate based on the solid volume, exemplified here as first-order kinetics with saturation capacities $q_{\text{sat},i}$.

If simple adsorption behavior can be considered (e.g., which is often not the case in bioseparations), the assumption of local adsorption equilibrium (Section 6.2.2.7) is normally valid. In this case Equation 6.31 formally reduces to an isotherm relationship connecting pore concentration a and solid loading a (Equation 6.34). As mentioned in Section 6.2.2.7, the two balances in the stationary phase (Equations 6.5 and 6.6) and the adsorption kinetics (Equation 6.31) can then be replaced by an isotherm equation (Equation 6.34) and the overall material balance for one particle (Equation 6.7). The latter can be derived analogous to Equations 6.81 and 6.82, leading to

$$\varepsilon_p \frac{\partial c_{p,i}}{\partial t} + (1-\varepsilon_p)\frac{\partial q_i}{\partial t} = \frac{1}{r^2} \cdot \frac{\partial}{\partial r}\left[r^2 \cdot \left(\varepsilon_p \cdot D_{\text{pore},i} \cdot \frac{\partial c_{p,i}}{\partial r} + (1-\varepsilon_p) \cdot D_{\text{solid},i} \cdot \frac{\partial q_i}{\partial r}\right)\right] \tag{6.83}$$

Gu, Tsai, and Tsao (1990a) and Gu et al. (1990b) proposed an even more reduced general rate model, which only considers pore diffusion inside the particles (pore diffusion model). Then, Equation 6.83 is replaced by Equation 6.84:

$$\varepsilon_p \frac{\partial c_{p,i}}{\partial t} + (1-\varepsilon_p)\frac{\partial q_i}{\partial t} = \varepsilon_p \cdot \frac{1}{r^2} \cdot \frac{\partial}{\partial r}\left(r^2 \cdot D_{\text{pore},i} \cdot \frac{\partial c_{p,i}}{\partial r}\right) \tag{6.84}$$

Not considering an established adsorption equilibrium between $c_{p,i}$ and q_i, adsorption kinetics were included later (Gu, Tsai, and Tsao, 1991, 1993; Gu, 1995). The approach by Gu et al. neglecting surface diffusion is justified for so-called "low-affinity" adsorbents in adsorption chromatography, where transport by free pore diffusion dominates surface diffusion (Furuya, Takeuchi, and Noll, 1989). Only for adsorbents with a pronounced micropore system ("high-affinity" adsorbents) surface diffusion can dominate over pore diffusion, and the more comprehensive model consisting of Equations 6.80–6.82 must be used. In such high-affinity adsorbents, the loadings are several orders of magnitudes higher than in particles applied in "normal" adsorption chromatography. The resulting high loading gradients can lead to a dominance of surface diffusion (Ma, Whitley, and Wang, 1996), although the transport coefficient D_{solid} is lower by orders of magnitude compared to D_{pore} (Suzuki, 1990).

Note that Equations 6.83 and 6.84 are identical if the pore diffusion coefficient is taken as a lumped concentration-dependent parameter that includes pore and surface diffusion:

$$D_{\text{app,pore},i} = D_{\text{pore},i} + \frac{1 - \varepsilon_p}{\varepsilon_p} \cdot D_{\text{solid},i} \cdot \frac{\partial q_i}{\partial c_{p,i}} \tag{6.85}$$

Applying general rate models, boundary conditions for the adsorbent phase are necessary in addition to boundary conditions at the column inlet and outlet (Section 6.2.7). The choice of appropriate boundary conditions is mathematically subtle and often a cause for discussion in the literature. The following boundary condition can be frequently used in solving the "complete" general rate model (Ma, Whitley, and Wang, 1996).

Owing to symmetry, the concentration and loading gradients vanish at the particle center:

$$\left.\frac{\partial c_{p,i}}{\partial r}\right|_{r=0} = \left.\frac{\partial q_i}{\partial r}\right|_{r=0} = 0 \tag{6.86}$$

The links between liquid, pore, and solid phase are given by mass balances at the particle boundary. Equation 6.29 connects the external mass transfer rate and the diffusion inside all particles, which after insertion of Equations 6.23, 6.24 and 6.28 results in

$$k_{\text{film},i} \cdot (c_i - c_{p,i}(r = r_p)) = \varepsilon_p \cdot D_{\text{pore},i} \left.\frac{\partial c_{p,i}}{\partial r}\right|_{r=r_p} + (1 - \varepsilon_p) \cdot D_{\text{solid},i} \left.\frac{\partial q_i}{\partial r}\right|_{r=r_p} \tag{6.87}$$

The set of boundary conditions is completed by recognizing that no surface diffusion exists outside the particles and, therefore, the gradient of the corresponding flux is zero:

$$\left.\frac{\partial \dot{m}_{\text{diff,solid},i}}{\partial r}\right|_{r=r_p} = \left.\frac{\partial}{\partial r}\left(\frac{\partial q_i}{\partial r}\right)\right|_{r=r_p} = \left.\frac{\partial^2 q_i}{\partial r^2}\right|_{r=r_p} = 0 \tag{6.88}$$

Considering Equation 6.82 the corresponding boundary condition is:

$$\left.\frac{\partial q_i}{\partial t}\right|_{r=r_p} = D_{\text{solid},i} \cdot \frac{3}{r_p} \cdot \left.\frac{\partial q_i}{\partial r}\right|_{r=r_p} + \psi_i|_{r=r_p} \tag{6.89}$$

6.2.7
Initial and Boundary Conditions of the Column

Mathematically, all models form a system of partial differential and algebraic equations. For the solution of such systems initial and boundary conditions for the chromatographic column are necessary. The initial conditions for the concentration and the loading specify their values at time $t=0$. Generally, not preloaded columns (zero values) are assumed:

$$c_i = c_i(t=0) = 0$$
$$c_{p,i} = c_{p,i}(t=0, r) = 0 \qquad (6.90)$$
$$q_i = q_i(t=0, r) = 0$$

Since no adsorbent enters or leaves the column, suitable inlet and outlet boundary conditions have to be provided only for the mass balance of the mobile phase (Equation 6.3).

One condition frequently applied at the column inlet is the classical "closed boundary" condition for dispersive systems derived by Danckwerts (1953):

$$c_i(t, x=0) = c_{\text{in},i} - \frac{D_{\text{ax}}}{u_{\text{int}}} \cdot \frac{\partial c_i(t, x=0)}{\partial x} \qquad (6.91)$$

In general, the overall balance for the mass transport streams (Equations 6.21 and 6.22) at the column inlet and outlet has to be fulfilled. In Equation 6.91 the closed boundary condition is obtained by setting the dispersion coefficient outside the column equal to zero. In open systems, the column stretches to infinity and in these limits concentration changes are zero.

For real chromatographic systems, the dispersion coefficient is usually small as the number of stages is very high ($N > 100$) and convection dominates. Therefore, Equation 6.91 may be simplified to

$$c_i(t, x=0) = c_{\text{in},i}(t) \qquad (6.92)$$

which is the same condition as for open systems.

A common condition of the inlet function (the injection profile) is a rectangular pulse (Section 6.3.2.1), that is, an injection of a constant feed concentration c_{feed} for a given injection time period, t_{inj}:

$$c_{\text{in},i}(t) = \begin{cases} c_{\text{feed},i}, & t \leq t_{\text{inj}} \\ 0, & t > t_{\text{inj}} \end{cases} \qquad (6.93)$$

For the outlet boundary condition typically a zero gradient of the fluid concentration is assumed (Danckwerts, 1953):

$$\frac{\partial c_i(t, x=L_c)}{\partial x} = 0 \qquad (6.94)$$

Notably, Equation 6.92 is in any case the "correct" inlet condition for all models without axial dispersion and is therefore often convenient to use. In practice, the difference between the solutions for different boundary conditions is typically irrelevant for highly efficient columns (Guiochon et al., 2006). Within numerical simulations, the effect of different boundary conditions can be easily evaluated.

6.2.8
Models of Chromatographic Reactors

Modeling approaches for so-called chromatographic reactors are typically based on the same models used to describe chromatographic separations. To extend these models for chromatographic reactors homogeneous or heterogeneous reactions have to be taken into account additionally. Owing to the typically high degrees of dilution, thermal effects due to the heats of reaction can be frequently neglected.

Mass transfer kinetics are given by simple linear driving force models. The fraction of the catalyst within the whole fixed bed is described by the factor X_{cat}. Assuming reaction occurs homogeneously in the liquid phase and heterogeneously at the surface of the catalyst, the differential mass balance for component i can be written for the liquid phase as

$$\frac{\partial c_i}{\partial t} + u_{int} \cdot \frac{\partial c_i}{\partial x} + -(1 - X_{cat}) \cdot \frac{1-\varepsilon}{\varepsilon} \cdot \left[\varepsilon_{ads,p} \frac{\partial c_{p,ads,i}}{\partial t} + (1 - \varepsilon_{ads,p}) \frac{\partial q_{ads,i}}{\partial t}\right]$$

$$- X_{cat} \cdot \frac{1-\varepsilon}{\varepsilon} \cdot \left[\varepsilon_{cat,p} \frac{\partial c_{p,cat,i}}{\partial t} + (1 - \varepsilon_{cat,p}) \frac{\partial q_{cat,i}}{\partial t}\right] = D_{ax} \cdot \frac{\partial^2 c_i}{\partial x^2} + v_i \cdot r_{hom,i}$$

(6.95)

and for both solid phases

$$\varepsilon_{ads,p} \frac{\partial c_{p,ads,i}}{\partial t} + (1 - \varepsilon_{ads,p}) \frac{\partial q_{ads,i}}{\partial t} = k_{eff,ads,i} \cdot \frac{3}{r_{p,ads}} \cdot (c_i - c_{p,ads,i}) \quad (6.96)$$

$$\varepsilon_{cat,p} \frac{\partial c_{p,cat,i}}{\partial t} + (1 - \varepsilon_{cat,p}) \frac{\partial q_{cat,i}}{\partial t} = k_{eff,cat,i} \cdot \frac{3}{r_{p,cat}} \cdot (c_i - c_{p,cat,i}) + v_i \cdot r_{het,i}$$

(6.97)

Equations 6.95–6.97 represent the mass balances based on the transport-dispersive model in their most general form. For certain applications they reduce to simpler versions.

6.2.9
Stage Models

An entirely different powerful approach to describe front propagation phenomena in chromatographic columns should also be briefly introduced here. This approach is based on considering the column as a sequence of connected equilibrium stages leading to the class of equilibrium stage or plate models. Instead of formulating dynamic microscopic balances, the column is modeled as a sequence of a finite number N of similar stages. Each stage is filled with liquid and solid and the two phases are completely mixed. Two general groups of equilibrium stage models have been reported. The so-called Craig model is based on the assumption of a constant residence time in each stage, which is sufficient to achieve equilibrium, and a subsequent discrete exchange of the liquid phase in the flow direction (Craig, 1944). Another stage model concept was introduced by Martin and Synge (1941). It is equal to the concept of a stirred tank cascades common in reaction engineering. Just the latter model is used here to illustrate the application of stage models to simulate chromatographic separation.

A constant flow of mobile phase passes continuously a cascade of N ideally stirred tanks (C.S.T.). Each tank has a total volume equal to V_c/N. Inside each tank, a fraction $(1 - \varepsilon_t)$ is occupied by the solid phase and the concentration inside the liquid is the same in the bulk and in the pore phase. This leads to the following mass balance for the kth tank, where overall accumulation is equal to the difference between the inlet and the outlet streams:

$$\frac{V_c}{N} \cdot \left(\varepsilon_t \cdot \frac{\partial c_i^k}{\partial t} + (1 - \varepsilon_t) \cdot \frac{\partial q_i^k}{\partial t} \right) = \dot{V} \cdot (c_i^{k-1} - c_i^k) \tag{6.98}$$

Now, in addition, equilibrium between the two concentrations c and q is assumed.

Band broadening effects such as dispersion and mass transfer resistance are represented by the number of tanks (or stages) N. This can be explained by evaluating the moments of the analytical solution of Equation 6.98. For linear isotherms and the injection of an ideal Dirac pulse of one component, this equation yields a gamma density function for the concentration profile. With the retention time $t_{R,\text{lin},i}$ of Equation 6.49 one obtains the elution profile as the time dependence of the concentration in the last tank ($k = N$):

$$c_N(t) = \frac{m_{\text{inj}}}{\dot{V}} \frac{N}{t_{R,\text{lin},i}} \cdot \left(\frac{N \cdot t}{t_{R,\text{lin},i}} \right)^{N-1} \cdot \frac{1}{(N-1)!} \cdot \exp\left\{ -\frac{N \cdot t}{t_{R,\text{lin},i}} \right\} \tag{6.99}$$

The injected amount m_{inj} is given by Equation 6.61.

Calculation of the first moment μ_t (Equation 2.38) and the second moment σ_t^2 (Equation 2.39) results in the following:

$$\mu_t = t_{R,\text{lin},i} \tag{6.100}$$

$$\sigma_t = \frac{t_{R,\text{lin},i}}{\sqrt{N}} \tag{6.101}$$

$$\Leftrightarrow N = \left(\frac{t_{R,\text{lin},i}}{\sigma_t} \right)^2 = \left(\frac{\mu_t}{\sigma_t} \right)^2 \tag{6.102}$$

Equation 6.102 is identical to Equation 2.40 and provides the definition of the number of stages N (Van Deemter, Zuiderweg, and Klinkenberg, 1956), which is equal to the number of tanks in this model. As the second moment (variance) is directly related to the "width" of the peak (Section 2.4.2), the stage number is an appropriate value to describe band broadening. This model's behavior is consistent with the requirement that a high number of stages leads to a highly efficient separation (low band broadening). For high N the gamma density function can be approximated by a Gaussian distribution (Equation 6.60).

In principle, different numbers of stages must be used for each component to account for individual band broadening. This is a clear disadvantage of stage models, making it impossible to get a proper description of the elution behavior of every component in a multicomponent separation. However, if N is large, this disadvantage might be negligible.

6.2.10
Assessment of Different Model Approaches

All models presented in the previous sections fulfill the general requirement of compatibility with different modes of operation (by specifying the corresponding initial and boundary conditions) and adsorption behaviors (by specifying the corresponding equilibrium functions). Also, appropriate analytical or numerical methods are available to solve all model equations. Because of the many combinations of solutes and mobile and stationary phases, it might be difficult to determine a priori which model is most suitable for a certain specific problem. A first guideline for model selection is provided by the number of stages N (Golshan-Shirazi and Guiochon, 1992; Seidel-Morgenstern, 1995; Guiochon et al., 2006), which can be obtained either from the column manufacturer or by a few simple preliminary experiments (Sections 2.4.2 and 6.5.3.1).

The ideal model provides good accuracy only if the column efficiency is very high ($N \gg 1000$). If band broadening is significant, the ideal model as well as the stage models, which can only reproduce the number of stages for a single component, are not suited to describe multicomponent separations accurately.

The equilibrium-dispersive model offers an acceptable accuracy only for $N \gg 100$, which may be sufficient for many practical cases. However, sometimes the number of stages is lower and more sophisticated models have to be applied.

Ma, Whitley, and Wang (1996) and Whitley, Van Cott, and Wang (1993) have provided methods to decide if effects such as pore and surface diffusion or adsorption kinetics have to be considered in a model. Their approach is based on the qualitative assessment of measured breakthrough curves, which result from a step input for different feed concentrations and flow rates. When the physical parameters are known or can be estimated, the value of dimensionless parameters defined in the mentioned publications may be used to select a model.

Since the main focus of this book is on adsorption chromatography (low-affinity adsorbents), influences of finite adsorption kinetics as well as of surface diffusion can be neglected. Thus, the reaction models and models including surface diffusion are no more considered in the following.

Consequently, the two remaining groups considering columns with a low number of stages are the lumped and the general rate models. Under the simplification mentioned above, general rate models take into account the individual mass transfer resistance in the liquid film/boundary layer (Equation 6.80) and inside the pore system of the particles (Equation 6.84). Based on theoretical studies, Ludemann-Hombourger, Bailly, and Nicoud (2000) concluded that, even for very small particle diameters ($d_p = 2$ μm) and extremely low fluid velocities, pore diffusion can be the limiting mass transfer step in adsorption chromatography (Section 6.5.8). For most practical applications it is therefore not necessary to use two parameters to describe mass transfer and a lumped rate model using only one parameter (e.g., Section 6.2.5.1) is sufficient.

The decision for a certain model also has to include considerations regarding methods to measure or estimate the model parameters. Concerning the general rate model (Equations 6.80 and 6.84) it is not possible to derive independently different transport parameters such as D_{pore} and k_{film} for a given column from a small number of measured chromatograms. Therefore, one of the two parameters has to be provided independently (e.g., by correlations; Section 6.5.8) and used to determine subsequently the other analyzing elution profiles.

The transport-dispersive model (Section 6.2.5.1) appears to be appropriate to simulate systems with considerable band broadening (Section 6.6), using two different parameters to characterize packing properties (D_{ax}) and mass transfer (k_{eff}). Kaczmarski and Antos (1996) provide rules, in which cases both TDM and GRM give identical results.

As increasing computational power and sophisticated numerical solvers are now available, it is no longer necessary to use oversimplified models to reduce computing time.

Table 6.2 gives recommendations for the application of the models discussed above.

The rest of this chapter will consider just the transport-dispersive model.

Table 6.2 Fields of application of different models.

Model	Recommended application
Ideal model	• Considers thermodynamic effects only ($N \gg 1000$) • Useful for rapid process evaluation and design
Equilibrium-dispersive model	• Adsorption chromatography for products with low molecular weights • High accuracy only if $N \gg 100$
Transport-dispersive model	• Adsorption chromatography for products with low molecular weights • Generally high accuracy • Useful to quantify chiral separation
General rate model	Chromatographic separation of solutes with complex mass transfer and adsorption behavior (e.g., bioseparations or ion exchange chromatography)
Equilibrium stage model	• Adsorption chromatography for products with low molecular weights • Accurate only for single components or it there are only small differences in N_i for all components

6.2.11
Dimensionless Model Equations

To reduce the number of parameters and to analyze their interdependence it is recommended that the model equations as well as the boundary conditions be converted into a dimensionless form.

Feed concentrations $c_{\text{feed},i}$ should be selected as a reference for concentrations and loadings:

- Dimensionless fluid concentration:

$$C_{\text{DL},i} = \frac{c_i}{c_{\text{feed},i}} \Rightarrow \partial C_{\text{DL},i} = \frac{1}{c_{\text{feed},i}} \cdot \partial c_i \qquad (6.103)$$

- Dimensionless concentration in the particle:

$$C_{p,\text{DL},i} = \frac{c_{p,i}}{c_{\text{feed},i}} \Rightarrow \partial C_{p,\text{DL},i} = \frac{1}{c_{\text{feed},i}} \cdot \partial c_{p,i} \qquad (6.104)$$

- Dimensionless loading (as one possibility):

$$Q_{\text{DL},i} = \frac{q_i}{c_{\text{feed},i}} \Rightarrow \partial Q_{\text{DL},i} = \frac{1}{c_{\text{feed},i}} \cdot \partial q_i \qquad (6.105)$$

A dimensionless axial coordinate Z is obtained by division through the column length:

$$Z = \frac{x}{L_c} \Rightarrow \partial x = L_c \cdot \partial Z \qquad (6.106)$$

and a dimensionless time τ can be defined with the residence time $t_{0,\text{int}}$ (Equation 6.42) based on the interstitial velocity:

$$\tau = \frac{t}{t_{0,\text{int}}} = \frac{u_{\text{int}}}{L_c} \cdot t \Rightarrow \partial \tau = \frac{u_{\text{int}}}{L_c} \cdot \partial t \qquad (6.107)$$

Dimensionless parameters represent ratios of different mass transport and reaction phenomena. One example is the axial Péclet number, which is the ratio of convection rate to axial dispersion:

$$Pe = \frac{u_{\text{int}} \cdot L_c}{D_{\text{ax}}} \qquad (6.108)$$

Additional parameters arise according to the model selected and the physical effects that are taken into account. Parameters appropriate for the general rate model are given, for example, by Berninger et al. (1991) and Ma, Whitley, and Wang (1996).

Introducing Equations 6.103–6.107 into the equations of the transport-dispersive model leads to the following dimensionless mass balances:

$$\frac{\partial C_{DL,i}}{\partial \tau} + \frac{\partial C_{DL,i}}{\partial Z} + \frac{1-\varepsilon}{\varepsilon} \cdot \left(\varepsilon_p \frac{\partial C_{DL,i}}{\partial \tau} + (1-\varepsilon_p) \frac{\partial Q_{DL,i}}{\partial \tau} \right) = \frac{1}{Pe} \cdot \frac{\partial^2 C_{DL,i}}{\partial Z^2}$$

(6.109)

$$\varepsilon_p \frac{\partial C_{DL,i}}{\partial \tau} + (1-\varepsilon_p) \frac{\partial Q_{DL,i}}{\partial \tau} = \frac{L_c}{u_{int}} \cdot \frac{6}{d_p} \cdot k_{eff,i} \cdot (C_{DL,i} - C_{p,DL,i})$$

(6.110)

The dimensionless parameter in Equation 6.109 is the Péclet number (Equation 6.108). The remaining parameter in Equation 6.110 determines the ratio of effective mass transport to convection, which is defined by the modified (effective) Stanton number (St_{eff}):

$$St_{eff,i} = k_{eff,i} \cdot \frac{6}{d_p} \cdot \frac{L_c}{u_{int}}$$

(6.111)

This finally leads to a system of differential equations that depend only on these two dimensionless parameters:

$$\frac{\partial C_{DL,i}}{\partial \tau} + \frac{\partial C_{DL,i}}{\partial Z} + \frac{1-\varepsilon}{\varepsilon} \cdot \left(\varepsilon_p \frac{\partial C_{p,DL,i}}{\partial \tau} + (1-\varepsilon_p) \frac{\partial Q_{DL,i}}{\partial \tau} \right) = \frac{1}{Pe} \cdot \frac{\partial^2 C_{DL,i}}{\partial Z^2}$$

(6.112)

$$\varepsilon_p \frac{\partial C_{p,DL,i}}{\partial \tau} + (1-\varepsilon_p) \frac{\partial Q_{DL,i}}{\partial \tau} = St_{eff,i} \cdot (C_{DL,i} - C_{p,DL,i})$$

(6.113)

In addition to the mass balances, the isotherm equations and the boundary conditions have to be transformed into a dimensionless form. Equation 6.114, for example, presents the dimensionless form of the multicomponent Langmuir equation (Equation 2.57) for a binary mixture:

$$Q_{DL,i} = \frac{H_i \cdot C_{p,DL,i}}{1 + b_1 \cdot c_{feed,1} \cdot C_{p,DL,1} + b_2 \cdot c_{feed,2} \cdot C_{p,DL,2}}, \quad \text{with} \quad H_i = q_{sat,i} \cdot b_i$$

(6.114)

The dimensionless parameters of this equation are the Henry coefficients H_i and the dimensionless Langmuir parameters $b_1 c_{feed,1}$ and $b_2 c_{feed,2}$. This is the mathematical indication that the feed concentrations are parameters for a certain separation problem, since the H_i and b_i are constant for the chromatographic system.

Conversion of the boundary condition at the column inlet (Equation 6.93) leads to

$$C_{DL,in,i}(t) = \begin{cases} 1, & \tau \leq \dfrac{t_{inj}}{t_{0,int}} = \dfrac{t_{inj} \cdot \dot{V}}{\varepsilon \cdot V_c} \\ 0, & \tau > \dfrac{t_{inj}}{t_{0,int}} \end{cases}$$

(6.115)

where $t_{inj}/t_{0,int}$ is the dimensionless injection time.

In summary, chromatographic batch separation depends, besides the feed concentrations, on the following dimensionless parameters: Péclet and Stanton numbers, dimensionless injection time, Henry coefficients, and dimensionless isotherm (e.g. Langmuir) parameters.

6.3
Modeling HPLC Plants

6.3.1
Experimental Setup and Simulation Flow Sheet

Section 6.2 presents various models capable to describe front propagation phenomena in chromatographic columns. It has to be kept in mind that these models account only for effects occurring within the packed bed. A HPLC plant, however, consists of several additional equipment and fittings besides the column. Therefore, the effect of this extra column equipment has to be accounted for to obtain reasonable agreement between experimental results and process simulation. Peripheral equipment (e.g., pipes, injection system, pumps, and detectors) causes dead times and mixing. Thus, it can contribute considerably to the band broadening measured by the detector.

So-called "plant dispersion" or "extra column effects" have to be taken into account by additional mathematical models rather than including them indirectly in the model parameters of the column, for example, by altering the dispersion coefficient. The combination of peripheral and column models can be implemented in a modular simulation approach. In a flow-sheeting approach the boundary conditions of different models are connected by streams (node balances) and all material balances are solved simultaneously.

Figure 6.5a shows the standard setup of an HPLC plant. Injections of rectangular pulses are performed via a three-way valve and a subsequent pump or a six-port valve with a sample loop. The feed passes the connecting pipes and a flow distributor before entering the packed bed. At the exit there are another flow distributor and connecting pipes before a detector records the chromatogram. Because of these additional elements, an exact rectangular pulse will not enter the chromatographic column and the detected chromatogram is not identical to the concentration profile at the column exit.

As mentioned in Chapters 2 and 4, peak distortion is caused by nonideal equipment not only outside the column but also inside the column. Although sophisticated measurements such as NMR (Tallarek, Bayer, and Guiochon, 1998) allow the investigation of the packed bed only, from a practical viewpoint the observable performance of a column always includes the effects of walls, internal distributors, and filters. Using the method described in this chapter, these are always contained in certain model parameters (e.g., in D_{ax}). The column manufacturers have to ensure a proper bed packing and flow distribution (Chapter 4) and, thus, negative influence of imperfections on column performance, for example, on D_{ax}, can be

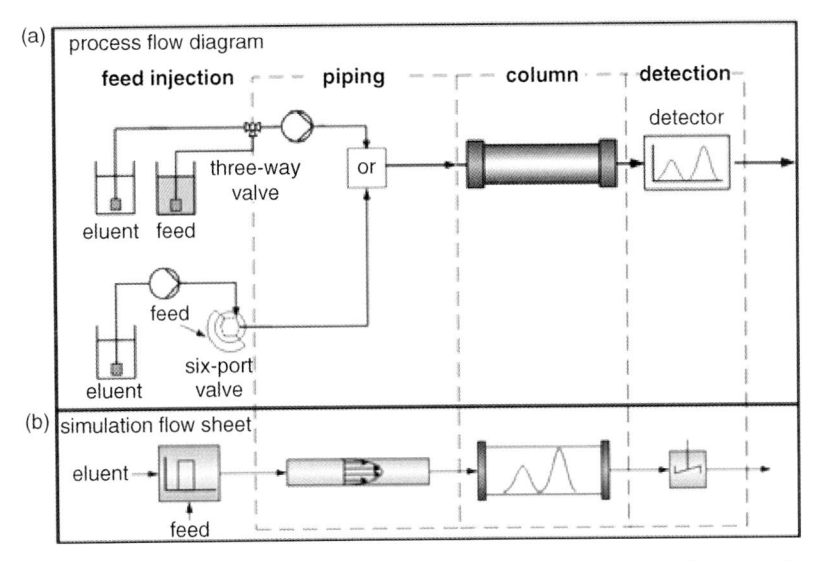

Figure 6.5 Comparison between (a) process flow diagram and (b) simulation flow sheet for an HPLC plant.

assumed to be small. However, the aspect of extra column effects should always be kept in mind when doing scale-up.

Transformation of the process diagram into a corresponding simulation flow sheet is illustrated in Figure 6.5b. In principle, all plant elements may be represented by separate submodels. For practical applications, though, it is typically sufficient to take into account only a time delay as well as the dispersion of the peaks until they enter the column. This can be achieved by a pipe flow model that includes axial dispersion. Large volume detectors (including some connecting pipes) can be approximated, if required, by a stirred tank model.

6.3.2
Modeling Extra Column Equipment

6.3.2.1 Injection System

The injection system is sufficiently described by setting the appropriate boundary conditions at the entry of the pipe, for example, a rectangular pulse:

$$c_{pipe,i}(x=0, 0 \leq t \leq t_{inj}) = c_{feed,i}$$
$$c_{pipe,i}(x=0, t > t_{inj}) = 0$$
(6.116)

Injection volume and the injection time are related by

$$V_{inj} = \dot{V} \cdot t_{inj}$$
(6.117)

6.3.2.2 Piping

If the piping only contributes to the dead time of the plant, the delay can be described by a pipe model assuming an ideal plug flow:

$$\frac{\partial c_{pipe,i}}{\partial t} = -u_{0,pipe} \cdot \frac{\partial c_{pipe,i}}{\partial x} \tag{6.118}$$

When the plant behavior without the column shows non-negligible backmixing, a dispersed plug flow model might be used (Equation 6.119):

$$\frac{\partial c_{pipe,i}}{\partial t} = D_{ax,pipe} \cdot \frac{\partial^2 c_{pipe,i}}{\partial x^2} - u_{0,pipe} \cdot \frac{\partial c_{pipe,i}}{\partial x} \tag{6.119}$$

The fluid velocity u_0 inside the pipe is given by the continuity equation:

$$u_{0,pipe} = \frac{\dot{V}}{A_{pipe}} \tag{6.120}$$

It is not necessary to model individual pipes if the cross-sections are represented by a typical diameter. In any case the primary parameter of interest is the volume (Equation 6.121) or the dead time (Equation 6.122) of the unit:

$$V_{pipe} = A_{pipe} \cdot L_{pipe} \tag{6.121}$$

$$t_{0,pipe} = \frac{V_{pipe}}{\dot{V}} \tag{6.122}$$

If $t_{0,pipe}$ is determined from experiments and A_{pipe} is set, all other parameters are known from Equations 6.120–6.122.

6.3.2.3 Detector

Detectors contain measuring cells that exhibit a backmixing behavior that dominates the influence of the pipe system behind the chromatographic column. Therefore, the whole system behind the column is modeled as an ideal continuously stirred tank (C.S.T.):

$$\frac{\partial c_{tank,i}}{\partial t} = \frac{\dot{V}}{V_{tank}} \cdot (c_{in,tank,i} - c_{tank,i}) \tag{6.123}$$

The dead time of the tank (Equation 6.124) is

$$t_{0,tank} = \frac{V_{tank}}{\dot{V}} \tag{6.124}$$

Note that the overall dead time of the plant (Equation 6.125) is the sum of the dead times of both the pipes and the detector (tank):

$$t_{plant} = t_{0,pipe} + t_{0,tank} \tag{6.125}$$

6.4
Calculation Methods

6.4.1
Analytical Solutions

For the most simple column models under certain simplifying conditions there are analytical solutions of the model equations available. Related to the equilibria this holds for problems where all components of interest are characterized by linear isotherm equations, in which the Henry constants are not affected by the presence of other component. Then all kinetic effects causing band broadening can be described by a single lumped parameter, for example, the number of theoretical plates. Consequently, the usage of two or more kinetic parameters is not justified. This field of linear or analytical chromatography has been extensively studied and is quite mature (Guiochon et al, 2006; Snyder, Kirkland, and Dolan, 2010).

For a number of nonlinear and competitive isotherm models analytical solutions of the mass balance equations can be provided for only one strongly simplified column model. This is the "ideal model" of chromatography, which considers just convection and neglects all mass transfer processes (Section 6.2.3). Using the method of characteristics within the elegant equilibrium theory, analytical expressions were derived capable to calculate single elution profiles for single components and mixtures (Helfferich and Klein, 1970; Helfferich and Carr 1993; Helfferich and Whitley 1996; Helfferich 1997; Rhee, Aris, and Amundson, 1970; Rhee *et al.*, 1986; Rhee, Aris, and Amundson, 1989; Guiochon *et al.*, 2006). However, these predictions are typically too optimistic since real columns are not characterized by an infinite efficiency. In many case they are however close to real profiles and offer the chance for rapid process design and easy process evaluation.

6.4.2
Numerical Solution Methods

6.4.2.1 General Solution Procedure
The balance equations described in the previous sections include both space and time derivatives. Apart from a few simple cases, the resulting set of coupled partial differential equations (PDE) cannot be solved analytically. The solution (the concentration profiles) must be obtained numerically, using either self-developed programs or commercially available dynamic process simulation tools. The latter can be distinguished in general equation solvers, where the model has to be implemented by the user, or special software dedicated to chromatography. Some providers are given in Table 6.3.

The generalized numerical solution procedure typically involves the following steps:

1) transformation of the PDE system into ordinary differential equations (ODE) with respect to time by discretization of the spatial derivatives;

Table 6.3 Examples of dynamic process simulation tools.

Examples for dynamic process simulation tools	
Aspen Engineering Suite™ (e.g., Aspen Custom Modeler®) (Aspen Technology, Inc., USA)	http://www.aspentech.com
gPROMS® (Process Systems Enterprise Limited (PSE), UK)	http://www.psenterprise.com
Examples for application software for liquid chromatography	
Aspen Chromatography® (Aspen Technology, Inc., USA)	http://www.aspentech.com
ChromSim® for Windows™ (Wissenschaftliche Gerätebau Dr. Ing. Herbert Knauer GmbH)	http://www.knauer.net
BatchChromLehrstuhl für Anlagentechnik, Universität Dortmund	http://atwww.bci.uni-dortmund.de
SMBOptLehrstuhl für Anlagensteuerungstechnik, Universität Dortmund	http://astwww.bci.uni-dortmund.de
Chromulator (Tingyue Gu's Chromatography Simulation)	http://www.ent.ohiou.edu/~guting/
VERSE (The Bioseparations Group, School of Chemical Engineering, Purdue University, West Lafayette)	http://atom.ecn.purdue.edu/~biosep/research.html
ChromWorks® (Burlington/MA, USA)	http://chromworks.com/wp/products/cw2012/

2) solving the ODEs using numerical integration routines, which generally involve another discretization into a nonlinear algebraic equation system and subsequent iterative solution.

The stability, accuracy, and speed of the solution process depend on the choice and parameter adjustment of the individual mathematical methods in each step as well as their proper combination. Notably, in simulating a batch column, computational time is hardly an issue using today's PC systems.

Detailed discussion of discretization techniques and numerical solution methods is beyond the scope of this book. Numerical methods are discussed in detail by, for example, Finlayson (1980), Davis (1984), and Du Chateau and Zachmann (1989). Summaries of different discretization methods applied in the simulation of chromatography are given by Guiochon, Golshan-Shirazi, and Katti (1994) and Guiochon and Lin (2003). For an introduction into numerical programming procedures see, for example, Press *et al.* (2002, http://www.nr.com) or Ferziger (1998).

6.4.2.2 Discretization

Discretization "replaces" the continuous space–time domain by a rectangular mesh or grid of discrete elements and points (Figure 6.6). Note that initial and boundary conditions of the system must also be considered.

As a result of the discretization, numerical solutions are only approximations of the "true" continuous solutions at discrete points of the space–time domain. The quality of the numerical solution depends on the structure of the discretization method and the number of discretes.

An example of the "errors" introduced by the approximation is the effect of "numerical dispersion," which leads to an additional artificial band broadening.

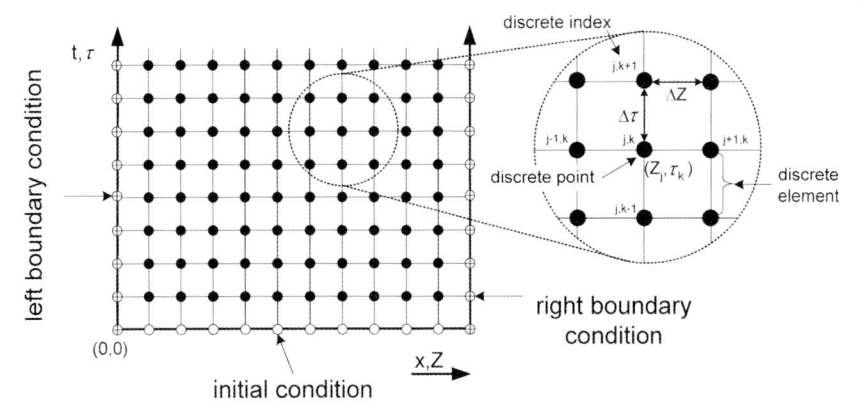

Figure 6.6 Scheme of discretized space–time domain using a uniform grid.

Another example is the occurrence of nonphysical oscillations in the profile. Typically holds that the finer the grid is, the more accurate the numerical approximation becomes and, finally, approaches the true solution. Naturally, smaller grid spacings mean more equations and longer calculation times. Practically, the grid is "fine enough" if the simulated elution profiles do not change noticeably on using a smaller grid size. Other simple tests for accuracy involve comparison with analytical solutions in special cases or control of the mass balance. Numerical integration of the simulated peak (Equation 6.61) area should not deviate by more than 1% from the injected amount (Guiochon and Lin, 2003).

For brevity, further discussion is restricted to the spatial discretization used to obtain ordinary differential equations. Often, the choice and parameter selection for this method is left to the user of commercial process simulators, while the numerical (time) integrators for ODEs have default settings or sophisticated automatic parameter adjustment routines. For example, using finite difference methods for the time domain, an adaptive selection of the time step is performed that is coupled to the iteration needed to solve the resulting nonlinear algebraic equation system. For additional information concerning numerical procedures and algorithms the reader is referred to the special literature.

Finite difference methods (FDM) are directly derived from the space–time grid. Focusing on the space domain (horizontal lines in Figure 6.6), the spatial differentials are replaced by discrete difference quotients based on interpolation polynomials. Using the dimensionless formulation of the balance equations (Equation 6.109), the convection term at a grid point j (Figure 6.6) can be approximated by assuming, for example, the linear polynomial:

$$\frac{\partial C_{DL}(Z=Z_j)}{\partial Z} \approx \frac{C_{DL,j+1} - C_{DL,j}}{\Delta Z}, \quad \text{with} \quad Z_j = j \cdot \Delta Z \tag{6.126}$$

Figure 6.7 illustrates this approach.

Equation 6.126 is a forward FDM, as grid points "after" j (here $j+1$) are used for evaluation. It is a first-order scheme, as only one point is used. Higher order

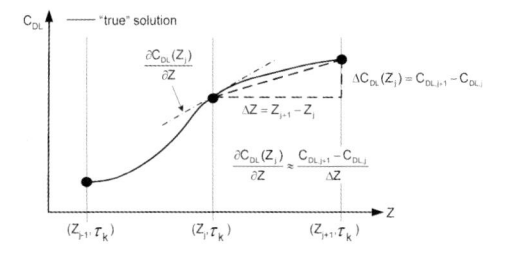

Figure 6.7 Approximation of spatial derivatives by difference quotients.

schemes involve more neighboring points for the approximation and are generally more accurate and have greater numerical stability. Other FD schemes involve grid points with lower indexes (backward FDM) or points on both sides of j (central FDM). Equation 6.126 illustrates only the principle of discretization. The continuous coordinates are replaced by the corresponding numbers of grid points (n_p) and the continuous profile $C_{DL}(Z)$ by a vector, with n_p elements $C_{DL,j}$.

The major drawback of (simple) FD schemes is that a rather fine grid, resulting in a large number of equations and hence high computational time, is necessary to obtain a stable and accurate solution, as chromatography is a convection-dominated process. Fewer grid points and shorter computational times can be realized by using UPWIND schemes for the convection term (Du Chateau and Zachmann, 1989; Shih, 1984). These are either higher order backward FDM or more complex schemes (Leonard, 1979).

FD methods are point approximations, because they focus on discrete points. In contrast, finite element methods focus on the concentration profile inside one grid element. As an example of these segment methods, orthogonal collocation on finite elements (OCFE) is briefly discussed below.

The collocation method is based on the assumption that the solution of the PDE system can be approximated by polynomials of order n. In this method the whole space domain (the column), including the boundary conditions, is approximated by one polynomial using $n+2$ collocation points. Polynomial coefficients are determined by the condition that the differential equation must be satisfied at the collocation points. This approach allows the spatial derivatives to be described by the known derivatives of the polynomials and transforms the PDE into an ODE system.

Mathematical methods exist that guarantee an "optimal" placement of the collocation points. In orthogonal collocation (OC), the collocation points are equal to the zero points of the orthogonal polynomials.

The OCFE method (Villadsen and Stewart, 1967) is derived straightforwardly by dividing the column into n_e elements (Figure 6.8). For each element an own orthogonal solution polynomial is used, leading to an ODE system for the column with $(n+2)n_e$ equations. The element boundaries are connected by setting equal values and slopes for the adjacent polynomials at the boundary points, which guarantees the continuity of the concentration profile.

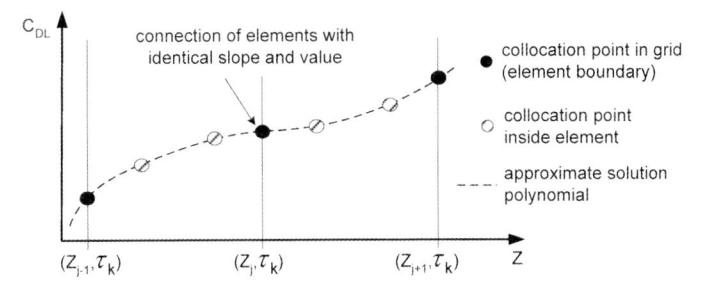

Figure 6.8 Representation of the numerical solution by the OCFE method for a second-order polynomial.

Finite element methods proved to be a good choice for various column models (Guiochon, Golshan-Shirazi, and Katti, 1994; Seidel-Morgenstern, 1995; Kaczmarski et al., 1997; Lu and Ching, 1997; Berninger et al., 1991). Comparisons show that they are, for the type of models discussed here, to be preferred to difference methods, especially if high accuracy is needed (Dünnebier, 2000; Finlayson, 1980; Guiochon and Lin, 2003).

Newer works report successful application of spatial discretization using finite volumes and the weighted essentially nonoscillatory (WENO) method (von Lieres and Andersson, 2010), high-resolution finite volume schemes (Javeed et al., 2011a), and a discontinuous Galerkin method (Javeed et al., 2011b).

6.5
Parameter Determination

After a suitable model for process simulation has been chosen and the techniques to solve the corresponding model equations are available, the model parameters have to be determined.

6.5.1
Parameter Classes for Chromatographic Separations

6.5.1.1 Design Parameters

The models presented in Section 6.2 contain sets of independent and dependent parameters. Designing a chromatographic process starts with selecting the chromatographic system (Chapter 3) and, subsequently, the kind of the chromatographic process as well as the hardware required to realize it (Chapters 4 and 5). These selections lead to the specification of the design parameters:

- stationary phase (particle diameter d_p);
- mobile phase (type, composition);
- length (L_c) and diameter (d_c) of the chromatographic bed;
- maximum allowable pressure drop (Δp);

- temperature;
- further process-specific design parameters (e.g., number of columns per section in SMB processes).

Various design parameters specify the chromatographic plant. Hereby geometrical design parameters are typically also objects of optimization, if no already existing plants are used for the separation.

6.5.1.2 Operating Parameters
The second class of parameters consists of the operating parameters for a given plant, which can be changed during plant operation:

- flow rate ($\dot{V}_{feed}$);
- feed concentration (c_{feed});
- amount of feed (V_{inj}, m_{inj});
- additional degrees of freedom, for example, switch times (fraction collection) in batch chromatography or flow rates in each section of a SMB process.

Like geometrical design parameters, operating parameters are part of the degrees of freedom for optimizing chromatographic separations. Their impact will be discussed in Chapter 7.

6.5.1.3 Model Parameters
System inherent physical and chemical parameters that specify the chromatographic system within the column as well as the plant operation make up the third set of parameters. These model parameters are typically not known a priori:

- plant parameters (describing the residence time distribution of the plant peripheral);
- packing parameters (void fraction, porosity, and pressure drop);
- axial dispersion coefficients;
- equilibrium isotherms;
- mass transfer coefficients.

The above list refers to the transport-dispersive model (Section 6.2.5.1) and, if other models are selected, the model equations and hence the corresponding number of parameters are expanded or reduced.

Physical properties and especially the isotherms depend on temperature as well as eluent composition. Feed and eluent composition influence the viscosity and therefore the fluid dynamics. The latter effects have already to be taken into account when selecting the chromatographic system (Chapter 3). The operating temperature for preparative processes is commonly selected in liquid chromatography to be close to room temperature for cost and also stability reasons. Consequently, temperature and eluent composition are early fixed parameters that are often not explicitly considered during subsequent process design.

The time and money invested in model parameter determination has to be in balance with the aim of a certain chromatographic separation. Often only sample products are needed for further studies with the pure substances and, to evaluate a separation process, only a few milligrams of the feed mixture is available. Time is typically an important factor and "quick and dirty"/shortcut methods will be applied to find conservative operating parameter values for a safe separation. A quite different situation is the design of commercial large-scale production plants. Here, process economics are more essential. Therefore, more precise parameters are necessary for process optimization. As some parameters may change during operation, repeated determination may be necessary to maintain or reestablish optimal production.

Research has a third focus. Here, results have to be as accurate and reliable as possible. Consequently, more time and money are invested to acquire the data. Research aims are, for instance, to better understand the process and to improve methods for process design, optimization, and control. This cannot be done experimentally only. Therefore, verified models are needed to perform the investigations, with the aid of process simulation as virtual plant experiments.

Against this background, both precise and straightforward shortcut methods for model parameter determination are presented below. Based on the individual task users have to decide specifically which procedures are most adequate. In this context it must be mentioned that the adsorption equilibrium has clearly the largest influence on the position and shape of chromatographic profiles and, therefore, the isotherms deserve the greatest care.

6.5.2
Determination of Model Parameters

A method for consistent parameter determination is depicted in Figure 6.9. It was published by Altenhöner *et al.* (1997) to determine the free parameters of the transport-dispersive model. The basic idea is to start with the simplest parameters and to use them subsequently to determine the ones that are more difficult to determine. The procedure is structured into the following steps:

1) Determine extra column effects such as dead volume and the backmixing of the plant.
2) Detector calibration.
3) Determine void fraction, porosity, and axial dispersion of the packed bed as well as pressure drop parameters.
4) Determine the adsorption equilibrium for the pure component and mixtures.
5) Determine transport coefficients for the (adsorbable) solutes.

The experimental techniques are mainly based on the injection of small (peaks) or large pulses (breakthrough curves) and analysis of the obtained chromatogram. Although the presented approach tries to limit the impact of measurement errors, experimental conditions and, especially, the material (adsorbent, eluent) used should be the same or at least comparable to the envisaged preparative application to ensure reliable results. The effects of plant peripherals should be kept low. Another aspect that deserves care is the calibration of the detectors.

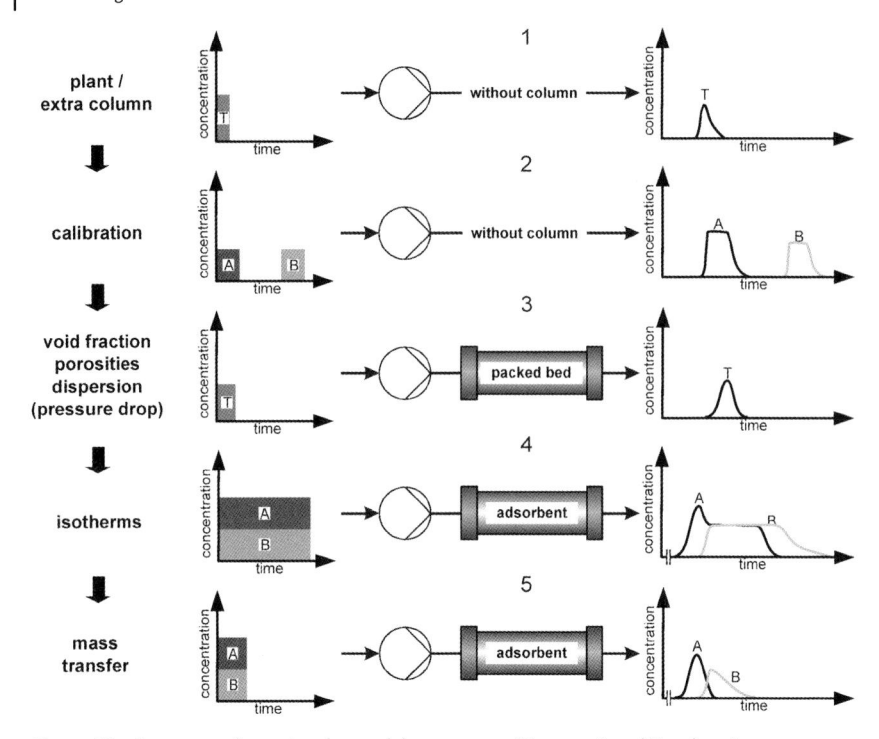

Figure 6.9 Concept to determine the model parameters (T, tracer; A and B, solutes).

Probably the simplest theoretical methods for determining the parameters from the experimental data involve the use of analytical solutions of simple column models and the analysis of the first moment of the elution profiles, for example, to determine the dead time or the Henry coefficient. In some cases where less accuracy is acceptable, parameters such as axial dispersion might be estimated by means of empirical correlations from literature. An overall kinetic coefficient can be also estimated analyzing the corresponding second moment of the elution profile.

To determine isotherms, methods based on numerical integration and differentiation of dynamic concentration profiles in combination with an analysis of mass balances can be used. A more versatile method is to use a parameter estimation tool to fit whole elution profiles ("peak fitting", "inverse method"), which is often the method of choice to obtain consistent and accurate estimates for many model parameters (e.g. Felinger and Guiochon, 2004). Parameter estimation routines are included in some commercially available simulation programs (Section 6.4) or can be linked to one's own simulation software. By solving an optimization problem these tools minimize the difference between the measured data and the simulation results by varying the model parameters. Thus, the result is an optimal set of parameter values. Since it is the preferred method, "parameter estimation" is typically used in the following, if not noted otherwise.

Table 6.4 gives an overview of different methods for parameter determination and illustrates the contradictory influences of accuracy and speed. Figure 6.10

Table 6.4 Parameter determination methods.

General remarks
Use for parameter estimation the same adsorbent as in the later applications and preferably "semipreparative" columns
Use one plant for all measurements and try to keep extra column effects low
Take care in calibrating the detectors
Check influence of volume flow to evaluate the magnitude of kinetic effects
With fluid mixtures as eluent check sensitivity to eluent composition fluctuations

---

Method	High accuracy and experimental effort	Medium accuracy and experimental effort	Initial guess ("quick and dirty")
Plant/ extra column	Experiment + moment analysis + parameter estimation: V_{pipe}, V_{tank} (t_{plant}), if necessary $D_{ax,pipe}$	Experiment + moment analysis: t_{plant}	Experiment + moment analysis: t_{plant}
Packing: void fraction and porosity	Experiment + moment analysis + parameter estimation: ε and ε_t	Experiment + moment analysis: ε_t, set ε	ε_t (from manufacturer), set ε
Packing: D_{ax}	Experiment + parameter estimation	Experiment + moment analysis	Use correlation
Isotherm	Elution profiles (peak fitting), frontal analysis, perturbation analysis	Measure using ECP and peak maximum method	Measure with ECP
Mass transfer: k_{eff}	Experiment (overload injection) + parameter estimation	Experiment + moment analysis (or HETP)	Use correlation

describes the work flow for parameter determination. Different methods for the individual parameters are discussed in detail below. First, some general methods are presented, which are helpful in evaluating a given chromatogram.

6.5.3
Evaluation of Chromatograms

The previous section gives an overview of the experiments necessary to determine the model parameters. This section describes general procedures for evaluating parameter values from experimental data. Model parameters are defined in Chapter 2, while Section 6.2 shows how they appear in the model equations. Based on the assumptions for these models it follows that all plant effects as well as axial dispersion, void fraction, and mass transfer resistance are independent of the adsorption/desorption processes occurring within the column.

The following methods are useful and required for the evaluation of model parameters: moment analysis, HETP plots, and peak fitting.

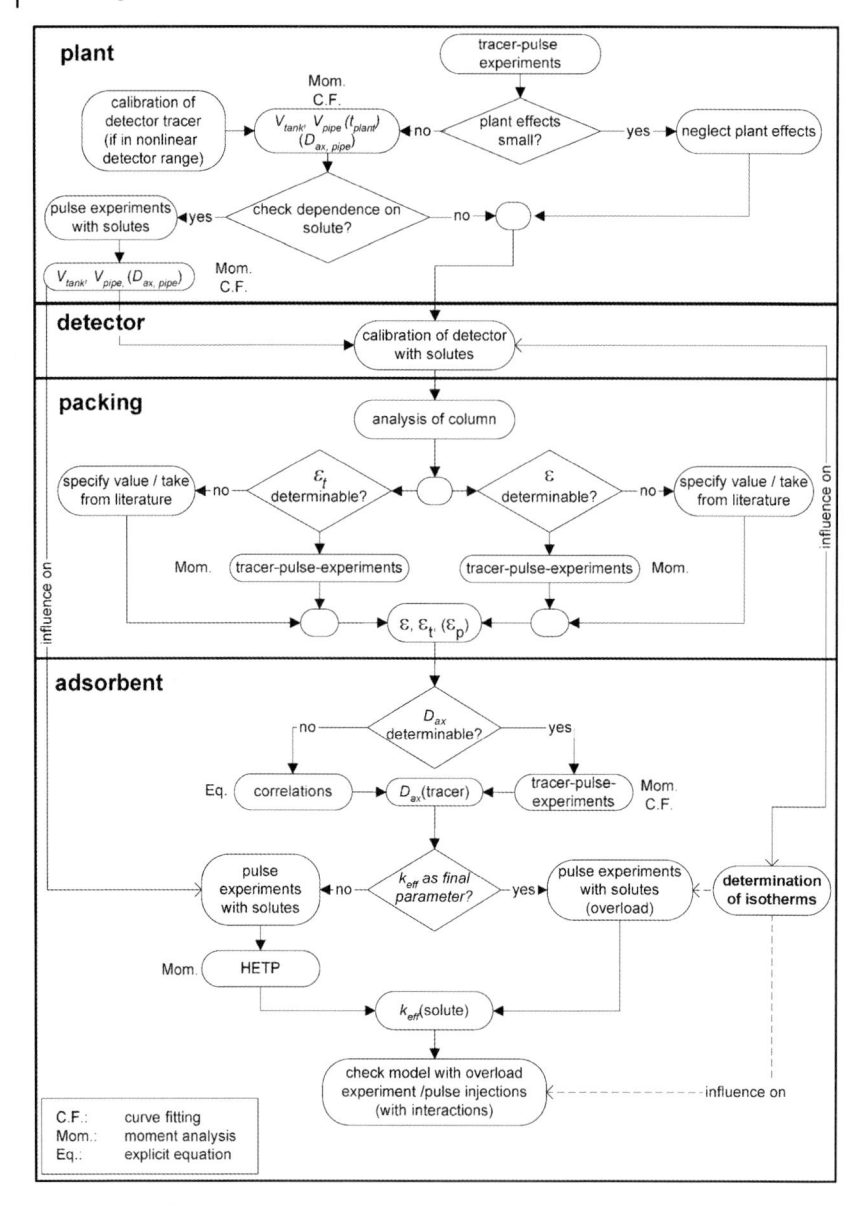

Figure 6.10 Different ways to determine various kinds of model parameters.

6.5.3.1 Moment Analysis and HETP Plots

Moment analysis and HETP plots were already described in Chapter 2. Here, it will be shown how they can be used in estimating the model parameters. If the detector signal is a linear function of concentration, analysis can be carried out directly with the detector signals without calibration. Since the concentration profiles are typically available as n_p discrete concentration versus time values, the first two moments (Equations 2.38 and 2.39) have to be calculated by numerical integration (summation):

$$\mu_t = \frac{\int_0^\infty t \cdot c(t) \cdot dt}{\int_0^\infty c(t) \cdot dt} \approx \frac{\sum_{j=1}^{n_p} t_j \cdot c_j \cdot \Delta t}{\sum_{j=1}^{n_p} c_j \cdot \Delta t} \tag{6.127}$$

$$\sigma_t^2 = \frac{\int_0^\infty (t - \mu_t)^2 \cdot c(t) \cdot dt}{\int_0^\infty c(t) \cdot dt} \approx \frac{\sum_{j=1}^{n_p} (t_j - \mu_t)^2 \cdot c_j \cdot \Delta t}{\sum_{j=1}^{n_p} c_j \cdot \Delta t} \tag{6.128}$$

The signal quality (=low detector noise) and sampling rate should be sufficiently high to obtain correct results, else postprocessing of the data is necessary, such as smoothing, baseline correction, and so on. Even then the result of the integration in Equation 6.128 depends very much on the extension of the baseline, and the obtained value of the second moment can be very inaccurate (Section 6.5.3.3).

To determine reliable model parameters, a minimum of three injection experiments with tracers and one tracer experiment for each solute of interest should be carried out. The evaluation of these experiments is sketched in Figure 6.11, together with the symbols of the measured first moments. The injected signal is assumed to be a rectangular pulse. The first tracer experiment detects the dead time of the plant while the column is replaced by a zero-volume connector. The other experiments are carried out with the column in place, using a tracer that cannot get into the pores (Tracer 1) and another one that penetrates the pores (Tracer 2). These experiments are necessary to determine the dead times $t_{0,\mathrm{int}}$ and t_0. Finally, a solute peak is analyzed, but it has to be noted that meaningful results

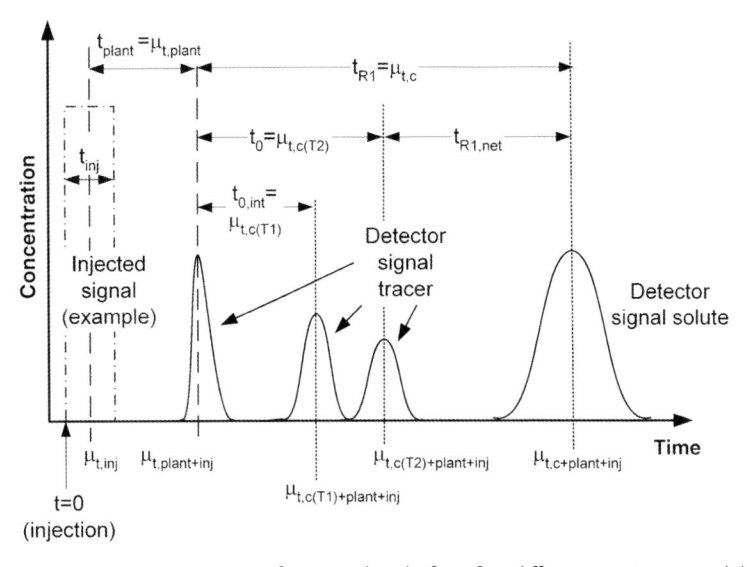

Figure 6.11 Representation of measured peaks from four different experiments and the characteristic times.

are obtained only if the concentrations are within the linear range of the isotherm. Figure 6.11 gives an overview of the different first moments and the respective times that are evaluated.

Contributions to the moments from all parts of the chromatographic plant (Section 6.3.1) are additive, as well known in linear chromatography (Ashley and Reilley, 1965). Assuming the hypothetical injection of a Dirac pulse "prior" to the injector (the injector "transforms" this into the rectangular pulse illustrated in Figure 6.11), the model parameters can be extracted from the measurements. For plant peripherals, equations for the resulting moments can be found in standard chemical engineering textbooks (e.g., Levenspiel, 1999; Baerns, Hofmann, and Renken, 1999):

Injector (Equation 6.116):

$$\mu_{t,inj} = \frac{t_{inj}}{2}, \quad \sigma^2_{t,inj} = \frac{t^2_{inj}}{12}, \quad t_{inj} = \frac{V_{inj}}{\dot{V}} \tag{6.129}$$

Pipe (Equation 6.119):

$$\mu_{t,pipe} = \frac{V_{pipe}}{\dot{V}}, \quad \sigma^2_{t,pipe} = 2\left(\frac{V_{pipe}}{\dot{V}}\right)^2 \cdot \frac{D_{ax,pipe}}{u_{0,pipe} \cdot L_{pipe}} \tag{6.130}$$

Detector (Equation 6.123):

$$\mu_{t,tank} = \frac{V_{tank}}{\dot{V}}, \quad \sigma^2_{t,tank} = \left(\frac{V_{tank}}{\dot{V}}\right)^2 \tag{6.131}$$

Plant:

$$\mu_{t,plant} = \mu_{t,pipe} + \mu_{t,tank}$$
$$\sigma^2_{t,plant} = \sigma^2_{t,pipe} + \sigma^2_{t,tank} \tag{6.132}$$

As the parameters for the injector in Equation 6.129 are known, the plant characteristics can be calculated from the experimentally determined values of $\mu_{t,plant+inj}$ and $\sigma_{t,plant+inj}$:

$$t_{plant} = \mu_{t,plant} = \mu_{t,plant+inj} - \mu_{t,inj} = \mu_{t,plant+inj} - \frac{t_{inj}}{2}$$
$$\sigma^2_{t,plant} = \sigma^2_{t,plant+inj} - \sigma^2_{t,inj} = \sigma^2_{t,plant+inj} - \frac{t^2_{inj}}{12} \tag{6.133}$$

If axial dispersion in the piping can be neglected, Equations 6.129–6.133 allow all plant parameters to be determined from one experiment only. Plant characterization is further discussed in Section 6.5.5.

Experimental determination of the first and second moments of column elution profiles is straightforward, using a pulse injection with and without the column and the parameters previously determined:

$$\mu_{t,c} = \mu_{t,c+plant+inj} - \mu_{t,plant+inj} = \mu_{t,c+plant+inj} - \mu_{t,plant} - \frac{t_{inj}}{2}$$
$$\sigma^2_{t,c} = \sigma^2_{t,c+plant+inj} - \sigma^2_{t,plant+inj} = \sigma^2_{t,c+plant+inj} - \sigma^2_{t,plant} - \frac{t^2_{inj}}{12} \tag{6.134}$$

Note that Equation 6.134 is valid for tracer and solute injections.

In the following the calculated values for the moments are related to the model parameters (e.g., ε, ε_t, D_{ax}, H, and k_{eff}) of the column. For this the theoretical solutions for the first and second moments derived for different column models are presented. Besides the sole purpose of parameter determination, this approach further allows us to demonstrate a justification for the application of "lumped parameter" models.

Only systems in the linear range of the isotherm are discussed based on classical work published by Kucera (1965) and Ma, Whitley, and Wang (1996).

For a general rate model including axial dispersion mass transfer (Equation 6.80), (apparent) pore diffusion (Equations 6.82, 6.84 and 6.85), and linear adsorption kinetics (Equations 6.32 and 6.33), Kucera (1965) derived the moments by Laplace transformation, assuming the injection of an ideal Dirac pulse. If axial dispersion is not too strong ($Pe \gg 4$), the equations for the first and second moments can be simplified to (Ma, Whitley, and Wang, 1996)

$$\mu_{t,c,GRM} = t_{R,lin,i} = \frac{L_c}{u_{int}}\left[1 + \frac{1-\varepsilon}{\varepsilon}(\varepsilon_p + (1-\varepsilon_p)\cdot H)\right] = \frac{L_c}{u_{int}}(1+\tilde{k}')$$

(6.135)

$$\sigma^2_{t,c,GRM} = 2\left(\frac{L_c}{u_{int}}\right)^2 \cdot \left(\frac{D_{ax}}{u_{int}\cdot L_c}(1+\tilde{k}')^2 + \tilde{k}'^2 \cdot \frac{\varepsilon}{1-\varepsilon}\cdot\left(\frac{u_{int}\cdot r_p^2}{15\cdot\varepsilon_p\cdot D_{app,pore}\cdot L_c} + \frac{r_p\cdot u_{int}}{3\cdot k_{film}\cdot L_c}\right)\right)$$

$$+ 2\left(\frac{L_c}{u_{int}}\right)^2 \tilde{k}'^2 \cdot \frac{\varepsilon}{1-\varepsilon} \cdot \left(\frac{((1-\varepsilon_p)/\varepsilon_p)H_i}{1+((1-\varepsilon_p)/\varepsilon_p)H_i}\right)^2 \cdot \frac{u_{int}}{(1-\varepsilon_p)\cdot q_{sat}\cdot k_{ads}\cdot L_c}$$

(6.136)

Note that the first moment does not depend on any mass transfer and dispersion coefficients and is equivalent to the retention time derived for the ideal model (Equation 6.49). The factor $\tilde{k}'$ is a modified retention factor that is zero for non-pore-penetrating tracers. It is connected to k' (Equation 2.1) by

$$\tilde{k}' = \frac{1-\varepsilon}{\varepsilon}(\varepsilon_p + (1-\varepsilon_p)\cdot H) = \frac{\varepsilon_t}{\varepsilon}(1+k') - 1$$

(6.137)

Whereas, in principle, simple experiments with tracers and one for each solute (explained below) allow the determination of ε, ε_t, and the component specific H from the experimentally determined $\mu_{t,c}$, the other model parameters cannot be simply extracted from the second moment. Dispersion (D_{ax}), liquid film mass transfer (k_{film}), diffusion inside the particles ($D_{app,pore}$), and adsorption kinetics (k_{ads}) contribute in a complex manner jointly to the overall band broadening as described by $\sigma_{t,c}$ (Equation 6.136). Therefore, an independent determination of these four parameters is not possible from Equation 6.136 only. In principle, additional equations could be obtained from higher moments (Kucera, 1965; Kubin, 1965). However, as the effect of detector noise on the accuracy of the moment value strongly increases the higher the order of the moment, a meaningful measurement of the third, fourth, and fifth moments is practically impossible. Equation 6.136 is thus not directly suited for parameter determination, but establishes the important connection between the model parameters of transport and their influence on band broadening.

The corresponding equations for the first and second moments for the simple transport-dispersive model (Lapidus and Amundson, 1952; Van Deemter, Zuiderweg, and Klinkenberg, 1956) are

$$\mu_{t,c,TDM} = t_{R,lin} = \frac{L_c}{u_{int}}(1 + \tilde{k}') \tag{6.138}$$

$$\sigma^2_{t,c,TDM} = 2\left(\frac{L_c}{u_{int}}\right)^2 \cdot \left[\tilde{k}'^2 \cdot \frac{\varepsilon}{1-\varepsilon} \cdot \frac{r_p \cdot u_{int}}{3 \cdot k_{eff} \cdot L_c} + \frac{D_{ax}}{u_{int} \cdot L_c}(1+\tilde{k}')^2\right] \tag{6.139}$$

The result for the first moment (Equation 6.138) is identical to Equation 6.135 for the GRM. Comparison of Equations 6.136 and 6.139 shows that both models describe similar band broadening if k_{eff} is given by

$$\frac{1}{k_{eff}} = \frac{r_p}{5\varepsilon_p \cdot D_{app,pore}} + \frac{1}{k_{film}} + \left(\frac{((1-\varepsilon_p)/\varepsilon_p)H}{1+((1-\varepsilon_p)/\varepsilon_p)H}\right)^2 \frac{3}{r_p \cdot (1-\varepsilon_p) \cdot q_{sat} \cdot k_{ads}} \tag{6.140}$$

Equation 6.140 defines a formal connection between the effective mass transport and the film transport, the pore diffusion, and the adsorption rate coefficient. It illustrates that k_{eff} is a "lumped parameter," composed of several transport effects connected in series. This also gives reasons for the use of lumped rate models as it proves that in linear chromatography the impact of the lumped parameters on the most important peak characteristics, retention time and peak width, is identical to the effect described by general rate model parameters.

The second moment (Equation 6.139) still consists of two transport parameters, k_{eff} and D_{ax}, but with additional assumptions about their dependence on the flow rate, Equation 6.139 can be used to determine both (see below).

For completeness, a similar deduction can be drawn for the equilibrium-dispersive model (EDM) (Section 6.2.4.1). The first moment (Equation 6.62) is again identical to Equation 6.135 and the second moment of the EDM is given by Equation 6.63. By comparing Equations 6.63 and 6.139, the following relationship between the apparent dispersion coefficient (D_{app}), k_{eff}, and D_{ax} can be derived:

$$D_{app} = D_{ax} + \left(\frac{\tilde{k}'}{1+\tilde{k}'}\right)^2 \cdot \frac{\varepsilon}{1-\varepsilon} \cdot \frac{r_p \cdot u_{int}^2}{3 \cdot k_{eff}} \tag{6.141}$$

Equation 6.141 illustrates the meaning of the apparent dispersion coefficient as a lumped parameter.

In addition to the interpretation of lumped parameters, Equations 6.135–6.141 may be used to calculate the model parameters from experimentally determined moments $\mu_{t,c}$ and $\sigma_{t,c}$. This method is described here for the TD model but can also be applied to the ED model.

Table 6.5 Determination of parameters of the transport-dispersive model based on moment analysis.

Experiment	Measurement	Equations	Unknown
Defined by experimentalist	t_{inj} (V_{inj})	6.129 (6.117)	$\mu_{t,inj}$, $\sigma_{t,inj}$
Plant effects	$\mu_{t,plant+inj}$, $\sigma_{t,plant+inj}$	6.133	$\mu_{t,plant}$ (t_{plant}), $\sigma_{t,plant}$
Tracer 1 (does not enter pores)	$\mu_{t,c(T1)+plant+inj}$	6.134 (2.9)	ε
Tracer 2 (does enter pores)	$\mu_{t,c(T2)+plant+inj}$	6.134 (2.9)	ε_t
Tracer 1 at different flow rates	$\sigma_{t,c(T1)+plant+inj} = f(u_{int})$ or HETP–u_{int} plot	6.134 (2.9)	$D_{ax} = f(u_{int})$ (with $k_{eff} = \infty$)
Solutes (linear isotherm)	$\mu_{t,c+plant+inj}$	6.134, 6.138 (2.9)	H
Solutes (linear isotherm); at different flow rates	$\sigma_{t,c+plant+inj}$ or HETP–u_{int} plot	6.134, 6.139 (2.9)	k_{eff}

Based on tracer and solute experiments (Figure 6.11) the model parameters are determined step by step, beginning with the void fraction, total porosity, and the axial dispersion coefficient (Section 6.5.6). All experimental data must be corrected for plant effects (Equation 6.134).

Table 6.5 summarizes the different steps of parameter estimation from the point of view of moment analysis. Owing to the limitations mentioned above, the derived parameters often possess just approximate character. The experimental procedure is listed in chronological order, where each step uses data from the previous ones.

If D_{ax} is assumed to be a linear function of the fluid velocity (Section 6.5.6.2), its value is derived from experiments with different volume flows. Subsequently, the Henry (Section 6.5.7.2) and mass transfer coefficients (linear isotherm range; Section 6.5.8) are calculated based on experiments with the solutes. In the framework of Figure 6.9, k_{eff} is used as the final fitting parameter and may differ from the value determined for the linear case (Antos et al., 2003; Section 6.5.8).

Table 6.5 also includes the application of the classical HETP–u_{int} plot as an alternative to direct analysis of the second moment. Transport parameters are derived from this plot by the following method.

Using the definition of the number of stages N (Equation 2.37) and the equations for the first and second moments (Equations 6.138 and 6.139) one obtains for the transport-dispersive model

$$\frac{1}{N_{TDM}} = \left(\frac{\sigma_{t,c,TDM}}{\mu_{t,c,GRM}}\right)^2 = 2 \cdot \left(\frac{\tilde{k}'}{1+\tilde{k}'}\right)^2 \cdot \frac{\varepsilon}{1-\varepsilon} \cdot \frac{r_p \cdot u_{int}}{3 \cdot k_{eff} \cdot L_c} + 2\frac{D_{ax}}{u_{int} \cdot L_c} \quad (6.142)$$

The corresponding simpler equation for the equilibrium-dispersive model originating from setting k_{eff} to infinity was given already by Equation 6.64.

Assuming D_{ax} is proportional to u_{int}, $D_{ax}=0.5(A*u_{int}+B)$ (Section 6.5.6.2), and k_{eff} is independent of the velocity (Section 6.5.6.2), HETP can be expressed as a function of the model parameters using Equations 2.42 and 6.142:

$$\text{HETP}_{TDM} = \frac{L_c}{N_{TDM}} = 2\frac{D_{ax}}{u_{int}} + 2 \cdot \left(\frac{\tilde{k}'}{1+\tilde{k}'}\right)^2 \cdot \frac{\varepsilon}{1-\varepsilon} \cdot \frac{r_p \cdot u_{int}}{3 \cdot k_{eff}}$$
$$= A + \frac{B}{u_{int}} + C \cdot u_{int} \quad (6.143)$$
$$\approx A + C \cdot u_{int}$$

The second row in Equation 6.143 represents the famous Van Deemter equation (Equation 2.42) and shows how the model parameters A, B, and C are related to D_{ax} and k_{eff}. For preparative chromatography the B term can be often neglected (Section 6.5.6.2).

Applying these simplifications, Equation 6.143 can be used to estimate D_{ax} and k_{eff}. Using the classical the Van Deemter analysis the parameters A and C must be determined experimentally analyzing HETP values measured for different flow rates (Figure 6.12).

As pointed out in Table 6.5, the HETP plot is useful to analyze peak widths measured for tracers and solutes. With nonadsorbable tracers the theoretical HETP plot would be a straight horizontal line ($C=0$) allowing to determine A and, thereby, D_{ax}/u_{int}. To fit within the simplified framework presented here, for every solute only the slope C needs to be determined and k_{eff} can be subsequently evaluated. Basically, A can also be derived from these plots, but in practice different axis intercepts A might be observed for solutes and tracers. Because of the frequently limited precision in determining second moments and the application of the linearized HETP plot, the accuracy of the derived parameters is rather limited and the values for A and C (D_{ax}/u_{int} and k_{eff}) can be seen as "initial guesses"

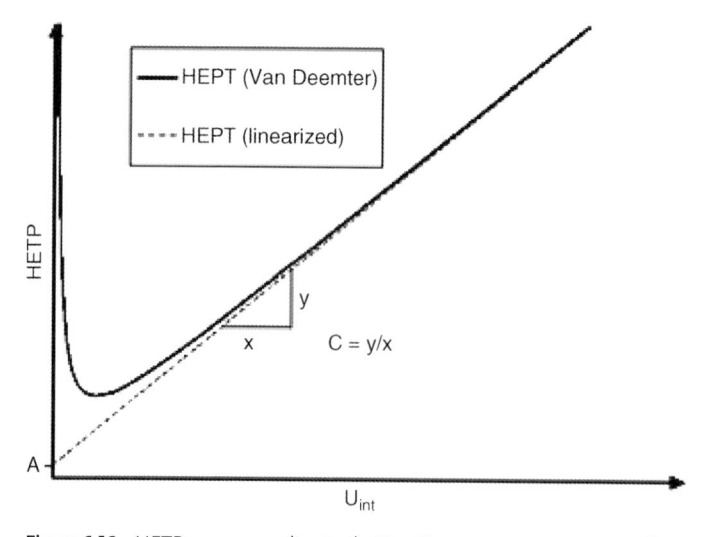

Figure 6.12 HETP curve according to the Van Deemter approximation and in simple linearized form.

only (Section 6.5.3.3). Thus, the differences in the observed axis intercepts are often negligible. For further discussions of HETP plots see, for example, Van Deemter, Zuiderweg, and Klinkenberg (1956), Grushka, Snyder, and Knox (1975), and Weber and Carr (1989).

6.5.3.2 Parameter Estimation

The analysis of two moments offers to estimate only two characteristic parameters that characterize a peak but not the whole peak shape. By "fitting" of either analytical equations (Section 6.5.3.3) or simulation results to the complete peak observed this drawback may be overcome. It also allows a direct comparison between the calculated and measured concentration profiles.

As suitable analytical solutions are not available for most of the column models, simulation-based parameter estimation using simulation software can be recommended as a powerful tool. This method is very versatile in terms of the number and complexity of models that can be handled. Estimation tasks can be solved, for example, with the gEST tool included in the gPROMS program package (PS Enterprise, UK). An additional advantage of the simulation-based approach is the consistency of the obtained data, if the same models and simulation tools are used for subsequent process analysis and optimization.

The fitting procedure results in a set of model parameters, minimizing the difference between theoretical and measured concentration profiles. A discussion of statistical-based objective functions and optimization procedures is beyond the scope of this book. For further information see, for example, Lapidus (1962), Barns (1994), Korns (2000), and Press et al. (2002, http://www.nr.com/).

Objective functions O are often based on least squares methods. They can be, for example, a function of the absolute or relative squared error of n_p measured concentration values, c_{exp}, and the theoretical values, c_{theo}:

$$O = f\left(\sum_{j=1}^{n_p} (c_{exp,j} - c_{theo,j})^2\right) \quad \text{or} \quad f\left(\sum_{j=1}^{n_p} \left(\frac{c_{exp,j} - c_{theo,j}}{c_{exp,j}}\right)^2\right) \qquad (6.144)$$

In general, objective functions can capture more than one set of experimental data.

To solve the arising optimization problem, the solution (in this case the shape of the simulated chromatogram) must be sensitive to the variable in question; otherwise no meaningful parameter set can be obtained. It is not recommended to determine more than two or maximal three parameters simultaneously, because this increases the chances of being trapped in local minima or in finding more than one set of parameters that fulfills optimization criteria. The concept depicted in Figure 6.9 follows this consideration as not all column parameters are obtained from one measured chromatogram. Instead, estimation is performed step by step, using different experimental data and increasing the complexity of the model required in each step.

Another aspect to be considered is the quality of the measurement, as all errors (e.g., detector noise) affect the accuracy of the determined parameters. Notably, in contrast to these statistical errors, systematic measurement errors cannot be minimized by repeated measurements.

It should be further remarked that initial guesses of the parameters must be provided by the users, coming from initial tests, experience, or shortcut calculations. If the equations are nonlinear, the initial guesses influence the result of the optimization and it is recommended to test the sensitivity of the obtained parameters with respect to the initial guesses.

In any case, the simulated profiles must be compared with the measured data to check the validity of the determined data as well as the assumed model. Statistical methods to quantify the goodness of the fit are given, for example, in Lapidus (1962), Barns (1994), Korns (2000), and Press et al. (2002).

6.5.3.3 Peak Fitting Functions

A simplified approach to peak fitting is to use a suitable analytical function to approximate the measured peak. As in the previous section, the function parameters are adjusted to obtain an optimal fit to the experimental data. In a next step the adjusted function (or its parameters) is used to calculate the first and second moments. This procedure may, for example, help to overcome the inaccuracy of moment analysis in case of asymmetrical peaks. It also offers the benefit that standard software such as spread sheets can be used instead of special parameter estimation systems.

The simplest method is the representation of an ideal chromatographic peak by a Gaussian function according to Equation 6.145:

$$c_g(x = L_c, t) = \frac{m_{inj}}{\dot{V} \cdot t_g} \cdot \frac{t_g}{\sigma_g \sqrt{2\pi}} \cdot \exp\left\{-\frac{t_g^2[(t/t_g) - 1]^2}{2 \cdot \sigma_g^2}\right\} \tag{6.145}$$

The retention time $t_{R,lin}$ and the second moment for the Gaussian profile (Equation 6.60) have been replaced by variables indexed with "g." The parameters t_g and σ_g must be optimized by curve fitting. Equation 6.145 is suitable only for symmetrical peaks. Analytical solutions of, for example, the transport-dispersive model (which describes asymmetrical band broadening only for a very low number of stages) are not suited to describing the asymmetry often encountered in practical chromatograms. Thus, many different, mostly empirical, functions have been developed for peak modeling. A recent extensive review by Marco and Bombi (2001) lists over 90 of them.

One most frequently used equation is called the exponentially modified Gauss (EMG) function (Jeansonne and Foley, 1991, 1992; Foley and Dorsey, 1983, 1984). It is defined as a Gaussian peak (Equation 6.145) superimposed by an exponential decay function h_{exp}:

$$h_{exp}(t) = \begin{cases} \dfrac{1}{\tau_{EMG}} \cdot \exp\left\{-\dfrac{t}{\tau_{EMG}}\right\} & (t \geq 0) \\ 0 & (t < 0) \end{cases} \tag{6.146}$$

where τ_{EMG} is used to specify the peak skew in the EMG function. The exponential decay function Equation 6.146 is equivalent to the residence time distribution of a stirred tank. Notably, the use of any empirical function to describe peak tailing does not provide any additional physical information about the nature of this phenomenon. Thus, it is recommended to apply these equations to increase the accuracy in moment analysis in the case of linear chromatography. In fact, the equations are an approximate correlation derived from the EMG model (Foley and Dorsey, 1983).

Another common application of fitting functions is in deconvoluting partially resolved peaks (Marco and Bombi, 2001).

The superposition (convolution) of Equations 6.145 and 6.146 gives the EMG function:

$$c_{EMG}(t) = \int_0^t h_{exp}(t - t') \cdot c_g(t') dt' \tag{6.147}$$

Solution of Equation 6.147 employs the error function (erf):

$$\text{erf}(t) = \frac{2}{\sqrt{\pi}} \cdot \int_0^t \exp\{-y^2\} dy \tag{6.148}$$

and is given by

$$c_{EMG}(t) = \frac{m_{inj}}{\dot{V} \cdot t_g} \cdot \frac{t_g}{\tau_{EMG}} \cdot \exp\left\{\frac{1}{2} \cdot \left(\frac{\sigma_g}{\tau_{EMG}}\right)^2 - \frac{t - t_g}{\tau_{EMG}}\right\}$$
$$\cdot \left[1 + \text{erf}\left\{\frac{1}{\sqrt{2}}\left(\frac{t - t_g}{\sigma_g} - \frac{\sigma_g}{\tau_{EMG}}\right)\right\}\right] \tag{6.149}$$

This function is easily implemented in standard software tools. To approximate erf (x) polynomial approximations exist (Foley and Dorsey, 1984).

The moments of this function are listed below:

$$\text{Area}_{(zeroth\ moment)} = \frac{m_{inj}}{\dot{V}} \tag{6.150}$$

$$\mu_t(x = L_c) = t_g + \tau_{EMG} \tag{6.151}$$

$$\sigma_t^2(x = L_c) = \sigma_g^2 + \tau_{EMG}^2 \tag{6.152}$$

Due to mass balance requirements the area (Equation 6.150) is identical to that of a Gaussian peak. The parameters t_g, σ_g, and τ_{EMG} can be determined by curve fitting and, thus, the values of the moments are directly available.

To illustrate the peak shape resulting from the EMG function, two peaks are shown in Figure 6.13. One plot represents a Gaussian peak with $N_g = 500$ stages while the other stands for the EMG function with $\tau_{EMG}/t_g = 0.05$, resulting in $N_{EMG} = 245$. This EMG peak is calculated for the same t_g and σ_g of the Gaussian peak and both have the same area. The concentrations are normalized by the maximum concentration of the Gaussian peak (Equation 6.145) and the time axis is scaled with t_g. For the EMG function considerable peak tailing occurs and the position of the maximum differs from that of the Gaussian peak.

Figure 6.14 compares different methods of calculating the number of stages N.

The black squares represent the data calculated by Equations 2.36 and 2.40. In this case it is a general problem to determine the peak width or variance from a

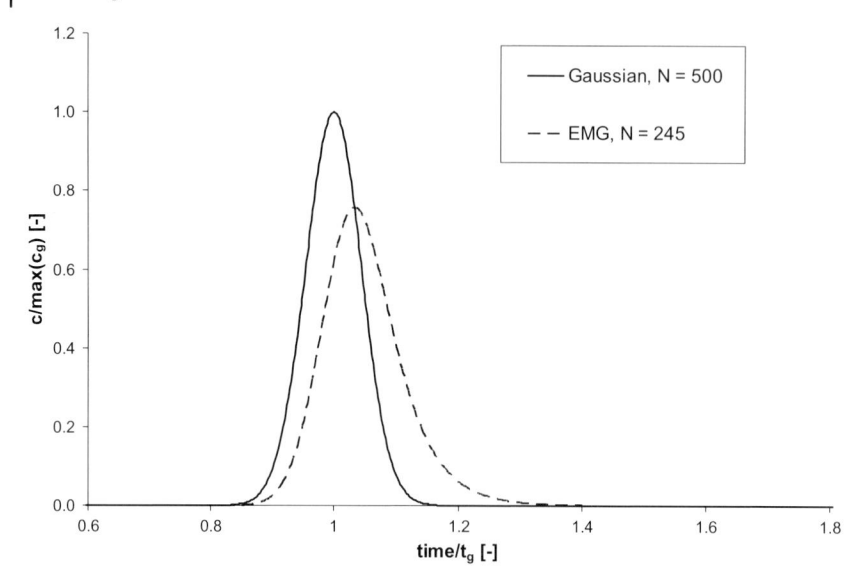

Figure 6.13 Comparison of peak profiles resulting from the Gaussian equation ($N_g = 500$) and the derived EMG function ($N_g = 500$, $\tau_{EMG}/t_g = 0.05$, $N_{EMG} = 245$) (concentrations normalized by the maximum concentration of the Gaussian peak).

measured chromatogram. Especially, if the second central moment has to be calculated, the result depends very much on the extension of the baseline, which has to be chosen for the integration interval. As an example, Table 6.6 shows the number of stages for different integration intervals, which are specified in Figure 6.15. Due

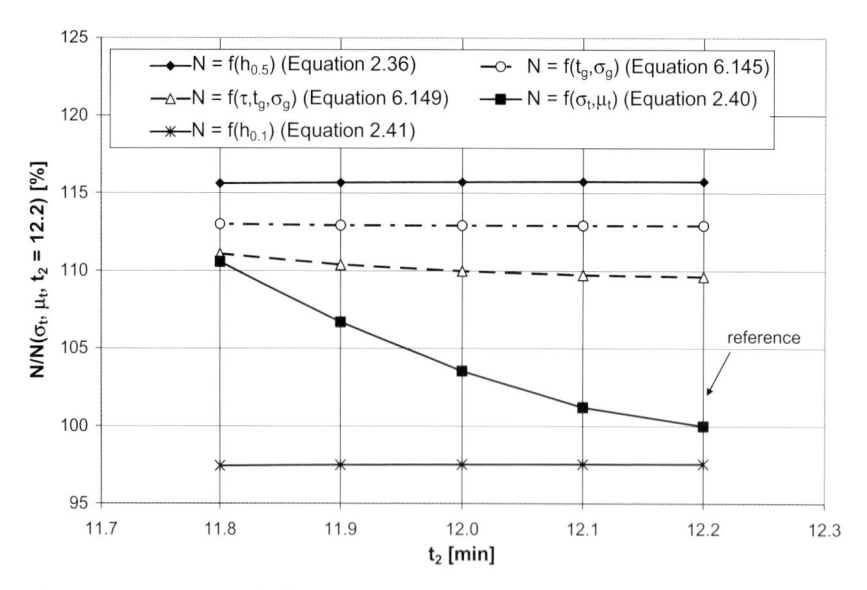

Figure 6.14 Comparison of different methods to determine the stage number.

Table 6.6 Residence time and plate number for different integration intervals.

	1	2	3	4	5
t_1 (min)	10.5	10.5	10.5	10.5	10.5
t_2 (min)	11.8	11.9	12	12.1	12.2
μ_t	11.20	11.20	11.20	11.21	11.21
σ_t^2	0.0400	0.0416	0.0428	0.0438	0.0444
$N_i = f(h_{0.5})$	3276	3277	3278	3279	3279
$N_i = f(\mu_t, \sigma_t)$	3138	3020	2933	2864	2828

to the numerical integration (Equation 6.128) the second moment decreases for increasing integration boundaries t_2. The value calculated for $t_2 = 12.2$ min is taken as a reference for comparison with the other procedures.

Additionally, Figure 6.14 presents the number of stages calculated by the peak width at a certain peak height (Equations 2.36 and 2.41) as well as peak fitting by the Gaussian function (Equation 6.145) and the EMG function (Equation 6.149). All functions are fitted to the peak given in Figure 6.15 for the different integration intervals. In contrast to the stage number determined from moment analysis, all other methods provide stage numbers nearly independent of the integration boundary but of quite different values. This confirms that the same calculation method should always be used for comparing different chromatographic systems. In view of the limited accuracy of any of the methods mentioned, a robust procedure should be used. In most cases this will be a method already implemented into

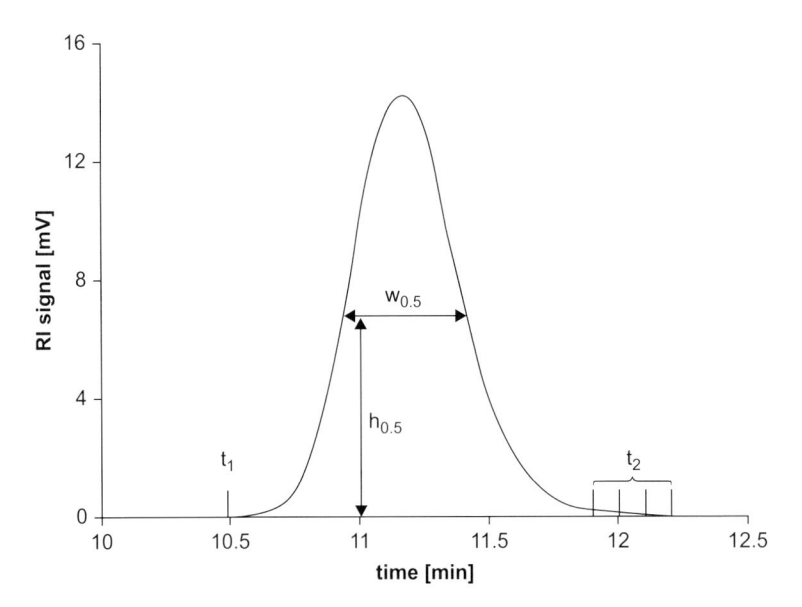

Figure 6.15 Determination of the plate number.

peak detection software of the equipment used or one that can be applied easily by taking the peak width at a certain peak height (Equations 2.36 and 2.41).

6.5.4
Detector Calibration

Careful detector calibration is crucial, as it directly influences the accuracy of the measured data. Hereby, a major practical problem is often the limited availability of pure components.

When adsorption isotherms are determined by breakthrough experiments (see below), the acquired data can be directly taken for the detector calibration. Breakthrough curves should be recorded after injecting well-characterized solutions into the plant without the column. This required considerably less sample to reach complete saturation. Since the plateau concentrations are the known injected feed concentrations, a plot of c_{feed} versus the detector signal u provides the sought-after relationship and a calibration function can be easily fitted to these values.

If the form of the calibration function

$$c = f_{cal}(u) \tag{6.153}$$

is known a priori, a pulse experiment may also be used to obtain the calibration curve. For a known amount of injected sample

$$m_{inj} = V_{inj} \cdot c_{feed} = \dot{V} \cdot t_{inj} \cdot c_{feed} \tag{6.154}$$

the mass balance (Equation 6.61) for the detected peak is

$$m_{inj} = \dot{V} \cdot \int_0^\infty c(L_c, t) dt = \dot{V} \cdot \int_0^\infty f(u(t)) dt \tag{6.155}$$

In the special case of a linear calibration function

$$c = F_{cal} \cdot u \tag{6.156}$$

Equations 6.154 and 6.155 together with the numerical integration of the detector signal allow the determination of the calibration factor F_{cal}:

$$F_{cal} = \frac{V_{inj} \cdot c_{inj}}{\dot{V} \cdot \int_0^\infty u(t) dt} = \frac{t_{inj} \cdot c_{inj}}{\int_0^\infty u(t) dt} \tag{6.157}$$

For nonlinear detector responses calibration curves may be also determined by curve fitting. For such more complex calibration functions with more than one unknown parameter, additional information (e.g., about the physical foundations of the measuring method) or pulse experiments with different concentrations are necessary. In any case, it is advised to check the validity of the calibration curves in the low and high concentration ranges by separate experiments with different feed concentrations.

6.5.5
Plant Parameters

Following the concept of Figure 6.9, the parameters describing the fluid dynamic and residence time distribution of the plant peripherals are determined by means of pulse experiments. For this purpose, a small amount of tracer is injected into the plant without the column and the output concentration is measured.

The overall retention time of the plant $\mu_{t,\text{plant}+\text{inj}}$ can be obtained by moment analysis of a chromatogram. Using the known volume of the injected sample V_{inj}, the dead volume V_{plant} and the dead time t_{plant} are evaluated using Equation 6.133:

$$V_{\text{plant}} = \mu_{t,\text{plant}+\text{inj}} \cdot \dot{V} - \mu_{t,\text{inj}} = \mu_{t,\text{plant}+\text{inj}} \cdot \dot{V} - \frac{V_{\text{inj}}}{2} = \mu_{t,\text{plant}} \cdot \dot{V} \quad (6.158)$$

$$t_{\text{plant}} = \mu_{t,\text{plant}} = \frac{V_{\text{plant}}}{\dot{V}} \quad (6.159)$$

Using a parameter estimation tool is another possible way to determine the parameters for the plant. Volumes of the pipe V_{pipe}, detector system V_{pipe}, and, if necessary, the axial dispersion coefficient $D_{\text{ax,pipe}}$ are estimated based on the model equations given in Sections 6.3.2 (Equations 6.119 and 6.123).

An initial estimate for the tank volume can also be obtained from moment analysis by Equations 6.131 and 6.133, if axial dispersion in the pipe is neglected and $\sigma_{t,\text{plant}+\text{inj}}$ is measured:

$$\sigma_{t,\text{tank}}^2 \approx \sigma_{t,\text{plant}}^2 = \sigma_{t,\text{plant}+\text{inj}}^2 - \sigma_{t,\text{inj}}^2$$

$$\Leftrightarrow \left(\frac{V_{\text{tank}}}{\dot{V}}\right)^2 \approx \sigma_{t,\text{plant}+\text{inj}}^2 - \frac{t_{\text{inj}}^2}{12} \quad (6.160)$$

$$\Leftrightarrow V_{\text{tank}} \approx \dot{V} \cdot \sqrt{\sigma_{t,\text{plant}+\text{inj}}^2 - \frac{t_{\text{inj}}^2}{12}}$$

As Equation 6.161 is always valid, exploiting Equations 6.132 and 6.125, an initial value for the pipe volume can be obtained from:

$$V_{\text{plant}} = V_{\text{pipe}} + V_{\text{tank}} \quad (6.161)$$

The dispersion coefficient and the parameter "pipe length" in Equation 6.121 are mainly adjusted parameters, especially since the actual cross-section of the pipe is not known exactly and therefore a representative value has to be chosen. For the axial dispersion coefficient, initial guesses using a very low value (e.g., 10^{-5} cm^2 s^{-1}) should be used and the pipe length is set according to Equation 6.121 and the specified cross-section.

Figure 6.16 illustrates the application of two different concepts for parameter estimation. The corresponding simulated concentration profiles are compared with the measurements. In one case ideal plug flow (Equation 6.118) and in the other case axial-dispersive flow (Equation 6.119) is assumed for the pipe system, while both models use the C.S.T. model (Equation 6.123) to describe the detector system. Figure 6.16 shows that the second model using axial dispersion provides an excellent fit for this setup, while the other cannot predict the peak deformation.

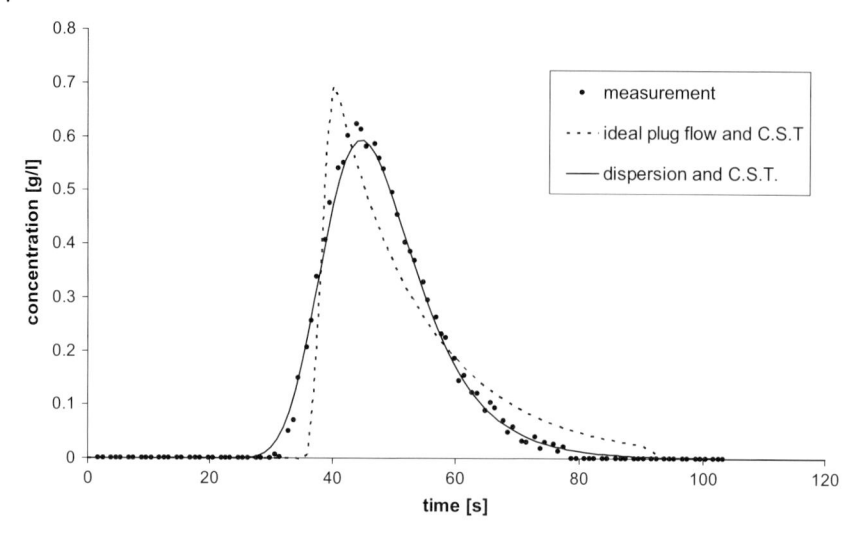

Figure 6.16 Comparison of experimental and simulated profiles (feed: R-enantiomer EMD53986 at 4 mg ml^{-1}, $\dot{V}$ = 20 ml min^{-1}, preparative scale; for additional data see Appendix A.1).

When comparing simulated profile with measured profiles, the desired level of accuracy has to be specified by the user taking into account the relevance of the plant effects in comparison to the column effects. The dead time of the plant should always be determined. The necessity of the other parameters depends on the design of the plant. Especially for analytical-scale plants, careful parameter estimation is recommended. Finally, influences of the volumetric flow rate on the plant parameters should be checked. In case of deviations it should be remembered that, on changing the fluid patterns in the piping and detector cells, the finite response time of the detector and/or changes of temperature might alter the measured signal and different parameter might be obtained.

6.5.6
Determination of Packing Parameters

When the model parameters of the plant are known, packing quality properties, that is, void fraction, pressure drop, and dispersion, should be evaluated (step 3 in Figure 6.9).

6.5.6.1 Void Fraction and Porosity of the Packing

To calculate the void fraction (Equation 2.6) and the (total) porosity (Equation 2.8) pulses of a nonpenetrating tracer (T1) and of a nonadsorbable tracer (T2), which penetrates into the pore system (t_0 marker), respectively, are injected. When the resulting measured elution profiles are evaluated by means of moment analysis (Section 6.5.3.1; Figure 6.11), the retention time μ_t has to be corrected by the dead time of the plant (Equation 6.134). Using the definition of the dead times (Equations 6.42 and 2.11) one obtains

$$\varepsilon = t_{0,\text{int}} \cdot \frac{\dot{V}}{V_c}, \quad \text{with} \quad t_{0,\text{int}} = \mu_{t,c(T1)+\text{plant}+\text{inj}} - t_{\text{plant}} - \frac{t_{\text{inj}}}{2} \tag{6.162}$$

$$\varepsilon_t = t_0 \cdot \frac{\dot{V}}{V_c}, \quad \text{with} \quad t_0 = \mu_{t,c(T2)+\text{plant}+\text{inj}} - t_{\text{plant}} - \frac{t_{\text{inj}}}{2} \tag{6.163}$$

If no suitable tracers are available, alternative methods (Section 2.2.2) may be used. In case of difficulties a void fraction of about $\varepsilon = 0.4$ can be estimated, since typical values for packings of spheres are between 0.32 and 0.42.

6.5.6.2 Axial Dispersion

The axial dispersion coefficient can be estimated from the concentration profile of a nonpenetrating tracer (T1). An approximation regarding its velocity dependence goes back to Van Deemter, Zuiderweg, and Klinkenberg (1956). The axial dispersion coefficient is seen as the sum of the contributions of molecular diffusion and eddy diffusion (Section 2.3.4; Ruthven, 1984):

$$D_{\text{ax}} = \gamma \cdot D_m + \lambda \cdot d_p \cdot u_{\text{int}} \tag{6.164}$$

where D_m is the molecular diffusion coefficient while γ and λ are the (external) tortuosity and a characterization factor of the packing, respectively. For typical values the following holds:

$$\gamma \approx 0.7, \quad \lambda \approx 1$$
$$D_m \approx 10^{-6} \text{ to } 10^{-5} \text{ cm}^2 \text{ s}^{-1}, \quad d_p \cdot u_{\text{int}} \approx (10^{-3} \times 0.1) \text{ cm}^2 \text{ s}^{-1} = 10^{-4} \text{ cm}^2 \text{ s}^{-1} \tag{6.165}$$

For typical chromatographic conditions, the contribution of molecular diffusion is relatively small and D_{ax} approximately becomes a linear function of the velocity:

$$D_{\text{ax}} \approx \lambda \cdot d_p \cdot u_{\text{int}} \tag{6.166}$$

Several methods for determining D_{ax} have been described in the literature. If a suitable tracer is available, chromatograms at different flow rates should be measured. If parameter estimation is performed to obtain the axial dispersion coefficient or the factor λ, the complete model (including the plant peripherals and the column) as well as all parameters determined so far have to be taken into account. Another possibility is offered by a moments analysis (Section 6.5.3.1, Equation 6.134), which is based on exploiting the connection between the axial dispersion coefficient and the second moment (Section 6.5.3.1):

$$\lambda = \text{constant} \approx \frac{D_{\text{ax}}}{u_{\text{int}} \cdot d_p} = \frac{\sigma^2_{t,c(T1)}}{\mu^2_{t,c(T1)}} \cdot \frac{L_c}{2 \cdot d_p} = \frac{\sigma^2_{t,c(T1)}}{t^2_{0,\text{int}}} \cdot \frac{L_c}{2 \cdot d_p} \tag{6.167}$$

$$D_{\text{ax}} = \frac{\sigma^2_{t,c(T1)}}{t^2_{0,\text{int}}} \frac{L_c u_{\text{int}}}{2} \tag{6.168}$$

All moments have to be corrected by the dead time of the plant and the injection volume (Section 6.5.3.1). Alternatively, λ can be estimated from the A-term evaluating a HETP plot as illustrated in Figure 6.12:

$$\lambda = \frac{A}{2 \cdot d_p} = \frac{\text{HETP}_{T1}(u_{int} = 0)}{2 \cdot d_p} \tag{6.169}$$

Chung and Wen (1968) and Wen and Fan (1975) proposed a dimensionless equation using the dependency of the dispersion coefficient on the (particle) Reynolds number Re (Equation 6.171) for fixed and expanded beds. It is an empirical correlation, based on published experimental data and correlations from other authors, which covers a wide range of Reynolds numbers Re. Owing to two different definitions of the Reynolds number, actual forms of the correlations vary in the literature. Since the particle diameter d_p is the characteristic value of the packing, Equation 6.170 based on the (particle) Péclet number Pe_p (Equation 6.172) is given as follows:

$$Pe_p = \frac{0.2}{\varepsilon} + \frac{0.011}{\varepsilon} \cdot (\varepsilon \cdot Re)^{0.48} \Rightarrow 10^{-3} \leq Re \leq 10^3 \tag{6.170}$$

$$Re = \frac{u_{int} \cdot d_p \cdot \varrho_l}{\eta_l} = \frac{u_{int} \cdot d_p}{\nu_l} \tag{6.171}$$

$$Pe_p = \frac{u_{int} \cdot d_p}{D_{ax}} \tag{6.172}$$

Notably, the particle Péclet number differs from the axial Péclet number defined in Equation 6.108 by the ratio of particle diameter to column length:

$$Pe = \frac{u_{int} \cdot L_c}{D_{ax}} = Pe_p \frac{L_c}{d_p} \tag{6.173}$$

Transforming Equation 6.170 leads to Equation 6.174:

$$D_{ax} = \frac{u_{int} \cdot d_p \cdot \varepsilon}{0.2 + 0.011 \cdot (\varepsilon \cdot Re_p)^{0.48}} \tag{6.174}$$

which allows a rough estimation of the axial dispersion coefficient, especially for high-quality packings. For such packings, Re has little influence as the value is under chromatographic conditions typically considerably smaller than 1.

6.5.6.3 Pressure Drop

Pressure drop and flow rate are usually linearly related – expressed by Darcy's law (Section 2.3.5). Pressure drops are usually measured for different volume flows, once with column in the plant and once without the column, using preferably a "zero-volume" connector. The difference between the two values yields the pressure drop characteristic Δp_c of the column alone. By plotting Δp_c versus u_{int} the unknown coefficient k_0 is readily determined from the slope of the curve by rearranging Equation 2.25:

$$\Delta p_c = \frac{\eta L_c}{k_0 d_p^2} \cdot u_{int}$$

$$\Leftrightarrow k_0 = \frac{u_{int}}{\Delta p_c} \cdot \frac{\eta L_c}{d_p^2} \qquad (6.175)$$

6.5.7
Isotherms

6.5.7.1 Determination of Adsorption Isotherms

The adsorption isotherms have the most pronounced influence on the courses of the chromatograms. Consequently, single-component and multicomponent isotherms should be determined with high accuracy in order to achieve a good agreement between simulations and experiments, including all model parameters measured so far (Figure 6.9).

Characteristic parameters of nearly all types of adsorption isotherm models are the Henry coefficients as well as the saturation capacities valid for large concentrations. In general, it is advisable to check the validity of the identified single-component isotherm equation before further considering the determination of additional multicomponent interaction parameters. In general, the decision on a certain isotherm equation should be made on the basis of the ability to predict experimentally observed overloaded concentration profiles. In any case, consistency with the Henry coefficients determined from initial pulse experiments with very low sample amounts must be assured.

This chapter discusses some of the most frequently used methods to determine isotherms and the corresponding model parameters. For further information, see reviews by Seidel-Morgenstern and Nicoud (1996) and Seidel-Morgenstern (2004) as well as the monograph by Guiochon *et al.* (2006). Table 6.7 summarizes the most important methods.

These techniques do not differ only in terms of accuracy and experimental effort. Their application is also limited by the availability of the required equipment. Not all methods are suitable for columns with low efficiency, because peak deformations will affect the isotherm parameters. The actual setup and experimental conditions should be as close as possible to the operating conditions for later production purposes. However, to save time and solutes, the column dimensions are usually only of smaller (analytical) scale, or "semipreparative" at maximum. Preparative chromatography requires typically the application of adsorbents, with larger particle sizes compared to analytical applications. In isotherm measurements temperature control is crucial, since adsorption may show significant temperature dependence. Sometimes the costs of solutes prohibit the use of *methods that need larger amounts of samples.*

The graphical guideline in Figure 6.17 illustrates the general steps of isotherm determination.

Table 6.7 Methods and steps in isotherm determination.

General aspects
Use directly the preparative adsorbents and, if possible, "semipreparative" columns
Use one plant for all measurements and try to keep extra column effects low
Determine all dead volumes of plant and column
Control column temperature within 1 °C
Calibrate the detectors carefully
For multicomponent systems: try to apply component-specific detectors
Favor measurement methods with packed columns
Evaluate concentration range of the application process
Check applicability of measurement method
Determine single-component parameters first
Check consistency with standard pulse tests of linear chromatography, especially Henry coefficients
Select carefully isotherm model equation
Try to simulate mixture experiments with single-component isotherms
Determine component interactions only if necessary
Check agreement between theoretical and experimental chromatograms

Static methods

Name	Methods	Isotherm type	Comments
Batch method (closed vessel)	• Immersion of adsorbent in solution • Mass balance	Single-component and multicomponent	• No packed column • No detector calibration necessary (only HPLC analytics) • Tedious • Limited accuracy
Adsorption/desorption method	• Loading and unloading of column • Mass balance	Single-component and multicomponent	• No detector calibration necessary (only HPLC analytics) • Tedious but accurate
Circulation method	• Closed system • Stepwise injection and circulation of components till equilibrium • Mass balance	Single-component and multicomponent	• Lower amount of sample • Accumulation of errors for the injection • Possibility for automation

Dynamic methods

Name	Methods	Isotherm type	Comments
Frontal analysis	• Integration of step response (breakthrough curves)	Single-component and multicomponent	• High accuracy

(continued)

Table 6.7 (Continued)

	• Numerical integration and mass balance		• Easy automation
			• Detector calibration necessary (directly obtainable from breakthrough experiments)
			• Component-specific detectors or fractionated analysis necessary for multicomponent isotherms only
			• High amounts of sample
			• Suitable for low efficient columns
			• No kinetic errors
			• Impurities must not cause significant detector signals
Analysis of disperse fronts: elution by characteristic point (ECP), frontal analysis by characteristic point (FACP)	• Pulse or step injection (high concentration)	Single component	• Small sample amounts
	• Slope of dispersive front		• Highly efficient columns and small plant effects necessary
			• Phase equilibrium is required (sensitive to kinetics)
			• Precise detector calibration necessary
Peak maximum	• Peak injection with systematic overload	Single component	• Small sample amounts
	• Peak maximum equal to retention time		• Less sensitive to low-efficiency columns than ECP/FACP
			• For special cases no detector calibration necessary
Perturbation method (minor disturbance method)	• Determination of the retention times of small (minor) disturbances	Single-component and multicomponent	• No detector calibration necessary

(continued)

Table 6.7 (Continued)

	• Column with different loading states		• Can deal with small impurities
			• Equilibrium is required • Can deal with low efficient columns • Isotherm models necessary to calculate multicomponent isotherms • Potential for automation
Curve fitting (peak fitting) of chromatograms	• Parameter estimation	• Single and binary (difficult!)	• Column and plant model necessary
		• Multicomponent not feasible	• Isotherm model necessary • Isotherm parameters must be sensitive

6.5.7.2 Determination of the Henry Coefficient

Henry coefficients are generally determined independently from other isotherm parameters analyzing the response to pulse injections performed with very low amounts of solutes to ensure linear isotherm behavior (Section 6.5.3.1). The linearity can be tested by comparing two or three pulse responses belonging to different concentrations. If the results for the determined Henry coefficients are identical, the system is linear. H is calculated by moment analysis using the measured $\mu_{t,c+inj+plant}$ and Equations 6.134 and 6.137:

$$H = \frac{1}{1-\varepsilon_t}\left(\mu_{t,c}\frac{\dot{V}}{V_c} - \varepsilon_t\right), \quad \text{with} \quad \mu_{t,c} = t_{R,lin} = \mu_{t,c+plant+inj} - t_{plant} - \frac{t_{inj}}{2} \quad (6.176)$$

6.5.7.3 Static Isotherm Determination Methods

Static methods (Figure 6.18) determine phase equilibrium data based on overall mass balances. They are often more time consuming and less accurate than dynamic methods described below:

Batch method: A known amount of adsorbent V_{ads} with initially empty pores is added to a solution of the volume V_l containing the solute with the concentration $c_{0,i}$. The mixture is then agitated in a closed vessel until equilibrium is reached. The final concentration in the solution ($c_{eq,i}$) is determined by standard analytical methods. From the following mass balance the appropriate equilibrium loading, $q(c_{eq})$, is calculated:

$$V_l \cdot c_{0,i} - V_l \cdot c_{eq,i} + V_{ads}(1-\varepsilon_p)q(c_{eq,i}) \quad (6.177)$$

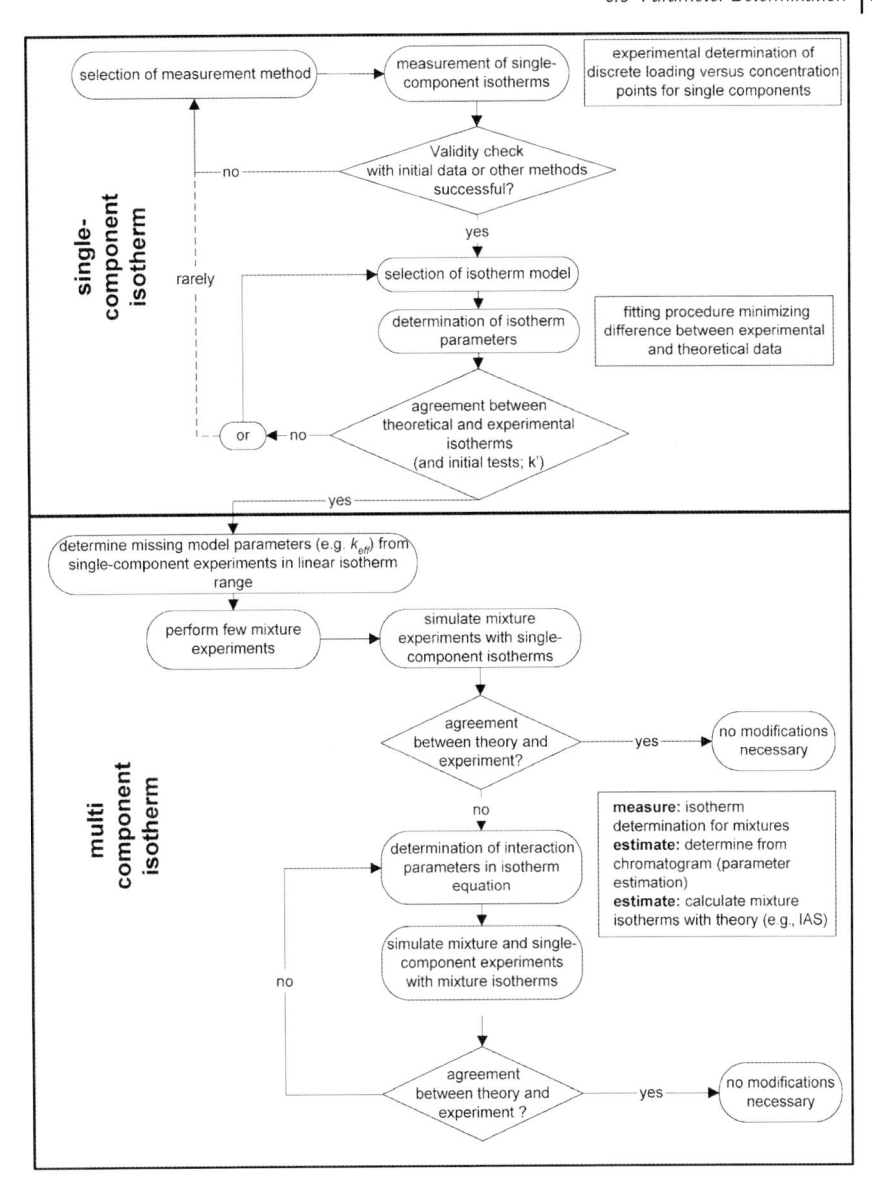

Figure 6.17 Graphical guideline for isotherm determination.

Using different initial concentrations or adsorbent amounts, a relevant concentration range can be covered. The method can be easily expanded to multi-component mixtures, where the loadings are functions of all components present. Drawbacks are the time-consuming preparations of the different mixtures and the transferability of the results to packed columns (e.g., due to uncertainties in phase ratio/porosity). Because of the numerous steps of manual work

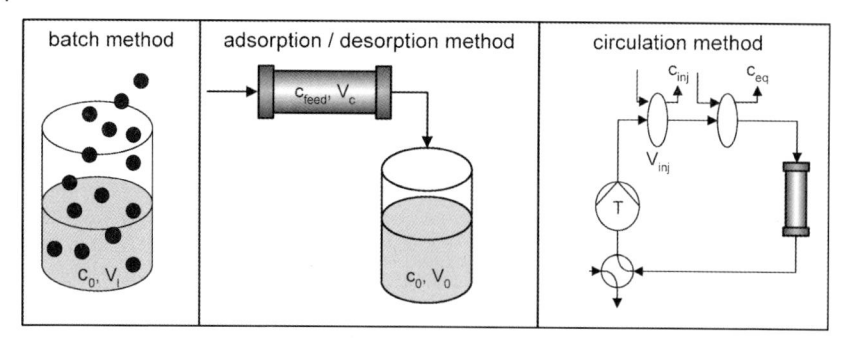

Figure 6.18 Principle of different static methods for isotherm determination.

and an uncertainty when equilibrium is fully reached, the accuracy of the method is not too high.

Adsorption–desorption method: An initially unloaded ($q = 0$) column is equilibrated by a feed concentration, c_{feed}, which may be a single-component or multicomponent mixture. Equilibrium is achieved by pumping a sufficient quantity of feed through this column. The plant is then flushed without the column to remove the solute solution. Afterwards, all solute is eluted from the column, collected, and analyzed to obtain the desorbed amount $m_{\text{des},i}$. The equilibrium loading $q(c_{\text{feed}})$ for each component i can be calculated according to

$$m_{\text{des},i} = \varepsilon \cdot V_c \cdot c_{\text{feed},i} + (1 - \varepsilon) \cdot V_c \cdot [\varepsilon_p \cdot c_{\text{feed},i} + (1 - \varepsilon_p) \cdot q(c_{\text{feed},i})] \tag{6.178}$$

The experimental effort to include different concentrations is considerable but the obtained equilibrium values are typically rather reliable.

Circulation method: Another static method is based on a closed fluid circuit that includes the chromatographic column. A known amount m_{inj} of solute or a mixture of several solutes is injected into this circuit and pumped around until equilibrium is established. Samples are taken and analyzed to determine the resulting equilibrium concentration c_{eq}. The mass balance for the equilibrium loading accounts for the holdup of the complete plant:

$$m_{\text{inj},i} = V_{\text{plant}} \cdot c_{\text{eq},i} + \varepsilon \cdot V_c \cdot c_{\text{eq},i} + (1 - \varepsilon) \cdot V_c \\ \cdot [\varepsilon_p \cdot c_{\text{eq},i} + (1 - \varepsilon_p) \cdot q(c_{\text{feed},i})] \tag{6.179}$$

with

$$m_{\text{inj},i} = V_{\text{inj}} \cdot c_{\text{inj},i} \tag{6.180}$$

Subsequently, new injections are made to change the concentration step-by-step. The successive nature of this method saves solute, but unavoidable inaccuracies accumulate.

6.5.7.4 Dynamic Methods

Dynamic methods extract information about the isotherms from the measured transient concentration profiles. The basic principle is to inject disturbances in an equilibrated column and to analyze the column response. Based on the type of disturbance three major groups can be distinguished. Injection of a large sample, with a concentration different from the existing equilibrated state, results in a breakthrough curve (frontal analysis, method of step response). The complementary case induces small disturbances to an equilibrated chromatographic system. Another possibility is the injection of overloaded pulses with concentrations in the nonlinear range of the isotherm, but for injection volumes small enough not to reach breakthrough. The last two methods are sometimes also referred to as pulse response techniques.

6.5.7.5 Frontal Analysis

Frontal analysis is one of the most popular methods to determine isotherms. At $t = 0$ a step signal with concentration c^{II} is injected into the column until $t_{inj} = t_{des}$, where the feed concentration is once again lowered to the initial feed concentration. The outgoing concentration profiles are detected. The injected volume has to be large enough to reach a plateau concentration, resulting in a concentration profile as depicted in Figure 6.19 for a pure component and a column equilibrated initially with the concentration c^{I}. A component index i is omitted for brevity.

It takes a certain time for the outlet profile to reach a plateau. During this adsorption period a new equilibrium is established, with a liquid concentration being the feed concentration c^{II}. Likewise, during the desorption step the initial equilibrium is restored with a delay of t_{des}. This experimental procedure is easy to implement and to automate if a gradient delivery system is available.

A very effective way to evaluate the equilibrium is to calculate the overall mass balances by numerical integration. Area "A" in Figure 6.19 is equivalent to the solute accumulated inside the plant and the column, which is split into two parts accumulated in the liquid and the adsorbent phase. The integral mass balance

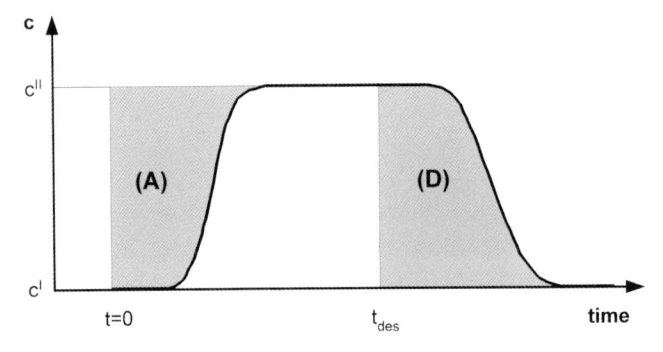

Figure 6.19 Typical breakthrough curve for adsorption and desorption of a pure component (reproduced from Seidel-Morgenstern, 1995).

allows the calculation of the loading, $q^{II} = q(c^{II})$, provided the status I and the total porosity are known:

$$V_{\text{plant}} \cdot (c^{II} - c^{I}) + V_c \cdot [\varepsilon_t \cdot (c^{II} - c^{I}) + (1 - \varepsilon_t) \cdot (q(c^{II}) - q(c^{I}))]$$
$$= \dot{V} \cdot \int_0^{t_{\text{des}}} (c^{II} - c(t)) dt \qquad (6.181)$$

If the solute is injected in the plant without the column, the dead time of the plant can be estimated at the inflection point of the breakthrough curve. The finally observed plateau signal also allows one to verify that the linear range of the detector was not exceeded.

Frontal analysis is straightforward when starting from an unloaded column ($c^{I} = 0$). A modification to reduce the amount of solute is the stepwise increase of the feed concentration, starting from the unloaded column. This results in successive plateaus. Desorption steps are obtained after the highest concentration plateau has passed through the column, if the concentrations are reduced inversely to the adsorption steps. To consume even less feed mixture, this procedure can be performed in closed loop or circulation operation (Figure 6.18).

Independent of the method applied, the gray area "D" in Figure 6.19 has to be equal to the area "A". A comparison may serve as a consistency check. Another possibility to verify the results is to stop the flow after the plateau is reached and analyze the desorption front according to the adsorption–desorption method (Section 6.5.7.3).

An extension to multicomponent systems is straightforward, by injecting more than one component. A prerequisite is the measurement of the concentration profile of each solute during the elution. This can be achieved either by using solute-specific detectors or by collection of multiple fractions and subsequent chemical analysis. Solute-specific detection may be performed, for example, by using different wavelengths for UV detection or multidetector setups. For a binary separation, the latter must provide two independent signals. A typical breakthrough curve of a binary mixture for convex (Langmuir-type) isotherms and interaction of the solutes is given in Figure 6.20. The concentration profile of the weaker adsorbed

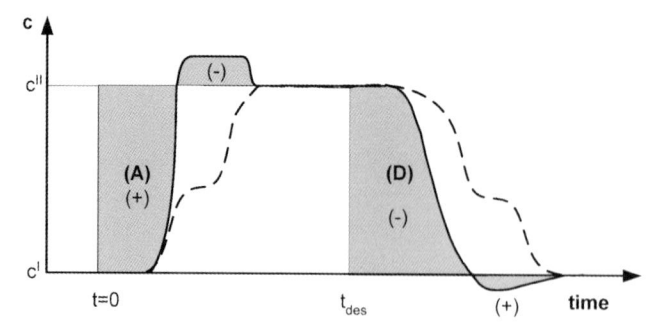

Figure 6.20 Typical breakthrough curve for adsorption and desorption of a binary mixture component (solid line, weaker retained component; dashed line, stronger retained component) (reproduced from Seidel-Morgenstern, 1995).

component is denoted by a solid line, while the more strongly adsorbed component is represented by the dashed line.

The increase above feed concentration c^{II} for the first eluting component is due to the displacement by the stronger adsorbed second component (Section 2.6.2). Calculation of the equilibrium loading is done for each component using Equation 6.181. For the first eluting component, the areas (+) and (−) illustrate the positive and negative contributions, respectively, in the integral. As with single components, calculations can be done using the adsorption profile (A) or the desorption profile (D). For $c^I = 0$, no positive contribution can be found in the desorption profile.

With compressive fronts (Section 2.2.3) the integration of the steep front may be simplified, if the kinetic effects are low (large number of stages). In the case of a Langmuir-type (convex, self-sharpening) isotherm, the front "A" is very steep, originating from the concentration shock (Section 6.2.3). Using Equation 6.50 and taking t_R as the time at the inflection point of the front profile, the loading q^{II} is obtained by Equation 6.182:

$$t_R = t_{plant} + t_0 \cdot \left(1 + \frac{1-\varepsilon_t}{\varepsilon_t} \cdot \frac{q(c^{II}) - q(c^I)}{c^{II} - c^I}\right) \tag{6.182}$$

Parameters of the isotherm equation can be determined from a set of experimental data using a least squares approximation (Section 6.5.7.11).

As an example, Figure 6.21 gives the experimental breakthrough curve for a racemic mixture of EMD53986 using Chiralpak AD ($d_p = 20\,\mu m$, Daicel) as adsorbent and ethanol as eluent. Here, a two-detector setup of polarimeter and UV detector was used, permitting the solute-specific detection of both components (Mannschreck, 1992; Jupke, 2004; Epping, 2005).

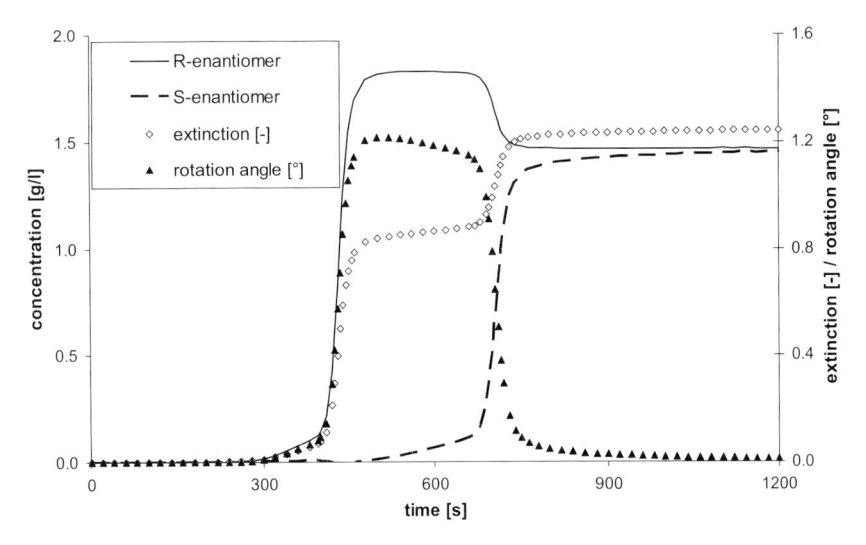

Figure 6.21 Experimental breakthrough curves of the mixture of R- and S-enantiomers of EMD53986 (lines, concentrations; symbols, detector signals; $c_{feed} = 3\,g\,l^{-1}$, $\dot{V} = 20\,ml\,min^{-1}$; for additional data see Appendix A.1).

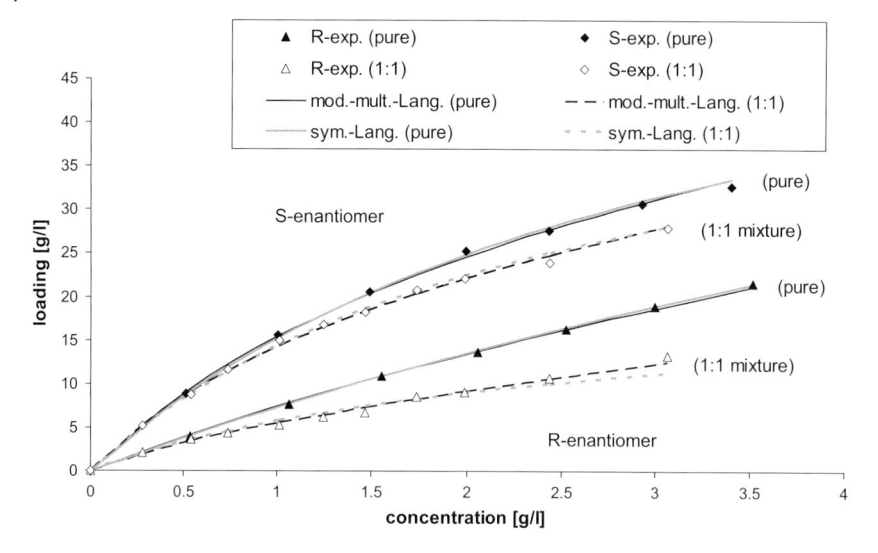

Figure 6.22 Comparison of experimental data (rhombuses and triangles) and fitted isotherm equations (lines) for R- and S-enantiomers of EMD53986 (pure components and the 1:1 mixture on Chiralpak AD in ethanol at 25 °C; mod-multi-Langmuir, Equation 2.58; sym-Langmuir, Equation 2.57; for additional data see Appendix A.1).

Breakthrough curves (lines in Figure 6.21) are calculated from the detector signals (symbols). The rotation angle detected by the polarimeter is zero if both enantiomers are present ($t > 800$ s) while the UV signal is additive and has the highest value there.

The two breakthrough curves of both components are described by convex isotherms with competitive adsorption and show a typical displacement effect for the weaker retained R-enantiomer.

The corresponding equilibrium data for pure components and different mixtures are represented in Figure 6.22 by filled rhombuses and triangles. Also shown are the results of different isotherm equations (solid and dashed lines) that have been fitted to the experimental data. In this case the modified multicomponent Langmuir isotherm (Equation 2.58)

$$R\text{-enantiomer}: \quad q_R = 2.054 \cdot c_R + \frac{5.847 \cdot c_R}{1 + 0.129 \cdot c_R + 0.472 \cdot c_S}$$

$$S\text{-enantiomer}: \quad q_S = 2.054 \cdot c_S + \frac{19.902 \cdot c_S}{1 + 0.129 \cdot c_R + 0.472 \cdot c_S}$$

(6.183)

and the classical Langmuir isotherm (Equation 2.57) for the multicomponent system were used. Isotherms for the mixtures result in lower loadings than those of the pure components and the interaction is stronger for the more weakly retained component, which is characteristic for competitive adsorption.

Other examples are isotherms for the isomers fructose and glucose. Figure 6.23 shows that the resulting isotherms exhibit an upward curvature and the slope of the

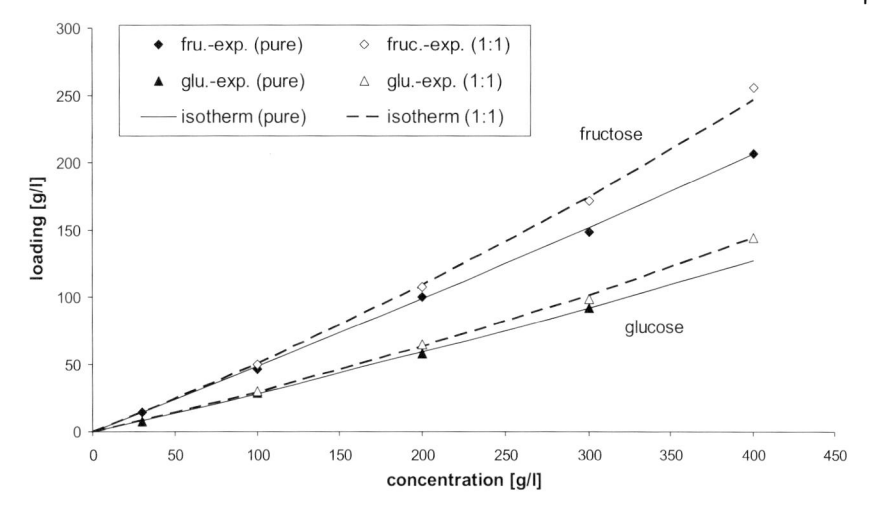

Figure 6.23 Measured and calculated isotherms for pure components and mixtures of glucose and fructose (adsorbent: ion exchange resin Amberlite CR1320 Ca, $d_p = 325$ µm, eluent: water; for additional data see Appendix A.3).

isotherm is increased in the case of mixtures. This anti-Langmuir behavior is explained from the specific interaction between the hydrated solute molecules and the eluent (water) (Saska et al., 1991; Saska, Clarke, and Iqbal, 1992, Nowak et al., 2007). These isotherms are expressed by the following empirical correlations:

$$q_{glu} = 0.27 \cdot c_{glu} + 0.000122 \cdot c_{glu}^2 + 0.103 \cdot c_{glu} \cdot c_{fru}$$
$$q_{fru} = 0.47 \cdot c_{fru} + 0.000119 \cdot c_{fru}^2 + 0.248 \cdot c_{glu} \cdot c_{fru}$$
(6.184)

Another aspect to keep in mind when performing frontal analysis is that severe changes in the general shape of the breakthrough curve for different concentrations may hint at mass transfer or adsorption kinetic effects. As similar changes occur when changing the flow rate, the variation of these two parameters can be used to further discriminate kinetic effects (Ma, Whitley, and Wang, 1996). Although analogue changes are observed for peak injection too, breakthrough curves are easier to analyze because of the defined conditions for the equilibrium plateau.

Finally, a few characteristics of the frontal analysis method should be summarized:

- Only equilibrium states are evaluated; thereby, errors due to kinetic effects are widely eliminated.
- Amount of solutes and experimental efforts are high, as only one equilibrium point is determined per injection. This is increasingly important if kinetic effects are strong and thus the time needed to reach the phase equilibrium is rather long. Closed loop operation may be used to reduce the consumption of solutes.

- Frontal analysis typically provides a high accuracy compared with the other methods mentioned.
- Detector calibration is needed but directly supported by the signal values collected at the plateaus.
- Multicomponent isotherms can be determined if suitable off-line analytical methods or solute-specific detectors are available.

6.5.7.6 Analysis of Disperse Fronts (ECP/FACP)

Analysis of disperse fronts exploits the equilibrium theory of chromatography (Section 6.2.3). It generally reduces the experimental effort as well as the sample amount needed compared with frontal analysis. Because of the complex mathematical solutions in the case of mixtures, it is only suitable for single-component isotherms. Two different forms are described in literature. After injection of a pulse that is not wide enough to cause a concentration breakthrough, the disperse part of the overloaded concentration profile is analyzed. For Langmuir-type isotherms, this is the rear part of the peak. This method is called "elution by characteristic point" (ECP). If the injected volume is high enough to get a breakthrough, it is termed "frontal analysis by characteristic points" (FACP).

The principle of both methods is explained in Figure 6.24.

According to equilibrium theory, the retention time t_R at a characteristic concentration c^+ correlates to the slope of the isotherm at this specific concentration. Rearranging Equation 6.48, the slope can be calculated by

$$\left.\frac{dq}{dc}\right|_{c^+} = \frac{t_R(c^+) - t_0}{t_0} \frac{\varepsilon_t}{1 - \varepsilon_t} \tag{6.185}$$

The measured retention times must be corrected by the dead time t_{plant} of the plant and for Langmuir-shaped isotherms additionally by the injection time t_{inj} (Equation 6.51).

If this analysis is repeated at different positions of the profile, the whole isotherm can be estimated based on analyzing a single chromatogram. The concentration range is limited up to the value at the peak maximum. Isotherm parameters can be determined from the slope versus concentration values in two ways.

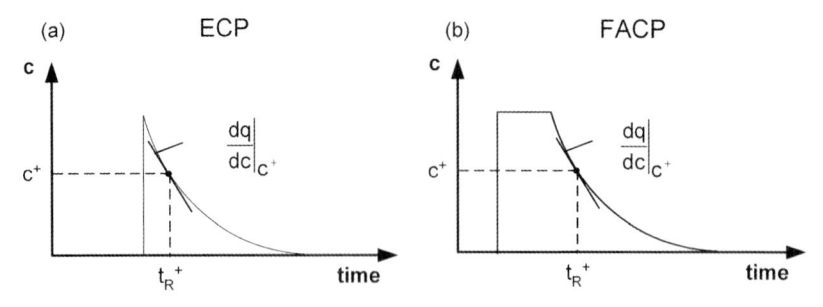

Figure 6.24 Illustration of the principle of the ECP and FACP methods.

First, numerical integration leads to the desired relationship between loading and concentration according to

$$q(c^+) = \int_0^{c^+} \frac{dq}{dc} dc = \int_0^{c^+} \frac{t_R(c) - t_0}{t_0} \frac{\varepsilon_t}{1 - \varepsilon_t} dc \qquad (6.186)$$

The isotherm equation is then fitted as described in Section 6.5.7.11.

Second, the derivative of a selected isotherm model equation can be fitted directly to the measured slope values. Thus, in the optimization problem the errors of the loading values are replaced by the errors of the slopes.

Regardless of which method is used, the initial slope of the isotherm must be determined with great care. Either the lower bound for integration must be approximated with the lowest recorded concentration or the slope for zero concentration must be determined separately. As the latter task is identical to the determination of the Henry constant, this value can be obtained from pulse experiments with very low injection concentrations using momentum analysis (Section 6.5.7.2).

ECP and FACP are limited to single-component systems. They save time and sample as the whole isotherm is derived from one experiment only. The most severe limitation of the method is the fact that highly efficient columns are required, because kinetic effects are completely neglected. This limits the applicability to systems where the plate number is above 1000. Experimental errors hampering the application of the method are

- Imprecise detector calibration directly influences the concentration profile and thus the isotherm.
- Peak deformation through extra column effects of the plant cannot be accounted for.
- Determination of the slopes from retention times for very low concentrations causes typically errors and provides Henry constants differing from values obtained exploiting pulse experiments.

Regarding the FACP method, complementary frontal analysis might be used to check the determined isotherms at the breakthrough concentration.

6.5.7.7 Peak Maximum Method

A possibility to reduce the influence of limited column efficiencies on the results obtained by the ECP method is to detect just the positions of the peak maxima. The method based on this is called consequently the peak maximum or retention time method. Chromatograms as shown in Figure 6.24a are acquired by measuring a series of pulse response injections for different injection concentrations. The concentration and position of the maximum is strongly influenced by the adsorption equilibrium due to the compressive nature of either the front or the rear of the peak (Section 2.2.3). Thus, the obtained values are less sensitive to kinetic effects than in the case of the ECP method. The isotherm parameters can be evaluated in the same way as described in Section 6.5.7.6, but the same limitations have to be kept in mind. For some isotherm equations, analytical solutions of the ideal model can be used to evaluate the retention times of the concentration maxima. (Golshan-Shirazi and Guiochon, 1989; Guiochon et al., 2006). Thus, only retention times must be considered and detector calibration can be omitted in these cases.

6.5.7.8 Minor Disturbance/Perturbation Method

The minor disturbance or perturbation method relies on equilibrium theory too and was suggested, for example, by Reilley, Hildebrand, and Ashley (1962). As known from linear chromatography and exploited above already frequently, the retention time of the response to a small pulse injected into a column filled with pure eluent can be used to obtain the initial slope of the isotherm. This approach can be expanded to cover the whole isotherm range. For the example of a single-component system the procedure is as follows (Figure 6.25): the column is equilibrated with a concentration c_a and, once the plateau is established, a small pulse is injected at a time $t_{start,a}$ and a pulse of a different concentration is detected at the corresponding retention time $t_{R,a}$.

The injected concentration can be either higher or lower than the plateau concentration, c_a. However, to maintain equilibrium conditions inside the column, the concentration should not deviate too much from c_a and the injected volume should be small. Care has to be taken to assure that the resulting peak is large enough to be distinguishable from the signal noise. Since for that reason the injected amount must be comparatively high, it is recommended to average results obtained for more or less concentrated injections. In practice, pure eluent with very small injected volumes often provides sufficient accuracy.

According to equilibrium theory, elution of the small disturbance depends on the isotherm slope at the plateau concentration. Because the perturbation peak is almost Gaussian, the time at the peak maximum (respectively, minimum) can be taken to estimate in a simple manner local isotherm slopes by Equation 6.48:

$$\left.\frac{dq}{dc}\right|_{c_a} = \frac{t_{R,a} - t_0}{t_0} \frac{\varepsilon_t}{1 - \varepsilon_t} \tag{6.187}$$

The characteristic retention times have to be corrected again by the dead time of the plant and eventually by the injected time (Section 6.5.3.1). Using values at different

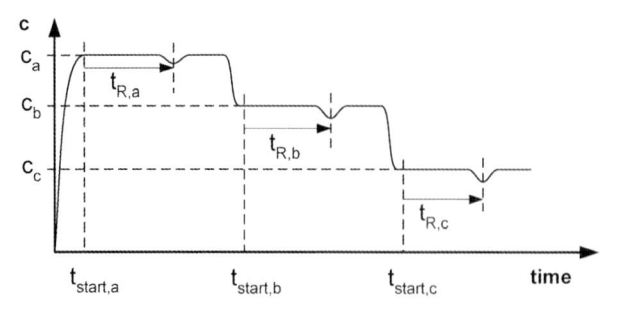

Figure 6.25 Principle of the minor disturbance method for a single-component system (lower concentrated sample injection).

plateau concentrations, the isotherm parameters are derived similar as described in Section 6.5.7.6.

To determine multicomponent isotherms, the column has to be preequilibrated at well-defined mixture concentrations. Perturbations now trigger several recordable peaks. For example, an injection of pure mobile phase on a column preequilibrated with a two-component mixture results in two peaks. Their retention times are the experimental information. For exploitation a competitive isotherm model must be assumed. Using Equation 6.46 and the definition of the retention times (Equation 6.48), for two-component systems together with the coherence condition (Equation 6.53), the two measured retention times can be used to calculate the four partial differentials of Equation 6.52 of the assumed isotherm model at the plateau concentrations. The complexity of these calculations increases rapidly with increasing number of components. Additionally, detector noise can make it difficult to clearly distinguish the earlier and later eluting peaks.

An advantage of this method compared with ECP is that the perturbation method is not as sensitive to the number of stages of a column. Additionally, detector calibration is not necessary and the analysis does not require solute-specific detectors.

The method described can be advantageously combined with the frontal analysis method, which also requires a concentration plateau and thus shares the disadvantage of high sample consumption. As indicated in Figure 6.25, the measurement procedure starts at maximum concentration. This concentration plateau is reduced step-by-step by diluting the solution. To reduce the amount of samples needed for the isotherm determination the experiments can be done in a closed loop arrangement (Figure 6.18, Blümel et al., 1999). It is easily possible to automate this procedure.

As an example, Figure 6.26a gives the results of the isotherm determination for Tröger's base enantiomer on Chiralpak AD ($d_p = 20\,\mu m$) from perturbation measurements (Mihlbachler et al., 2001). Theoretical retention times for the pure components and racemic mixtures (lines) were fitted to the measured data (symbols) by means of Equation 6.187 to determine the unknown parameter in Equation 6.188. Total differentials for the mixture (Equation 6.52) were evaluated using the coherence condition (Equation 6.53), providing parameters of the isotherm equation (Equation 6.188). Note that the Henry coefficients were independently determined by pulse experiments and were fixed during the fitting procedure:

$$\text{R-enantiomer}: \quad q_R = \frac{0.0311 \cdot c_R \cdot (54 + 0.732 \cdot c_S)}{1 + 0.0311 \cdot c_R} + \frac{0.732 \cdot 0.0365 \cdot c_R \cdot c_S}{1 - 0.0365 \cdot c_R}$$

$$\text{S-enantiomer}: \quad q_S = \frac{27 \cdot c_S \cdot (0.1269 + 2 \cdot 0.0153 \cdot c_S)}{1 + 0.1269 \cdot c_S + 0.0153 \cdot c_S^2}$$

(6.188)

The resulting isotherms are illustrated as lines in Figure 6.26b. The stronger adsorbing S-enantiomer isotherm exhibits an inflection point and can be roughly assumed to be independent of the R-enantiomer concentration. The R-enantiomer isotherm shows typical Langmuir behavior and minor interaction with the

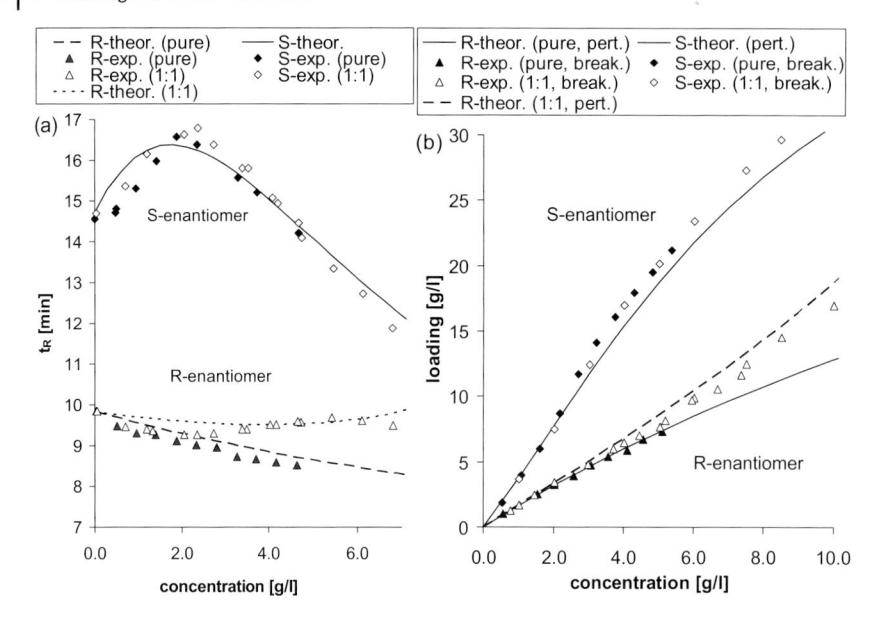

Figure 6.26 (a) Results for isotherm determination for Tröger's base on Chiralpak AD for pure components and mixtures (symbols, experimental data; lines, retention times calculated by Equations 6.189 and 6.190 ($t_0 = 5.144$ min, $\dot{V} = 1$ ml min^{-1}); for additional data see Appendix A.2). (b) Comparison of Equation 6.190 with experimental data from frontal analysis. (Data taken from Mihlbachler et al., 2001.)

S-enantiomers. The unusual behavior of the R-enantiomer can be explained e.g. with multilayer adsorption processes (Mihlbachler et al., 2001). A more detailed discussion related to consequences of inflection points in isotherm courses is given by Arnell and Fornstedt (2006).

Figure 6.26b also shows the isotherm data obtained from frontal analysis experiments (symbols), which agree sufficiently well with those obtained using the perturbation analysis.

The determined competitive equilibrium data could be more precisely successfully fitted to a second-order model capable to describe concave isotherms with saturation (Hill, 1960; Lin et al., 1989; Mihlbachler et al., 2001; Jupke, 2004):

$$R\text{-enantiomer}: q_R = \frac{54 \cdot c_R \cdot (0.035 + 0.0046 \cdot c_S)}{1 + 0.035 \cdot c_R + 0.062 \cdot c_S + 0.0046 \cdot c_R \cdot c_S + 0.0052 \cdot c_S^2}$$

$$S\text{-enantiomer}: q_S = \frac{54 \cdot c_S \cdot (0.062 + 0.0046 \cdot c_R + 2 \cdot 0.0052 \cdot c_S)}{1 + 0.035 \cdot c_R + 0.062 \cdot c_S + 0.0046 \cdot c_R \cdot c_S + 0.0052 \cdot c_S^2}$$

(6.189)

6.5.7.9 Curve Fitting of the Chromatogram

The approach of parameter estimation exploiting dynamic process models can also be applied to determine isotherm parameters. In the framework of Figure 6.9 this requires the selection of the column model (Section 6.2) and the plant model (Section 6.3.2) together with an (assumed) adsorption equilibrium model (i.e., an isotherm equation). If the plant parameters, the packing parameters, and detector calibration curves are determined, transport coefficients (Section 6.5.8) can be estimated from analyzing in the linear range of the isotherm responses to injected pure component pulses. These standard experiments provide also the Henry coefficient (using Equation 6.176 and momentum analysis, Section 6.5.3.1). The remaining unknown isotherm parameters can then be estimated by matching model predictions to the peak shapes observed during overloading experiments (large pulses). To determine significant branches of the isotherms, the injection and thus elution concentrations should be as "high" as possible. This curve fitting method is frequently called "inverse method" (or sometimes just peak fitting). It was first applied systematically by James et al. (1999) in order to determine the competitive adsorption isotherms of the ketoprofen enantiomers on a cellulose-based chiral stationary phase. A systematic comparison of the results achieved using this method with the results of frontal analysis was published by Vajda, Felinger, and Cavazzini (2010). Cornel et al. (2010) suggested recently an attractive extension, designated as direct inverse method, which avoids the calibration of the detector.

In order to identify the correct isotherm model, the analysis has to be repeated for several potential candidate models. In each case realistic initial estimates for the free parameters have to be provided in order to facilitate convergence of the nonlinear optimization procedure required. A drawback of this curve fitting approach is that all errors of the assumed column and plant models have an effect on the quality of the isotherm parameters estimated. Thus, this approach is in particular recommended to get relatively fast a first idea about the thermodynamic properties of the chromatographic system investing only small amounts of sample.

6.5.7.10 Prediction of Mixture Behavior from Single-Component Data

As mentioned in Section 2.5.2.3, theoretical methods exist to calculate data for multicomponent isotherms just based on the knowledge regarding the corresponding single-component equations. When the resulting accuracy is acceptable, this eliminates the need to determine experimentally additional mixture parameters. As an example, the experimental data shown in Figure 6.22 are used to derive the multicomponent isotherm based on the single-component equation and IAS theory (Section 2.5.2.3). Experimental data of the EMD system (rhombuses for S-enantiomer and triangles for R-enantiomer) as well as the data calculated by the IAS theory (straight lines) are plotted in Figure 6.27. Data for the modified multi-Langmuir isotherm (Figure 6.22) are added for reference as dashed lines. Figure 6.27 shows a relative satisfying agreement between IAS theory predictions and experimental data, although the deviation increases for higher concentrations due to the required extrapolation of pure-component data in the IAS calculations.

As with the selection of a suitable single-component adsorption isotherm model, a final decision about the accuracy and validity of this attractive approach to predict mixture isotherms has to be based on the results of a critical comparison between simulated and measured concentration profiles.

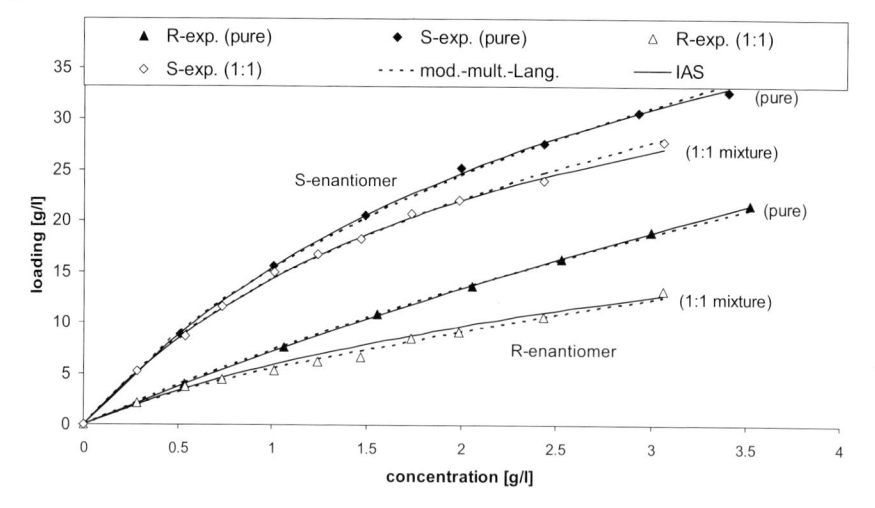

Figure 6.27 Multicomponent isotherm derived from single-component data using the IAS theory as well as reference data for the EMD system (for other data see Figure 6.22 and Appendix A.1).

6.5.7.11 Data Analysis and Accuracy

When dynamic methods are applied, the measured profiles should include enough points and sufficiently low detector noise to perform numerical calculations (e.g., integration). This is discussed in another context in Sections 6.5.3.2 and 6.5.3.3. The number of data points in the measured profile can be increased by changing the flow rate of the pump or the sample rate.

Through repetition of the same experiment and subsequent evaluation of the equilibrium data, the accuracy may be increased by averaging the values, if no systematic errors occur. Too high a deviation between equivalent measurements indicates problems in either the data evaluation or the experiment itself. In the latter case, it should be checked if the pumps deliver a constant flow rate and that the temperature is constant in the range of a few tenths of 1 °C. If the eluent consists of a fluid mixture, possible influences of slight changes in eluent composition must be critically evaluated (Section 6.5.7.1).

All plant and packing parameters, as well as the calibration curve of the detector, must be determined with care to give proper results. Simulation and experiment might still agree sufficiently if the "wrong" void fractions and porosities are determined and, consequently, the isotherm parameters are inaccurate. This is because not the isotherm but rather the isotherm multiplied by the porosities appears in the model equations and thus both inaccuracies cancel each other out to a certain degree.

Most determination methods finally lead to discrete loading versus concentration data that have to be fitted to a continuous isotherm equation. For this purpose it is advised to use a least squares method to obtain the parameters of the isotherm. Nonlinear optimization algorithms for such problems are implemented in standard spreadsheet programs. To select a suitable isotherm equation and obtain a meaningful fit, the number of data points should not be too low and their distribution should be such that changes in the slope or curvature are properly represented. A suitable objective function to minimize the overall error Δ includes the sum of the weighed relative error:

$$\delta = \sqrt{\sum_{j=1}^{n_p} f_j \cdot \left(\frac{q_{\exp,j} - q_{\text{theo},j}}{q_{\exp,j}}\right)^2} \qquad (6.190)$$

where $q_{\text{theo},j}$ is the value obtained from the isotherm equation, $q_{\exp,j}$ the experimental value, and f_j describes a weighting factor. Often, f_j is a constant and depends only on the number of experimentally determined points (n_p) and the number of isotherm parameters (n_{para}):

$$f_j = \frac{1}{n_p - n_{\text{para}}} \qquad (6.191)$$

This allows a comparison of the fitting quality for certain isotherm equations with different numbers of adjustable parameters. Further equations for statistical analysis can be found, for example, in Barns (1994) and Press et al. (2002). The use of relative instead of absolute errors is necessary to increase the fit in the low concentration region of the isotherm. Otherwise, the isotherm is inaccurate in the low concentration region and the calculated band profile is inaccurate at the rear boundary.

A problem often encountered in nonlinear optimization is the necessity to provide suitable initial guesses for the free parameters to be estimated. Therefore, it should be tested if different initial guesses lead to different final sets of the parameters. This is connected to the sensitivity problem, which is pronounced in the case of multicomponent systems where several parameters need to be fitted at once. Substantial initial guesses are often difficult to find and the sensitivity is often low, which demands much experimental data or leads to the selection of other isotherm equations.

For the linear part of the isotherm, the Henry coefficient may be determined separately by pulse experiments (Section 6.5.7.2). If a significant deviation from the value obtained with the isotherm equation is encountered, additional experiments in the low concentration range should be carried out. These can be used to clarify whether the isotherm equation (e.g., adding an extra linear term as in Equation 2.58) or the method of isotherm determination must be changed.

As the overall aim of parameter determination is mostly the (simulation-based) prediction of the process behavior, the final decision about the suitability of the isotherm equation can only be made by comparing experimental and theoretical elution profiles (Section 6.6). Depending on the desired accuracy, this may involve iteration loops for the selection of isotherm equations or in some cases even the methods (Figure 6.17). In this context it should be remembered that, according to the model equations (e.g., Equation 6.46), the migration velocities and thus the positions of the profiles are functions of the isotherm slopes. Therefore, it is most important for the reliability of process simulation that the slopes of the measured and calculated elution profiles fit well to each other.

6.5.8
Mass Transfer

According to the approach summarized in Figure 6.9 for the transport-dispersive model, the transport coefficient is the last parameter to be determined. All prior experimental errors and model inaccuracies are lumped now into this final parameter.

The parameter k_{eff} could be determined by a simulation-based parameter estimation that needs to be now performed with the full set of model equations, using the parameters for the plant as well as the column (Figure 6.5). The estimated value of k_{eff} might be verified by measurements at different volumetric flow rates and injection amounts.

An initial guess of k_{eff} can be obtained from moment analysis (Section 6.5.3.1) or empirical correlations. For peak injections in the linear range of the isotherm k_{eff} is calculated by Equation 6.139 if the axial dispersion coefficient is already known (Section 6.5.6.2). Another rough estimation is based on the slope C of the simplified HETP curve (Equation 6.143):

$$k_{eff} = 2 \cdot \left(\frac{\tilde{k}'}{1+\tilde{k}'}\right)^2 \cdot \frac{\varepsilon}{1-\varepsilon} \cdot \frac{r_p}{3} \cdot \frac{1}{C} \tag{6.192}$$

If mass transfer in the film and diffusion inside the pores are taken into account, the effective mass transfer coefficient is given as a series connection of the internal $(1/k_{pore})$ and external $(1/k_{film})$ mass transfer resistances:

$$\frac{1}{k_{eff}} = \frac{d_p}{10 \cdot \varepsilon_p \cdot D_{pore}} + \frac{1}{k_{film}} = \frac{1}{k_{pore}} + \frac{1}{k_{film}} \tag{6.193}$$

For the film transfer coefficient k_{film}, Wilson and Geankoplis (1966) developed the following correlations:

$$Sh = \frac{1.09}{\varepsilon} \cdot (\varepsilon \cdot Re)^{0.33} \cdot Sc^{0.33} \quad (0.0015 < \varepsilon \cdot Re < 55)$$
$$Sh = \frac{1.09}{\varepsilon} \cdot (\varepsilon \cdot Re)^{0.33} \cdot Sc^{0.33} \quad (55 < \varepsilon \cdot Re < 1050) \tag{6.194}$$

where the Sherwood number Sh and the Schmidt number Sc are defined as

$$Sh = \frac{k_{film} \cdot d_p}{D_m} \tag{6.195}$$

$$Sc = \frac{\eta_l}{\varrho_l \cdot D_m} = \frac{\nu_l}{D_m} \tag{6.196}$$

The (particle) Reynolds number is given by Equation 6.171:

$$Re = \frac{u_{int} \cdot d_p \cdot \varrho_l}{\eta_l} = \frac{u_{int} \cdot d_p}{\nu_l} \tag{6.171}$$

Using the same assumptions as in Section 6.5.6.2 and typical viscosity values ν_l of about $0.01 \, cm^2 \, s^{-1}$, the Schmidt number for small molecules is of the order of 1000. The Reynolds number is about 0.01 so that the first equation in

Equation 6.194 is applicable. For a typical void fraction ε of 0.37–0.4, the Sherwood number is about 4.5 and the film transfer coefficient is then calculated by

$$k_{\text{film}} = \frac{1.09}{\varepsilon} \cdot \frac{D_m}{d_p} \cdot \left(\varepsilon \cdot \frac{u_{\text{int}} \cdot d_p}{D_m} \right)^{0.33} \tag{6.197}$$

According to the approximations mentioned above, the film transfer coefficient k_{film} is about 4.5×10^{-2} cm s^{-1} for a 10 μm particle.

The intraparticle (pore) diffusion coefficient defined in Section 6.2.2.4 may be estimated by the Mackie–Meares correlation (Mackie and Meares, 1955):

$$D_{\text{pore}} = \frac{\varepsilon_p}{(2 - \varepsilon_p)^2} \cdot D_m \tag{6.198}$$

The factor in front of D_m is an approximation of the internal tortuosity factor. For a porosity of 0.9–0.5, D_{pore} is five times smaller than D_m or even lower, so that the contribution to the effective mass transfer coefficient (Equation 6.193) is

$$k_{\text{pore}} = \frac{10 \cdot \varepsilon_p \cdot D_{\text{Pore}}}{d_p} = \frac{10}{d_p} \frac{\varepsilon_p^2}{(2 - \varepsilon_p)^2} \cdot D_m \tag{6.199}$$

which gives a value for k_{pore} of about 1×10^{-3} cm s^{-1}. Intraparticle transport is an order of magnitude slower than film transfer and is, therefore, the rate-limiting step (see also Ludemann-Hombourger, Bailly, and Nicoud, 2000).

Thus, k_{eff} is mainly defined by Equation 6.199. This simplified analysis provides also a justification to take the effective transfer coefficient as independent of volumetric flow rate and inversely proportional to particle diameter.

6.5.9
Identification of Isotherms and Mass Transfer Resistance by Neural Networks

Standard approaches to determine isotherms, such as frontal analysis or the perturbation method, detect equilibrium data and fit parameters of isotherm model equations, as, for example, Langmuir-type or empirical equations to these data. Generally, these methods are very precise, but they are also time consuming and require an amount of product that is not always negligible. Other methods such as ECP or FACP need only a small amount of product but are applicable for single components only and are of restricted accuracy. Generally, the experimental efforts increase with increasing interactions of the components and the demand on accuracy of the model. Another common problem is the limited amount of samples available, especially for studies during the first steps of process development.

Schlinge et al. (2011) presented an alternative method that requires a lot of preliminary computer simulations but estimates isotherm parameters as well as mass transfer coefficients on the basis of only a few experiments and very small amounts of samples in short time. The basic idea is to simulate HPLC chromatograms for given isotherm equations and to vary the isotherm parameters within the range of practical applications. The calculated data are then used to train and validate neural

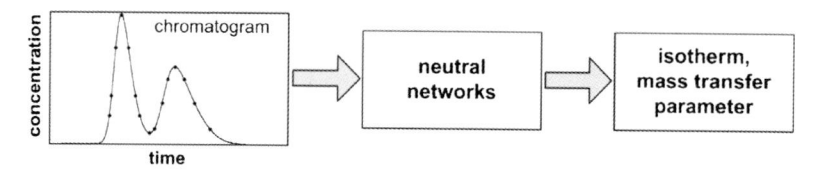

Figure 6.28 Determination of isotherm parameters by neural networks.

networks (NN) for the estimation of the parameters of the isotherm equation. The accuracy improves if for each isotherm parameter a separate NN is trained. After this preliminary work is done the NN can repeatedly be used to estimate isotherm parameters and mass transfer coefficients for new separation problems based on a few experimentally measured chromatograms (Figure 6.28).

Schlinge et al. (2011) used the transport-dispersive model (Equations 6.72 and 6.73) for process simulation. Only selected characteristic points of the calculated chromatograms (e.g., peak maximum and the time for concentrations between 90% and 10% in steps of 10–20%) are necessary to train and validate the NN for the estimation of the parameters of an isotherm equation (Figure 6.28) and a mass transfer coefficient. The operating data as well as the design data of the HPLC plant are always kept constant (i.e., interstitial velocity, particle diameter, porosity, axial dispersion, and column length). For the measurement of chromatograms it is essential to use a well-designed HPLC plant where dead times as well as mixing are reduced as much as possible and to verify the flow rate of the pump. All plant parameter (Sections 6.3 and 6.5.5) of the HPLC have to be determined in advance in order to use them for the simulations.

After having developed the NN for certain types of isotherms the estimation of the isotherm data for a new chromatographic system is done in following steps:

- Measure chromatograms for the pure components and the mixture and determine characteristic points of the chromatograms.
- Estimate the type of an isotherm.
- Calculate the isotherm parameters as well as mass transfer coefficient for the pure components by NN.
- Determine additional isotherm parameters for the real mixture based on the parameters for the pure components.

It has to be kept in mind that the development of the NN and especially the number of computer simulations increase exponentially with the bandwidth and the number of parameters, but this costs computer time and not additional man power. For some parameters such as axial dispersion, particle diameter and porosity mean values should be chosen and kept constant. The resulting deviations should be acceptable if they are balanced with the low experimental effort and small amount of samples necessary.

As an example for the efficiency of the NN method, Schlinge et al. (2011) published a comparison between simulated and measured chromatograms for a racemic mixture. The NN were validated for a column of 4 × 300 mm and a flow rate of

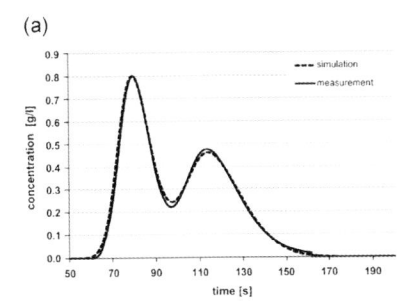

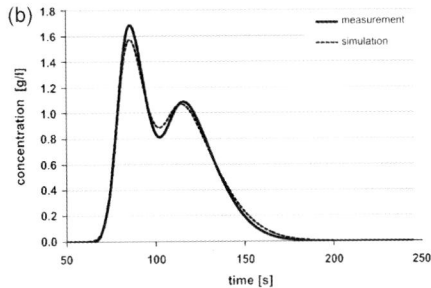

Figure 6.29 Comparison of experimental and simulated results: (a) $L_c = 100$ mm and $\dot{V} = 1.5$ ml min^{-1} and (b) $L_c = 300$ mm and $\dot{V} = 3.5$ ml min^{-1} (reproduced from Schlinge et al., 2011).

$\dot{V} = 1.5$ ml min^{-1}. For the chromatograms shown in Figure 6.29 different values are chosen. The isotherms are described by the multi-Langmuir equation (Equation 2.57).

6.6
Experimental Validation of Column Models

After all parameters for the plant (Section 6.5.5) and column (Sections 6.5.6–6.5.8) models are determined, either from experimental data or from empirical correlations, the validity of a model should be checked using different experiments than those analyzed during parameter determination. In particular, the correctness of predicting the positions of the adsorption and desorption fronts is an important indicator for the reliability of a model.

Below, we will shortly demonstrate the degree of agreement between experimental and simulation results that can be achieved using the approach described in this chapter. Hereby, we will focus first on classical isocratic batch elution, since the corresponding single-column model can be seen as the building block for describing also the more sophisticated process concepts described in Chapter 5. Extensions essentially require respecting the corresponding specific initial and boundary conditions. Then an illustration will be given with respect to the periodically operated multicolumn SMB process introduced in Section 5.2.4. For this process also an additional simpler steady-state model is shortly introduced, which assumes a hypothetical true countercurrent between the two phases involved.

6.6.1
Batch Chromatography

The following examples illustrate a few effects encountered in model validation based on our research (Epping, 2005; Jupke, 2004). All process simulations are based on the transport-dispersive model. Model equations were solved by the gPROMS$^\text{R}$ Software (PS Enterprise, UK) using the OCFE method (Section 6.4).

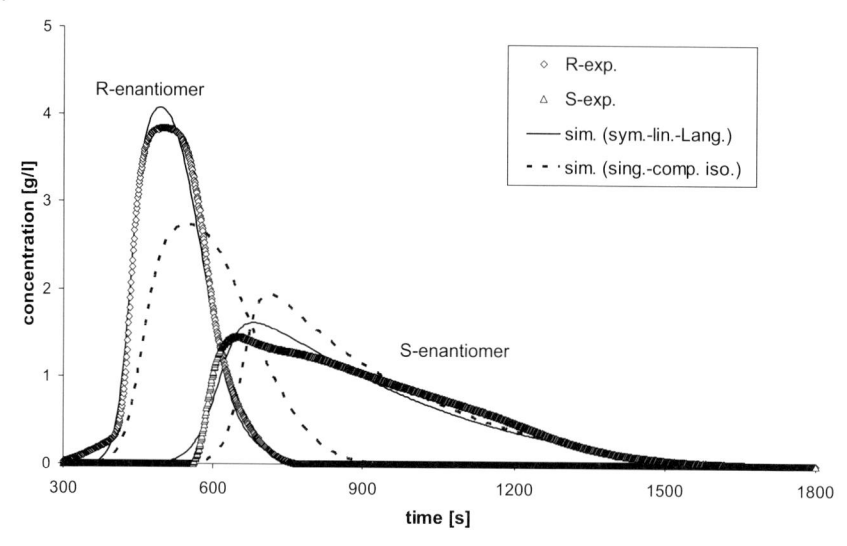

Figure 6.30 Comparison of experimental and simulated profiles for the separation of EMD53986 racemic mixture using single- and multicomponent isotherms (exp., experimental; sim., simulated; $c_{feed} = 6\,g\,l^{-1}$, $\dot{V} = 20\,ml\,min^{-1}$, $V_{inj} = 80\,ml$, $V_c = 54\,ml$; for additional data see Appendix A.1).

Figure 6.30 compares measured and simulated profiles for the batch separation of EMD53986. Very good agreement between theory (solid lines) and experiment (symbols) is achieved using the multicomponent modified Langmuir isotherm (Figure 6.22). Also shown are the simulation results neglecting component interaction by using only the single-component isotherms (dashed line), which deviate strongly from the observed mixture behavior. Typical for competitive adsorption is the displacement of the weaker retained R-enantiomer and the peak expansion of the stronger adsorbed S-enantiomer.

Finally, selected simulated profiles using an experimentally determined multicomponent isotherm and the respective data calculated according to the IAS theory are compared (Figure 6.27). Figure 6.31 shows the very close agreement between these methods for this enantiomer system. Especially when considering the effort necessary to measure the multicomponent isotherms, IAS theory or its extensions may provide a good estimate for the component interaction in the case of competitive adsorption. Therefore, it is advisable to simulate elution profiles using the IAS theory after single-component isotherms have been measured. These calculations should then be compared with a few separation experiments to decide if additional measurements of the multicomponent isotherm are still necessary.

The validity of the transport-dispersive model was further successfully confirmed by experiments with other chromatographic systems such as Tröger's base (Mihlbachler et al., 2001; Jupke, 2004), the WMK-Keton (Epping, 2005), and fructose–glucose (Jupke, 2004).

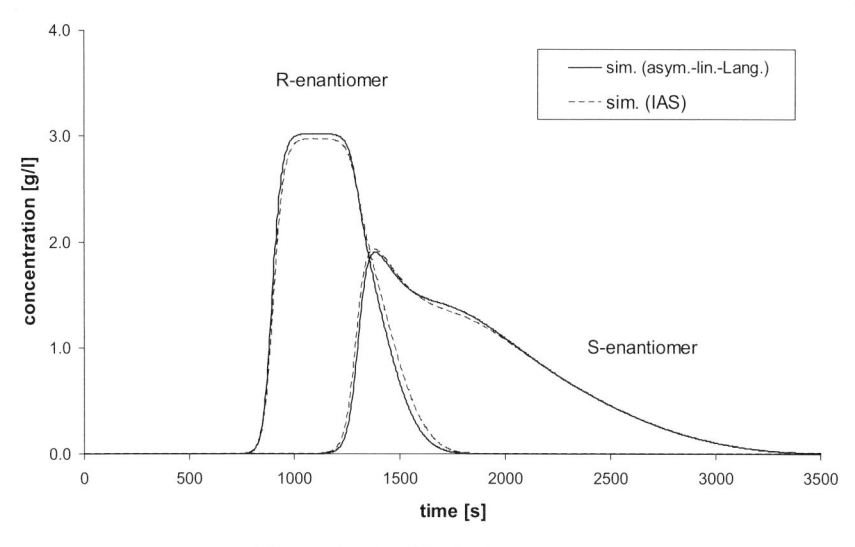

Figure 6.31 Comparison of the simulated profiles for the modified multicomponent Langmuir isotherm and the IAS equation (Figure 6.27) ($c_{feed} = 4.4\,\text{g}\,\text{l}^{-1}$, $\dot{V} = 10\,\text{ml}\,\text{min}^{-1}$, $V_{inj} = 120\,\text{ml}$, $V_c = 54\,\text{ml}$; for additional data see Appendix A.1).

As one example, Figure 6.32 shows the results for Tröger's base using isotherm data determined from breakthrough curves (Equation 6.189).

Even for the fructose–glucose isomer system, with liquid phase concentrations that are an order of magnitude higher than those of the enantiomer system, the transport-dispersive model is valid. As shown in Figure 6.33, the model approach

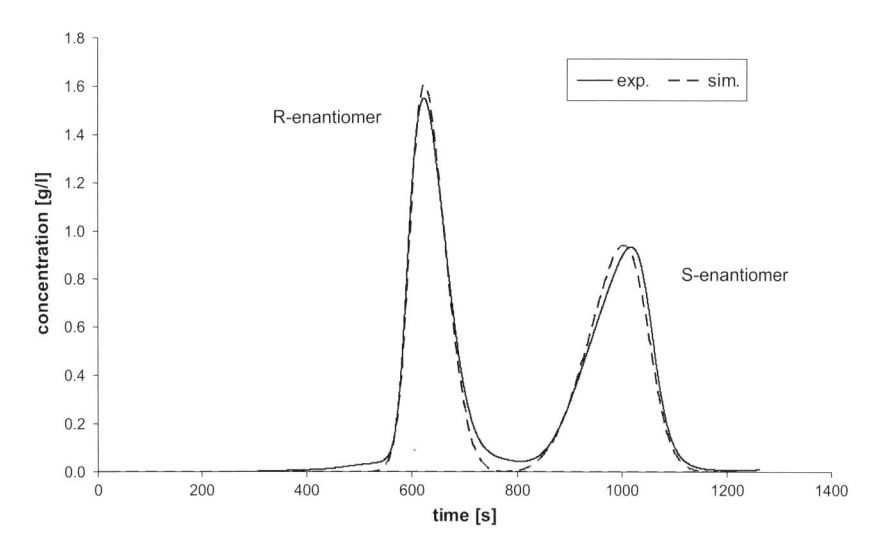

Figure 6.32 Measured and simulated pulse experiment for the Tröger's base racemate ($\dot{V} = 1\,\text{ml}\,\text{min}^{-1}$, $V_{inj} = 1\,\text{ml}$, $c_{feed} = 2.2\,\text{g}\,\text{l}^{-1}$, $V_c = 7.9\,\text{ml}$; for additional data see Appendix A.2).

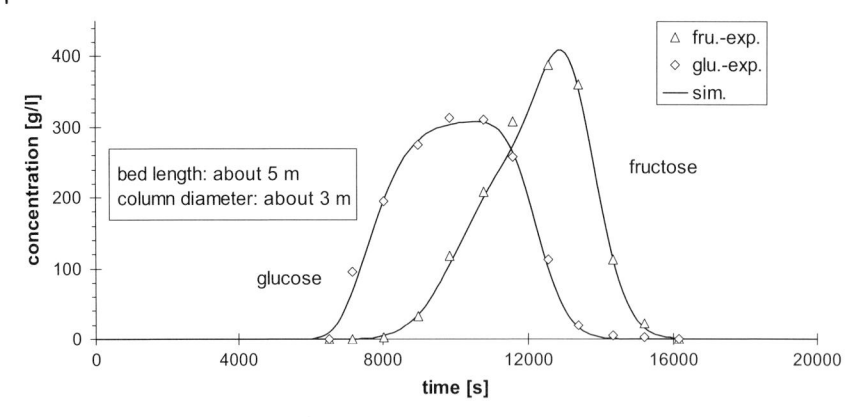

Figure 6.33 Measured and simulated pulse experiment for the glucose–fructose mixture on an industrial plant (see Appendix A.3).

is in good agreement with experimental data for an industrial-scale plant with a column diameter of about 3 m. In this case the extra column dead volumes were especially important to account for.

6.6.2
SMB Chromatography

6.6.2.1 Model Formulation and Parameters

In SMB chromatography a countercurrent movement of the liquid and solid phases is achieved by shifting in a cascade of connected columns the positions of the inlet and outlet streams in the direction of the fluid by valve switching (Section 5.2.4; Figure 6.34). The process reaches after a certain number of shifts a cyclic steady state. Assuming identical columns, the special profiles are identical after each shifting time, just shifted by one column length. The complex dynamics, cyclic operation, and numerous influence parameters make a purely empirical design of SMB process difficult or even impossible. Since the introduction of this technology, models with different levels of details have been used to obtain the operating parameters (Ruthven and Ching, 1989; Barker and Ganetsos, 1993; Storti et al., 1993; Chu and Hashimoto, 1995; Strube, 1996; Zhong and Guiochon, 1996; Dünnebier, 2000).

In case of linear isotherms there are analytical solutions available (Ruthven and Ching, 1989; Storti et al., 1993; Zhong and Guiochon, 1996). More complex isotherm equations require the application of numerical methods for the single-column processes (Strube, 1996; Pais, Loureiro, and Rodrigues, 1998; Migliorini, Mazzotti, and Morbidelli, 2000; Dünnebier, 2000). The SMB simulation approach takes into account the shifting of the inlet and outlet streams just as in the real plant and the process reaches a cyclic steady state (Hashimoto et al., 1983; Storti, Masi, and Morbidelli, 1988; Strube, 1996; Zhong and Guiochon, 1996; Pais, Loureiro, and Rodrigues, 1998; Dünnebier, 2000).

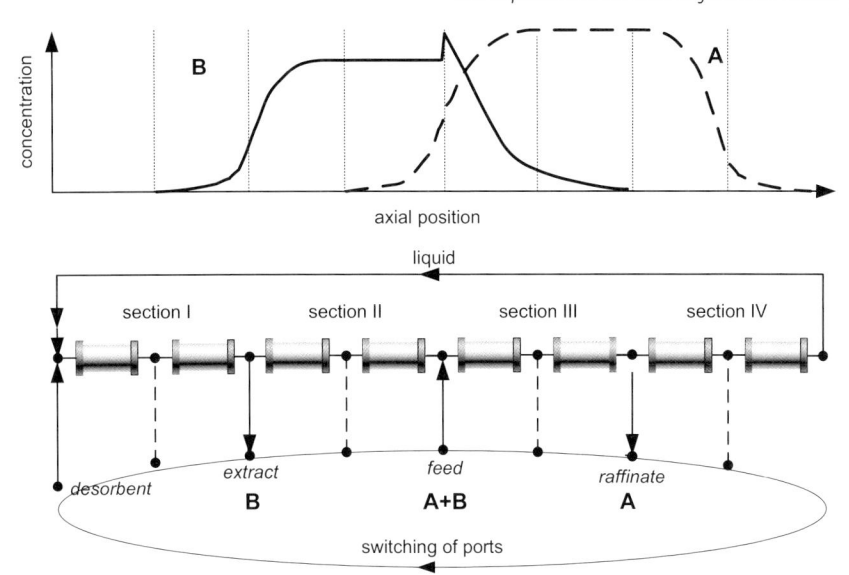

Figure 6.34 Simplified axial concentration profile and flow sheet for an SMB process (standard configuration).

An SMB plant consists, for instance, of eight chromatographic columns connected by pipes (Figure 6.35). The piping also includes valves for attaching external streams as well as measurement devices or pumps.

The overall model of an SMB process is developed by linking the models of individual chromatographic columns (Section 6.2). As with the chromatographic batch process, the plant setup of Figure 6.35 is converted into a simulation flow sheet. Figure 6.36 shows the SMB column model.

Mathematically, a SMB model can be set up by connecting the boundary conditions of each column model, including nodes represented by material balances of splitting or mixing models. These so-called node models (Ruthven and Ching, 1989) are given for a component i in the sections I–IV by

- Desorbent node:

$$\dot{V}_{des} = \dot{V}_{I} - \dot{V}_{IV}$$
$$c_{des,i} \cdot \dot{V}_{des} = c_{in,I,i} \cdot \dot{V}_{I} - c_{out,IV,i} \cdot \dot{V}_{IV}$$
(6.200)

- Extract draw-off node:

$$\dot{V}_{ext} = \dot{V}_{I} - \dot{V}_{II}$$
$$c_{ext,i} = c_{out,I,i} = c_{in,II,i}$$
(6.201)

- Feed node:

$$\dot{V}_{feed} = \dot{V}_{III} - \dot{V}_{II}$$
$$c_{feed,i} \cdot \dot{V}_{feed} = c_{in,III,i} \cdot \dot{V}_{III} - c_{out,II,i} \cdot \dot{V}_{II}$$
(6.202)

6 Modeling and Model Parameters

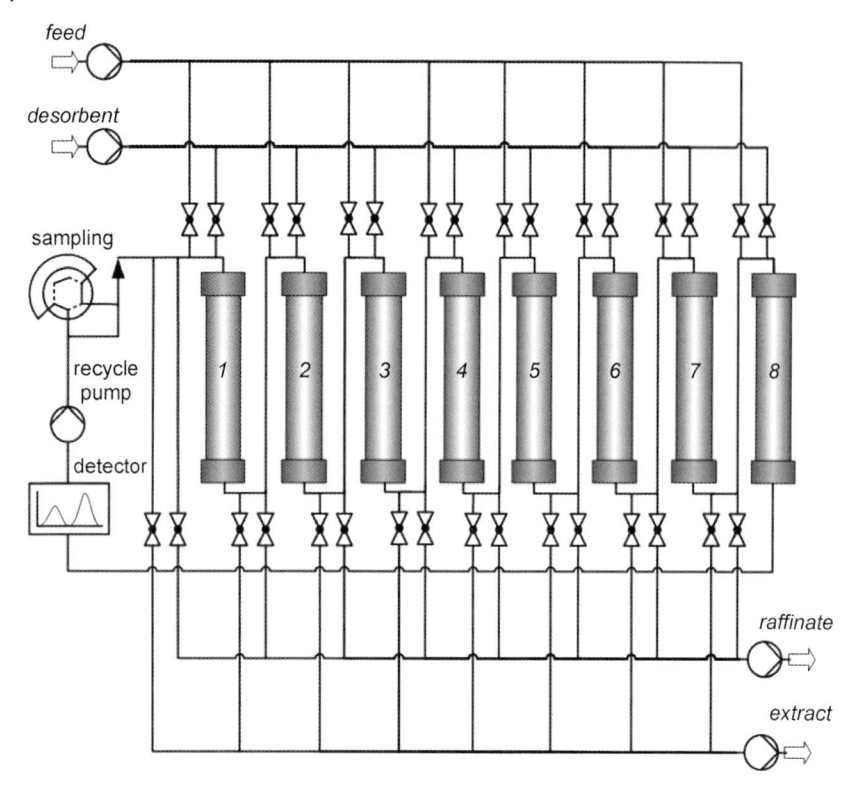

Figure 6.35 Principle setup of an SMB plant including detector systems in the recycle stream (eight-column standard configuration).

- Raffinate draw-off node:

$$\dot{V}_{raf} = \dot{V}_{III} - \dot{V}_{IV}$$
$$c_{raf,i} = c_{out,III,i} = c_{in,IV,i}$$
(6.203)

Concentrations c_{in} are the inlet boundary conditions (Equation 6.91 or 6.92) of the columns at the beginning of each section while c_{out} are the outlet

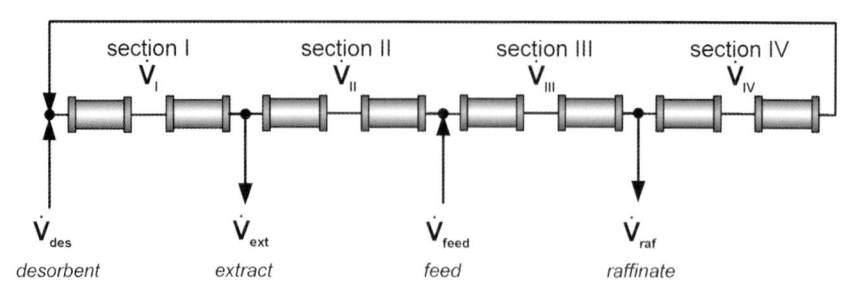

Figure 6.36 Simulation flow sheet of the SMB process ("SMB column model").

concentrations calculated at the end of each section. Intermediate node balances consist of setting equal volume flows and assigning the outlet concentration to the inlet concentration of the subsequent column. Since SMB is a periodic process, the boundary conditions for every individual column are changed after a switching period t_{shift}. If SMB modifications such as Varicol, Modicon, and so on (Chapter 5), are used, the boundary conditions have to be modified accordingly.

Further extensions of the SMB model are necessary to account for the fluid dynamic effects of piping and other peripheral equipment such as measurement devices (Figure 6.35), especially for plants with large recycle lines (Jupke, 2004). This is achieved by adding stirred tanks and pipe models (Section 6.3.2) to the simulation flow sheet, resulting in the extended SMB model given in Figure 6.37. If the distribution of the dead volumes in the process is uneven, an asynchronous shifting (Chapter 5) of the inlet and outlet ports is required (Hotier and Nicoud, 1995; Migliorini, Mazzotti, and Morbidelli, 1999). Sources for the dead volume between the columns are the connecting pipes as well as the switch valves. Dead volume in the recycle stream additionally includes the recycle pump and the measurement systems.

If the number of columns is increased, the process characteristics gradually approach to the characteristics of a hypothetical process, in which the solid and liquid phases are moving continuously in countercurrent directions. This hypothetical process is designated as a true moving bed (TMB) process. It reaches a real steady state that can be described mathematically much easily compared to the approach described above based on exploiting a dynamic model (Liapis and Rippin, 1979; Ruthven and Ching, 1989; Barker and Ganetsos, 1993).

Because of the strong analogy between simulated and true countercurrent flows, TMB models are frequently used to design SMB processes, more specifically to find suitable operating parameters (inlet and outlet flow rates and shifting times). The simpler TMB process can be elegantly analyzed with the equilibrium theory using the method of characteristics mentioned in Section 6.2.3. For applications of the equilibrium theory see Sections 7.4.1 and 7.6.

As an example of how the basic equations of a TMB model can be formulated, the transport-dispersive model for batch columns can be modified by adding an

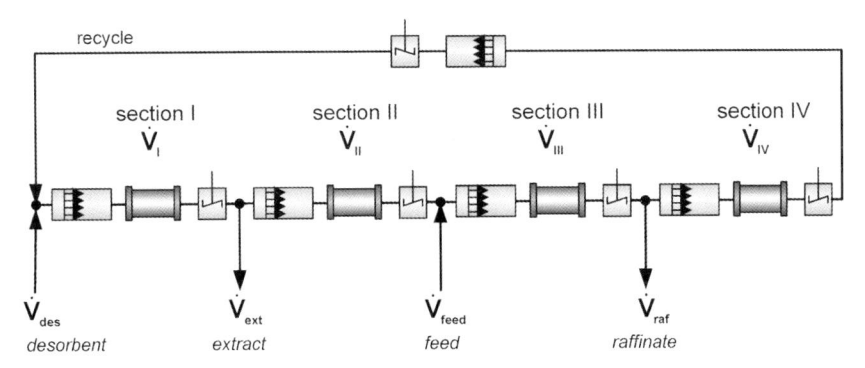

Figure 6.37 Simulation flow sheet for the "extended SMB model."

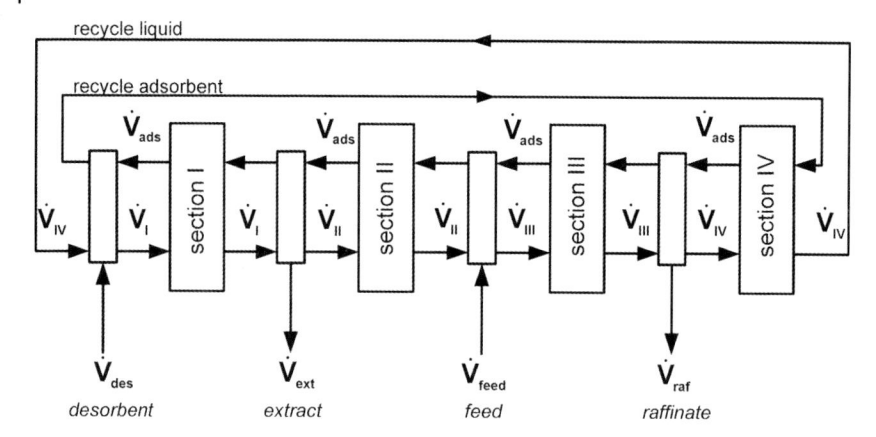

Figure 6.38 Node model for the TMB process.

adsorbent volume flow in the opposite direction $\dot{V}_{ads}$ (Figure 6.38), which results in a convection term in the solid phase mass balance with the velocity u_{ads}.

The dynamic mass balances for component i and section j in the liquid phase are

$$\frac{\partial c_{j,i}}{\partial t} = -u_{int,j,TMB} \cdot \frac{\partial c_{j,i}}{\partial x} + D_{ax,j} \cdot \frac{\partial^2 c_{j,i}}{\partial x^2} - \frac{1-\varepsilon}{\varepsilon} \cdot k_{eff\,j,i} \cdot \frac{3}{r_p} \cdot (c_{j,i} - c_{p,j,i}) \quad (6.204)$$

and in the adsorbent phase are

$$\varepsilon_p \frac{\partial c_{p,j,i}}{\partial t} + (1-\varepsilon_p) \frac{\partial q_{j,i}}{\partial t} = +u_{ads} \cdot \left(\varepsilon_p \frac{\partial c_{p,j,i}}{\partial x} + (1-\varepsilon_p) \frac{\partial q_{j,i}}{\partial x}\right) + k_{eff\,j,i} \cdot \frac{3}{r_p} \cdot (c_{j,ij} - c_{p,j,i}) \quad (6.205)$$

For process evaluation, just the stationary form of Equations 6.204 and 6.205 is needed, in which the accumulation terms are zero. To transfer the results of TMB to SMB, the interstitial velocity in the SMB process must be equal to the relative velocity of fluid and adsorbent in the TMB process:

$$u_{int,SMB} = u_{int,TMB} + u_{ads} \quad (6.206)$$

u_{ads} is positive and the direction inverse to u_{int} is specified in Equation 6.205 by a positive sign in front of the convection term.

Liquid phase velocities are related to the volume flows in each section while the adsorbent movement in the case of SMB is equal to the column volume moved per shifting time:

$$u_{int,j,SMB} = \frac{\dot{V}_j}{\varepsilon \cdot A_c} \quad (6.207)$$

$$u_{ads} = \frac{\dot{V}_{ads}}{A_c \cdot (1-\varepsilon)} = \frac{V_c}{A_c \cdot t_{shift}} = \frac{L_c}{t_{shift}} \quad (6.208)$$

Below, a short illustration is given to illustrate the similarity between predictions generated by TMB and SMB models.

Figure 6.39 gives an example of the difference in axial concentration profiles between TMB and SMB models, where the number of columns per SMB section is varied while the overall bed length is kept constant. The operating parameters are taken from an optimized TMB process with 99.9% purity of the product streams. Clearly, the end-cycle SMB profiles approach the TMB profile only for a high number of columns.

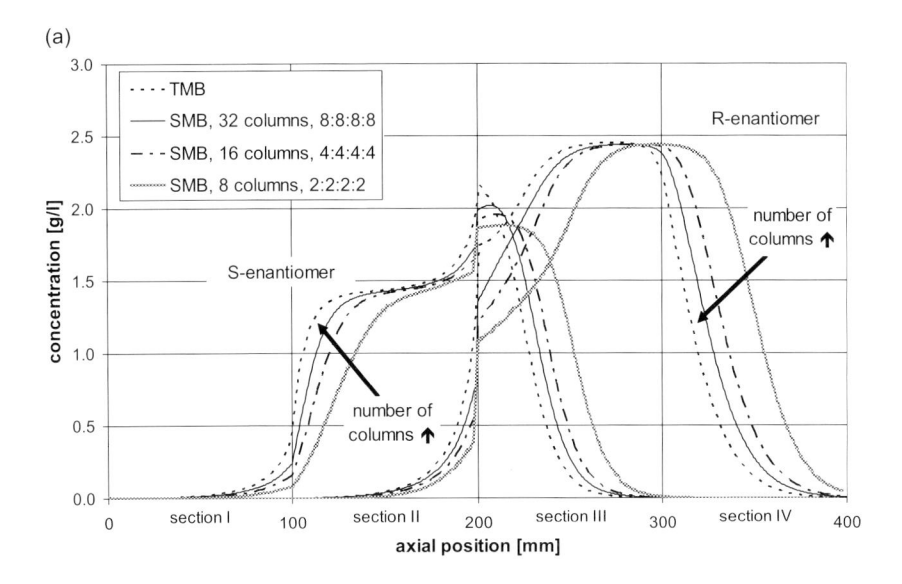

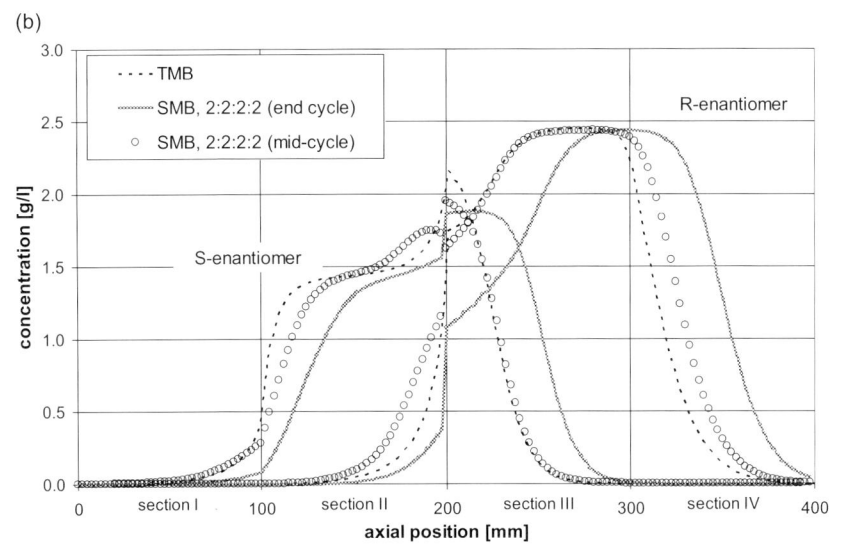

Figure 6.39 Axial concentration profile for a) TMB and SMB processes with different number of columns per section (end-cycle profiles); b) TMB and SMB processes with 8 columns and profiles at end-cycle and mid-cycle (separation of EMD53986, Appendix A.1, equal operating parameters).

Quite similar concentration profiles generated with the two models were reported frequently for three or more columns per section by Ruthven et al. (1989), Lu and Ching (1997), and Pais, Loureiro, and Rodrigues (1998). However, due to high investment costs SMB plants often have fewer columns. Then the real SMB plant behavior starts to differ considerably from the TMB process (Chapter 7; Chu and Hashim, 1995; Strube, Schmidt-Traub, and Schulte, 1998; Pais, Loureiro, and Rodrigues, 1998). Nevertheless, initial process design can be efficiently performed based on true moving bed assumptions (Storti et al., 1993; Charton and Nicoud, 1995; Ma and Wang, 1997). This will be outlined further in Chapter 7.

To simulate an SMB process, the model parameters of the individual columns (Section 6.5) and, if necessary, the dead volumes (Figure 6.37) must be known.

It is advisable to use nearly identical columns (concerning bed length and packing structure), which is easily checked by comparing their outgoing signals from batch experiments with small injected amounts and determining the Henry coefficient. Comparison of the product peaks for the individual columns may also be used as a test for the packing. If strong deviations occur, the packing procedure must be repeated and checked for errors.

Model parameters should then be obtained only for one column, as these should be the same for all. Column parameters are determined by batch experiments or are known from previous tests. Finally, the dead volume inside an SMB plant (Figure 6.37) can be determined from tracer pulse experiments by connecting the respective part of the plant directly to the pump and a detector.

For the operating parameters, it is necessary to specify five independent variables. A common method is to specify the four internal flow rates ($\dot{V}_{I-IV}$) and the switching time (SMB model) or solid flow (TMB model). Note that the four external flow rates have to fulfill the overall mass balance and only three flow rates are independent.

6.6.2.2 Experimental Validation of SMB Models

6.6.2.2.1 Equipment and Measurements
Ever since the development and application of mathematical models for the design of SMB processes, beginning in the 1980s, efforts have been made to validate these models by comparing measured and simulated data. SMB and TMB models of different complexity have been used for this task, for example, the ideal and equilibrium-dispersive SMB model as well as TMB and SMB transport-dispersive models. A common approach is to compare the internal concentration profile with the simulation results. Another way is to use the product concentrations at extract and raffinate and their temporal evolution. Some characteristic points of the internal profiles can be obtained from taking samples (see sample valve in Figure 6.35) and off-line analysis. Classical examples include those from applications in the petrochemical and sugar industries (Ruthven and Ching, 1989; Hashimoto, Adachi, and Shirai, 1993; Ching et al., 1993; Ma and Wang, 1997; Deckert and Arlt, 1997; Beste et al., 2000). Example applications in enantioseparation are

given by Pais, Loureiro, and Rodrigues [1997a, 1997b], Heuer et al. (1998), Kniep, Blümel, and Seidel-Morgenstern (1998), Kniep et al. (1999) and comprehensively by Rajendran et al. (2009). One disadvantage of taking samples is the normally low data density. Often, only one sample per shifting interval is taken to reduce experimental effort and to limit the impact of withdrawn sample on the internal concentration profile. As the shifting interval may be up to several thousand seconds, a detailed and meaningful model validation is difficult.

Chromatographic separations with product purities exceeding 99% and high separation costs require precise prediction of the optimal process design. This demands carefully validated models, especially in the separation sections of the SMB plant. Consequently, methods are needed to increase the number of measured data points.

This can be achieved by online analysis of the concentration profile. As depicted in Figure 6.35, one or more detectors are placed in the recycle stream. They are positioned in front of the recycle pump, because some detectors are sensitive to high pressure. Due to the shifting of external streams the detector "travels" through every process section during one cycle (see below).

Yun, Zhong, and Guiochon (1997a, 1997b) placed one UV detector in the recycle stream to measure the fronts in sections I and IV for the separation of phenylethanol and phenylpropanol. The use of only one detector allows the measurement of the pure components in each regeneration section, but the concentration of the mixture in the separation section cannot be determined. Jupke et al. used a multi-detector setup for binary separation, which gave the possibility of measuring the concentration profiles of all components in all sections individually (Jupke, 2004; Epping, 2005; Mannschreck, 1992; Jupke, Epping, and Schmidt-Traub, 2002; Mihlbachler et al., 2002).

The general relation between the data obtained in the recycle stream, which can be used for model validation, and the axial profile, which can be used for process analysis, is given in Figure 6.40. The setup is identical to Figure 6.35. In Figure 6.40a, the simulated temporal profile measured behind column 8 during one complete cycle is shown. The concentration values after each shifting interval are given as symbols. In the ideal case they are identical to those of the samples taken for off-line analysis. As the process is in cyclic steady state, the axial profiles after each switching time are identical; only their position is shifted by one column. Thus, the symbols marked above can be translated into points of the axial profile at the end of one shifting period (Figure 6.40b). Simulated curves are added for comparison. Although it is recommended to use the end-of-interval axial profile for process characterization (Chapter 7), the situation at mid-interval may be determined when taking the samples at these specific times. If no asymmetrical distribution of dead volumes is present in the plant, the end-of-shifting-interval profiles are identical to those at the end of a cycle.

The axial position in Figure 6.40b is normalized to the length of each column and, therefore, column 1 lies between the coordinates 0 and 1. The temporal positions 0, 1, 2, . . . , 8 correspond to the axial positions 8, 7, . . . , 0. The feed is injected in front of column 5 at the start of the cycle and the last feed position at

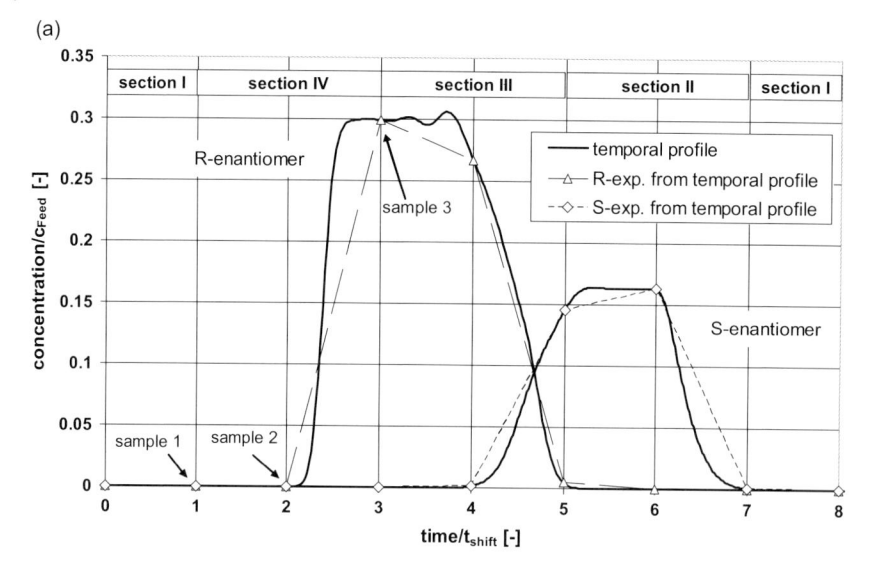

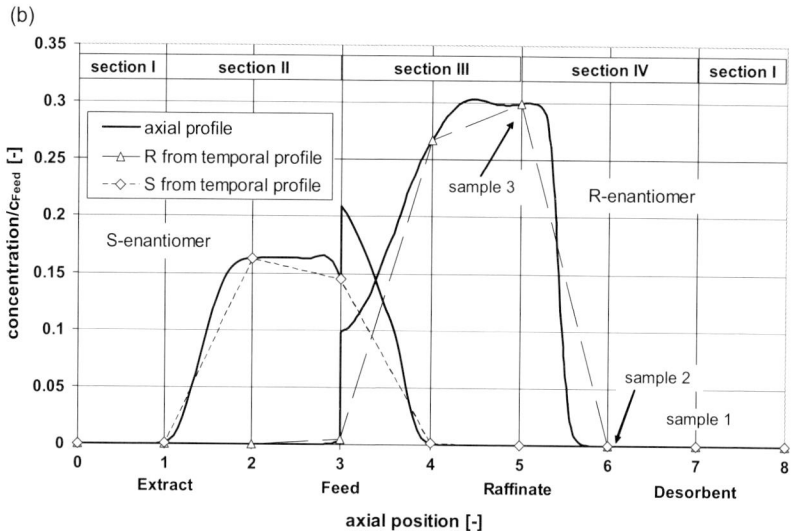

Figure 6.40 Relationship between the temporal profile measured behind the eighth column (a) and the axial profile at the end of cycle just before the feed position is switched in front of the fifth column (b); symbols represent the identical points.

the end is in front of column 4 (Figure 6.40). For clarity, no graphical shift is performed to have the feed position in front of column 5 as, for example, in Figure 6.34.

Importantly, sampling gives only a limited number of points in the axial profile. The complete curve cannot be reconstructed from any temporal measurement, as the concentration fronts inside the columns change from the beginning to the end of the shifting interval. However, increasing the number of experimental samples per interval, of course, allows a comparison of the simulated temporal profile on a broader base without online measurement.

6.6.2.2.2 Isocratic SMB In the following some examples based on our own research are given for the validation of the SMB transport-dispersive model, using an online detection system in the recycle stream. All flow sheet, column, and plant models were implemented in the gPROMS (PS Enterprise, UK) simulation tool and solved with OCFE methods (Section 6.4).

A more detailed discussion of the experiments and simulations for various operating points and conditions can be found in Jupke (2004) and Mihlbachler *et al.* (2002). Example results for the systems EMD53986, Tröger's base, and fructose–glucose are discussed. Pulse tests with the solutes proved that all columns for each SMB setup behaved identically within small deviations. Most of the experiments were performed on a commercial plant "Licosep Lab" (Figure 6.41) from Novasep (France).

In all cases good agreement between simulation and experiment was found.

Simulations were performed considering the extended SMB configuration (Figure 6.37), which included dead volumes and synchronous as well as asynchronous port switching. The latter accounts for the dead volume in the recycle stream, which is the dominant contribution (Hotier and Nicoud, 1995; Migliorini, Mazzotti, and Morbidelli, 1999). As mentioned in the previous section, the temporal concentration profiles are shown and the partitions of the time axis mark the switching time. The additional time added after the eighth period is the asynchronous switching time.

Figure 6.42 shows the simulated and measured concentration profiles for EMD53986. Good agreement can be observed for this system, which is characterized by its strong coupled and nonlinear adsorption behavior. Only slight deviation in the position of the fronts and the maximum height is observed. Especially, the

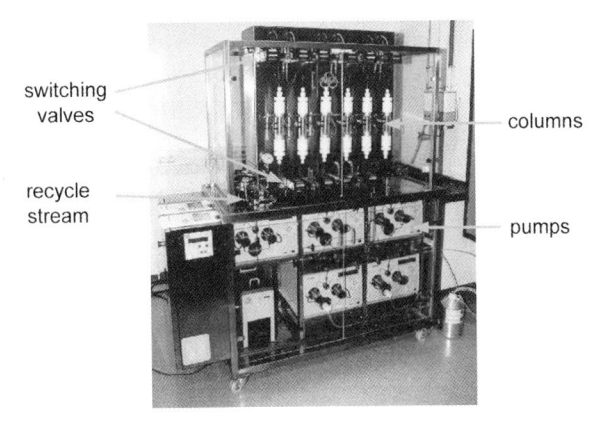

Figure 6.41 Photograph of an SMB plant "Licosep Lab" from Novasep.

414 | *6 Modeling and Model Parameters*

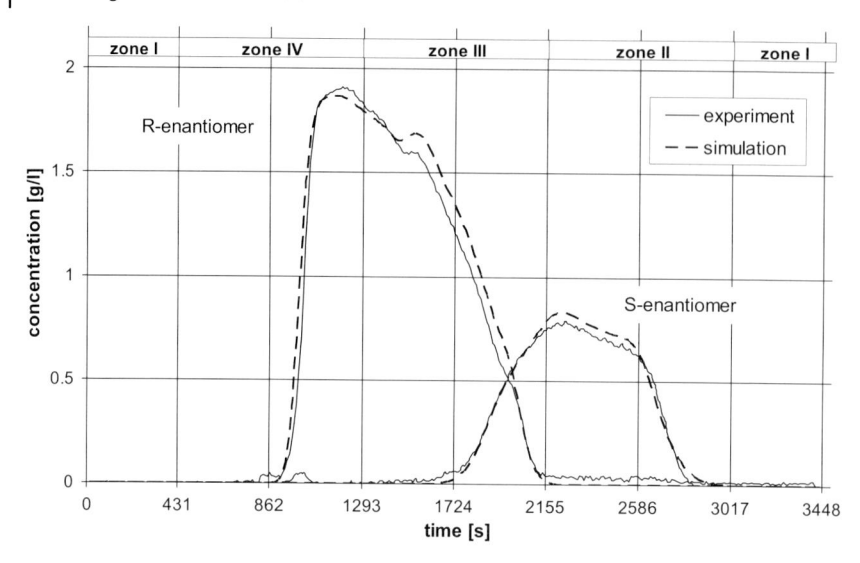

Figure 6.42 Measured and simulated concentration profiles in the SMB for EMD53986 (cycle 8, $c_{feed} = 5\,g\,l^{-1}$; for additional data see Appendix A.1; reproduced from Jupke, Epping, and Schmidt-Traub, 2002).

steepness of the fronts is reproduced very well. The model can also predict the start-up of the SMB plant (Figure 6.43) correctly. Deviations in the first cycle are presumably due to the pressure fluctuations often encountered directly after start-up. Good agreement between simulation and experiment was observed for other operating conditions, too.

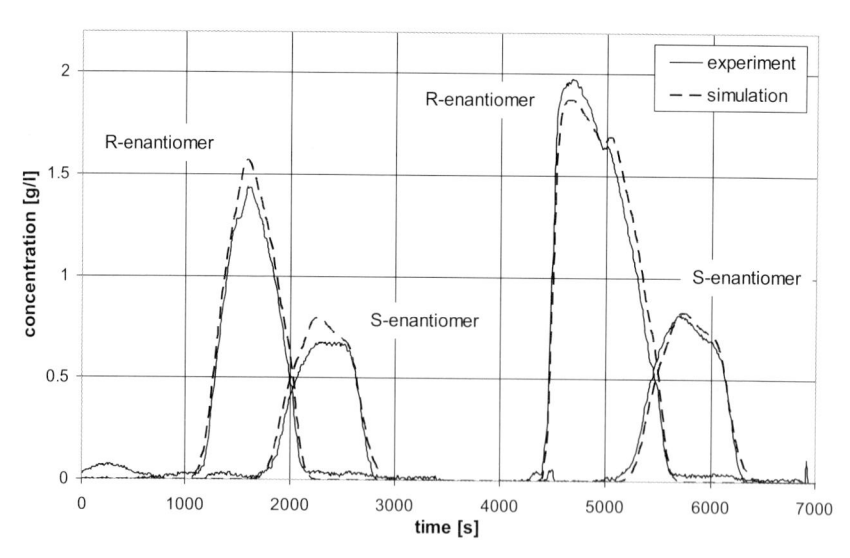

Figure 6.43 Measured and simulated concentration profiles in the SMB for EMD53986 during start-up (one and two cycles; for other data see Figure 6.39).

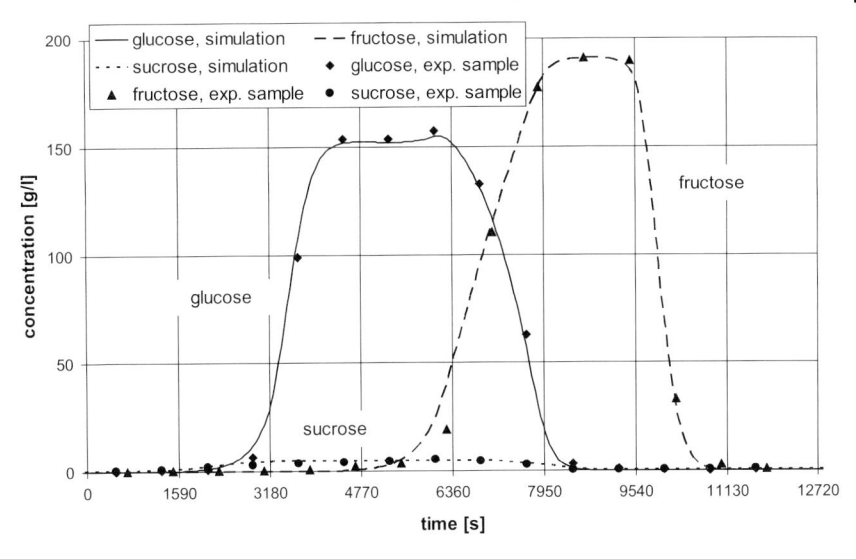

Figure 6.44 Simulated and measured (sampled) concentration profiles in the SMB for fructose–glucose ($c_{feed} = 300$ g l^{-1}) and sucrose ($c_{feed} = 18$ g l^{-1}) (eighth cycle; for additional data see Appendix A.3).

The separation of an industrial feed mixture of fructose–glucose ($c_{feed} = 300$ g l^{-1}) containing 6% sucrose as an impurity is displayed in Figure 6.44. Once again, good agreement between experiment and simulation is found. The individual concentration profiles of this three-component mixture cannot be determined by two-detection systems. Therefore, additional samples (two per shifting interval) were taken and analyzed by HPLC.

Isotherm data for sucrose in this concentration range were determined from pulse experiments and represented by a linear isotherm. Because of the low adsorptivity, this impurity is collected at the raffinate together with the glucose.

The examples for experimental validation of the SMB model are based on the "extended model" (Figure 6.37) that takes into account the fluid dynamic effect of piping, especially recycle lines and other peripheral equipment such as measurement devices. From point of process simulation these are additional elements of the plant that have to be regarded within the overall flow sheet.

6.6.2.2.3 Gradient SMB

Gradients are known to be attractive also in preparative chromatography (Seidel-Morgenstern, 2005). Standard SMB processes have two inlet streams. Therefore, there is an interesting option to improve their performance by using step gradients as described in Section 5.2.6.

The gradient concept can be applied also in SMB chromatography. This approach was first suggested for separations using supercritical fluids (Clavier, Nicoud, and Perrut, 1996). It is capable of increasing productivity and product concentrations, as well as of reducing eluent consumption. Theoretical and experimental studies reveal that it is particularly attractive to apply a solvent as desorbent that has stronger elution strength than the solvent containing the feed. An increasingly important application of gradient SMB chromatography is the separation of proteins by ion exchange or hydrophobic interaction chromatography using salt gradients. Examples of successful application of gradient SMB were given by, for example, Jensen et al. (2000), Antos and Seidel-Morgenstern (2001), Houwing et al. (2002), Wekenborg, Susanto, and Schmidt-Traub (2005), Keßler et al. (2007), and Gueorguieva et al. (2011). Contributions to the analysis and design were made, for example, by Abel, Mazzotti, and Morbidelli (2002, 2004), Antos and Seidel-Morgenstern (2002), and Beltscheva, Hugo, and Seidel-Morgenstern (2003).

Wekenborg (2009) realized the elution gradient by a stepwise change of the salt concentration and investigated the separation of the proteins β-lactoglobulin A and B. Corresponding to the strongly adsorbed products the elution strength in zones 1 and 2 is increased by a higher salt concentration than in zones 3 and 4. The model parameters are documented in Appendix A.5. Process simulation was performed using the extended SMB model based on the transport-dispersive model for the columns (Section 6.2.5.1) and the node balances at the inlet and outlet ports, respectively. The only difference between isocratic and gradient SMB concerning modeling is given by the fact that salt as an additional component has to be taken into account. The isotherms for β-lactoglobulin A and B as well as salt are described by SAM isotherms (Section 2.5.2.4); their parameters are given in Appendix B.5. The salt concentration $c_{S,I,II}$ in zones I and II results from mixing of the recycle stream and the desorbent that enters the process with a given salt concentration. At the feed node this concentration is diluted into the lower concentration $c_{S,III,IV}$ in order to reduce the eluent strength in zones III and IV. The balance for this node is given by Equation 6.209:

$$[m_{II}(1-\varepsilon_p)+\varepsilon_p](c_{S,II}-c_{S,\text{feed}}) = [m_{III}(1-\varepsilon_p)+\varepsilon_p](c_{S,III}-c_{S,\text{feed}}) \quad (6.209)$$

The process model regards the salt as a component that does not take part in adsorption.

The comparison of experimental and simulated data for the separation of β-lactoglobulin A and B is shown in Figure 6.45. During the experiments the salt concentration (Figure 6.45a) and the total protein concentration (Figure 6.45b) were measured continuously. The concentrations of β-lactoglobulin A and B were analyzed from samples that were taken periodically (Figure 6.45c).

All in all, the simulated data agree very well with the experimental data so that the extended SMB model is approved also for further theoretical investigation and optimization of gradient SMB processes.

The concentration profiles of the salt deviate considerably from the intended step gradient. These fluctuations correlate directly with the concentration of the proteins. The reason is a "carryover" at the beginning of each switching period. The column that is switched from zone 1 to 4 contains a higher salt concentration

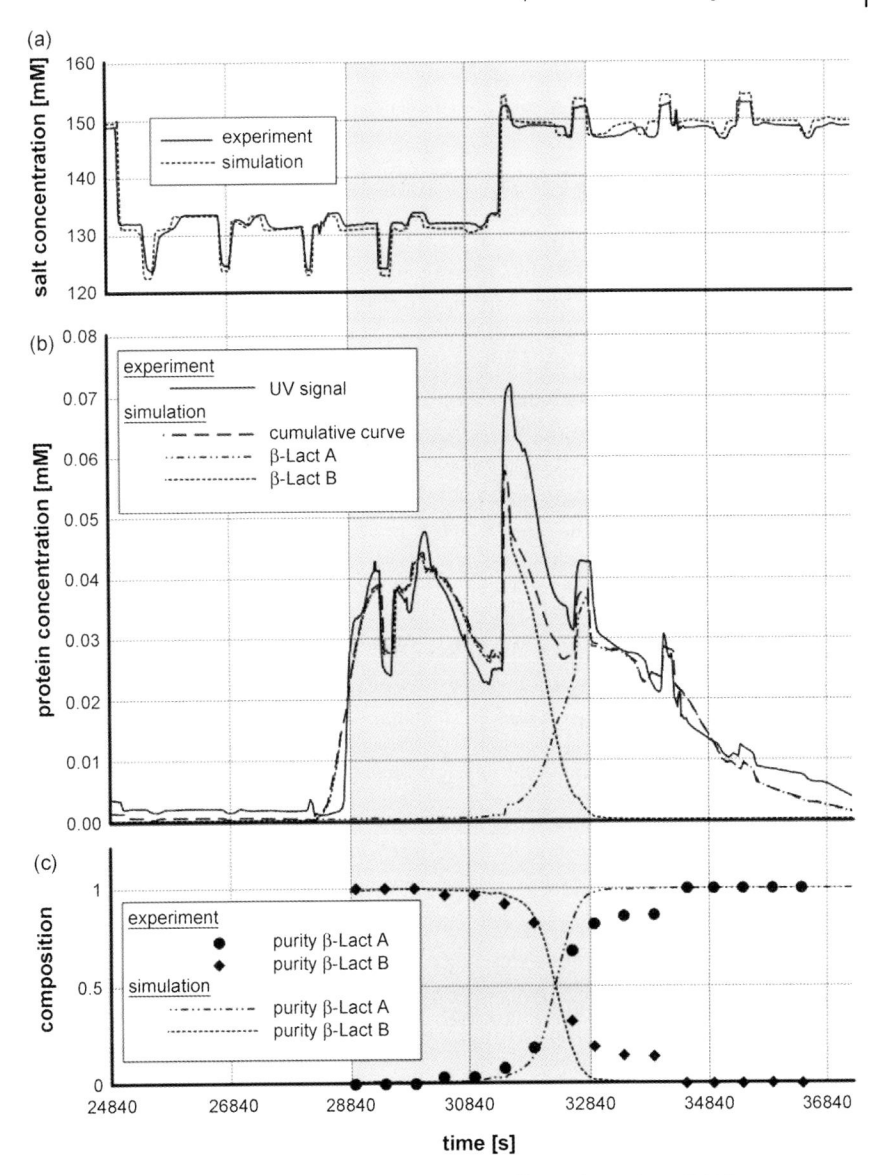

Figure 6.45 Measured and simulated concentration profiles in the gradient SBM for separation of β-lactoglobulin A and B (reproduced from Wekenborg, 2009).

compared to the theoretical profile of the step gradient and causes a concentration peak. Vice versa, the concentration in the column that enters zone 2 is too low and induces a steep decline of the concentration. During the switching period these fluctuations are compensated and start again at the next switch.

References

Abel, S., Mazzotti, M., and Morbidelli, M. (2002) Solvent gradient operation of simulated moving beds. I. Linear isotherms. *J. Chromatogr. A*, **944**, 23–39.

Abel, S., Mazzotti, M., and Morbidelli, M. (2004) Solvent gradient operation of simulated moving beds. II. Langmuir isotherms. *J. Chromatogr. A*, **1026**, 47–55.

Altenhöner, U., Meurer, M., Strube, J., and Schmidt-Traub, H. (1997) Parameter estimation for the simulation of liquid chromatography. *J. Chromatogr. A*, **769**, 59–69.

Antos, D., Kaczmarski, K., Wojciech, P., and Seidel-Morgenstern, A. (2003) Concentration dependence of lumped mass transfer coefficients – linear versus non-linear chromatography and isocratic versus gradient operation. *J. Chromatogr. A*, **1006**, 61–76.

Antos, D. and Seidel-Morgenstern, A. (2001) Application of gradients in simulated moving bed processes. *Chem. Eng. Sci.*, **56**, 6667–6682.

Antos, D. and Seidel-Morgenstern, A. (2002) Two-step solvent gradients in simulated moving bed chromatography: numerical study for linear equilibria. *J. Chromatogr. A*, **944**, 77–91.

Arnell, R. and Fornstedt, T. (206) Validation of the tracer-pulse method for multicomponent liquid chromatography, a classical paradox revisited. *Anal. Chem.*, **78**, 4615–4623.

Ashley, J.W. and Reilley, C.N. (1965) De-tailing and sharpening of response peaks in gas chromatography. *Anal. Chem.*, **37** (6), 626–630.

Baerns, M., Hofmann, H., and Renken, A. (1999) *Chemische Reaktionstechnik in Lehrbuch der Technischen Chemie Band 1*, Georg Thieme Verlag, Stuttgart.

Barker, P.E. and Ganetsos, G. (1993) *Preparative and Production Scale Chromatography*, Chromatographic Science Series 61, Marcel Dekker Inc., New York.

Barns, J.W. (1994) *Statistical Analysis for Engineers and Scientists*, McGraw-Hill.

Bellot, J.C. and Condoret, J.S. (1991) Liquid chromatography modelling: a review. *Process Biochem.*, **26**, 363–376.

Beltscheva, D., Hugo, P., and Seidel-Morgenstern, A. (2003) Linear two-step gradient counter-current chromatography: analysis based on a recursive solution of an equilibrium stage model. *J. Chromatogr. A*, **989**, 31–45.

Berninger, J.A., Whitley, R.D., Zhang, X., and Wang, N.-H. (1991) Versatile model for simulation of reaction and nonequilibrium dynamics in multicomponent fixed-bed adsorption processes. *Comput. Chem. Eng.*, **15** (11), 749–768.

Beste, Y.A., Lisso, M., Wozny, G., and Arlt, W. (2000) Optimization of simulated moving bed plants with low efficient stationary phases: separation of fructose and glucose. *J. Chromatogr. A*, **868**, 169–188.

Blümel, C., Hugo, P., and Seidel-Morgenstern, A. (1999) Quantification of single solute and competitive adsorption isotherms using a closed-loop perturbation method. *J. Chromatogr. A*, **865**, 51–71.

Borren, T. (2007) Untersuchungen zu chromatographischen Reaktoren mit verteilten Funktionalitäten, Fortschritts-Berichte VDI: Reihe 3 Nr. 876, VDI Verlag GmbH, Düsseldorf.

Charton, F. and Nicoud, R.-M. (1995) Complete design of a simulated moving bed. *J. Chromatogr. A*, **702**, 97–112.

Ching, C.B., Chu, K.H., Hidajat, K., and Ruthven, D.M. (1993) Experimental study of a simulated counter-current adsorption system – vii effects of non-linear and interacting isotherms. *Chem. Eng. Sci.*, **48** (7), 1343–1351.

Chu, K. and Hashim, M. (1995) Simulated countercurrent absorption processes: a comparison of modelling strategies. *Chem. Eng. J.*, **56**, 59–65.

Chung, S.F. and Wen, C.Y. (1968) Longitudinal dispersion of liquid flowing through fixed and fluidised beds. *AIChE J.*, **14** (6), 857–866.

Clavier, J.-Y., Nicoud, R.-M., and Perrut, M. (1996) A new efficient fractionation process: the simulated moving bed with supercritical eluent, in *High Pressure Chemical Engineering* (eds Ph. Rudolf von Rohr and Ch. Trepp), Elsevier, London.

Cornel, J., Tarafder, A., Katsuo, S., and Mazzotti, M. (2010) The direct inverse method: a novel approach to estimate adsorption isotherm parameters. *J. Chromatogr. A*, **1217**, 1934–1941.

Craig, L.C. (1944) Identification of small amounts of organic compounds by distribution studies. II: separation by counter-current distribution. *J. Biol. Chem.*, **155**, 519–534.

Danckwerts, P.V. (1953) Continuous flow systems – distribution of residence times. *Chem. Eng. Sci.*, **2** (1), 1–13.

Davis, M.E. (1984) *Numerical Methods and Modeling for Chemical Engineers*, John Wiley & Sons, Ltd.

De Vault, D. (1943) The theory of chromatography. *J. Am. Chem. Soc.*, **65**, 532.

Deckert, P. and Arlt, W. (1997) Pilotanlage zur simulierten Gegenstromchromatographie-Ergebnisse für die Trennung von Fructose und Glucose. *Chem. Ing. Tech.*, **69**, 115–119.

Du Chateau, P. and Zachmann, D. (1989) *Applied Partial Differential Equations*, Harper & Row.

Dünnebier, G. (2000) Effektive Simulation und mathematische Optimierung chromatographischer Trennprozesse. Dissertation. Universität Dortmund, Shaker Verlag.

Epping, A. (2005) *Modellierung, Auslegung und Optimierung chromatographischer Batch-Trennung*, Shaker-Verlag, Aachen.

Ferziger, J.H. (1998) *Numerical Methods for Engineering Applications*, 2nd edn, Wiley-Interscience.

Finlayson, B. (1980) *Numerical Analysis in Chemical Engineering*, McGraw-Hill, New York.

Foley, J.P. and Dorsey, J.G. (1983) Equations for calculation of chromatographic figures of merit for ideal and skewed peaks. *Anal. Chem.*, **55**, 730–737.

Foley, J.P. and Dorsey, J.G. (1984) A review of the exponentially modified Gaussian (EMG) function: evaluation and subsequent calculation of universal data. *J. Chromatogr. Sci.*, **22**, 40–46.

Fricke, J. (2005) Entwicklung einer Auslegungsmethode für chromatographische SMB-Reaktoren, Fortschritts-Berichte VDI: Reihe 3 Nr. 844, VDI Verlag GmbH, Düsseldorf.

Furuya, F., Takeuchi, Y., and Noll, K.E. (1989) Intraparticle diffusion of phenols within bidispersed macrorectangular resin particles. *J. Chem. Eng. Jpn.*, **22** (6), 670.

Glueckauf, E. (1949) Theory of chromatography: VII. The general theory of two solutes following non-linear isotherms. *Discuss. Faraday Soc.*, **7**, 12.

Glueckauf, E. and Coates, J.I. (1947) Theory of chromatography, part 4: the influence of incomplete equilibrium on the front boundary of chromatogram and on the effectiveness of separation. *J. Chem. Soc.*, 1315–1321.

Golshan-Shirazi, S. and Guiochon, G. (1989) Experimental characterization of the elution profiles of high concentration chromatographic bands using the analytical solution of the ideal model. *Anal. Chem.*, **61**, 462–467.

Golshan-Shirazi, S. and Guiochon, G. (1992) Comparison of the various kinetic models of non-linear chromatography. *J. Chromatogr. A*, **603**, 1–11.

Gomes, P.S., Minceva, M., and Rodrigues, A.E. (2006) Simulated moving bed technology: old and new. *Adsorption*, **12**, 375–392.

Grushka, E.S., Snyder, L.R., and Knox, J.H. (1975) Advances in band spreading theories. *J. Chromatogr. Sci.*, **13**, 25–37.

Gu, T. (1995) *Mathematical Modeling and Scale-Up of Liquid Chromatography*, Springer-Verlag, New York.

Gu, T., Tsai, G.-J., and Tsao, G.T. (1990a) New approach to a general nonlinear multicomponent chromatography model. *AIChE J.*, **36** (5), 784–788.

Gu, T., Tsai, G.-J., and Tsao, G.T. (1991) Multicomponent adsorption and chromatography with uneven saturation capacities. *AIChE J.*, **37** (9), 1333–1340.

Gu, T., Tsai, G.-J., and Tsao, G.T. (1993) Modeling of nonlinear multicomponent chromatography, in *Advances in Biochemical Engineering/Biotechnology*, vol. **49** (ed. A. Fiechter), Springer-Verlag, New York, pp. 45–71.

Gu, T., Tsai, G.-J., Tsao, G.T., and Ladisch, M.R. (1990b) Displacement effect in multicomponent chromatography. *AIChE J.*, **36** (8), 1156–1162.

Gueorguieva, L., Palani, S., Rinas, U., Jayaraman, G., and Seidel-Morgenstern, A. (2011) Recombinant protein purification using gradient assisted simulated moving bed hydrophobic interaction chromatography. Part II: process design and experimental validation. *J. Chromatogr. A*, **1218**, 6402–6411.

Guiochon, G., Felinger, A., Shirazi, D.G., and Katti, A.M. (2006) *Fundamentals of Preparative and Nonlinear Chromatography*, Elsevier, Amsterdam Press.

Guiochon, G., Golshan-Shirazi, S., and Katti, A. (1994) *Fundamentals of Preparative and Nonlinear Chromatography*, Academic Press, Boston.

Guiochon, G. and Lin, B. (2003) *Modeling for Preparative Chromatography*, Elsevier, Amsterdam.

Hashimoto, K., Adachi, S., Noujima, H., and Maruyama, H. (1983) Models for separation of glucose–fructose mixtures using a simulated moving bed adsorber. *J. Chem. Eng. Jpn.*, **16** (4), 400–406.

Hashimoto, K., Adachi, S., and Shirai, Y. (1993) Operation and design of simulated moving-bed adsorbers, in *Preparative and Production Scale Chromatography* (eds G. Ganetsos and P.E. Barker), Marcel Dekker Inc., New York.

Helfferich, F.G. (1997) Non-linear waves in chromatography III. Multicomponent Langmuir and Langmuir-like systems. *J. Chromatogr. A*, **768**, 169–205.

Helfferich, F.G. and Carr, P.W. (1993) Non-linear waves in chromatography, I. Waves, shocks, and shapes. *J. Chromatogr.*, **629**, 97–122.

Helfferich, F.G. and Klein, G. (1970) Multicomponent chromatography, in *A Theory of Interferences*, Marcel Dekker, New York.

Helfferich, F.G. and Whitley, R.D. (1996) Non-linear waves in chromatography II. Wave interference and coherence in multicomponent systems. *J. Chromatogr. A*, **734**, 7–47.

Heuer, C., Küsters, E., Plattner, T., and Seidel-Morgenstern, A. (1998) Design of the simulated moving bed process based on adsorption isotherm measurements using a perturbation method. *J. Chromatogr. A*, **827**, 175–191.

Hill, T.L. (1960) *An Introduction to Statistical Thermodynamics*, Addison-Wesley, Reading, MA.

Hotier, G. and Nicoud, R.M. (1995) Separation by simulated moving bed chromatography with dead volume correction by desynchronization of periods. Europäisches Patent EP 688589 A1.

Houwing, J., van Hateren, S.H., Billiet, H.A.H., and van der Wielen, L.A.M. (2002) Effect of salt gradients on the separation of dilute mixtures of proteins by ion-exchange in simulated moving beds. *J. Chromatogr. A*, **952**, 85–98.

James, F., Sepúlveda, M., Charton, F., Quinones, I., and Guiochon, G. (1999) Determination of binary competitive equilibrium isotherms from the individual chromatographic band profiles. *Chem. Eng. Sci.*, **54**, 1677–1696.

Javeed, S., Qamar, S., Seidel-Morgenstern, A., and Warnecke, G. (2011a) Efficient and accurate numerical simulation of nonlinear chromatographic processes. *Comp. Chem. Eng.*, **35**, 2294–2305.

Javeed, S., Qamar, S., Seidel-Morgenstern, A., and Warnecke, G. (2011b) A discontinuous Galerkin method to solve chromatographic models. *J. Chromatogr. A*, **1218**, 7137–7146.

Jeansonne, M.S. and Foley, J.P. (1991) Review of the exponentially modified Gaussian chromatographic peak model since 1983. *J. Chromatogr. Sci.*, **29**, 258–266.

Jeansonne, M.S. and Foley, J.P. (1992) Improved equations for calculation of chromatographic figures of merit for ideal and skewed peaks. *J. Chromatogr.*, **594**, 1–8.

Jensen, T.B., Reijns, T.G.P., Billiet, H.A. H., and van der Wielen, L.A.M. (2000) Novel simulated moving-bed method for reduced solvent consumption. *J. Chromatogr. A*, **873**, 149–162.

Jupke, A. (2004) Experimentelle Modellvalidierung und modellbasierte Auslegung von Simulated Moving Bed (SMB) Chromatographieverfahren, Fortschrittbericht VDI: Reihe 3 Nr. 807, VDI Verlag GmbH, Düsseldorf.

Jupke, A., Epping, A., and Schmidt-Traub, H. (2002) Optimal design of batch and SMB chromatographic separation processes. *J. Chromatogr. A*, **944**, 93–117.

Kaczmarski, K. and Antos, D. (1996) Modified Rouchon and Rouchon-like algorithms for solving different models of multicomponent preparative chromatography. *J. Chromatogr. A*, **756**, 73–87.

Kaczmarski, K., Mazzotti, M., Storti, G., and Morbidelli, M. (1997) Modeling fixed-bed adsorption columns through orthogonal collocation on moving finite elements. *Comp. Chem. Eng.*, **21**, 641–660.

Keßler, L.C., Gueorguieva, L., Rinas, U., and Seidel-Morgenstern, A. (2007) Step gradients in 3-zone simulated moving bed chromatography: application to the purification of antibodies and bone morphogenetic protein-2. *J. Chromatogr. A*, **1176**, 69–78.

Klatt, K.U. (1999) Modellierung und effektive numerische Simulation von chromatographischen Trennprozessen im SMB-Betrieb. *Chem. Ing. Tech.*, **71**, 6.

Kniep, H., Blümel, C., and Seidel-Morgenstern, A. (1998) Efficient design of the SMB process based on a perturbation method to measure adsorption isotherms and on a rapid solution of the dispersion model. Oral presentation at *SPICA* 98, Strasbourg, 23–25 August.

Kniep, H., Mann, G., Vogel, C., and Seidel-Morgenstern, A. (1999) Enantiomerentrennung mittels Simulated Moving Bed-Chromatographie. *Chem. Ing. Tech.*, **71**, 708–713.

Korns, G.A. (2000) *Mathematical Handbook for Scientists and Engineers*, Dover Publications Inc.

Kubin, M. (1965) Beitrag zur theorie der chromatographie. *Collect. Czech. Chem. Commun.*, **30**, 1104–1118.

Kucera, E. (1965) Contribution to the theory of chromatography/linear non-equilibrium elution. *J. Chromatogr.*, **19** (2), 237–248.

Lapidus, F.L. (1962) *Digital Computation for Chemical Engineers*, McGraw-Hill, New York.

Lapidus, L. and Amundson, N.R. (1952) A descriptive theory of leaching: mathematics of adsorption beds. *J. Phys. Chem.*, **56**, 984–988.

Leonard, B. (1979) A stable and accurate convective modelling procedure based on quadratic upstream procedure. *Comp. Methods Appl. Mech. Eng.*, **19**, 59–98.

Levenspiel, O. (1999) *Chemical Reaction Engineering*, 3rd edn, John Wiley & Sons, Inc., New York.

Levenspiel, O. and Bischoff, K.B. (1963) Patterns of flow in chemical process vessels. *Adv. Chem. Eng.*, **4**, 95 ff.

Liapis, A.I. and Rippin, D.W.T. (1979) The simulation of binary adsorption in continuous counter-current operation and a comparison with other operation modes. *AIChE J.*, **25** (3), 455–460.

Lin, B., Ma, Z., Golshan-Shirazi, S., and Guiochon, G. (1989) Study of the representation of competitive isotherms and of the intersection between adsorption isotherms. *J. Chromatogr. A*, 475.

Lu, Z. and Ching, C.B. (1997) Dynamics of simulated moving-bed adsorption separation processes. *Sep. Sci. Technol.*, **32**, 1118–1137.

Ludemann-Hombourger, O., Bailly, M., and Nicoud, R.-M. (2000) Design of a simulated moving bed: optimal size of the stationary phase. *Sep. Sci. Technol.*, **35** (9), 1285–1305.

Ma, Z. and Wang, N.-H. (1997) Standing wave analysis of SMB-chromatography: linear systems. *AIChE J.*, **43**, 2488–2508.

Ma, Z., Whitley, R.D., and Wang, N.-H. (1996) Pore and surface diffusion in multicomponent adsorption and liquid chromatography systems. *AIChE J.*, **42** (5), 1244–1262.

Mackie, J.S. and Meares, P. (1955) The diffusion of electrolytes in a cation-exchange resin membrane. *Proc. R. Soc. Lond. A*, **267**, 498–506.

Mannschreck, A. (1992) Chiroptical detection during liquid chromatography. *Chirality*, **4**, 163–169.

Marco, V. and Bombi, G. (2001) Mathematical functions for representation of chromatographic peaks. *J. Chromatogr. A*, **931**, 1–30.

Martin, A.J.P. and Synge, R.L.M. (1941) A new form of chromatogram employing two liquid phases: a theory of chromatography. 2. Application to the micro-determination of the higher monoamino-acids in proteins. *Biochem. J.*, **35**, 1358 ff.

Migliorini, C., Mazzotti, M., and Morbidelli, M. (1999) Simulated moving-bed units with extra-column dead volume. *AIChE J.*, **45**, 1411–1421.

Migliorini, C., Mazzotti, M., and Morbidelli, M. (2000) Robust design of countercurrent adsorption separation processes: 5. Nonconstant selectivity. *AIChE J.*, **46**, 1384–1399.

Mihlbachler, K., Fricke, J., Yun, T., Seidel-Morgenstern, A., Schmidt-Traub, H., and Guiochon, G. (2001) Effect of the homogeneity of the column set on the performance of a simulated moving bed unit, part I: theory. *J. Chromatogr. A*, **908**, 49–70.

Mihlbachler, K., Jupke, A., Seidel-Morgenstern, A., Schmidt-Traub, H., and Guiochon, G. (2002) Effect of the homogeneity of the column set on the performance of a simulated moving bed unit, part II: experimental study. *J. Chromatogr. A*, **944**, 3–22.

Nowak, J., Gedicke, K., Antos, D., Piątkowski, W., and Seidel-Morgenstern, A. (2007) Synergistic effects in competitive adsorption of carbohydrates on an ion-exchange resin. *J.Chromatogr. A*, **1164**, 224–234.

Pais, L., Loureiro, J., and Rodrigues, A. (1997a) Separation of 1,1-bi-2-naphthol enantiomers by continuous chromatography in simulated moving bed. *Chem. Eng. Sci.*, **52**, 245–257.

Pais, L., Loureiro, J., and Rodrigues, A. (1997b) Modeling, simulation and operation of a simulated moving bed for continuous chromatographic separation of 1,1-bi-2-naphthol enantiomers. *J. Chromatogr. A*, **769**, 25–35.

Pais, L., Loureiro, J., and Rodrigues, A. (1998) Modelling strategies for enantiomers separation by SMB chromatography. *AIChE J.*, **44** (3), 561–569.

Press, W.H., Flannery, B.P., Teukolsky, S. A., and Vetterling, W.T. (2002) *Numerical Recipes in C^{++}: The Art of Scientific Computing*, 2nd edn, Cambridge University Press, Cambridge.

Rajendran, A., Paredes, G., and Mazzotti, M. (2009) Simulated moving bed chromatography for the separation of enantiomers. *J. Chromatogr. A*, **1216**, 709–738.

Reilley, C.N., Hildebrand, G.P., and Ashley Jr, J.W. (1962) Gas chromatographic response as a function of sample input profile. *Anal. Chem.*, **34**, 1198–1213.

Rhee, H.K., Aris, R., and Amundson, N.R. (1970) On the theory of multicomponent chromatography. *Philos. Trans. R. Soc. Lond. A*, **267**, 419–455.

Rhee, H.K., Aris, R., and Amundson, N.R. (1986) First-order partial differential equations, in *Theory and Application of Hyperbolic Systems of Quasilinear Equations*, vol. **I**, Prentice-Hall, New Jersey.

Rhee, H.K., Aris, R., and Amundson, N.R. (1989) First-order partial differential equations, in *Theory and Application of Hyperbolic Systems of Quasilinear Equations*, vol. **I**, Prentice-Hall, New Jersey.

Ruthven, D.M. (1984) *Principles of Adsorption and Adsorption Processes*, John Wiley & Sons, Inc., New York.

Ruthven, D.M. and Ching, C.B. (1989) Review article no. 31: counter-current and simulated counter-current adsorption separation processes. *Chem. Eng. Sci.*, 44 (5), 1011–1038.

Saska, M., Clarke, S.J., and Iqbal, K. (1992) Continuous separation of sugarcane molasses with a simulated-moving-bed adsorber: absorption equilibria, kinetics and applications. *Sep. Sci. Technol.*, 27, 1711–1732.

Saska, M., Clarke, S.J., Wu, M.D., and Iqbal, K. (1991) Applications of continuous chromatographic separation in the sugar industry, part I. Glucose/fructose equilibria on Dowex Monosphere 99 Ca resin at high sugar concentrations. *Int. Sugar J.*, 93, 1115.

Schlinge, D., Krasberg, N., Burghoff, B., and Schembecker, G. (2011) Determination of isotherm and mass transfer coefficients applying neural networks. *J. Chromatogr. A*, submitted for publication.

Seidel-Morgenstern, A. (1995) *Mathematische Modellierung der präparativen Flüssigchromatographie*, Deutscher Universitätsverlag, Wiesbaden.

Seidel-Morgenstern, A. (2004) Experimental determination of single solute and competitive adsorption isotherms. *J. Chromatogr. A*, 1037 (1–2), 255–272.

Seidel-Morgenstern, A. (2005) Preparative gradient chromatography. *Chem. Eng. Technol.*, 28, 1265–1273.

Seidel-Morgenstern, A. and Nicoud, R.-M. (1996) Adsorption isotherms: experimental determination and application to preparative chromatography. *Isolation Purif.*, 2, 165–200.

Shih, T.M. (1984) *Numerical Heat Transfer*, Series in Computational Methods in Mechanics and Thermal Sciences, Springer-Verlag, New York.

Siitonen, J. and Sainio, T. (2011) Explicit equations for the height and position of the first component shock for binary mixtures with competitive Langmuir isotherms under ideal conditions. *J. Chromatogr. A.*, 1218, 6379–6389.

Snyder, L.R., Kirkland, J.J., and Dolan, J.W. (2010) *Introduction to Modern Liquid Chromatography*, 3rd edn, John Wiley & Sons, Inc., Hoboken, NJ.

Storti, G., Masi, M., and Morbidelli, M. (1988) *Optimal design of SMB adsorption separation units through detailed modelling and equilibrium theory*. Proceedings of NATO Meeting, Vimeiro, Portugal. Research and Technology Agency, Neuilly sur Seine, France.

Storti, G., Mazzotti, M., Morbidelli, M., and Carrà, S. (1993) Robust design of binary countercurrent adsorption separation processes. *AIChE J.*, 39, 471–492.

Strube, J. (1996) Simulation und Optimierung kontinuierlicher Simulated-Moving-Bed (SMB)-Chromatographie-Prozesse. Dissertation. Universität Dortmund.

Strube, J., Schmidt-Traub, H., and Schulte, M. (1998) Auslegung, Betrieb und ökonomische Betrachtung chromatographischer Trennprozesse. *Chem. Ing. Tech.*, 70 (10), 1271–1279.

Suzuki, M. (1990) *Adsorption Engineering*, Elsevier, Amsterdam.

Tallarek, U., Bayer, E., and Guiochon, G. (1998) Study of dispersion in packed chromatographic columns by pulsed field gradient nuclear magnetic resonance. *J. Am. Chem. Soc.*, 120 (7), 1494–1505.

Thomas, H. (1944) Heterogeneous ion exchange in a flowing system. *J. Am. Chem. Soc.*, 66, 1664–1666.

Vajda, P., Felinger, A., and Cavazzini, A. (2010) Adsorption equilibria of proline in hydrophilic interaction chromatography. *J. Chromatogr. A*, 1217, 5965–5970.

Van Deemter, J.J., Zuiderweg, F.J., and Klinkenberg, A. (1956) Longitudinal diffusion and resistance to mass transfer as causes of nonideality in chromatography. *Chem. Eng. Sci.*, 5, 271–289.

Villadsen, J. and Stewart, W.E. (1967) Solution of boundary-value problems by orthogonal collocation. *Chem. Eng. Sci.*, 22, 1483 ff.

von Lieres, E. and Andersson, J. (2010) A fast and accurate solver for the general rate model of column liquid chromatography. *Comp. Chem. Eng.*, **34**, 1180–1191.

Vu, T.D. and Seidel-Morgenstern, A. (2011) Quantifying temperature and flow rate effects on the performance of a fixed-bed chromatographic reactor. *J. Chromatogr. A*, **1218**, 8097–8109.

Weber, S.G. and Carr, P.W. (1989) The theory of the dynamics of liquid chromatography, in *High Performance Liquid Chromatography* (eds P.R. Brown and R.A. Hartwick), John Wiley & Sons, Inc., New York.

Wekenborg, F.K. (2009) Kontinuierliche Trennung von Proteinen durch nicht-isokratische SMB-Chromatographieprozesse, Shaker Verlag GmbH, Aachen.

Wekenborg, K., Susanto, A., and Schmidt-Traub, H. (2005) Modelling and validated simulation of solvent-gradient simulated moving bed (SG-SMB) processes for protein separation, in *Computer Aided Chemical Engineering*, Elsevier vol. 20 (eds L. Puigjaner and A. Espuña), pp. 313–318.

Wen, C.Y. and Fan, L.T. (1975) *Models for Flow Systems and Chemical Reactors*, Marcel Dekker, New York.

Whitley, R.D., Van Cott, K.E., and Wang, N.-H. (1993) Analysis of nonequilibrium adsorption/desorption kinetics and implications for analytical and preparative chromatography. *Ind. Eng. Chem. Res.*, **32**, 149–159.

Wicke, E. (1939) *Kolloid Z.*, **86**, 285 ff.

Wilson, E.J. and Geankoplis, C.J. (1966) Liquid mass transfer at very low Reynolds numbers in packed beds. *Ind. Eng. Chem. Fundam.*, **5**, 9.

Yun, T., Zhong, G., and Guiochon, G. (1997a) Simulated moving bed under linear conditions: experimental vs. calculated results. *AIChE J.*, **43** (4), 935–945.

Yun, T., Zhong, G., and Guiochon, G. (1997b) Experimental study of the influence of the flow rates in SMB chromatography. *AIChE J.*, **43** (11), 2970–2983.

Zhang, W., Shan, Y., and Seidel-Morgenstern, A. (2006) Breakthrough curves and elution profiles of single solutes in case of adsorption isotherms with two inflection points. *J. Chromatogr. A*, **1107**, 216–225.

Zhong, G. and Guiochon, G. (1996) Analytical solution for the linear ideal model of simulated moving bed chromatography. *Chem. Eng. Sci.*, **51** (18), 4307–4319.

7
Model-Based Design, Optimization, and Control

Henner Schmidt-Traub, Malte Kaspereit, Sebastian Engell, Arthur Susanto[*],
Achim Epping[*], and Andreas Jupke[*]

Previous chapters deal with the selection of chromatographic systems, process concepts, as well as formulation and validation of process models. The present chapter focuses on process design by rigorous simulations using validated models as well as shortcut methods. In order to determine optimal process conditions, it is necessary to define performance indicators as well as objective functions and to take into account how design and process parameters influence the process performance. The dimensionless representation of these parameters minimizes the degrees of freedom of the decision space and therefore simplifies optimization procedures. Dimensionless parameters are also helpful to scale up or down experimentally or theoretically proven processes to the projected scale. TMB shortcut methods are important tools to estimate operating regions for SMB processes and to decide upon the values of process parameters that have to be chosen in order to meet certain process specifications. These process parameters are also good starting points to determine optimal process conditions by mathematical optimization or case studies based on rigorous simulation. If decisions have to be made in short time or suitable optimization tools are not available, shortcut results including safety margins may also be applied to run a production plant. General design procedures for batch and SMB processes are presented below and explained by example calculations. Additionally, enhancements of standard isocratic batch and SMB processes using variable process conditions including solvent gradients are outlined. Finally, methods for advanced model-based process control are presented.

Properties of the chromatographic systems that are used for example calculations are documented in Appendix A.

7.1
Basic Principles and Definitions

7.1.1
Performance, Costs, and Optimization

Performance and economic criteria – such as yield, productivity, or total cost – are indicators of efficiency and quality of preparative chromatographic separations.

[*]These authors have contributed to the first edition.

They depend on the operating and design parameters of a plant. Hereby, they are potential variables for any optimization problem and the basis for the formulation of objective functions. The definitions of objective functions are often identical for diverse chromatographic processes (e.g., batch or SMB), but different operating and design parameters must be applied for their calculation.

7.1.1.1 Performance Criteria

The yield Y_i of a process is the ratio between the amount of a product (component i) that can be collected in the outlet and the amount that was introduced into the system through the feed stream within a batch cycle. Due to the continuous operation mode in an SMB plant, both collected and introduced amounts of component i are calculated within a shifting interval:

$$Y_{i,\text{Batch}} = \frac{m_{\text{out},i}}{\dot{V} \cdot c_{\text{feed},i} \cdot t_{\text{inj}}} \tag{7.1}$$

$$Y_{i,\text{SMB}} = \frac{m_{\text{out},i}}{\dot{V}_{\text{feed}} \cdot c_{\text{feed},i} \cdot t_{\text{shift}}} \tag{7.2}$$

The amount of component i in the outlet stream can be calculated for a batch chromatographic process using Equation 7.3, whereby $t_{1,i}$ and $t_{2,i}$ are the beginning and the end, respectively, of the fraction collection for pure component i:

$$m_{\text{out},i} = \dot{V} \cdot \int_{t_{1,i}}^{t_{2,i}} c_{i,\text{out}}(t) \cdot dt \tag{7.3}$$

In comparison with batch chromatography t_1 and t_2 for an SMB plant are the beginning and the end, respectively, of a shifting interval. Furthermore, component i can be collected from either raffinate or extract stream:

$$m_{\text{out},i} = \dot{V}_{\text{raf/ext}} \cdot \int_{t_1}^{t_2} c_{i,\text{raf/ext}}(t) \cdot dt \tag{7.4}$$

Using the purified amount of product i within a batch cycle, the average mass flow of purified product i (also called as production rate $\dot{m}_{\text{prod},i}$) in batch chromatography can be expressed as

$$\dot{m}_{\text{prod},i} = \frac{m_{i,\text{out}}}{t_{\text{batch-cycle}}} \tag{7.5}$$

For an SMB plant the production rate must be determined within a shifting interval instead:

$$\dot{m}_{\text{prod},i} = \frac{m_{i,\text{out}}}{t_{\text{shift}}} \tag{7.6}$$

The efficiency of a chromatographic separation can be expressed by comparing the purified amount of product i with the time taken and the resources necessary to produce it. This expression is referred to as productivity and can be applied in terms of

- volume-specific productivity VSP_i (production rate over total solid adsorbent volume);
- cross section-specific productivity ASP_i (production rate over free cross section of the column):

$$ASP_{i,\text{Batch}} = \frac{\dot{m}_{\text{prod},i}}{A_c \cdot \varepsilon} \tag{7.7}$$

$$ASP_{i,\text{SMB}} = \frac{\dot{m}_{\text{prod},i}}{N_{\text{COL}} \cdot A_c \cdot \varepsilon} \tag{7.8}$$

$$VSP_{i,\text{Batch}} = \frac{\dot{m}_{\text{prod},i}}{V_{\text{solid}}} = \frac{\dot{m}_{\text{prod},i}}{V_c \cdot (1 - \varepsilon_t)} \tag{7.9}$$

$$VSP_{i,\text{SMB}} = \frac{\dot{m}_{\text{prod},i}}{V_{\text{solid,total}}} = \frac{\dot{m}_{\text{prod},i}}{N_{\text{COL}} \cdot V_c \cdot (1 - \varepsilon_t)} \tag{7.10}$$

Here, the number of columns N_{COL} in SMB has to be considered.

In many cases the cost of eluent (i.e., solvent) is not negligible and in some cases it even represents the greatest contribution to the total separation cost. Therefore, it is advisable in these cases to observe the eluent consumption or, better still, the efficiency of eluent usage during the chromatographic separation. This can be characterized by specific eluent consumption EC_i, which means the amount of eluent required to purify a certain amount of product i:

$$EC_{i,\text{Batch}} = \frac{\dot{V}}{\dot{m}_{\text{prod},i}} \tag{7.11}$$

Since in an SMB plant the eluent is introduced into the system through the desorbent port as well as feed port, both flow rates have to be considered in the calculation:

$$EC_{i,\text{SMB}} = \frac{\dot{V}_{\text{feed}} + \dot{V}_{\text{des}}}{\dot{m}_{\text{prod},i}} \tag{7.12}$$

Notably, eluent consumption is calculated by neglecting the influence of solute concentration since the change in liquid volume during the dissolution of solid components can generally be neglected.

Another performance criterion that is also often used as a boundary condition for a separation problem is the purity of a product i, Pu_i. Product purity in a chromatographic process can be calculated by Equation 7.13, where N_{COMP} is the number of components in the solution:

$$Pu_i = \frac{m_{i,\text{out}}}{\sum_{i=1}^{N_{\text{COMP}}} m_{i,\text{out}}} \tag{7.13}$$

7.1.1.2 Economic Criteria

Katti and Jagland (1998) give a complete description of the contribution of different kinds of costs to the total separation costs for batch separation. Also, the influence of different physical properties of the separation system and the plant design is discussed. The authors demonstrated that the cost structure of the separation problem changes with the scale of the separation. For smaller production amounts the contribution of capital, labor, and maintenance costs to total cost is significantly higher than for bigger production rate. On this account, it is very important to consider the total separation cost, because only consideration of the separation cost will lead to an economic optimal process for both batch and SMB chromatography. Yield, productivity, eluent consumption, and purity are clearly defined performance criteria, while total separation cost and other economic criteria depend on the individual company and are based on site-related parameters. Therefore, the calculation of total separation cost is complex due to various influencing parameters and cost structures. This chapter proposes the following cost functions, which can be easily adapted to specific conditions:

1) **Separation problem independent costs (fixed costs):** These can be considered as fixed costs and they are specific for each company. They can be divided into:
   - Annual operating cost ($C_{operating}$): Operating cost includes overhead costs as well as wages and maintenance cost.
   - Annual depreciation ($C_{depreciation}$): Annual depreciation is the allocation of investment cost over the depreciation years.
2) **Separation problem dependent costs (variable costs):** These costs are directly correlated to a given separation problem and the desired production rate. They are the annual costs for eluent, adsorbent, and lost feed (also known as crude loss). All of these costs can be calculated by applying other performance criteria that characterize the efficiency of the separation process (e.g., yield, productivity). Further requirements for the calculation of separation problem dependent costs are the production rate or the annual produced amount $\dot{m}_{prod, annual}$ and prices for eluent f_{el} (€ per l eluent), for adsorbent f_{ads} (€ per kg adsorbent), and for feed f_{feed} (€ per kg feed). Additionally, to calculate the annual adsorbent cost C_{ads}, the lifetime of the adsorbent t_{life} has to be known or estimated:

$$C_{el} = EC_i \cdot \dot{m}_{prod, annual} \cdot f_{el} \tag{7.14}$$

$$C_{ads} = \frac{1}{VSP_i} \cdot \dot{m}_{prod, annual} \cdot \frac{\varrho_{ads} \cdot f_{ads}}{t_{life}} \tag{7.15}$$

$$C_{crudeloss} = \frac{1 - Y_i}{Y_i} \cdot \dot{m}_{prod, annual} \cdot f_{feed} \tag{7.16}$$

The total annual separation cost, C_{total}, can then be determined by adding up the fixed costs as well as the variable costs:

$$C_{total} = C_{operating} + C_{depreciation} + C_{ads} + C_{el} + C_{crudeloss} \tag{7.17}$$

However, to be able to compare the total annual separation costs for different production scales, the production-specific cost (cost per product amount) should be used instead:

$$C_{\text{spec,total}} = \frac{C_{\text{total}}}{\dot{m}_{\text{prod,annual}}} \tag{7.18}$$

7.1.1.3 Objective Functions

Objective functions define the aim of an optimization process that is by convention stated as a minimization problem. A general mathematical formulation is, for example:

$$\min f(x^*) \leq f(x), \quad \text{for } \|x - x^*\| < \delta \tag{7.19}$$

where x is the degrees of freedom for the given problem. Generally, objective functions have also to fulfill certain constraints:

$$g(x) \leq 0 \tag{7.20}$$

Related to preparative chromatography, diverse objective functions appear. For instance, they may focus on the economics of new plant for a certain product, or on maximizing the productivity VSP_i as a function of the injection volume under the constraints of minimal purities and yields. In the latter case a general formulation of the optimization task can be

$$\min f(x^*) = \frac{1}{\text{VSP}_i} \tag{7.21}$$

under the constraints

$$\text{Pu}_i \geq \text{Pu}_{i,\min} \quad \text{and} \quad Y_i \geq Y_{i,\min} \tag{7.22}$$

Performance and economic criteria as defined in the previous sections or their combinations are potential objective functions, while the design and operating parameters represent the degrees of freedom for the optimization task. In order to reduce the complexity of an optimization task, it is recommended to minimize the number of parameters by introducing dimensionless parameters, as described in the following section.

The solution of an optimization problem depends very much on the complexity of the decision space and the available tools. At first, overall decisions based on experience, for example, the selection of a certain process flow sheet, might be necessary to reduce the number of degrees of freedom. An optimal solution can be approached by trial and error or case studies based on experiments and/or theoretical simulations. Other possibilities are mathematical procedures for a rigorous determination of an optimal solution of the objective function. Direct mathematical optimizations are essential for advanced process control (Section 7.8), but in case of process optimization it is recommended that the design engineer also checks the robustness of the solution. This will usually require parametric studies in order to verify that within the operation plane the gradients, which surround the evaluated optimum, are not too steep. Goal-oriented case studies based on rigorous simulation and additional shortcut calculations are beneficial to get a better understanding of the process performance (see Sections 7.2 and 7.4).

Mathematical single-objective as well as multiobjective optimizations are out of the scope of this book; for certain applications in preparative chromatography the interested reader is referred to citations in the following sections.

7.1.2
Degrees of Freedom

7.1.2.1 Optimization Parameters

Models for a chromatographic column consist of many parameters that have already been classified into three groups in Section 6.5.1:

1) **Operating parameters:** Alterable parameters during the operation of the plant, for example:
   - Flow rate.
   - A number of operating parameters in batch chromatography result from the batch operation mode: switch times.
   - To characterize the operation of an SMB plant precisely, the following additional parameters are necessary: flow rate in each SMB section and shifting time.

2) **Design parameters:** Design parameters define the appearance of a plant and cannot be changed during line operation, for example:
   - column geometry (length and diameter);
   - diameter of adsorbent particle;
   - maximum pressure drop;
   - additional degree of freedom in an SMB plant: column configuration (number of columns in each SMB section).

3) **Model parameters:** Model parameters are system inherent parameters that result from the choice of chromatographic system. They describe, for example, the following phenomena:
   - thermodynamics;
   - fluid dynamics;
   - dispersion effects;
   - mass transfer resistance.

A good process description based on validated models is needed to predict the position of the optimal state accurately. Since model parameters are determined experimentally on the chromatographic system, which is generally predefined before optimization starts, model parameters remain unchanged during the optimization. On this basis, all model-based optimization strategies in the following sections apply only to operating and design parameters.

7.1.2.2 Dimensionless Operating and Design Parameters

The chromatographic process contains many design and operating parameters. Hence, the optimization of all parameters requires a great amount of resources (e.g., faster computer) and more complex optimization algorithms. However, this can be reduced by summarizing the numerous design and operating parameters into a smaller number of dimensionless parameters, thus reducing the number of optimization parameters and the complexity of the optimization problem.

Dimensionless formulation of model equations for a chromatographic column can be found in Chapter 6. For clarity, the dimensionless parameters and the phenomena described by them will also be mentioned here:

1) **Axial dispersion and convection:** Péclet number:

$$Pe_i = \frac{\text{convection}}{\text{dispersion}} = \frac{u_{int} \cdot L_c}{D_{ax,i}} \quad (7.23)$$

2) **Mass transfer and convection:** modified Stanton number:

$$St_{eff,i} = \frac{\text{mass transfer}}{\text{convection}} = k_{eff,i} \cdot \frac{6}{d_p} \cdot \frac{L_c}{u_{int}} \quad (7.24)$$

3) **Adsorption:** for linear chromatography, adsorption behavior is determined by the dimensionless Henry coefficient H_i:

$$H_i = \frac{Q_{DL,i}}{C_{p,DL,i}} \quad (7.25)$$

In most cases (nonlinear chromatography), however, the feed concentration plays a major role in the adsorption.

For example, the Langmuir isotherm:

$$Q_{DL,i} = \frac{H_i \cdot C_{p,DL,i}}{1 + (c_{feed,i} \cdot b_i) \cdot C_{p,DL,i}} \quad (7.26)$$

Here, the additional dimensionless number (product of $c_{feed,i} \cdot b_i$) must also be taken into account.

4) **Process-related boundary conditions:**
   - **For batch processes:** dimensionless injection time $t_{inj}/t_{0,int}$.

   In this chapter another dimensionless number is used to represent the injection time, namely, the loading factor LF_i, which was first introduced by Guiochon and Golshan-Shirazi (1989) and is defined as

$$LF_i = \frac{\text{injected feed amount}}{\text{saturation capacity}} = \frac{c_{i,feed} \cdot \dot{V} \cdot t_{inj}}{q_{sat,i} \cdot V_c \cdot (1 - \varepsilon_t)} \quad (7.27)$$

The saturation concentration ($q_{sat,i}$) can be easily determined for all isotherm types with saturation concentration, for example, the Langmuir type:

$$q_{sat,i} = \lim_{c_i \to \infty} \frac{H_i \cdot c_i}{1 + b_i \cdot c_i} = \frac{H_i}{b_i} \quad (7.28)$$

For other isotherm types with no saturation concentration (e.g., linear type) it is advisable to use the solid phase concentration at maximum feed concentration in the separation problem:

$$q_{sat,i} = q_i(c_{feed,max}) \tag{7.29}$$

The use of loading factor as degree of freedom is equivalent to the use of dimensionless injection time, as can be seen in Equation 7.30:

$$LF_i = \frac{c_{i,feed}}{q_{sat,i}} \cdot \frac{\varepsilon}{1-\varepsilon_t} \cdot \left(\frac{t_{inj}}{t_{0,int}}\right) \tag{7.30}$$

Since the loading factor is a more common parameter in practice, in this chapter it is preferred to the dimensionless injection time.

- **For SMB processes:** dimensionless shifting times $t_{shift}/t_{0,int,j}$.

Once again, other dimensionless numbers are used to represent the shifting boundary conditions. The periodic behavior of SMB processes is induced by the stepwise switching of the inlet and outlet ports. To describe this operation as a quasi-stationary process comparable to the TMB process the following flow rates are introduced:

$$\text{Net liquid flow rate:} \quad \dot{V}_{j,net} = \dot{V}_j - \frac{V_c \varepsilon_t}{t_{shift}} \tag{7.31}$$

$$\text{Simulated solid flow rate:} \quad \dot{V}_{ads} = \frac{V_c(1-\varepsilon_t)}{t_{shift}} \tag{7.32}$$

The ratio of these flow rates results in the following dimensionless parameter that is essential for the design of countercurrent chromatographic separation (Storti et al., 1993):

$$m_j = \frac{\text{net fluid flow rate in section } j}{\text{simulated solid flow rate}} = \frac{\dot{V}_j - [(V_c \cdot \varepsilon_t)/t_{shift}]}{[V_c \cdot (1-\varepsilon_t)]/t_{shift}} \tag{7.33}$$

Similar to the injection time in batch processes, the use of flow ratio m_j as a degree of freedom is equivalent to the use of dimensionless shifting time. Modifying Equation 7.33 leads to Equation 7.34:

$$m_j = \frac{\varepsilon}{1-\varepsilon_t} \cdot \left(\frac{t_{shift}}{t_{0,int,j}} - \frac{\varepsilon_t}{\varepsilon}\right) \tag{7.34}$$

After transforming the model equations and boundary conditions into their dimensionless form, there still remain quite a number of parameters that have to be taken into account for process optimization. In case of HPLC chromatography the complexity of a design/optimization problem can often be reduced without significantly reducing the accuracy of the solution by simplifications as discussed below. But the validity of these simplifications should be checked for each practical application.

The first important simplification is based on the assumption of linear isotherms (Guiochon et al., 2006). In this case the influence of mass transfer and axial dispersion are added to each other (Equation 6.142), and the Péclet (Pe_i) and modified Stanton ($St_{eff,i}$) numbers are combined into the number of stages N_i, which is the more common parameter in practice (Equation 7.35). A detailed derivation of the number of stages for a transport-dispersive model in dependency of Pe_i and $St_{eff,i}$ can be found in Section 6.5.3.1:

$$N_i = \frac{2}{Pe_i} + \frac{2\varepsilon}{1-\varepsilon} \left(\frac{\tilde{k}'_i}{1+\tilde{k}'_i} \right)^2 \frac{1}{St_{eff,i}} = f(Pe_i, St_{eff,i}) \qquad (7.35)$$

Equation 7.35 defines the number of stages for a transport-dispersive model. It indicates that for the assumptions mentioned above the influence of Pe_i and $St_{eff,i}$ is summarized by the number of plates that reduces the number of parameters for process optimization. From Equation 7.35 follows the equation for HETP:

$$HETP_i = \frac{L_c}{N_i} = \frac{2 \cdot D_{ax,i}}{u_{int}} + \frac{2\varepsilon}{1-\varepsilon} \cdot \left(\frac{\tilde{k}'_i}{1+\tilde{k}'_i} \right)^2 \cdot \frac{d_p}{6 \cdot k_{eff,i}} \cdot u_{int} \qquad (7.36)$$

Please note that Equation 7.36 is not used to calculate the number of stages during the optimization. Instead of this, the number of HETP is calculated by Equation 7.37:

$$HETP_i = A_i + \frac{B_i}{u_{int}} + C_i \cdot u_{int} \qquad (7.37)$$

The coefficients in Equation 7.37 (A_i, B_i, and C_i) are determined from experiment (see also Chapter 6).

For preparative liquid chromatographic processes the HETP equation can be simplified further by taking a closer look at the axial dispersion coefficient, $D_{ax,i}$, which is generally described as the sum of the contributions of eddy diffusion and molecular diffusion:

$$D_{ax,i} = \lambda \cdot d_p \cdot u_{int} + \gamma \cdot D_{m,i} \qquad (7.38)$$

where $D_{m,i}$ is the molecular diffusion coefficient for component i, and λ and γ the irregularity in the packed column and the obstruction factor for the diffusion in the packed column, respectively. In preparative liquid chromatography the contribution of molecular diffusion can be neglected due to high flow rates and relatively small molecular diffusion coefficients in liquids:

$$D_{ax,i} \approx \lambda \cdot d_p \cdot u_{int} \qquad (7.39)$$

Therefore, Pe_i is independent of interstitial velocity:

$$Pe_i = \frac{u_{int} \cdot L_c}{D_{ax,i}} \approx \frac{L_c}{\lambda \cdot d_p} \neq f(u_{int}) \qquad (7.40)$$

The HETP equation for calculating number of stages in preparative liquid chromatography can then be simplified to

$$HETP_i = 2 \cdot \lambda \cdot d_p + \frac{2\varepsilon}{1-\varepsilon} \cdot \left(\frac{\tilde{k}'_i}{1+\tilde{k}'_i} \right)^2 \cdot \frac{1}{k_{eff,i}} \cdot \frac{d_p}{6} \cdot u_{int} = A_i + C_i \cdot u_{int} \qquad (7.41)$$

In general, the substitution of two degrees of freedom by a single degree of freedom is mathematically not correct. However, in many cases the influence of one of these dimensionless parameters (Pe_i or $St_{eff,i}$) on chromatographic processes can be neglected, thus reducing the degrees of freedom into a single dimensionless parameter and justifying the simplification into the number of stages N_i. Cases in which the influence of axial dispersion (Pe_i) or mass transfer resistance ($St_{eff,i}$) can be neglected are as follows:

1) **Dominant mass transfer resistance (characterized by $St_{eff,i} \ll Pe_i$):** Compared with Pe_i, which is independent of the interstitial velocity (Equation 7.40), $St_{eff,i}$ is inversely proportional to interstitial velocity (Equation 7.24). This means that the influence of mass transfer resistance will grow and surpass the influence of axial dispersion at high interstitial velocity, which is almost always the case for preparative chromatographic processes. In some extreme cases, where the mass transfer coefficients are small and the chromatographic column is operated at high flow rates, the HETP equation for the calculation of N_i can even be simplified further to

$$HETP_i \approx C_i u_{int}$$

2) **Dominant axial dispersion (characterized by $Pe_i \ll St_{eff,i}$):** In chromatographic systems with relatively large effective mass transfer coefficients $k_{eff,i}$ (i.e., low mass transfer resistance) the influence of axial dispersion, especially eddy diffusion, dominates the concentration profile. $HETP_i$ and N_i are then independent of the interstitial velocity:

$$HETP_i \approx A_i \neq f(u_{int})$$

Since column efficiency is usually expressed by the number of stages, the flow rate through the column can be increased without changing the column efficiency, thus increasing the production rate of the column. The optimal flow rate for such chromatographic systems is therefore limited by the maximum allowed pressure drop.

In other words, the simplification of using the number of stages for process optimization is best applied if either mass transfer or dispersion dominates the peak broadening. Therefore, the optimization strategies discussed later in this chapter apply a validated transport-dispersive model, which can flexibly consider mass transfer and/or dispersion effect. Here, the number of stages is used as an independent variable for the optimization criteria such as productivity or eluent consumption. Other possible approaches would be the use of simplified simulation models such as transport model or equilibrium-dispersive model.

Another important simplification, which makes the scale-up of chromatographic plants easier, is to always use the same feed concentration. Since higher feed concentration also results in higher productivity, feed concentration is fixed at the maximum allowed concentration, which depends on the chromatographic system and the solubility of the feed components. However, in SMB processes, feed concentration affects the size of the operating parameter region for a total separation and, therefore, should not be fixed at too high values in order not to hinder the ability to get "robust" operating parameters (Section 7.4). Thus, the remaining and applied degrees of freedom for the design and optimization methods for batch as well as SMB processes in this chapter are

- **for batch processes:** the number of stages N_i, respectively, the height of a theoretical plate $HETP_i$ (representing the flow rate), and the loading factor LF_i (representing the injection conditions);
- **for SMB processes:** total number of stages $N_{tot,i}$ and dimensionless flow ratios m_j.

7.1.3
Scaling by Dimensionless Parameters

One of the advantages of the presented optimization method is the use of only a reduced number of optimization parameters, although nearly all design and operating parameters are considered. Furthermore, chromatographic processes with identical dimensionless parameters will have similar behavior. In other words, the values of many performance criteria (e.g., specific productivity, specific eluent consumption) for those processes will be the same and their dimensionless concentration profiles will also be identical. Therefore, no new optimization must be done after upscaling (or downscaling) the chromatographic plant, because the optimal dimensionless parameters from the old plant can be easily adapted to the new plant. This means

- For batch processes:

$$N_{i,new} \geq N_{i,old} \tag{7.42}$$
$$LF_{i,new} = LF_{i,old} \tag{7.43}$$

- For SMB processes:

$$N_{tot,i,new} \geq N_{tot,i,old} \tag{7.44}$$
$$m_{j,new} = m_{j,old} \tag{7.45}$$

Since the adoption of the number of stages from the old chromatographic column into the new column can never be done exactly, in Equations 7.42 and 7.44 greater than or equal to signs are placed instead of equal to signs. This should guarantee that the separation performance of the new column is equal to or better than the performance of the old column even if the concentration profiles and the values of performance criteria are not exactly the same. The reason of this difference will be illustrated in the following sections with the aid of numbers of stages for two different chromatographic plants (an old, optimized plant and a new plant):

$$N_{i,old} = \frac{L_{c,old}}{HETP_{i,old}} \tag{7.46}$$

$$N_{i,new} = \frac{L_{c,new}}{HETP_{i,new}} \tag{7.47}$$

7.1.3.1 Influence of Different HETP Coefficients for Every Component

Two different plants will behave similarly only if the numbers of stages *for all components* are equal. If the equality condition is valid for the first component, then

$$N_{1,\text{new}} = N_{1,\text{old}} \Leftrightarrow L_{c,\text{new}} = L_{c,\text{old}} \cdot \frac{\text{HETP}_{1,\text{new}}}{\text{HETP}_{1,\text{old}}} \tag{7.48}$$

Substituting this equation into the new number of stages for component 2 will result in

$$N_{2,\text{new}} = \frac{L_{c,\text{old}}}{(\text{HETP}_{1,\text{old}} \cdot \text{HETP}_{2,\text{new}})/\text{HETP}_{1,\text{new}}} \tag{7.49}$$

The equality condition will therefore only be valid for the second component if

$$N_{2,\text{new}} = N_{2,\text{old}} \Leftrightarrow \frac{\text{HETP}_{1,\text{old}} \cdot \text{HETP}_{2,\text{new}}}{\text{HETP}_{1,\text{new}}} = \text{HETP}_{2,\text{old}} \tag{7.50}$$

or as general expression

$$\frac{\text{HETP}_{i,\text{new}}}{\text{HETP}_{i,\text{old}}} = \frac{A_i + C_i \cdot u_{\text{int,new}}}{A_i + C_i \cdot u_{\text{int,old}}} = \text{constant} \tag{7.51}$$

Therefore, the equality condition for the number of stages is only fulfilled exactly by all components if the ratios between HETP of the old column and that of the new column are *equal for every component*. Due to the different HETP coefficients for each component the equality condition can only be exactly fulfilled if

1) **Both columns have the same length but different cross section areas:** Due to the identical column length the interstitial velocity in both columns must be kept constant in order to have the same column efficiency (i.e., same number of stages). Both columns have, however, different flow rates due to different cross section areas:

$$u_{\text{int,new}} = u_{\text{int,old}} \Leftrightarrow \frac{\text{HETP}_{i,\text{new}}}{\text{HETP}_{i,\text{old}}} = 1 = \text{constant} \tag{7.52}$$

2) **Dominant eddy diffusion:** For chromatographic systems with negligible mass transfer resistance the eddy diffusion is the dominant effect. Thereby, HETP values are constant and independent of the interstitial velocity:

$$\frac{\text{HETP}_{i,\text{new}}}{\text{HETP}_{i,\text{old}}} \approx 1 = \text{constant} \tag{7.53}$$

3) **Dominant mass transfer resistance:** In this case the approximated HETP value is directly proportional to the interstitial velocity. Thus, the HETP ratios between

old and new columns for all components are identical: they are always equal to the ratio of interstitial velocities:

$$\frac{\text{HETP}_{i,\text{new}}}{\text{HETP}_{i,\text{old}}} \approx \frac{u_{\text{int,new}}}{u_{\text{int,old}}} = \text{constant} \tag{7.54}$$

Whenever those cases apply, the number of stages in the new and old columns for other components will remain the same if the number of stages of one single component is not varied. Therefore, the column can be optimized by considering the loading factor and the number of stages of only one component (i.e., the reference component). Generally, however, the equality condition for all the number of stages cannot be fulfilled exactly, so that a certain deviation has to be accepted. The magnitude of the deviation depends on the magnitude of the contributions to the HETP. The more dominant the influence of mass transfer resistance or eddy diffusion, the smaller the deviation between the concentration profiles. However, if the modified equality condition (Equations 7.42 and 7.44) is applied, the separation performance of the new column can be guaranteed to be equal to or better than that of the old column.

7.1.3.2 Influence of Feed Concentration

Actually, the definition of number of stages and HETP originates from linear chromatography, although it can also be formulated for nonlinear chromatography. In nonlinear chromatography the number of stages and HETP are affected by the varying gradient of the adsorption isotherm; therefore, their value depends on the concentration range in the column. Due to the concentration dependency in the nonlinear case, the number of stages loses its original practical meaning as a measure of column efficiency.

For preparative chromatography, where we almost always have to deal with high feed concentrations and nonlinear adsorption isotherms, the following approach to the appropriate choice of feed concentration during model-based optimization is recommended:

1) Determination of HETP coefficients in the linear adsorption region (i.e., low feed concentration) as explained in Chapter 6.
2) Execution of model-based optimization using the same feed concentration that will be applied later in the preparative column. Generally, a high feed concentration is applied, which is close to the solubility limit of the feed components. Feed concentration is, therefore, not an optimization parameter and should not be varied during optimization. The number of stages, however, is calculated by applying the known HETP coefficients from the linear adsorption region.

Application of concentration-independent HETP coefficients from the linear isotherm region in calculating the number of stages in a nonlinear region is only a formalism. However, since feed concentration is not varied during the optimization, the calculated number of stages can nevertheless be used further as it is a characteristic value for the specific chromatographic separation. Scale-up of the

7.1.3.3 Examples for a Single Batch Chromatographic Column

To demonstrate the equality of concentration profiles, results from simulation of a transport-dispersive model with different design and operating parameters but identical number of stages and loading factor are compared. Here, it has to be kept in mind that these model calculations agree with experiments within measurement accuracy (Chapter 6).

The first exemplary problem is the separation of a racemic mixture EMD53986 (R: S-enantiomers $= 1:1$) on the chiral stationary phase Chiralpak AD from Daicel (20 μm particle diameter) (Table 7.1).

Please note that only the number of stages and loading factor of the R-enantiomer are taken into account, since R-enantiomer is used as the *reference component* in this separation problem. Furthermore, the number of stages is calculated using the experimentally determined HETP coefficients (Equation 7.37). All applied model parameters (isotherms, mass transfer coefficients, etc.) for the simulation of this enantiomeric separation are already experimentally validated. The simulation results (i.e., simulated chromatograms) are presented in Figure 7.1 in dimensionless form using the ratios $c_i/c_{feed,i}$ and $t/t_{0,int}$ instead of c_i and t, whereby $t_{0,int}$ is the mean residence time under nonadsorbing, nonpenetrating conditions (i.e., the residence time of large tracer molecules, e.g., dextran):

$$t_{0,int} = \frac{L_c}{u_{int}} \qquad (7.55)$$

As can be seen in Table 7.1, Pe_i far exceeds $St_{eff,i}$, clearly indicating that mass transfer resistance plays a major role in this separation problem. Therefore, the concentration profiles are completely identical despite the different Pe_i. This justifies simplification of the optimization method by using the number of stages instead of Pe_i and $St_{eff,i}$. Furthermore, the number of stages is calculated using the HETP coefficients determined in the linear isotherm region. However, even if the concentration range in the column no longer lies within the linear region of the adsorption isotherms, deviations of the concentration profiles are insignificant as long as the same feed concentration is used (Figure 7.1).

Table 7.1 Different design and operating parameters for a single column (EMD53986); $N_R = 25$; $LF_R = 1.25$; $c_{feed,R} = c_{feed,S} = 2.5\,g\,l^{-1}$.

	$\dot{V}$ (ml min^{-1})	u_{int} (cm s^{-1})	t_{inj} (s)	L_c (cm)	d_c (cm)	$St_{eff,R}$ (−)	Pe_R (−)
EMD-1	23.5	0.2247	208.0	10.8	2.5	21.63	3061.22
EMD-2	54.8	0.5237	206.6	25.0	2.5	21.48	7108.14
EMD-3	109.0	0.2605	207.7	12.5	5.0	21.59	3544.69

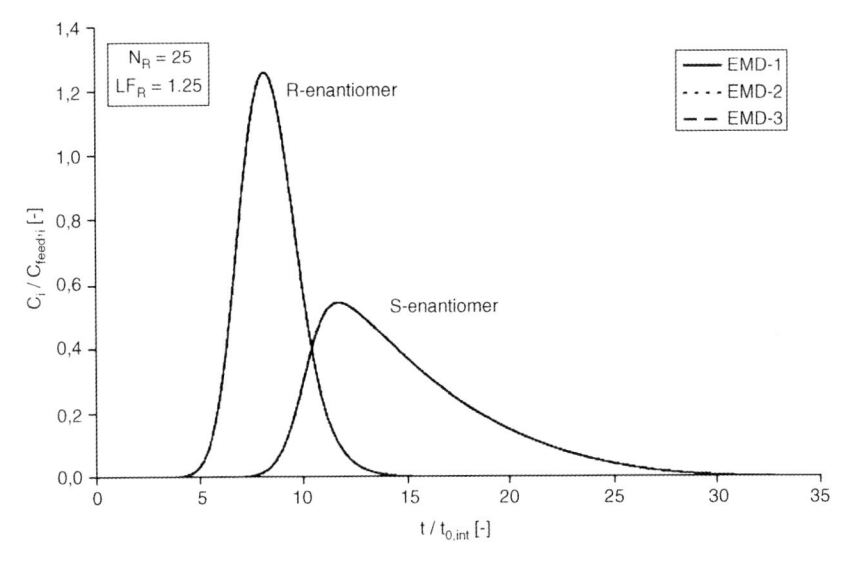

Figure 7.1 Equality of concentration profiles in a chromatographic column (EMD53986).

Dimensionless degrees of freedom do not always transfer to other column designs as perfectly as shown in Figure 7.1. In many cases some deviations in the concentration profiles have to be taken into account. These cases will now be demonstrated using a second exemplary separation problem: the chromatographic separation of 1 : 1 mixture of glucose and fructose on ion exchange resin Amberlite CR1320 Ca from Rohm and Haas (325 µm particle diameter) (Table 7.2).

Compared with the EMD53986 separation, Pe_i are smaller, although they are still much greater than $St_{eff,i}$. This indicates that, although mass transfer effects still dominate, axial dispersion now plays a small but significant role. One reason why the axial dispersion coefficient grows is surely the bigger particle diameter at 325 µm. A certain deviation during the scale-up has therefore to be taken into account (Figure 7.2).

Since the concentration profiles for this exemplary system are not perfectly identical, scale-up has to be done with caution. Especially, the "greater than or equal to" condition (Equation 7.42 or 7.44) has to be fulfilled for all components. Therefore, in this case it is not sufficient to look only at the number of stages of one component, but the number of stages of all components has to be taken into account.

Table 7.2 Different design and operating parameters for a single column (Glu/Fruc); $N_{Glu} = 100$; $LF_{Glu} = 0.5$; $c_{feed,Glu} = c_{feed,Fruc} = 300\,g\,l^{-1}$.

	$\dot{V}$ (ml min^{-1})	u_{int} (cm s^{-1})	t_{inj} (s)	L_c (cm)	d_c (cm)	$St_{eff,Glu}$ (−)	Pe_{Glu} (−)
FG-1	5.8	0.0490	328.1	56	2.6	14.76	950.15
FG-2	2.4	0.0201	428.0	30	2.6	19.25	505.52
FG-3	13.6	0.0313	367.7	40	5.0	16.54	676.07

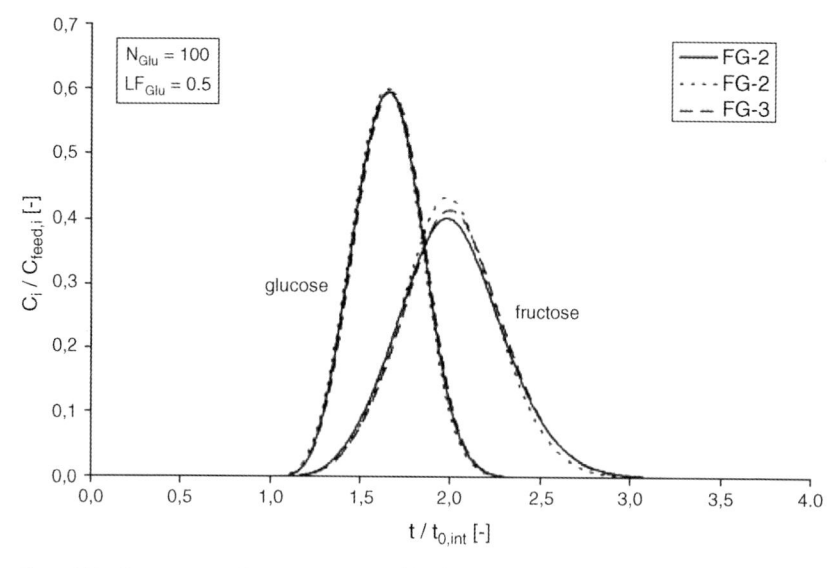

Figure 7.2 Comparison of concentration profiles from different columns (Glu/Fruc).

Table 7.3 and Figure 7.3 show the resulting operating points and concentration profiles if only the number of stages of glucose is considered during the scale-up. In this example, the column "FG-2" is scaled up by changing its length from 30 to 100 cm.

Table 7.3 shows that there are fewer stages of fructose in the new column ("FG-4") than in the old column ("FG-2"). This causes a peak broadening of fructose, thus worsening the separation of glucose and fructose (Figure 7.3). This can be avoided by taking fructose as reference component instead of glucose.

In summary the transport-dispersive model together with the simplified degrees of freedom N_i and LF_i is very useful for optimization and scale-up of chromatographic processes.

7.1.3.4 Examples for SMB Processes

An analogous procedure can also be applied to SMB processes (Table 7.4). The following compares the axial concentration profile of several eight-column SMB processes with different operating and design parameters but identical number of stages and dimensionless flow rate in each SMB section (m_j) (Table 7.5). For this purpose we use the same separation problem, a racemic

Table 7.3 Different design and operating parameters with $N_{i,new}$ less than or equal to $N_{i,old}$.

	$\dot{V}$ (ml min^{-1})	t_{inj} (s)	L_c (cm)	d_c (cm)	N_{Glu} (–)	N_{Fruc}	LF_{Glu} (–)	LF_{Fruc}
FG-2	2.4	428.0	30	2.6	100.0	48.2	0.5	0.3
FG-4	11.5	293.4	100	2.6	100.0	38.8	0.5	0.3

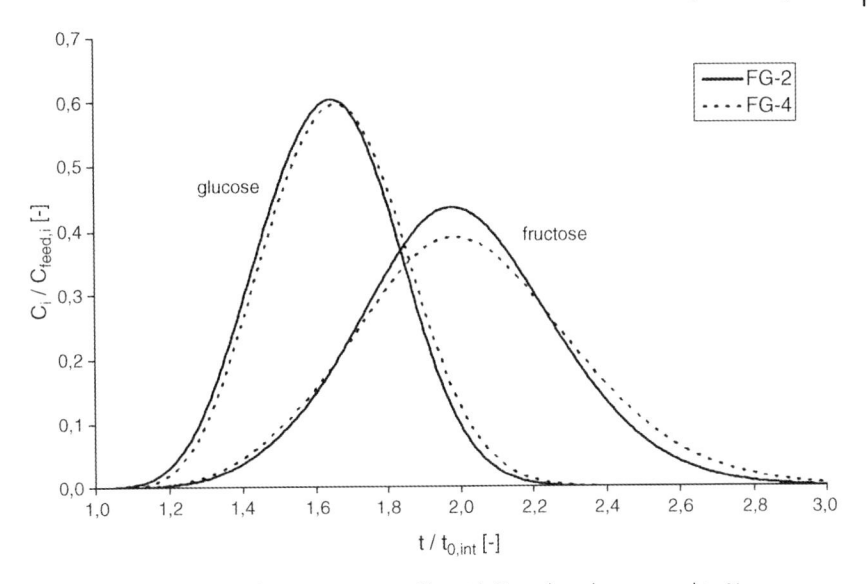

Figure 7.3 Comparison of concentration profiles with $N_{i,\text{new}}$ less than or equal to $N_{i,\text{old}}$.

Table 7.4 Identical dimensionless parameters for the SMB processes.

SMB-1, -2, and -3	
$N_{\text{tot},R}$	66.5
m_{I}	34.0
m_{II}	6.9
m_{III}	10.9
m_{IV}	5.7

Table 7.5 Different operating and design parameters for SMB plant.

	SMB-1	SMB-2	SMB-3
N_{COL} (−)	8 (2 per section)	8 (2 per section)	8 (2 per section)
L_c (cm)	4.5	5.2	7.6
d_c (cm)	2.5	2.5	2.5
t_{shift} (s)	153.7	153.5	153.0
$\dot{V}_{\text{des}}$ (ml min^{-1})	68.3	79.1	115.9
$\dot{V}_{\text{ext}}$ (ml min^{-1})	65.4	75.7	111.0
$\dot{V}_{\text{feed}}$ (ml min^{-1})	9.7	11.2	16.4
$\dot{V}_{\text{raf}}$ (ml min^{-1})	12.6	14.5	21.3

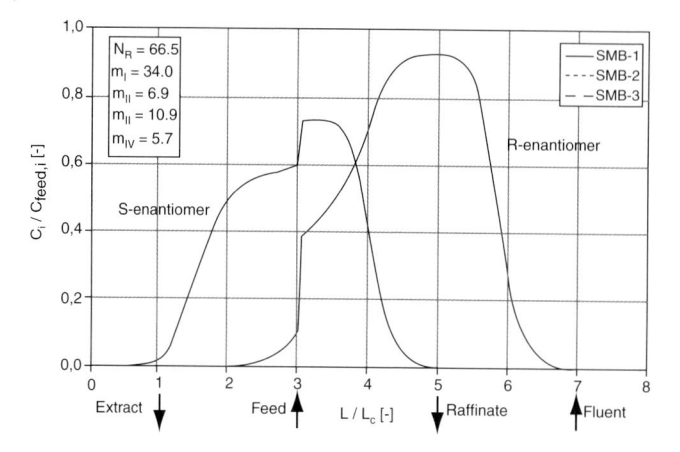

Figure 7.4 Equality of axial concentration profiles in SMB plant (EMD53986).

mixture EMD53986, and the same feed concentration (i.e., $2.5\,\mathrm{g\,l^{-1}}$) as above. Once again, R-enantiomer is used as reference component.

After the process simulation applying the experimentally validated model, all axial concentration profiles are plotted in a dimensionless diagram using $c_i/c_{feed,i}$ and L/L_c as axes. These profiles are taken at the end of a shifting interval from quasi-steady state. Figure 7.4 illustrates that all concentration profiles are completely identical.

Similar to batch processes, the transfer of number of stages is not always as perfect as in the case of enantiomeric separation of EMD53986. If significant deviations between the concentration profiles are observed, another component has to be taken as reference component so that the total number of stages of every component in the new plant is greater than or equal to the number of stages in the old plant (Equation 7.44).

7.2
Batch Chromatography

7.2.1
Fractionation Mode (Cut Strategy)

Within batch chromatography there exist many approaches for optimization. Basically, though, they can be classified into two approaches, which differ in their fractionation mode or cut strategy (Figure 7.5).

In cut strategy II the products are collected consecutively, that is, the next fraction collection starts immediately after the previous fraction. High purity demand can therefore only be satisfied by reducing the feed throughput. Furthermore, purity

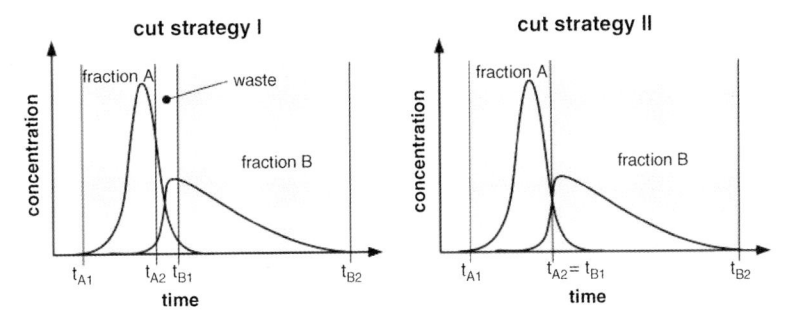

Figure 7.5 Cut strategies for a binary mixture.

and yield are coupled and cannot be varied independently. A high purity demand will consequently result in high yield. One limit of this cut strategy is the touching band or baseline separation. In this special case the concentration profiles do not overlap and 100% purity and yield are gained.

Cut strategy I allows a waste fraction between the product fractions, which can be either discarded or processed further (e.g., by recycling the waste fraction or by application of other separation steps, such as crystallization, to purify it). Due to the introduction of a waste fraction the optimization problem gains additional degrees of freedom, that is, times for the beginning and the end of waste collection. These additional degrees of freedom, however, will not increase the complexity of the optimization, because they are pinpointed automatically by the purity demand, which serves as a boundary condition for the optimization problem.

By use of cut strategy I the optimization problem can now be formulated with greater flexibility. For example, high purity demand can be satisfied either by reducing feed throughput or by increasing waste fraction, thus reducing the yield of separation. It is, however, essential in this cut strategy to consider the additional cost due to feed loss or cost for further treatment of waste fraction. Separation cost optimization with very high feed cost will often result in a very small waste fraction or even "baseline separation," which can be interpreted as the marginal case for cut strategy I if t_{A2} is equal to t_{B1} (Figure 7.5). However, if feed cost is comparable with other costs, then it is more favorable to apply a waste fraction, thus reducing the yield but maximizing feed throughput at constant purity. Strategies to determine suitable cut times for multicomponent mixtures are also discussed by Shan and Seidel-Morgenstern (2004).

Since cut strategy II is only the marginal case of cut strategy I and since the application of cut strategy I provides more flexibility and often leads to a better global optimum, only *cut strategy I* will be used for model-based optimization in this chapter.

7.2.2
Design and Optimization of Batch Chromatographic Columns

7.2.2.1 Design and Optimization Strategy

The main idea behind the design and optimization strategy for batch chromatographic columns is the equality of concentration profiles if their dimensionless parameters (i.e., number of stages and loading factor) are identical. As mentioned previously, column performance under this condition will also be identical (e.g., volume-specific productivity, specific eluent consumption, and yield at a defined purity). In other words, process performance will only vary if the number of stages and loading factor are varied. Therefore, it does not really matter which design or operating parameters are varied during the optimization as long as this variation does not involve a variation of number of stages and loading factor. Based on this main idea, a strategy can be developed that is suitable for the individual demand and boundary condition of the separation problem.

This chapter proposes a strategy that should be suitable for many separation problems. The following strategy should also be suitable for the optimization of operating parameters of an existing plant as well as the design of a new column:

- **Applied column model:** experimentally validated transport-dispersive model.
- **Applied cut strategy:** waste fraction is allowed (cut strategy I, Figure 7.5).
- **Objective function:** to be able to compare total separation costs independently of production rate, the total separation cost per product amount $C_{spec,total}$ is used. Since the consideration of fixed costs in the cost structure is usually handled differently and is subject to company policies, it does not make sense to consider the fixed costs in this book. Therefore, the optimization is focused on variable costs only (Section 7.1.1.2):

$$C_{spec,total} = C_{spec,el} + C_{spec,ads} + C_{spec,crudeloss}$$
$$C_{spec,total} = EC \cdot f_{el} + \frac{\varrho_{ads}}{VSP \cdot t_{life}} \cdot f_{ads} + \frac{1-Y}{Y} \cdot f_{feed} \qquad (7.56)$$

- **Boundary conditions:** purity, maximum pressure drop Δp_{max}, and desired production rate $\dot{m}_{prod}$. Since purity demand has to be fulfilled, the switch times for the fraction collection (t_{A1}, t_{A2}, t_{B1}, and t_{B2}) can be determined and they are not subjects of optimization.
- **Number of stages:** The number of stages plays a major role in this optimization strategy and it can be easily calculated using the HETP equation (Equation 7.37). Therefore, experimental HETP coefficients for every feed component, or at least for the reference component in the separation problem, have to be acquired before running this optimization strategy.

The first step in this design strategy is to assume any initial values for the column geometry (L_c and d_c) (Figure 7.6). If only the operating parameters of an existing chromatographic plant have to be optimized, L_c and d_c should be the length and diameter, respectively, of the current column. Using these initial values a series of dynamic simulations can be started to determine the values of following functions

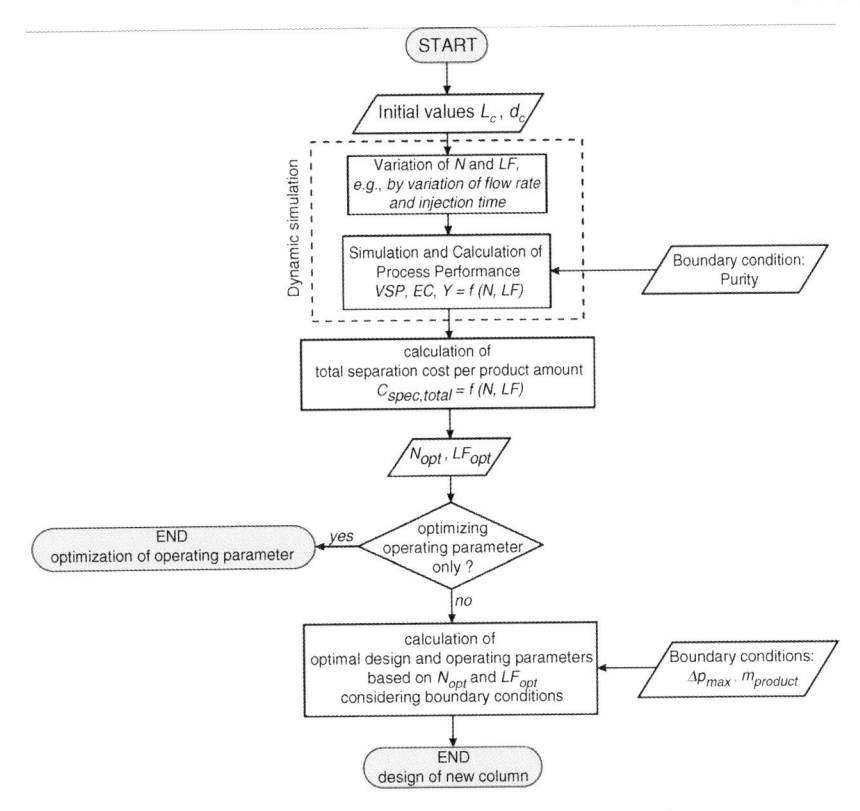

Figure 7.6 Design and optimization strategy for batch chromatographic column.

that characterize the column performance at different numbers of stages and loading factors:

- volume specific productivity VSP;
- specific eluent consumption EC;
- yield at predefined purity Y.

Provided prices for eluent, adsorbent, and feed loss are known, the total separation cost per product amount $C_{spec,total}$ at each number of stages and loading factor can also be calculated (Equation 7.56). As explained in Section 7.1.3.1, it is sufficient in most cases to take the number of stages and loading factor of a reference component into account. The number of stages and loading factor can be easily varied, for example, by varying the injection time t_{inj} and flow rate $\dot{V}$, whereby the maximum pressure drop can be neglected for the time being. However, during optimization of an existing plant it has to be taken into account that a very small number of stages (i.e., very large flow rates) cannot be realized due to the limiting pressure drop of the plant.

Afterwards, with the help of the results from dynamic simulation, the optimal number of stages N_{opt} and loading factor LF_{opt} can be determined where the value of objective function (i.e., total separation cost per product amount) is at its minimum. The applied total separation cost per product amount depends on productivity, eluent consumption, and yield, which in turn depend only on the number of stages and loading factor. Therefore, any process with an identical number of stages and loading factor will also have the same total separation cost per product amount. Please consider that this is only valid for variable costs. Fixed costs should not be expressed as functions of number of stages and loading factor, because they depend only on instrumental effort, not process performance. Therefore, consideration of fixed costs in the total separation cost would require an additional optimization loop to optimize the instrumental effort. However, in practice, fixed costs will hardly be an object of global optimization. In most cases there will be several options and the optimization task is reduced to several case studies.

When optimizing the operating parameters of an existing plant, there is no need to determine the optimal column design with the help of N_{opt} and LF_{opt}. Optimization procedures can be terminated at this point because the operation of the batch column already reaches its optimum here. However, during the design of a new column one more step is necessary, to determine the optimal column geometry and operating parameters based on N_{opt}, LF_{opt}, and the boundary conditions:

1) N_{opt} and Δp_{max}: Using optimal number of stages (N_{opt}), the unknown interstitial velocity, $u_{int,opt}$, and column length, $L_{c,opt}$, can be correlated as

$$N_{opt} = \frac{L_{c,opt}}{HETP(u_{int,opt})} = \frac{L_{c,opt}}{A + Cu_{int,opt}} \qquad (7.57)$$

where A and C are the HETP coefficients of the reference component. At the same time, the maximum limit for pressure drop should not be violated. If the Darcy equation (Equation 2.22) is applied, $L_{c,opt}$ and $u_{int,opt}$ can also be correlated as

$$\Delta p_{max} = \psi \cdot \frac{u_{int,opt} \cdot L_{c,opt} \cdot \eta_l}{d_p^2} \qquad (7.58)$$

It is very often assumed that $u_{int,opt}$ for a column is always the maximum allowed interstitial velocity limited by the pressure drop for a certain column length. But it has to be kept in mind that for every performance criterion there are numerous optimal combinations of interstitial velocities and column lengths, which are represented by one optimal number of stages, N_{opt}, provided mass transfer or axial dispersion is dominating (Section 7.1.2.2). Scherpian (2009) investigated this effect in case studies. He points out that in case of congruent dimensionless concentration profiles eluent consumption and yield are constant for all combinations of column length and interstitial velocity. But the specific productivity will vary depending on the A-term of the Van Deemter equation. Additionally, productivity will decrease if the column length is shortened and is limited by the maximum allowable pressure drop. The latter indicates that N_{opt} cannot be realized if the selected column length is too large, so that the corresponding interstitial velocity exceeds the maximum velocity due to the limiting pressure drop. Therefore, the column length should be so selected that the optimal number of stages can be realized at the maximum or less pressure drop. Both unknown variables ($L_{c,opt}$ and $u_{int,opt}$) can therefore be calculated, considering both N_{opt} and Δp_{max}:

$$L_{c,opt} \leq \frac{1}{2} \cdot \left(N_{opt} \cdot A + \sqrt{N_{opt}^2 \cdot A^2 + 4 \cdot N_{opt} \cdot C \cdot \frac{\Delta p_{max} \cdot d_p^2}{\psi \cdot \eta_1}} \right) \quad (7.59)$$

$$u_{int,opt} \leq \frac{\Delta p_{max} \cdot d_p^2}{\psi \cdot \eta_1 \cdot L_{c,opt}} \quad (7.60)$$

If the effect of mass transfer resistance on the HETP equation is very dominant, then Equation 7.59 can be simplified to

$$L_{c,opt} \leq \sqrt{N_{opt} \cdot C \cdot \frac{\Delta p_{max} \cdot d_p^2}{\psi \cdot \eta_1}} \quad (7.61)$$

2) Desired production rate, $\dot{m}_{product}$: Since values that characterize column performance (i.e., yield, EC, and productivity) at the optimal point are known, the required column diameter for a desired production rate can be easily determined using VSP at the optimal point (Equation 7.9):

$$d_{c,opt} = \sqrt{VSP_{opt} \cdot \frac{\dot{m}_{product}}{1 - \varepsilon_t} \cdot \frac{4}{\pi \cdot L_{c,opt}}} \quad (7.62)$$

The optimal flow rate can then be determined:

$$\dot{V}_{opt} = u_{int,opt} \cdot \varepsilon \cdot \frac{\pi}{4} \cdot d_{c,opt}^2 \quad (7.63)$$

3) Optimal loading factor, LF_{opt}: Using the optimal LF the last unknown operating parameter, the injection time, can be determined:

$$t_{inj,opt} = \frac{LF_{opt} \cdot q_{sat} \cdot \pi \cdot d_{c,opt}^2 \cdot L_{c,opt} \cdot (1 - \varepsilon_t)}{4 \cdot c_{feed} \cdot \dot{V}_{opt}} \quad (7.64)$$

7.2.2.2 Process Performance Depending on Number of Stages and Loading Factor

To improve the basic understanding of column performance in a batch chromatographic separation, the dependency of yield, eluent consumption, and productivity on the applied dimensionless parameters (i.e., number of stages and loading factor) will be investigated in here. Once again, the batch chromatographic separation of a racemic mixture EMD53986 on chiral stationary phase Chiralpak AD from Daicel (20 μm particle diameter) will be used as an exemplary chromatographic system. All the figures in this chapter result from the dynamic simulation of an experimentally validated column model.

Figure 7.7 illustrates the dependency of yield for 99% purity on the dimensionless parameters, showing that high yields can be realized at very low column loadings (which correspond to loading factors <1). Here, only a small amount of waste fraction is required. As the loading factor increases, the overlapping region in the concentration profile grows rapidly. This causes the amount of waste fraction to

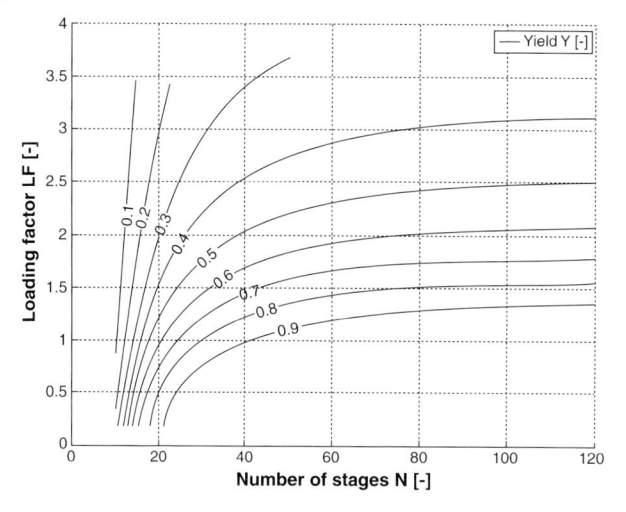

Figure 7.7 Dependency of yield for 99% purity on number of stages and loading factor.

increase in order to meet the purity demand, thus reducing the yield. However, the size of the overlapping concentration region, and consequently also the amount of waste fraction, can be reduced by increasing the number of stages, which also increases the peak resolution. Therefore, high yield at relatively higher loading factor can be realized in a column with a large number of stages.

Contrary to yield, the productivity has a maximum in the investigated parameter region (Figure 7.8). An increase of number of stages increases the yield due to better peak resolution. Separation becomes more efficient and a higher loading factor can be applied at constant yield, which, at first, results in a steep rise

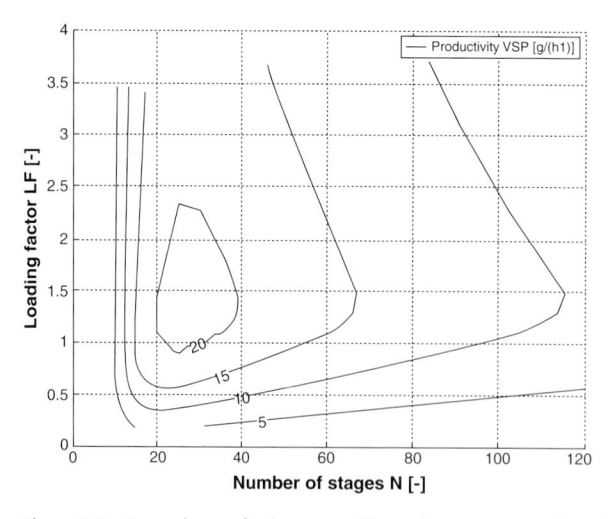

Figure 7.8 Dependency of volume-specific productivity on number of stages and loading factor.

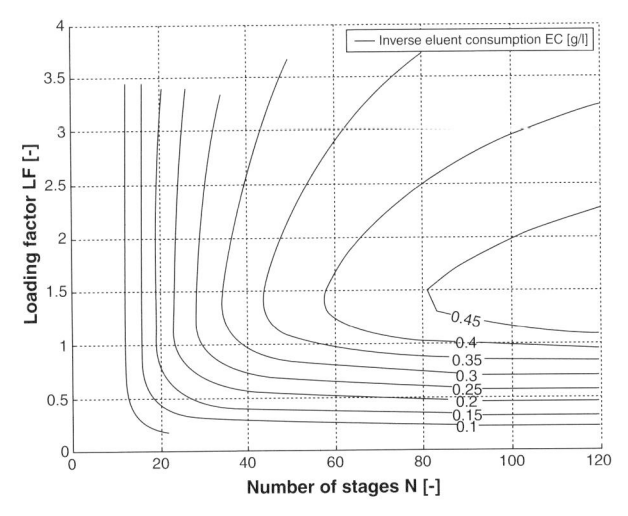

Figure 7.9 Dependency of specific eluent consumption on number of stages and loading factor.

of production rate and volume-specific productivity. But a higher number of stages also involves greater column length (i.e., bigger column) and/or reduction of interstitial velocity. The decrease of interstitial velocity increases the mean residence time of solutes in the column, thus increasing the batch cycle time. If the interstitial velocity falls below a certain point, the high batch cycle time will start to reduce production rate in spite of the higher loading factor. Therefore, a further increase in the number of stages will cause a smooth decrease in the volume-specific productivity, which is defined as production rate over required adsorbent volume.

The influence of loading factor on productivity is roughly similar to that of the number of stages. At first, the higher the loading factor, the higher is the production rate, and thus productivity also increases. However, as seen in Figure 7.7, the yield already drops rapidly at low loading factors (at about 1). Therefore, a further increase of loading factor will cause the production rate and productivity to decrease despite higher feed throughput since a considerable amount is lost in the waste fraction.

Figure 7.9 illustrates the *inverted* values of specific eluent consumption in dependency of dimensionless parameters. Higher values are therefore the more favorable values in Figure 7.9 as they indicate small eluent consumption. The dependency of specific eluent consumption on loading factor is similar to that of productivity, although the absolute eluent amount does not depend on the loading factor. However, in this chapter *specific* eluent consumption is used instead of just the amount of consumed eluent in order to compare the efficiency of eluent consumption. Since specific eluent consumption is defined as eluent per product amount (or as flow rate over production rate), a higher loading factor causes, at first, more favorable values, but further increase in loading factor shifts the specific eluent consumption to less favorable values due to decreasing production rate caused by an increasing amount of waste fraction.

Contrary to productivity, however, a high number of stages does not result in less favorable values for specific eluent consumption. As mentioned before, a high number of stages applies in columns with relatively low interstitial velocity. A definite increase in the number of stages causes a decrease of flow rate (eluent flow rate) at the same degree, whereas the production rate drops slowly. Further increase of their number will, therefore, continue to improve the specific eluent consumption, at least in the parameter region investigated in this chapter.

A very important conclusion from the case study in this chapter is that yield, productivity, and eluent consumption represent oppositional optimization goals during the design and optimization of batch chromatographic column. To demonstrate this, three particular examples within the N–LF diagram for a given column geometry will be compared. Each of these exemplary operating points represents one of the following three extreme cases:

- maximum productivity;
- high yield;
- low eluent consumption.

The positions in N–LF diagram and the values of the objective functions for each case are shown in Figure 7.10 and Table 7.6.

A closer look at the chromatogram at maximum productivity (Figure 7.11) reveals a large overlapping area (i.e., large waste fraction) between the concentration peaks of the pure components. This confirms the fact that maximum productivity can be reached in the region with relatively higher loading factor, where the yield already drops significantly below 100% (i.e., 60%). Furthermore, although productivity is at

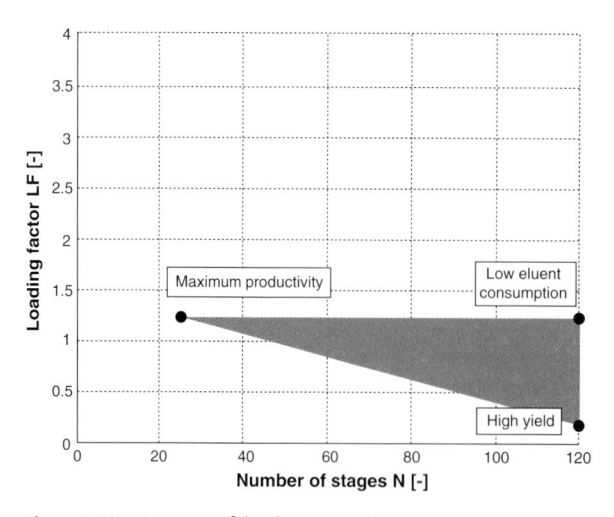

Figure 7.10 Positions of the three exemplary operating points.

7.2 Batch Chromatography

Table 7.6 Objective functions at the exemplary operating points.

	$\dot{V}$ (ml min^{-1})	t_{inj} (s)	N_R (−)	LF_R (−)	VSP (g l^{-1} h^{-1})	EC (l g^{-1})	Y (−)
Maximum productivity	23.5	208.0	25	1.25	23.36	4.07	0.60
Low eluent consumption	4.7	1045.2	120	1.25	9.89	1.92	0.94
High yield	4.7	167.2	120	0.20	1.80	10.53	1.00

Column geometry: $d_c = 2.5$ cm, $L_c = 10.8$ cm.

its maximum, there is still an optimization potential for the eluent consumption (Table 7.6).

High numbers of stages for a given column always represent low flow rate. Additionally, by increasing the number of stages the peak width will become smaller (Figure 7.11), thus reducing the batch cycle time but increasing the production rate. Both phenomena benefit the eluent consumption. Low eluent consumptions can therefore be reached in the region with very high number of stages, although the productivity already decreases significantly below its maximum value (Table 7.6).

As can be seen in Table 7.6 and Figure 7.12, a high yield can be reached by reducing the amount of feed introduced into the column (i.e., low loading factor) and by reducing the flow rate (i.e., high number of stages). However, at very low loading factors the productivity of the column and the efficiency of eluent usage will deteriorate seriously. Therefore, low loading factors benefit only yield of separation and not utilization of adsorbent and eluent.

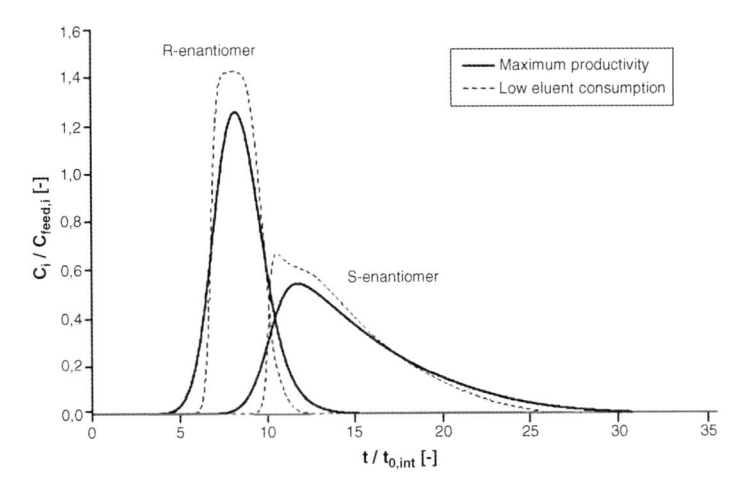

Figure 7.11 Comparison of chromatograms (maximum productivity; low eluent consumption).

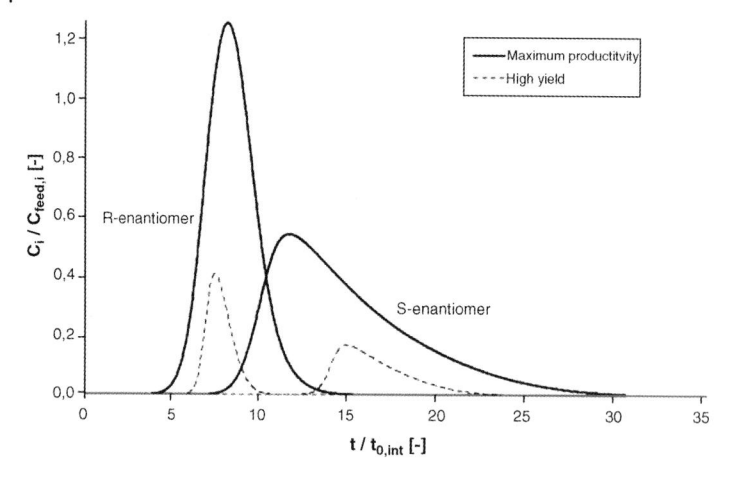

Figure 7.12 Comparison of chromatograms (maximum productivity; high yield).

The only evaluation method that unifies the contributions of these optimization goals is the total separation cost. The position of the optimum after a cost optimization depends heavily on the magnitude of each contribution relative to each other. If the price of adsorbent or fixed costs (maintenance and capital cost) is a dominant factor, then the cost optimum will coincide with the maximum productivity. For separation problems with low solubility of the components and/or very high eluent price, the cost optimum is approximately equal to the minimum of eluent consumption (i.e., very high number of stages). In other words, only the optimization of total separation cost leads to the "real" (economically) optimum of the separation problem (Figure 7.13).

7.2.2.3 Other Strategies

In many studies production rate is generally maximized while tolerating a decrease in yield, thus neglecting the potentially very high costs resulting from feed loss. To solve this problem Felinger and Guiochon (1996) proposed the product of yield and production rate as a new objective function. Other investigations of Felinger and Guiochon (1998) use this new objective function in dealing with the optimization and comparison of elution, displacement, and nonisocratic chromatography. This

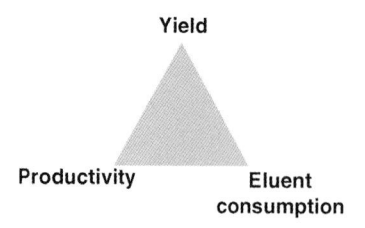

Figure 7.13 Triangle of oppositional optimization goals.

investigation covers a wide range of retention factors, but neglects the fact that the retention factor in real systems generally cannot be varied independently of other material-dependent parameters. Further information on economics of chromatographic separations is given in Guiochon et al., (2006).

Strube and Dünnebier applied a different approach by optimizing elution chromatography using cut strategy II. Strube used a transport-dispersive model for his simulation studies and optimized the productivity of batch (and SMB) separations (Strube, 2000). Dünnebier (2000) uses a general rate model and implements yield and purity demand as boundary conditions. However, since the "touching band" strategy (cut strategy II) is used, yield and purity are not independent, thus restricting the maximum achievable productivity.

Katti and Jagland (1998) have performed a systematic analysis of different parameters that influence the separation cost. They investigate cost contributions of eluent, feed loss, packing materials, equipment, and investment. Each cost contribution is also examined individually, depending on various free optimization parameters. Their work gives good insight into the general correlations and dependencies in a cost optimization.

Epping (2005) and Jupke, Epping, and Schmidt-Traub (2002) optimized batch (and SMB) processes, regarding not only the common process performance (eluent consumption, productivity, etc.) but also the total separation cost, which is the summation of various cost contributions. Based on design data of Dünnebier, Weirich, and Klatt (1998) and specific cost data from Jupke, Epping, and Schmidt-Traub (2002), Chan, Titchener-Hooker, and Sùrensen (2008) extended the economic evaluation of standard batch (and SMB chromatography) to single columns with recycling as well as recycling with peak shaving, but their findings are restricted to linear isotherms. For the economic optimization of large-scale plants (2000 kg a^{-1}) they investigated three scenarios: maximizing a weighted objective function (recovery·yield) proposed by Felinger and Guiochon (1996), minimizing production cost, and maximizing annual net profit. Their comparison of standard batch (and SMB) production affirms the general trend that batch production is favorable for short-term application while multicolumn processes are more economic for long-term production (>5 years).

7.3
Recycling Chromatography

The principles of closed loop as well as steady-state recycling chromatography have already been explained in Sections 5.1.3 and 5.1.4. In general, these process schemes may be chosen if for standard batch chromatography the number of plates has to be increased in order to get an efficient separation and the column length is restricted by the maximum allowable pressure drop. At the column outlet certain volumes of pure products are separated while the rest of the mixture is recycled to the column inlet. Through this a virtually longer column is achieved and maximum pressure drop is not exceeded.

Additionally to the operating and design parameters known for standard batch chromatography, cut or fraction times have to be determined. For the design of SSRC these are the cut times for the steady state while for CLRC the cut times of the last cycle (cut times of all cycles in case of peak shaving) have to be determined. As shown by Scherpian and Schembecker (2009), Kaspereit and Sainio (2011), or von Langermann *et al.* (2012), the performance of these processes can be theoretically calculated by rigorous process simulation based on the transport-dispersive model without decreasing accuracy inadmissibly because of accumulation of inaccuracies. For the design of new processes it is recommended first to determine the approximate operating map in order to estimate optimal process conditions. These data can be used to improve process design by rigorous process simulation or experiments. For a first approximate design the following design strategies can be applied. Section 7.3.1 describes shortcut methods for the approximate determination of operating conditions for SSRC while Section 7.3.2 outlines how for CLRC, for example, satisfying lab results are scaled up to production scale and how the process data can additionally be optimized based on experiments with the production plant.

7.3.1
Design of Steady-State Recycling Chromatography

The major challenge in designing any recycling chromatography process is to determine the cut times required to fulfill the given purity requirements. In most cases this task can be accomplished only using a dynamic process model or by experimental trial and error. However, for steady-state recycling in mixed recycle mode (MR-SSR, Section 5.1.4) simple shortcut design methods allow estimating a priori and rather accurately the cut times required in steady state – without performing any dynamic simulation. This facilitates a straightforward basic process design and rough performance evaluation, and is the basis for a subsequent more detailed design, optimization, and scale-up.

Two such shortcut approaches exist. The first neglects any band broadening effects and applies to systems with Langmuir isotherms (Equation 2.51), of which the parameters have to be known. Based on equilibrium theory (Section 6.2.3), Bailly and Tondeur (1982) derived explicit equations for designing MR-SSR processes for the case of complete separation, that is, 100% pure products. This approach was extended by Sainio and Kaspereit (2009) to any combination of purity requirements. In connection with a recent extension of equilibrium theory for Langmuir isotherms by Siitonen and Sainio (2011), these methods predict explicitly, for a given column geometry and injection volume, the cut times and the steady state of the process a priori. Dynamic simulation is not required.

The second methodology, which is explained in more detail below, is an extension of the former to systems with significant dispersion and favorable isotherms in general. Although it covers a broader range of problems, it is even simpler to

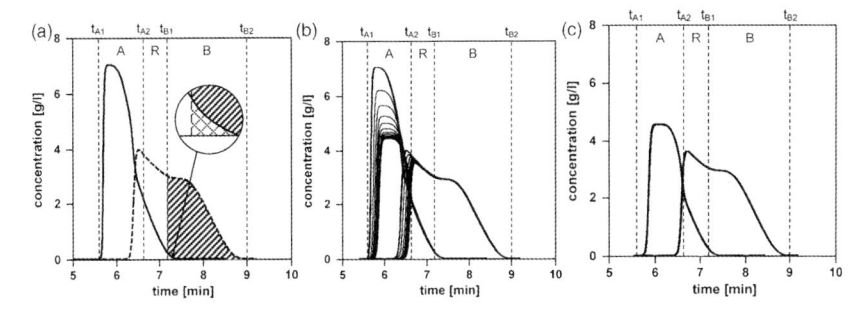

Figure 7.14 Shortcut design of SSRC according to Kaspereit and Sainio (2011). (a) Determination of cut times by integrating the rear of an initial batch chromatogram. (b) Transient behavior of the designed SSR process (overlay of 50 cycles, steady-state purities 95.6% and 99.4%, respectively). (c) Improved design and elimination of start-up phase by applying the shortcut method to the last chromatogram of (b) (overlay of 50 cycles, purities 96.0% and 99.3%).

apply. It requires only a measured and deconvoluted chromatogram. Neither a process model nor isotherm parameters are required.

Figure 7.14a illustrates the approach for a system with bi-Langmuir isotherms (Equation 2.53). The required purities are set to $Pu_{A,req} = Pu_{B,req} = 99\%$. Further details of the simulation results are given in the reference. For the injection volume to be investigated, an experimental chromatogram is analyzed to obtain the underlying individual concentrations. If available, also simulated data can be used (as in the figure). Based on the chromatogram, the four cut times required to achieve the purity requirements in steady state can be determined (Kaspereit and Sainio, 2011). During operation of an SSR process, these times are being held constant with respect to the start of each injection.

As for the four cut times, the end of fraction product fraction B, t_{B2}, is selected as the end of the chromatogram – for example, based on a low threshold concentration. If the initial slope of the isotherm of the stronger adsorbing component is known, Equation 7.65 might be used instead:

$$t_{B2} = t_{inj} + t_0 \left(1 + \frac{1 - \varepsilon_t}{\varepsilon_t} \frac{\partial q_2}{\partial c_2}\bigg|_{c_1 = c_2 = 0}\right) \quad (7.65)$$

where t_{inj} is the duration of the injection. The start of fraction B, t_{B1}, which defines the end of the recycle fraction R, is found by backwards integrating the dispersive rear of the profile, until the determined amounts m_{iB} ($i = 1, 2$) fulfill exactly the purity requirements, $Pu_{B,req}$, on this fraction:

$$t_{B1} = \max\left\{t < t_{B2} : Pu_B(t) = \frac{m_{2B}(t)}{m_{1B}(t) + m_{2B}(t)} = Pu_{B,req}\right\} \quad (7.66)$$

It can be shown that for favorable isotherms this rear part of the chromatogram does not change during SSR operation (Sainio and Kaspereit, 2009; Kaspereit and

Sainio, 2011). Thus, from a global material balance and the cycle-invariant amount of the target compound in fraction B, m_{2B}, an expression can be derived for the end time of fraction A, equal to the start of the recycle R:

$$t_{A2} = t_{B1} - t_{inj} + \frac{m_{2B}}{Y_2 \dot{V} c_{2F}} \tag{7.67}$$

where c_{2F} is the fresh feed concentration of this compound. Y_2 is its overall yield, which is already fully specified through the two given purity requirements:

$$Y_2 = \frac{Pu_{B,req}}{100\% - Pu_F} \frac{Pu_{A,req} - Pu_{B,req}}{Pu_{A,req} + Pu_{B,req} - 100\%} \tag{7.68}$$

Finally, for the start of fraction A, t_{A1}, some suitable value $t_{A1} \geq t_0$ is chosen. This time can be optimized later.

Figure 7.14b shows an overlay of the elution profiles of 50 SSR cycles performed using the cut times determined in Figure 7.14a, which is the reason why the elution profiles of the first cycle (largest concentrations) are identical to those in (a). The process reaches a steady state after about 10 cycles. Note that the second fraction is virtually invariant from cycle to cycle, while the purity for fraction A is fulfilled only after the process, which was started here with a full injection of fresh feed, reaches steady state. The purity requirement on A can be fulfilled also during start-up if the process is initiated using a lower injection concentration. However, the most elegant option is to eliminate the start-up completely, as shown in Figure 7.14c. As proposed by Bailly and Tondeur (1982), this is achieved when using for the first few injection(s) a prepared mixture that corresponds to the steady-state injections, and switching to the fresh feed after these eluted from the column.

This efficient approach was successfully validated experimentally by Kaspereit and Sainio (2011) for two cycloketones with bi-Langmuir isotherms, as well as for an enantioseparations for a pharmaceutically relevant compound with complex quadratic isotherms (von Langermann et al., 2012).

The design method was extended to processes with an additional solvent removal in the recycle (Siitonen, Sainio, and Kaspereit, 2011). Furthermore, although it does not directly apply to SSR processes operated in closed loop mode, it can be used to obtain initial guesses on cut times and performance of such process.

It is recommended to apply the shortcut method in a parametric study in order to optimize also the injection volume and column dimensions. The results should then be used in a detailed design study where all relevant design parameters are optimized against the performance criteria and objective functions discussed in Section 7.1.1. In particular, the mentioned transport-dispersive model is a good choice for such studies since it affords a high degree of accuracy when simulating recycling processes (Scherpian, 2009; Kaspereit and Sainio, 2011). Scale-up of the process can be performed using such model, preferably using dimensionless representations along the lines of the following section.

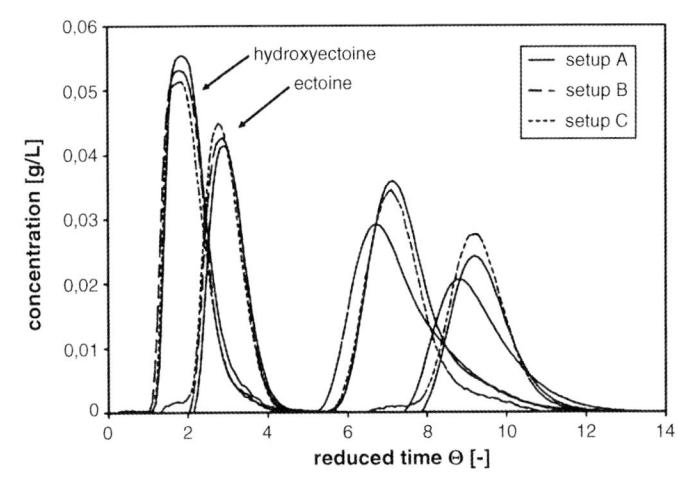

Figure 7.15 CLRC elution profiles of three different experiments for constant loading factors, number of plates, and recycle numbers (Scherpian and Schembecker, 2009).

7.3.2
Scale-Up of Closed Loop Recycling Chromatography

In Section 7.1.2 it is shown that in case of high efficient chromatography the number of dimensionless parameters reduces and instead of Stanton and Péclet numbers the number of plates is the dominating parameter. In order to proof these findings for CLRC, Scherpian and Schembecker (2009) measured the elution profiles for three different process setups for which the number of plates, the loading factor, and the recycle number are constant. The characteristic data of these experiments are presented in the reference.

Figure 7.15 presents the measured CLRC elution profiles. For the first cycle all profiles are in very good agreement and confirm the expected congruency. The deviations of setup B in the second and third cycles are due to dominating plant effects.

Comparable elution profiles for CLRC with peak shaving as depicted in Figure 7.16 are again in good congruence for types A and C. In case of peak shaving pure product is extracted from each recycle. The slight deviations of type B are caused by backmixing effects as mentioned above. The characteristic process data are the same as in Figure 7.15.

In Section 7.1 it has been confirmed that elution profiles of chromatographic processes are congruent if the parameters of the dimensionless transport-dispersive model are identical; based on these results Section 7.2 presents an optimization strategy for batch chromatography. A corresponding procedure is applicable for the theoretical optimization of recycle processes but it is very time consuming. An alternative for practical purposes is a scale-up based on experimentally optimized lab results or a practically satisfying process.

7 Model-Based Design, Optimization, and Control

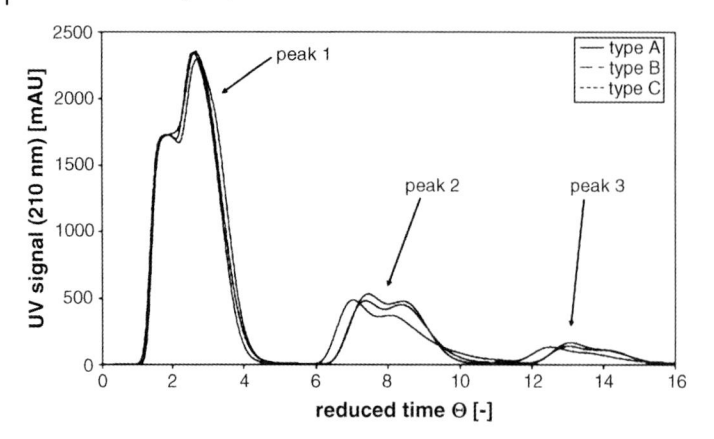

Figure 7.16 CLRC elution profiles with peak shaving for constant loading factors, number of plates, and recycle numbers (Scherpian and Schembecker, 2009).

Figure 7.17 depicts a strategy to scale-up CLCR processes. The idea is to start with experimental results that shall be transferred into a production-scale process. The same procedure can be applied if optimal figures for the dimensionless parameters have been evaluated by theoretical optimization.

Following the left-hand side of Figure 7.17:

- The scale-up strategy starts with a lab process that has been optimized experimentally.
- In order to determine the relevant parameters, the coefficients of the Van Deemter plot A_i and C_i are measured (Section 6.5.3.1). Additionally the void fraction ε (Section 2.2.2) and the resistance factor ψ (Section 2.3.5) are measured or estimated.
- The dimensionless parameters that have to be equal for the lab- and process-scale plants are defined as follows:

$$\text{Loading factor, } LF_i = \frac{c_{i,\text{feed}} \cdot \dot{V} \cdot t_{\text{inj}}}{q_{\text{sat},i} \cdot V_c \cdot (1 - \varepsilon_t)} \quad (7.69)$$

$$\text{Recycle number, } \Phi = \frac{t_{\text{plant}}}{t_{0,\text{int}}} = \frac{V_{\text{plant}}}{V_c \varepsilon} \quad (7.70)$$

$$\text{Number of plates, } N_i = \frac{L_c}{\text{HETP}_i} \approx \frac{L_c}{A_i + C_i u_{\text{int}}} \quad (7.71)$$

These parameters ensure congruent elution profiles. In order to achieve the same production rate, the yield has to be constant:

$$Y_i = \frac{m_{\text{prod,cyc}}}{c_{\text{feed},i} V_{\text{inj}}} = \frac{\dot{m}_{\text{prod}} \Theta_{\text{cyc}} t_{0,\text{int}}}{c_{\text{feed},i} V_{\text{inj}}} \quad (7.72)$$

where Θ_{cyc} is the reduced time of one injection cycle, that is, the time between two consecutive injections. In case of CLRC with peak shaving the dimensionless

7.3 Recycling Chromatography

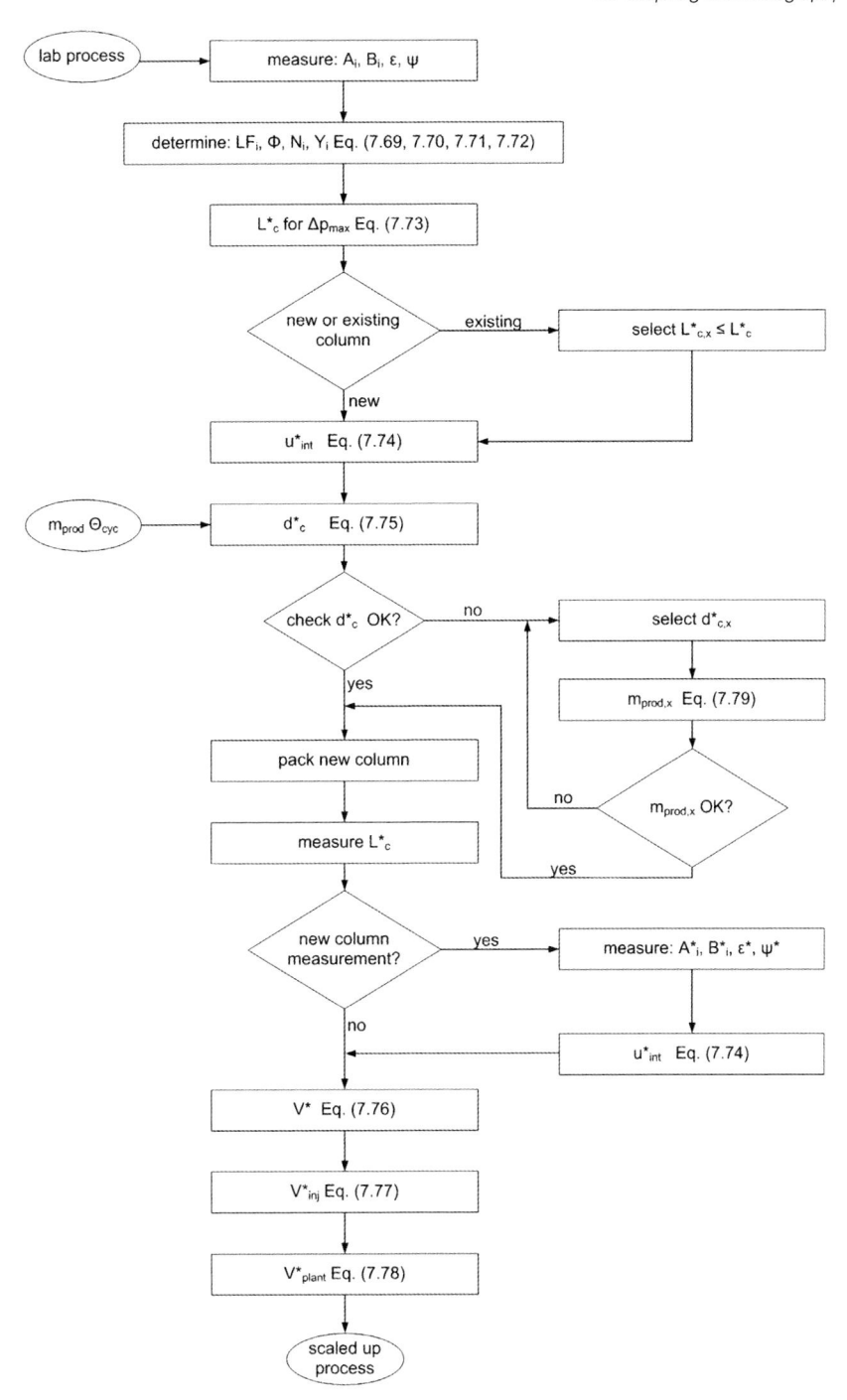

Figure 7.17 Scale-up strategy for closed loop recycling chromatography.

cutting times of the lab process are also the same for the scaled-up process (Scherpian, 2009).

Most parameters are different for each component. As already pointed out in Section 7.1.3, for all components the number of stages of the new column should be greater than or equal to the old one $N_i^* \geq N_i$. Hence, the key component is $N_i^* = N_i$.

- The maximum column length, which can be realized for the given number of stages and the maximum allowable pressure drop, is

$$L_c^* \leq \frac{1}{2}\left(N_i A_i + \sqrt{N_i^2 A_i^2 + 4 N_i C_i \frac{\Delta p_{max} d_p^2}{\psi \eta}}\right) \quad (7.73)$$

- The first decision is to get a new column of the calculated length or to take an already existing column of a given length.
- The interstitial velocity for the new column length is

$$u_{int}^* = \left(\frac{L_c^*}{N_i} - A_i\right)\frac{1}{C_i} \quad (7.74)$$

- The column diameter depends on the expected production rate and the cycle time of the process:

$$d_c^* = \sqrt{\frac{4}{\pi}\frac{\dot{m}_{prod,i}\Theta_{cyc}}{Y_i LF_i(1-\varepsilon_t)u_{int}^*}} \quad (7.75)$$

- After the dimensions are determined the column is packed with the same solid phase used for the lab process. The real column length is measured and it has to be decided if the parameters of the new packing shall be determined or not.
- Finally the volumetric flow rate, the injection volume, and volume of the plant are determined:

$$\text{Volumetric flow rate, } \dot{V}^* = \frac{\pi}{4}d_c^{*2}u_{int}^*\varepsilon \quad (7.76)$$

$$\text{Injection volume, } V_{inj}^* = \frac{LF_i q_{sat,i} V_c^*(1-\varepsilon_t)}{c_{feed,i}} \quad (7.77)$$

$$\text{Plant volume, } V_{plant}^* = \Phi V_c^* \varepsilon \quad (7.78)$$

The design strategy described so far is a theoretical approach. In practice, additional aspects will influence the scale-up strategy. They are characterized by three decision cycles that are depicted on the right-hand side of Figure 7.17:

- First, a column of a given length $L_{c,x}^*$ may be selected. It has to be equal to or smaller than the maximum allowable column length (Equation 7.73).

- Second, the calculated column diameter does not match with standard dimensions or an available column shall be used. In this case a certain diameter $d_{c,x}^*$ is selected and the resulting production rate is

$$\dot{m}_{\text{prod},i} = \frac{\pi}{4} d_c^2 \frac{Y_i \text{LF}_i (1 - \varepsilon_t) u_{\text{int}}}{\Theta_{\text{cyc}}} \tag{7.79}$$

- Third, the conditions for scale-up from lab to process plant are constant figures for the dimensionless parameters. But in practice it is not certain that the packing of the columns is always identical. Slight variations of the void fraction and HETP may occur. Additionally, differences in the fluid dynamics, especially at the column inlet and outlet, have to be taken into account. The theoretical scale-up strategy ignores these deviations. But in order to make sure that real numbers of plates of both plants are really the same, it is recommended to determine the Van Deemter plot, void fraction, and friction number for the new packing and to correct the interstitial velocity, the flow rate, and the injection volume.

Please note that the scale-up strategy is not influenced by dimensionless number Pe_{plant}. The "plant" that summarizes the effect of all equipment besides the column has two effects: the band broadening of the elution profile and the hold-up. Pe_{plant} stands for the band broadening that depends on the axial dispersion as well as back-mixing and has to be maximized by the equipment design. The hold-up determines the recycle time and is a boundary condition for the chromatographic column. Therefore, the recycle number has to be constant for scale-up.

Batch chromatography is a special case of the CLRC. Therefore, the same scale-up strategy is applicable if the reduced time of one injection and the recycle number are set:

$$\Phi = \frac{t_{\text{plant}}}{t_{0,\text{int}}} = \frac{V_{\text{plant}}}{V\varepsilon} = 0 \quad \text{and} \quad \Theta_{\text{cyc}} = 1 \tag{7.80}$$

7.4
Conventional Isocratic SMB Chromatography

Continuously operated chromatographic processes such as simulated moving beds (SMB) are well established for the purification of hydrocarbons, fine chemicals, and pharmaceuticals. They have proven ability to improve the process performance in terms of productivity, eluent consumption, and product concentration, especially for larger production rates. These advantages, however, are achieved with higher process complexity with respect to operation and layout. A purely empirical optimization is rather difficult and, therefore, the breakthrough for practical applications is linked to the availability of validated SMB models and shortcut methods based on the TMD model as described in Chapter 6.

Starting with the simplest model, the true moving bed (TMB) model, first it will be demonstrated how to determine parameters for the operation of SMB processes. Based on these TMB shortcut methods, a more detailed optimization of operating parameters applying complete SMB models will be presented. Finally, a strategy to determine the optimal design of an SMB unit will be illustrated, based on similar considerations to those already made for optimizing single-column batch chromatography.

7.4.1
Optimization of Operating Parameters

The operating point of an SMB unit is characterized by the flow rates of the liquid phase in the four sections, $\dot{V}_j$, as well as the switching time, t_{shift}. Another degree of freedom is the concentration of the solute in the feed stream – but according to Charton and Nicoud (1995) the feed concentration should be at its maximum to achieve best productivity. For the following considerations feed concentrations as well as temperatures are kept constant.

During operation of an SMB plant the propagation of components to be separated is influenced by the internal fluid flow rates in the different sections as well as the switching time that simulates movement of the solid. By appropriate choice of operating parameters the movement of the less retained component is focused on the raffinate port while the more retained component is collected in the extract stream. Figure 7.18 shows an optimal axial profile of the liquid concentrations at the end of a switching interval after the process has reached a periodic steady state. The adsorption and desorption fronts of both components have to start or stop at given points to achieve complete separation at maximum productivity.

For complete separation the desorption fronts of the two components must not exceed points 1 and 2, respectively, which are located one column downstream the desorbent and extract port. Since the concentration profile displayed in Figure 7.18 demonstrates the situation at the end of a switching interval, all ports will move one column downstream in the very next moment. In the case where the desorption front of component B does exceed point 2, the extract stream, meant to withdraw

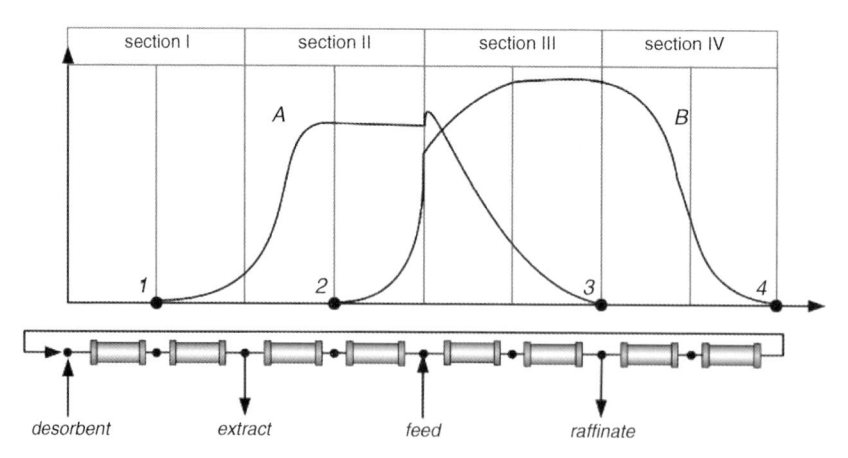

Figure 7.18 Optimal axial concentration profile at the end of a switching interval.

the more retained component A only, will be polluted with B after the ports have been switched. The same applies to point 1. If component A is shifted into section IV, the adsorbent will transfer it to the raffinate port and the raffinate will be polluted. For the adsorption fronts, components A and B must not violate points 3 and 4, respectively, at the end of the switching period. Otherwise, component A will pollute the raffinate, and component B will enter section I and pollute the extract.

To achieve 100% purity for both components it is necessary to fulfill the constraints mentioned before. In addition, it is favorable to push the fronts as far as possible toward points 1–4 to realize the highest throughput as well as the lowest eluent consumption.

If product purities lower than 100% are required, there are generally two possibilities to achieve this with a positive impact on other process parameters:

- A higher feed rate at constant eluent consumption can be realized when the adsorption front of component A exceeds point 3 and/or the desorption front of component B exceeds point 2.
- A lower eluent consumption at constant feed rate can be realized when the adsorption front of component B exceeds point 4 and/or the desorption front of component A oversteps point 1.

In these cases the raffinate will be polluted with the more strongly retained solute A and/or the extract will be polluted with the less retained component B. Depending on purity requirements and the objective function, any combination of these two cases can also be chosen.

Whenever dealing with the operation of SMB units one always has to consider the time the process needs to reach its periodic steady state. Depending on the chromatographic system, the process setup, and the column geometry it might take 15 or more complete cycles to reach steady state.

7.4.1.1 Process Design Based on TMB Models (Shortcut Methods)

Due to the similarity between TMB and SMB processes (Section 6.6.2) the TMB approach is quite often used to estimate the operating parameters of SMB units. Operating parameters for TMB processes are the liquid flow rates in the four sections, $\dot{V}_{j,\text{TMB}}$, and the volumetric flow rate of the adsorbent, $\dot{V}_{\text{ads}}$. To transfer these TMB parameters to SMB operating parameters ($\dot{V}_{j,\text{SMB}}$ and t_{shift}) the interstitial velocity in the SMB process must be equal to the relative velocity of fluid and adsorbent in the TMB process (Section 6.6.2.1). From this follow the relationships listed below:

$$\dot{V}_{\text{ads}} = \frac{(1-\varepsilon) \cdot V_c}{t_{\text{shift}}} \tag{7.81}$$

$$\dot{V}_{j,\text{SMB}} = \dot{V}_{j,\text{TMB}} + \frac{V_c \cdot \varepsilon}{t_{\text{shift}}} \tag{7.82}$$

7.4.1.1.1 Ideal Model, Linear Isotherm
Ruthven and Ching (1989), Hashimoto et al. (1983), and Nicoud (1992) have used analytical solutions of a TMB

process under the assumption of the ideal model (neglecting axial dispersion and mass transfer phenomena) and linear isotherms. In each section j of the TMB process the fluid phase moves with the flow rate $\dot{V}_{j,\text{TMB}}$ in one direction and the solid phase with $\dot{V}_{\text{ads}}$ in the opposite direction. Because, in the ideal model, equilibrium is assumed to be reached immediately, the mass flow $\dot{m}_{i,j}$ of component i depends on these two flow rates and the isotherm.

$$\dot{m}_{i,j} = (\dot{V}_{j,\text{TMB}} - \dot{V}_{\text{ads}} \cdot \varepsilon_p) \cdot c_{i,j} - \dot{V}_{\text{ads}} \cdot (1 - \varepsilon_p) \cdot q_{i,j} \qquad (7.83)$$

Figure 7.19 depicts the directions of migration of two components for a TMB separation.

Storti et al. (1993) introduced the dimensionless flow rate ratio m_j as the ratio between liquid and solid flows in every section:

$$m_j = \frac{\dot{V}_{j,\text{TMB}} - \dot{V}_{\text{ads}} \cdot \varepsilon_p}{\dot{V}_{\text{ads}} \cdot (1 - \varepsilon_p)} \qquad (7.84)$$

This was the impulse for the research groups of Morbidelli and Mazzotti at the ETH-Zurich to develop a method that is most instructive for the understanding of SMB separations and pioneered the design of SMB processes. The basic idea is to determine for sections II and III, where the chromatographic separation of a TMB process has to take place, an operating plane that indicates the area where complete separation of a two-component mixture is possible. Coordinates of this operating plane are the dimensionless flow ratios m_{II}–m_{III}. As shown in Figure 7.20 this plane is bordered by the diagonal and has in case of linear isotherms the shape of a triangle. Therefore, this method is often called triangle theory. For the m_{II}–m_{III} operating points within the

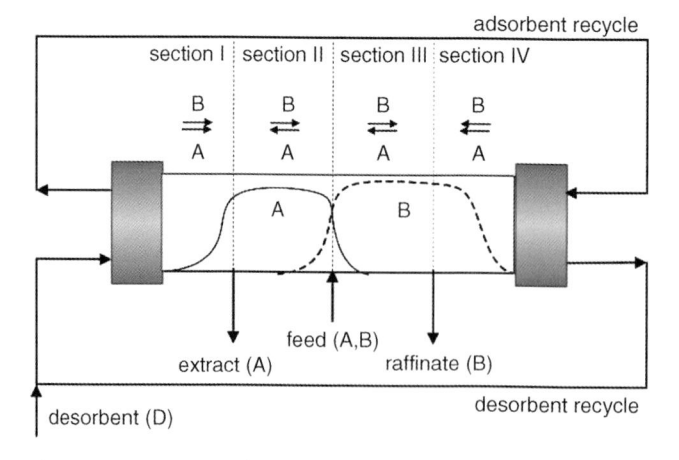

Figure 7.19 Directions of migration of two components in a TMB separation.

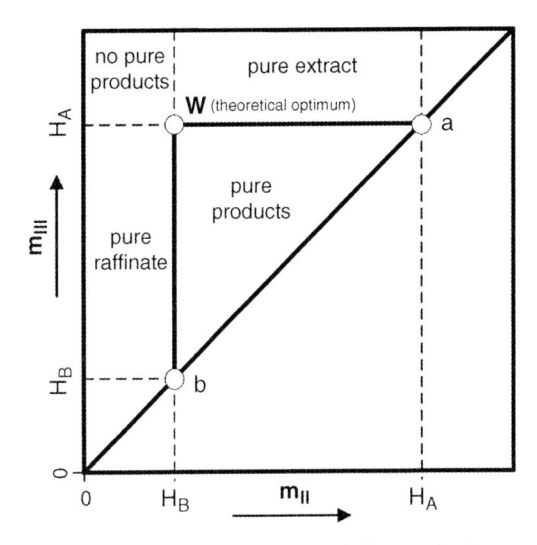

Figure 7.20 Operating plane or triangle diagram for linear isotherms.

triangle a two-component mixture is separated into pure products while outside the triangle only pure extract, raffinate, or impure products are achieved. The decisive point is that because of the similarity between TMB and SMB processes the TMB operating plane is also applicable for the estimation of SMB operating conditions if the TMB parameters are transferred into SMB parameters according to Equations 7.81 and 7.82. The basic principles of this method and the application for different types of isotherms are outlined in the following papers: Storti et al. 1993,1995 and Mazzotti, Storti, and Morbidelli (1994,1996,1997).

The analytical solution of the ideal model for linear isotherms to the following constraints for complete separation of a two-component feed mixture (A and B):

Section I : $\quad m_I \geq H_B, \quad m_I \geq H_A$ (7.85)
Section II : $\quad H_A \geq m_{II} \geq H_B$ (7.86)
Section III : $\quad H_A \geq m_{III} \geq H_B$ (7.87)
Section IV : $\quad m_{IV} \leq H_B, \quad m_{IV} \leq H_A$ and (7.88)
$\quad H_B \leq m_{II} \leq m_{III} \leq H_A$ (7.89)

As shown in Figure 7.20 the points a and b result from Henry coefficients. The theoretical optimum for process productivity is achieved at point w, where the difference between m_{II} and m_{III} is maximal and hence the feed flow rate for complete separation is as high as possible. Rigorous process simulations as well as experiments prove that SMB operating points for 100% purities are placed within this "triangle," but certain safety margins, especially to the vertex w, have to be met. The flow rate of fresh desorbent and, therefore, the eluent consumption can be minimized by choosing m_I as low and m_{IV} as high as possible.

7.4.1.1.2 Ideal Model, Nonlinear Isotherm
Application of the method described so far becomes more complicated when the isotherms are no longer linear and the migration velocities of the components are strongly influenced by fluid concentrations. One approach for determining the operating parameters is the explicit solution of the ideal TMB process model (Equations 6.204 and 6.205), as proposed by Storti, Mazzotti, and Morbidelli. By introducing the dimensionless time

$$\tau = \frac{t\,\dot{V}_{\text{ads}}}{V_c} $$

and the dimensionless axial position

$$z = \frac{x}{L_c}$$

as well as the flow rate ratio m_j, the balance for the TMB process under the assumption of ideal conditions is

$$\frac{d}{d\tau}[\varepsilon_t \cdot c_{i,j} + (1-\varepsilon_t) \cdot q_{i,j}] + (1-\varepsilon_p) \cdot \frac{d}{dz}[m_j \cdot c_{i,j} - q_{i,j}] = 0 \qquad (7.90)$$

This partial differential equation can be solved analytically for certain types of isotherms by applying the method of characteristics (Helfferich and Klein, 1970; Rhee, Aris, and Amundson, 1970). Following this approach direct solutions are available for isotherms with constant selectivity, such as the multi-Langmuir or the modified competitive multi-Langmuir isotherm (Storti et al., 1993,1995; Mazzotti, Storti, and Morbidelli, 1994,1996,1997). For multi-bi-Langmuir isotherms with nonconstant selectivities a numerical determination is necessary (Gentilini et al., 1998; Migliorini, Mazzotti, and Morbidelli, 2000).

As an example, the procedure for determining the operating diagram is explained for the EMD53986 system (Appendix A). The adsorption equilibrium of this system can be described by the multi-Langmuir isotherm (Equation 7.91):

$$q_i = \frac{H_i \cdot c_i}{1 + \sum b_i \cdot c_i}, \quad i = A, B \qquad (7.91)$$

Based on this isotherm, the nulls ω_G and ω_F ($\omega_G > \omega_F > 0$) have to be determined for the following quadratic equation:

$$(1 + a_A \cdot c_{A,F} + a_B \cdot c_{B,F}) \cdot \omega^2 - [H_A \cdot (1 + a_B \cdot c_{B,F}) + H_B \cdot (1 + a_A \cdot c_{A,F})] \cdot \omega \\ - H_A \cdot H_B = 0 \qquad (7.92)$$

With these data, the isotherm parameters for the EMD53986 and the feed concentration, which is $2.5\,\text{g}\,\text{l}^{-1}$ for each component, a plot as shown in Figure 7.21 can be created. The different points and lines are defined as follows:

$$\text{Line wf}: \quad [H_A - \omega_G \cdot (1 + b_A \cdot c_{\text{feed},A})] \cdot m_{\text{II}} + b_A \cdot c_{\text{feed},A} \cdot \omega_G \cdot m_{\text{III}} \\ = \omega_G \cdot (H_A - \omega_G) \qquad (7.93)$$

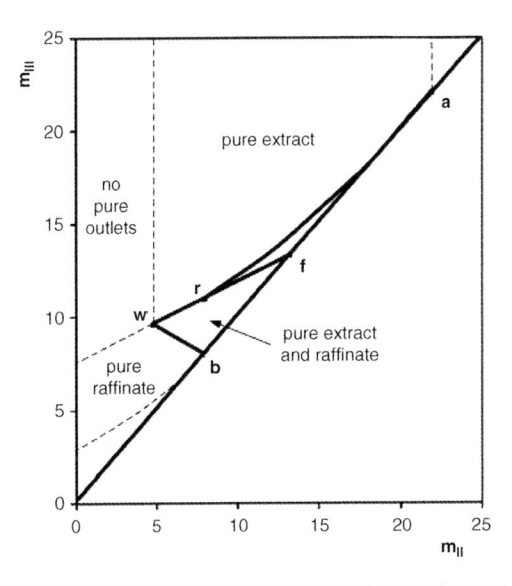

7.4 Conventional Isocratic SMB Chromatography

Figure 7.21 Operating plane or triangle diagram for nonlinear isotherms.

Line wb : $[H_A - H_A \cdot (1 + b_A \cdot c_{\text{feed},A})] \cdot m_{\text{II}} + b_A \cdot c_{\text{feed},A} \cdot H_B \cdot m_{\text{III}}$
$= H_B \cdot (H_A - H_B)$ (7.94)

Curve ra : $m_{\text{III}} = m_{\text{II}} + \dfrac{(\sqrt{H_A} - \sqrt{m_{\text{II}}})^2}{b_A \cdot c_{\text{feed},A}}$ (7.95)

Point a : (H_A, H_A),   point b : (H_B, H_B),   point f : (ω_G, ω_G) (7.96)

Point w : $\left(\dfrac{H_B \cdot \omega_G}{H_A}, \dfrac{\omega_G \cdot [\omega_F \cdot (H_A - H_B) + H_B \cdot (H_B - \omega_F)]}{H_B \cdot (H_A - \omega_F)} \right)$ (7.97)

Due to the strong nonlinearity of the isotherms, combined with the competitive adsorption behavior, the triangle is completely different from that for linear isotherms.

Since the slope of such isotherms depends strongly on the fluid concentration, the feed concentration has a remarkable influence on the shape of the separation region. Figure 7.22 illustrates the correlation between feed concentration and the operating diagram.

To complete the set of possible operating parameters for a four-section SMB unit, values for the dimensionless flow rate ratios m_{I} and m_{IV} have to be determined as well. In sections I and IV things are a little less complicated, since only the adsorption of single components is involved.

Because the solid phase is regenerated in section I, the more strongly retained component A has to be desorbed by the fluid flow. This can be guaranteed for all

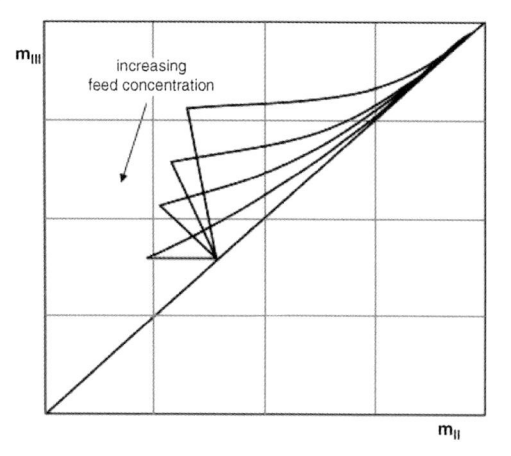

Figure 7.22 Influence of the feed concentration on the operating diagram.

Langmuir-type isotherms by the following condition:

$$m_{\mathrm{I}} \geq m_{\mathrm{I,min}} = H_A \tag{7.98}$$

In section IV the less retained component B has to be adsorbed and carried toward the raffinate port in order to regenerate the liquid phase. For the EMD53986 system with its multi-Langmuir isotherm, the corresponding constraint on the dimensionless flow rate ratio m_{IV} is

$$\frac{-\varepsilon_p}{1-\varepsilon_p} < m_{\mathrm{IV}} \leq m_{\mathrm{IV,max}} = \frac{1}{2}\Big[H_B + m_{\mathrm{III}} + b_B c_{\mathrm{feed,B}}(m_{\mathrm{III}} - m_{\mathrm{II}})$$
$$- \sqrt{(H_B + m_{\mathrm{III}} + b_B c_{\mathrm{feed,B}}(m_{\mathrm{III}} - m_{\mathrm{II}}))^2 - 4H_B m_{\mathrm{III}}}\Big] \tag{7.99}$$

The maximal m_{IV} is seen to be a function of flow rate ratios m_{II} and m_{III} as well as the feed concentration.

Table 7.7 lists the theoretical optimal values of all flow rate ratios for the separation of EMD53986 at a feed concentration of $2.5\,\mathrm{g\,l^{-1}}$. These values indicate the set of operating parameters where the productivity is at its maximum (point w), since the difference between m_{II} and m_{III} is biggest. In addition, eluent consumption reaches its minimum because the difference between m_{I} and m_{IV} is smallest.

Table 7.7 Operating parameters for EMD53986 obtained by the TMB shortcut method.

$m_{\mathrm{I,min}}$	19.75
m_{II} (at point w)	5.29
m_{III} (at point w)	10.57
$m_{\mathrm{IV,max}}$	6.28

7.4.1.1.3 Ideal Model, Shortcut for Langmuir Isotherms
The procedure described so far requires a detailed knowledge of the adsorption equilibrium. Naturally, the more accurately the isotherm parameters have been determined, the more reliable are the obtained operating parameters m_j. However, as discussed in Section 6.5.7, experimental determination of isotherm parameters might consume quite a lot of time and substance.

Therefore, the basic correlations for situations present in the different sections of the SMB process, combined with a simplified shortcut approach, will be presented. Without knowing the adsorption equilibria, an operating point can be generated with a minimum number of experiments. All considerations made in the following are valid only for Langmuir-type isotherms.

The retention time for a concentration c_i^+ of the dispersed desorption front can be described as a function of the isotherm derivative (Section 6.2.3). This implies that the maximum time for desorption or the minimum migration velocity is given by the derivative of the isotherms at lowest concentration ($c_i^+ \to 0$), which is equal to the Henry coefficient:

$$t_{R,i}(c^+) = t_0 \left(1 + \frac{1-\varepsilon_t}{\varepsilon_t} \cdot \frac{dq_i}{dc_i}\bigg|_{c^+ \to 0}\right) = t_0 \left(1 + \frac{1-\varepsilon_t}{\varepsilon_t} \cdot H_A\right) \quad (7.100)$$

For an adsorption shock front, the time for the front's breakthrough under Langmuir behavior can be calculated according to Equation 7.101, where the isotherm's secant at feed concentration is relevant. The smaller the ratio q_i/c_i, the higher is the velocity of the shock front:

$$t_{R,i,\text{shock}}(c^+) = t_0 \left(1 + \frac{1-\varepsilon_t}{\varepsilon_t} \cdot \frac{q_i}{c_i}\bigg|_{c^+ = c_{\text{Feed}}}\right) \quad (7.101)$$

These correlations can be transferred to continuous SMB or TMB chromatography where disperse as well as shock fronts are also present. The dimensionless flow rate ratios m_j can then be described as function of either the initial slope or the secant of the isotherm, depending on the situation in every zone.

Section I: In section I the more strongly retained component A has to desorb while no B is present. The minimum m_I is therefore the initial slope or the Henry coefficient of the isotherm:

$$m_{I,\text{min}} = \frac{dq_A}{dc_A}\bigg|_{c_A \to 0; c_B = 0} = H_A \quad (7.102)$$

As described in Section 6.5.7.2 the Henry coefficient can be determined by single pulse experiments at low concentrations of component A.

Section II: In section II the minimum value for the corresponding flow rate ratio m_{II} is fixed to the maximal possible initial isotherm slope of the less retained component, which has to be desorbed here. The highest initial slope for component B is reached if no A is around. Of course, A is always present in this zone (Figure 7.19). Since it is quite difficult to predict the concentration of A in this section, its presence causes a further decrease of the initial slope, and the worst case estimation of m_{II} is

$$m_{II,min} = \left.\frac{dq_B}{dc_B}\right|_{c_A=0;c_B\to 0} = H_B \qquad (7.103)$$

Again, the initial slope of the isotherm has to be determined by a pulse experiment of B.

Section III: The third section of an SMB process is characterized by the adsorption of more strongly retained component A. To achieve complete separation one has to ensure that the shock front of A does not exceed the raffinate port. The highest possible velocity of that front occurs when the q_A/c_A is smallest. This is the case when both components are present at their highest possible concentration. Again, it must be pointed out that it is difficult to determine the exact concentrations, but the feed concentrations are a good first guess, as well as the worst case estimation:

$$m_{III,max} = \left.\frac{q_A}{c_A}\right|_{c_A=c_{Feed};c_B=c_{Feed}} \qquad (7.104)$$

Experimentally, this parameter is determined by a breakthrough experiment where both components are fed at feed concentration to the chromatographic column.

Section IV: In the last section the eluent is regenerated by complete adsorption of component B. Again, the velocity of a shock front is limiting. In this case the highest possible velocity is determined by the lowest slope of the secant to the isotherm of component B. Since no A should be present, the minimum slope is reached when B appears at feed concentration. Therefore, the correlation for this section is as follows:

$$m_{IV,max} = \left.\frac{q_B}{c_B}\right|_{c_A=0;c_B=c_{Feed}} \qquad (7.105)$$

This parameter can be extracted from a breakthrough experiment with pure B.

Following this procedure a very first initial guess for the four dimensionless operating parameters can be obtained with a minimum number of experiments. The resulting parameters for the EMD53986 system are listed in Table 7.8.

Comparison of these results with the operating point obtained by the first shortcut method for nonlinear isotherms shows that, especially for the separation sections II and III, a point (m_{II}, m_{III}) has been found that is located within the area of complete separation (Figure 7.21).

Again, it must be pointed out that this procedure for determining the operating diagram is based on strong simplifications and represents something like a worst case scenario. Nevertheless, this can be done very quickly, leading to quite safe operating parameters. More detailed shortcut methods that also

Table 7.8 Operating parameters for EMD53986 obtained by the experimental shortcut method.

$m_{I,min}$	19.75
$m_{II,min}$	7.88
$m_{III,max}$	10.15
$m_{IV,max}$	6.53

need only rather small experimental effort have been published by Mallmann et al. (1998) and Migliorini, Mazzotti, and Morbidelli (2000). The advantage of these approaches is a more detailed estimation of the amount of components entering sections II and III. This can be achieved by hodographic analysis of breakthrough experiments.

Besides the description of the TMB process based on the ideal model, other model approaches can be used to follow the same goal of determining reasonable operating parameter for an SMB unit. One possibility is to utilize stage models as introduced in Section 6.2.9 for the batch column and extend these to the TMB setup (Ruthven and Ching, 1989; Pröll and Küsters, 1998). In addition to determining operating parameters, TMB stage models have also been used to optimize the design of SMB plant in terms of, for example, column geometry (Charton and Nicoud, 1995; Biressi et al., 2000). Heuer et al. (1998) have applied a TMB equilibrium-dispersive model for the determination of operating parameters. In this approach effects such as axial dispersion and mass transfer are lumped together in an apparent dispersion coefficient. Hashimoto et al. (1983) and Hashimoto, Adachi, and Shirai (1993) have used a TMB transport model to assist the layout.

The advantages of all approaches based on the stationary TMB process are the quite simple models and, in some cases, the possibility of direct analytical solutions. Without the need of dynamic simulations, much information with respect to the operation of SMB unit can be gathered. Nevertheless, these methods described so far can only provide the basis for a more or less proper guess of the operating parameters since the dynamics due to port switching as well as mass transfer and axial dispersion are neglected. In addition to this, of course, the quality of the isotherm plays an important role. Incomplete (e.g., neglecting competitive adsorption behavior) or wrong parameter determination influences significantly the quality of the operating point. Therefore, safety margins have always to be considered. Further optimization can be done afterwards by applying a more detailed model as presented below. The necessity of extra operating parameter optimization is shown below (Figure 7.26) where it can be seen that the theoretical optimum does not coincide with the real optimum of an SMB process.

7.4.1.2 Process Design Based on Rigorous SMB Models

Due to the simplifying assumptions made for the TMB approaches, accurate design and optimization of SMB processes is not possible. Several approaches based on SMB models have been suggested to improve the prediction and optimization of the SMB operation. Zhong and Guiochon (1996) have presented an analytical solution for an ideal SMB model and linear isotherms. The results of this ideal model are compared with concentration profiles obtained by an equilibrium-dispersive model as well as experimental data. Strube et al. (1997a, 1997b) used a transport-dispersive model in a systematic case study to optimize operating parameters. The strategy has been successfully tested on an industrial-scale enantioseparation (Strube et al., 1999). A similar approach has been published by Beste et al. (2000) to optimize the separation of sugars. A more detailed model, a general rate model (GRM), was used in Dünnebier and Klatt (1999) as well as in Dünnebier, Fricke, and Klatt (2000a) and Dünnebier, Jupke, and Klatt (2000b) to optimize SMB operating parameters.

This section presents a systematic strategy for the optimization of operating parameters, based on the transport-dispersive SMB model (Sections 6.6.2 and 6.2.5). A first approximation is performed by applying the triangle theory described above. Notably, the basic strategy of this procedure is not limited to a model-based design but can also be applied for the experimental optimization of a running separation process.

7.4.1.2.1 Operating Parameter Estimation

Knowing the isotherms, an initial guess of operating parameters can be performed by applying the shortcut methods described previously. With the triangle theory, one obtains values for the four dimensionless operating parameters m_j. Since simplifying assumptions have been made, a safety margin of at least 10% should always be considered. The four dimensionless parameters m_j have then to be transferred to a "real" SMB process where the number of independent operating parameters is five (four flow rates and one switching time). Thus, for a given plant setup, one out of these five parameters has to be fixed first. One possibility is to set the flow rate in section I to the maximum allowable pressure drop Δp_{max}, assuming that this flow rate is predominant in all sections. If Darcy's correlation for the description of the pressure drop is used, the flow rate in section I can be calculated by Equation 7.106:

$$\dot{V}_I = \frac{\Delta p_{max} \cdot A_c \cdot d_p^2}{\psi \cdot N_c \cdot L_c \cdot \eta_f} \tag{7.106}$$

Here, N_c stands for the total number of columns in the SMB plant. This procedure leads to higher pressure drops than in the real plant, since section I is the part of the process where, in general, the highest flow rate occurs. As demonstrated later, it is not necessary to fix the flow rate in section I to the maximum allowed pressure drop to achieve optimal process performance. However, for a given flow rate in section I and all dimensionless flow rate ratios m_j, the switching time can be calculated according to Equation 7.107:

$$t_{\text{shift}} = \frac{[m_I \cdot (1 - \varepsilon_t) + \varepsilon_t] \cdot V_c}{\dot{V}_I} \qquad (7.107)$$

The missing internal flow rates $\dot{V}_j$ are calculated by Equation 7.108:

$$\dot{V}_j = \frac{[m_j \cdot (1 - \varepsilon_t) + \varepsilon_t] \cdot V_c}{t_{\text{shift}}} \qquad (7.108)$$

The flow rates of all external streams such as desorbent, feed, extract, and raffinate are determined by the balances for the port nodes as given by Equations 6.200–6.203.

Applying this strategy, a first set of operating parameters is generated and it is possible to either run the SMB separation or make further optimizations using a more detailed model. Since the determination of these operating parameters is based on the TMB model with all its simplifying assumptions (neglecting all non-idealities such as axial dispersion, mass transfer, and port switching), the estimated operating point should always be handled with care.

7.4.1.2.2 **Optimization Strategy for Linear Isotherms** Figure 7.23 shows the internal axial concentration profiles at the end of a switching interval for a system of linear isotherms and no competitive interaction of the two components. After switching all ports downstream in the direction of the liquid flow, the initial concentration profile is represented by the upper gray line. In this case the extract will be polluted with component B because the desorption front of B violates point 2 (Figure 7.18).

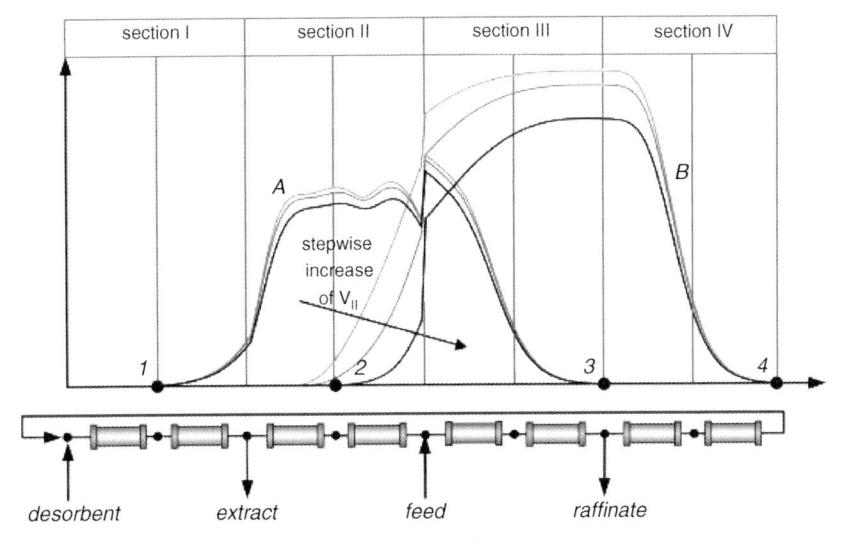

Figure 7.23 Axial concentration profiles for different flow rates in section II (system with linear isotherms).

To improve process performance and achieve a complete separation of the two components, the desorption front of component B has to be pushed in the direction of the feed port. Since the isotherms of both components are linear in this example and the isotherm of one component is not influenced by the presence of the other, the front can be moved in the desired direction by simply increasing the internal flow rate in section II, $\dot{V}_{II}$. This can be realized by decreasing the flow rates of extract and feed by the same amount. The effect of a stepwise decrease of these external flow rates with the associated increase of $\dot{V}_{II}$ is depicted in Figure 7.23.

Only the desorption front of component B is influenced by the increased flow rate in section II – all other fronts are not influenced and remain at their initial position. The decrease in feed flow rate has an impact on the total height of the concentration plateaus, especially in section III. This procedure of shifting the fronts is also applicable for all other sections of the SMB process, to optimize the internal concentration profile according to Figure 7.18 and improve the process performance with respect to purities, productivity, and eluent consumption.

7.4.1.2.3 **Optimization Strategy for Nonlinear Isotherms** Process optimization gets more difficult when the isotherms are no longer linear and the equilibrium of one component is strongly influenced by the presence of the other, as is the case for the EMD53986. Again, an initial guess based on the TMB model has been made for a given plant setup. The resulting internal concentration profile is plotted in Figure 7.24. Again, the desorption front of component B violates the optimization criterion for point 2, leading to pollution of the extract stream.

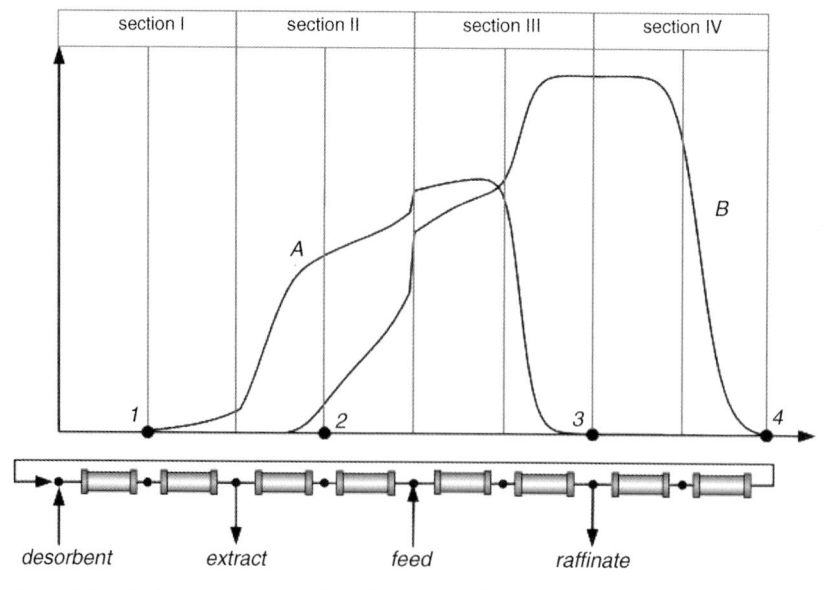

Figure 7.24 Axial concentration profile with pollution of the extract (system with nonlinear isotherms).

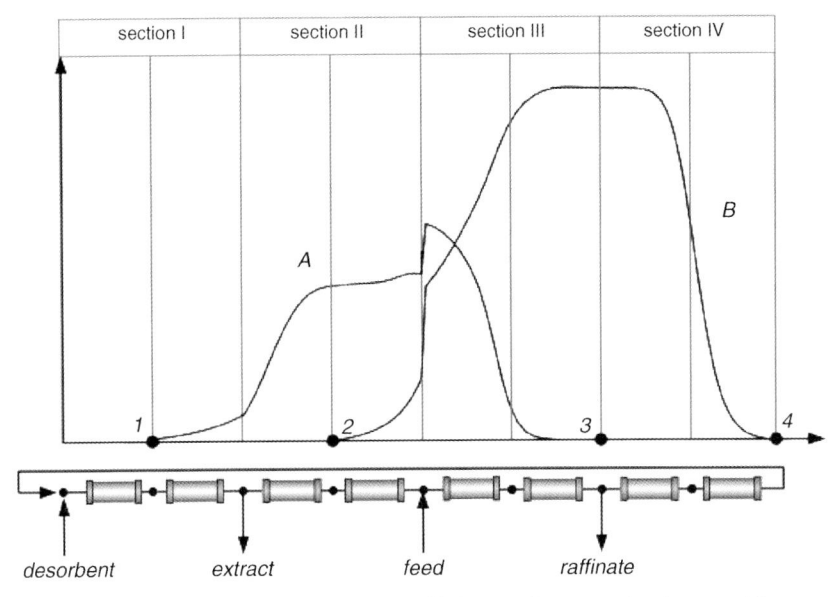

7.4 Conventional Isocratic SMB Chromatography | 475

Figure 7.25 Suboptimal axial concentration profile for complete separation (system with nonlinear isotherms).

Applying the same strategy as introduced for linear isotherms, an internal concentration profile as shown in Figure 7.25 is obtained. Again, only the flow rate in section II has been increased by decreasing both extract and feed flow rate. The desorption front of component B is again pushed in the correct direction. However, in this case a change of flow rate in section II does also effect the situation in other sections of the process. In this example, the adsorption front of component A has moved away from the raffinate port where, actually, the stopping point 3 (Figure 7.18) is located. Even though the two products can now be withdrawn with maximum purity, the whole process is not operated at its optimum with respect to productivity.

To achieve this goal a new optimization strategy has to be applied. The most difficult step is to adjust the flow rate ratios in sections II and III, especially for strongly nonlinear isotherms with competitive interaction. Therefore, the m_{II}–m_{III} plane is divided into segments by parallel lines to the diagonal. Each of these lines represents combinations of m_{II}–m_{III} values of constant feed flow – the so-called isofeed lines (Figure 7.26).

If, for instance, as in our example, the purity of the extract stream is too low while the adsorption front of component A is still far from stopping point 3, the operating point has to be varied along one isofeed line. In this case, m_{II} and m_{III} and, therefore, also the flow rates in sections II and III are increased by the same amount. This can be done by decreasing the extract flow rate and increasing the raffinate flow rate while the feed rate is held constant. When both the extract and the raffinate purity are higher than required, or the concentration profile can still be

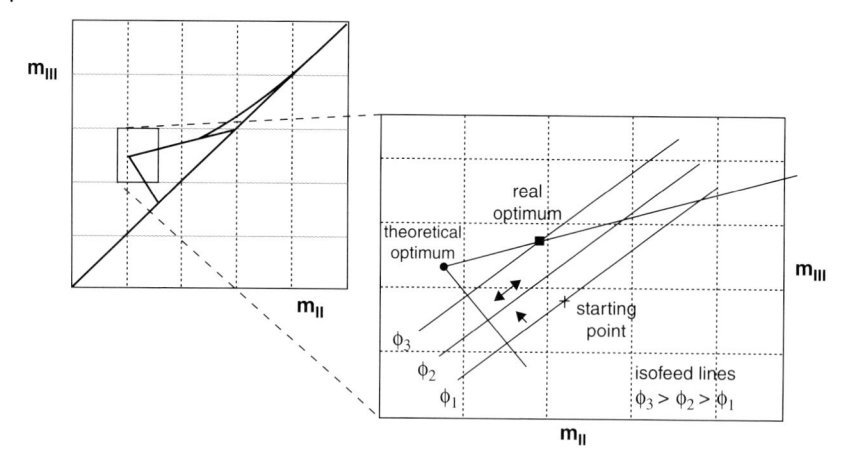

Figure 7.26 Strategy for the optimization of operating parameters (reproduced from Jupke, Epping, and Schmidt-Traub, 2002).

optimized according to Figure 7.18, the total feed flow can be increased by jumping to the next isofeed line closer to the vertex of the operating triangle. Otherwise, if both purities are too low, the feed flow has to be decreased. When the flow rates in sections II and III have been optimized to achieve the required purities and the highest productivity, sections I and IV have to be checked and optimized separately.

After adjusting the flow rates in all sections, process conditions for complete separation with highest productivity and lowest solvent consumption have been found. The resulting internal concentration profile exhibits all the constraints of an optimal concentration profile (Figure 7.18).

From this optimization with a detailed SMB model another very important aspect can be observed. Due to the simplifying assumption made to determine the operating diagram (Section 7.4.1.1), the "real" optimum of the SMB process does not coincide with the predicted theoretical optimum, as can be seen from Figure 7.26.

The mentioned optimizations of operating parameters can be performed for a given plant setup, where the total number of columns, their distribution over the different sections, and their geometry are given before. With such information and all operating parameters, the specific process parameters such as productivity $VSP_{i,SMB}$ (Equation 7.10) and eluent consumption $EC_{i,SMB}$ (Equation 7.12) can be calculated. But since these parameters are only valid for one given set of design parameters, a strategy to optimize these will be presented next.

7.4.2
Optimization of Design Parameters

Needless to say that, besides the operating parameters, the geometry and configuration of an SMB plant has a significant influence on the process performance in terms of various objective functions such as the VSP and cross section-specific productivity (VSP$_{i,SMB}$ and ASP$_{i,SMB}$) as well as the eluent consumption EC$_{i,SMB}$. Important design parameters for SMB processes are

- column geometry L_c and A_c;
- total number of columns N_c;
- distribution of columns over different sections j ($N_{c,j}$);
- maximum pressure drop;
- particle size.

Few published papers deal with the detailed optimization of design parameters. Charton and Nicoud (1995) and Nicoud (1998) have presented a strategy based on a TMB stage model to optimize the operation as well as the design of SMB processes. The basic influence of the number of stages and the particle diameter on productivity and eluent consumption is illustrated and compared with batch elution chromatography. Based on this work, Biressi et al. (2000) presented a method to optimize operating parameter and investigate the influence of the particle diameter as well as the feed concentration on process performance. Ludemann-Hombourger, Bailly, and Nicoud (2000a) have also analyzed the influence of particle diameter on the productivity.

In this contribution the focus will be on the optimization of column geometry for a fixed column distribution first, while the influence of the total number of columns as well as their distribution will then be shown for some examples. A more detailed description of this strategy has been published by Jupke (2004) and Jupke, Epping, and Schmidt-Traub (2002).

The basic idea for the optimization of process performance is, similar to batch chromatography, the summarization of all influencing parameters into dimensionless parameters. Relevant dimensionless parameters for an SMB process are the total number of plates (N_{tot}) in the system and the four dimensionless flow rate ratios m_j. Therefore, a single set of N_{tot} and m_j represents not only one unique SMB configuration but also a series of different SMB configurations with similar behavior. As demonstrated before, nearly identical concentration profiles (Figure 7.4) can be observed for different plant layouts if these parameters are kept constant. Also, since only the dimensionless parameters are used as optimization parameters, violation of the maximum pressure drop can be neglected during optimization. The maximum pressure drop is considered only at the end of the optimization in order to determine the "real" SMB configuration based on optimal values for the dimensionless parameters $N_{tot,opt}$ and $m_{j,opt}$.

Figure 7.27 illustrates the design procedure for a new SMB plant. For an initial column geometry and a fixed column distribution (e.g., 2/2/2/2) the optimal operating parameters, namely, the flow rate ratios m_j, can be obtained following the strategy described in the previous section. From these parameters the present dimensionless design parameter N_{tot} can be calculated, as well as the corresponding objective functions. Since the initial values for the column geometry, especially the column length, do not consequently represent the optimal design parameters, other plate numbers have to be examined. This can be done by choosing a different column length from the initial one, but to realize a different total number of

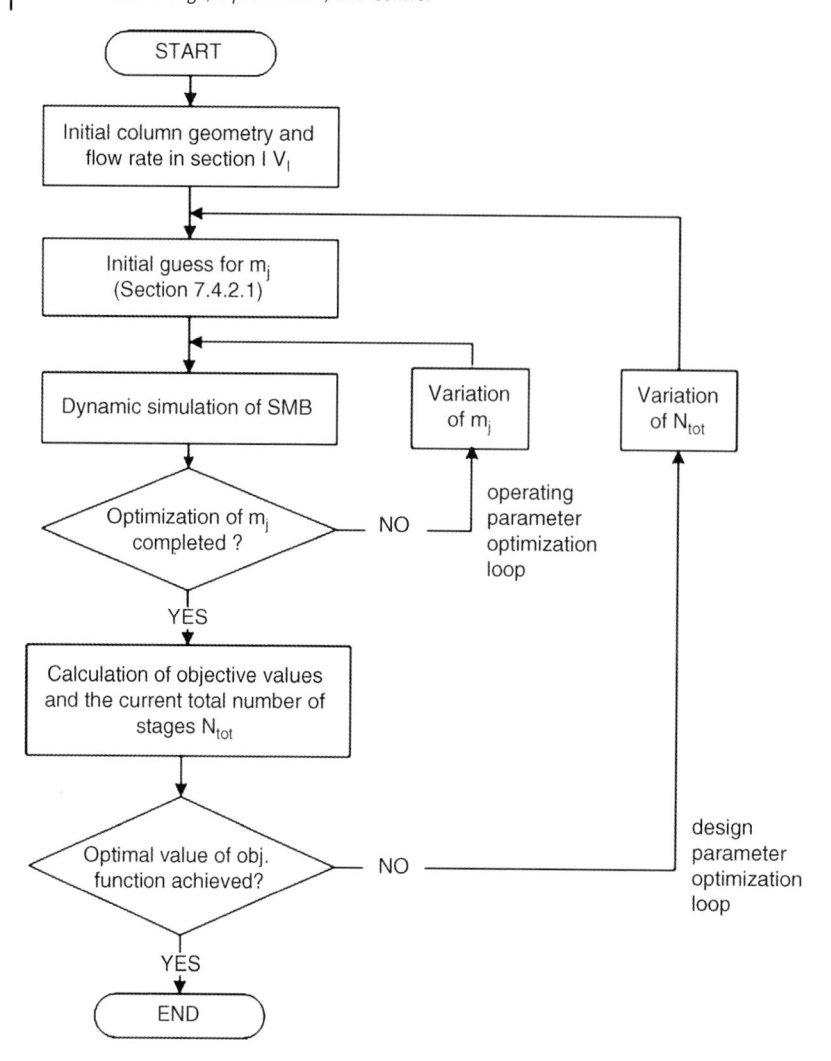

Figure 7.27 Algorithm for a complete SMB design optimization.

theoretical plates in the system, the flow rate in section I, as it had to be chosen before, should be held constant. Another possibility is to fix the column length to a constant value but to change the flow rate in section I. For the new setup again an optimization of operating parameters has to be performed to fulfill purity requirements, and so on.

The algorithm in Figure 7.27 is characterized by two optimization loops. The inner loop describes the operating parameter optimization based on the strategies introduced before. When for one setup (e.g., one column length) the optimized operating parameters have been found, a new setup (e.g., another column length) has to be chosen, represented by the outer design parameter optimization loop.

Table 7.9 Optimization of plate number for EMD53986.

Constant parameters							
N_c	8 (2/2/2/2)						
d_c (cm)	2.5						
$\dot{V}_I$ (ml min^{-1})	1.472						
$c_{feed,i}$ (g l^{-1})	2.5						
Parameters varied during optimization							
Step	1	2	3	4	5	6	7
L_c (cm)	20	10	6	5	4.5	4.26	4.0
N_{tot} (–)	231.3	124.6	83.3	71.9	65.6	62.3	57.1
t_{shift} (s)	500.7	275.5	187.7	163.4	151.2	145.1	136.3
m_I (–)	24.30	27.00	31.00	32.50	**33.50**	33.99	34.00
m_{II} (–)	5.55	5.89	6.30	6.57	**6.90**	7.15	8.00
m_{III} (–)	10.63	10.71	10.80	10.84	**10.95**	11.00	11.00
m_{IV} (–)	6.32	6.20	5.90	5.75	**5.70**	5.70	5.70
VSP (g h^{-1} l^{-1})	11.42	19.70	26.98	29.41	**30.13**	29.85	24.77
EC (l g^{-1})	1.81	2.12	2.63	2.91	**3.15**	3.34	4.17

This procedure is repeated until the optimum for the objective function (e.g., VSP, EC, or costs) has been found.

Results for the systematic optimization of the VSP for the separation of the EMD53986 system are summarized in Table 7.9.

In this example, starting with a given setup, the flow rate in section I was held constant while the column length has been changed to realize different plate numbers. Again, it must be pointed out that the pressure drop is not considered during the optimization.

According to Table 7.9 a maximum VSP is reached for a total number of plates of approximately 65 and the corresponding flow rate ratios m_j. This is more than a twofold increase over the productivity at a plate number of 231. But, in addition to the development of productivity, the specific EC has increased. As reported by Jupke et al., (2002) analogue correlations can be found for different particle diameters. With decreasing particle diameter an increase in terms of productivity can be observed, while the total number of stages remains nearly constant.

The separation problem has been transferred completely to the dimensionless space, where the best performance in terms of productivity VSP, eluent consumption (EC), and so on, can be achieved by one set of five parameters (N_{tot}, m_j). The next step is to transfer these parameters to an SMB process by considering a given production rate $\dot{m}_{prod,i}$, purity requirements, and pressure drop limitations.

For the desired production rate, $\dot{m}_{prod,i}$, together with the volume-specific productivity (VSP$_{i,SMB}$), as it had been introduced before (Equation 7.10), the optimal volume of one column $V_{c,opt}$ can be calculated according to Equation 7.109:

$$V_{c,opt} = A_{c,opt} \cdot L_{c,opt} = \frac{\dot{m}_{prod,i}}{(1 - \varepsilon_t) \cdot N_c \cdot VSP_{i,SMB}} \qquad (7.109)$$

This will lead to an infinite number of possible A_c/L_c ratios and, therefore, the column length has to be chosen according to the following equations.

The first equation represents the total number of theoretical plates of one component in the system, which has to be held constant to enable a proper scaling of the process:

$$N_{\text{tot},i} = \sum_{j=\text{I}}^{\text{IV}} \left(\frac{N_{c,j} \cdot L_{c,\text{opt}}}{A_i + C_i \cdot u_{\text{int},j}} \right) \tag{7.110}$$

But, in analogy to the procedure in batch chromatography, one has to ensure that for both components the total number of stages of the new process is equal to or greater than for the old one. This can be done, as introduced for the single column before, by choosing the correct reference component.

In addition to the plate number, the dimensionless flow rate ratios should remain constant:

$$m_{j,\text{opt}} = \frac{u_{\text{int},j} \cdot A_{c,\text{opt}} - [(V_{c,\text{opt}} \cdot \varepsilon)/t_{\text{shift}}]}{[V_{c,\text{opt}} \cdot (1 - \varepsilon)]/t_{\text{shift}}} \tag{7.111}$$

Also, as a final constraint, pressure drop limitations have to be considered, which are described by Equation 7.112:

$$\Delta p_{\text{max}} = \frac{\psi \cdot \eta_f}{d_p^2} \cdot \sum (u_{\text{int},j} \cdot N_{c,j} \cdot L_{c,\text{opt}}) \tag{7.112}$$

This procedure leads to one $L_{c,\text{opt}}/A_{c,\text{opt}}$ ratio that exactly fulfills the maximum pressure drop condition for the given feed stream. Of course, smaller L_c/A_c ratios are also applicable to reach the same optimal performance of the SMB plant but with a lower pressure drop. Notably, a lower pressure drop will decrease the investment cost of a chromatographic unit. However, the smaller that ratio becomes, and with that the lower the pressure drop is, the shorter the columns will be. Consequently, this leads to difficulty in ensuring a proper fluid distribution inside the column.

Finally, it is worth mentioning that, in the presented strategy, besides the total number of stages, the operating parameters in terms of the flow rate ratio were also held constant. Therefore, no further optimizations of the operating parameters have to be done.

Besides the column's geometry, of course, the total number of columns and their distribution over the unit are design parameters that influence the process performance. To compare different process configurations with the initial 2/2/2/2 setup, a complete optimization of the column geometry based on optimization of the number of stages has been performed for different configurations. Productivity again was the objective function to be optimized in these considerations. The results for selected configurations (2/3/2/2, 2/2/3/2, 2/3/3/2, and 3/3/3/3) are shown, relative to the data for the initial eight-column setup, in Figure 7.28.

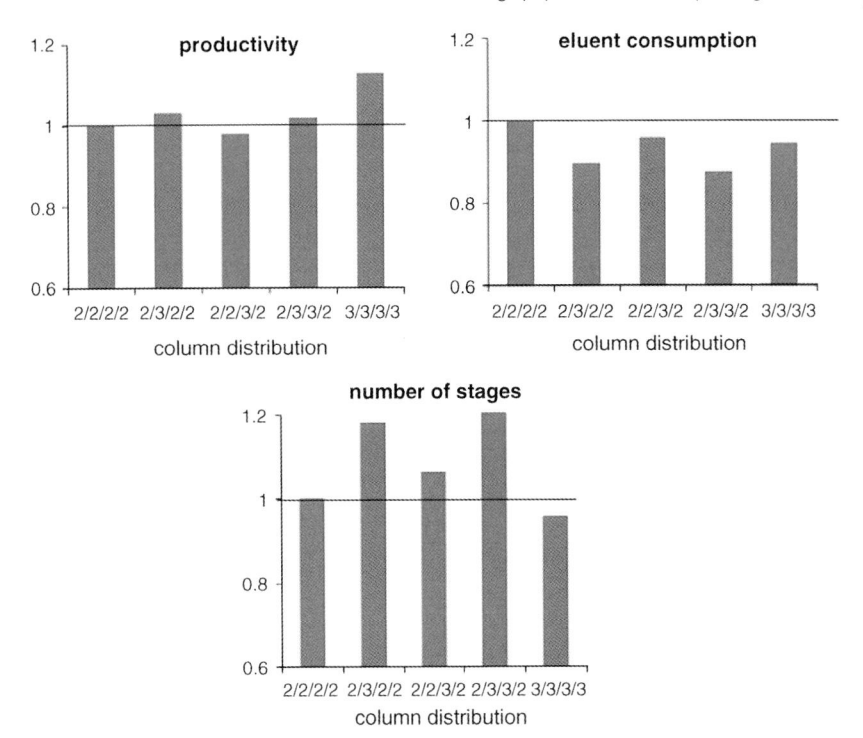

Figure 7.28 Influence of the number of columns and column distribution on the specific productivity and the specific eluent consumption.

This example indicates that the column configuration can be used to improve the process performance. For every configuration it is not enough just to change the number of columns and optimize the operating parameters, because every new configuration shows the best results in terms of productivity at its own optimal total number of stages.

Process configurations with a reduced number of columns, such as 1/2/2/2, 2/2/2/1, and 1/2/2/1, are not considered. Since the number of columns is reduced in the sections meant to regenerate the solid phase (section I) as well as the fluid phase (section IV), a large increase in eluent consumption will be the result. Nevertheless, these configurations should also be taken into account when fresh eluent is not the major block in the cost structure.

7.5
Isocratic SMB Chromatography under Variable Operating Conditions

Standard SMB processes are run under constant operating conditions during each shifting period. The first idea to abandon this constraint resulted in the Varicol process (Ludemann-Hombourger, Nicoud, and Bailly, 2000b). Shortly thereafter, the scientific world developed and analyzed other nonstandard SMB processes such as PowerFeed and Modicon (Section 5.2.5).

In a recent review Seidel-Morgenstern, Keßler, and Kaspereit (2008a, 2008b) gave an overview of new concepts to improve the performance of the standard SMB process. The research group in Magdeburg also investigated the application of Modicon processes (Schramm et al., 2003a, 2003b) as well as other SMB variants such as the fractionation and feedback scheme (FF-SMB) (Keßler and Seidel-Morgenstern, 2008; Li et al., 2010a, 2010b). In case of FF-SMB one or both outlet streams are fractioned into a product stream and an off-spec stream that is fed back to the process. The authors prove that the simultaneous fraction of both outlet streams is the most efficient operating scheme in terms of throughput while single fractionation modes can also be superior to the classical SMB process. More recently, a procedure for optimizing the variable operating conditions for starting up and shutting down SMB plants was proposed by Li et al. (2011).

The ETH groups of Morbidelli and Mazzotti (Zurich) compared Varicol, Power-Feed, and Modicon with SMB based on a detailed stage model and multiobjective optimization using a genetic algorithm and compared theoretical results with experiments (Zhang, Mazzotti, and Morbidelli, 2003a, 2003b, 2004a; Zhang, Morbidelli, and Mazzotti, 2004b).

The research group of Hidajat (Ontario) and Ray (Singapore) decided for the equilibrium-dispersive model and nonlinear isotherms to study the behavior of Varicol in comparison to SMB for enantioseparations and also for the large-scale Parex process using nondominating sorting genetic algorithms with jumping genes (NSGA-JG) (Zhang et al., 2002; Wongso, Hidajat, and Ray, 2004; Kurup, Hidajat, and Ray, 2005a, 2006; Zhang, Hidajat, and Ray, 2007, 2009). This group developed and improved this optimization method for several years and applied it also to other nonstandard and standard SMB processes with integrated chromatographic reactors (Subramani, Hidajat, and Ray, 2003; Subramani et al., 2004; Yu, Hidajat, and Ray, 2005; Kurup, et al. 2005b).

Another research group of Engell (Dortmund) developed alternative optimization methods and applied them to the design and especially control of SMB, Varicol, PowerFeed, and Modicon processes. For rigorous modeling they used the most detailed general rate model that additionally to other models regards pore diffusion. Process optimization based on the FFSQP algorithm highlights the advantage of the Varicol process. A general conclusion is that in comparison to SBM the productivity of Varicol is increased up to 40% for the same number of columns; alternatively, the number of columns can be reduced by 1–2 to achieve the same performance of SMB (Toumi, Engell, and Hanisch, 2002; Toumi et al., 2003; Engell and Toumi, 2005). For further process optimization the direct multiple shooting method, where the optimal state trajectory and corresponding operation parameters are determined simultaneously, was used (Toumi et al., 2007; Küpper and Engell, 2011).

Another approach is the simultaneous optimization of process setup and operating conditions as performed for various processes by the group around Biegler (Pittsburgh) and specifically for SMB processes by Kawajiri (Georgia Tech). The partial differential algebraic equations are fully discretized in temporal and spatial domains and the resulting mixed-integer nonlinear programming (MINLP) problem is handled by an interior-point solver IPOPT (Kawajiri and Biegler, 2006a, 2006b, 2006c). MINLP was also applied for the optimal synthesis of complete flow sheets based on SMB chromatography, crystallization, and chemical reactions (Kaspereit *et al.*, 2012).

Rodrigues and his group are known for theoretical and experimental research on various SMB processes. In context to this section it is worth mentioning an experimental procedure that allows measuring the performance of standard and nonstandard SMB processes with only one column (Rodrigues, Araújo, and Mota, 2007b; Rodrigues *et al.*, 2007a). The authors also investigated the robustness of optimized processes that is important for control and operability (Mota, Araújo, and Rodrigues, 2007).

In the following an overview of selected results of recent research is given and an attempt is made to summarize some recommendations concerning the applicability of nonstandard SMB processes.

In the first paper describing the Varicol process, Ludemann-Hombourger, Nicoud, and Bailly (2000b) investigated Varicol by theoretical case studies and experiments for a chiral separation. Varicol subdivides the switching period into subintervals (generally three or four) and alters the number of columns for each subinterval, so that the mean number of columns per zone for the switching period becomes noninteger (Section 5.2.5). The case studies are based on stage model calculations and competitive modified multicomponent Langmuir isotherms. The optimal internal flow rates for a given operating pressure are estimated by equilibrium theory and afterwards tuned to fully utilize the no leak condition in zones 1 and 4. The switching time is always the same. A comparison with standard SMB (Table 7.10) shows that for five and six columns Varicol outperforms SMB. As Varicol achieves comparable purities for fewer columns that offer the opportunity to increase productivity by increasing the feed rate until maximum allowable pressure drop is reached. Please keep in mind that these design data are determined by case studies and do not represent a global optimum.

These findings are proven by experimental results as shown in Table 7.11. Varicol with the same number of stages as SMB achieves higher purities, slightly increases

Table 7.10 Theoretical case study for SMB and Varicol (Ludemann-Hombourger, Nicoud, and Bailly, 2000b).

No. of columns	SMB		Varicol		Varicol–SMB
Total	Columns/zone	Raffinate/extract purity (%)	Columns/zone	Raffinate/extract purity (%)	Increase of average purity (%)
5	1/2/1/1	98.55/93.13	1.25/1.25/1.25/1.25	97.89/95.98	1.10
6	1/1/2/2	98.49/95.15	1.75/1.25/1.25/1.75	98.67/97.12	1.07
8	2/2/2/2	98.38/97.95	2.5/1.5/1.5/2.5	98.95/97.81	0.21

Table 7.11 Experimental case study for five-column SMB and Varicol processes (Ludemann-Hombourger, Nicoud, and Bailly, 2000b).

	Columns/zone	Extract purity (%)	Raffinate purity (%)	Increase in productivity (%)	Decrease in eluent (%)
SMB	1/2/1/1	91.2	96.3		
Varicol 1	1.25/1.25/1.25/1.25	92.7	96.9	1.8	10.7
Varicol 2	1.0/1.5/1.5/1.0	92.4	97.7	18.5	−0.3

productivity, and at the same time reduces the eluent consumption by more than 10%. According to the second Varicol configuration, it is possible to increase productivity by 18.5% compared to an optimized SMB process while the eluent consumption of both processes remains similar.

Zhang, Mazzotti, and Morbidelli (2004a) present the first systematic comparison of the optimal separation behavior of SMB, PowerFeed, Varicol, and Modicon for small number of columns. The nonstandard SMB processes subdivide the switching period into subintervals (generally three or four) and alter for each subinterval the feed flow rate, number of columns, and the feed concentration (Section 5.2.5). The authors' investigations are based on the following assumptions and constraints: three-, four-, and five-column processes, three subintervals for the nonstandard SMB processes, particle diameter rather large: 30 μm (i.e., columns not very efficient), and eluent consumption not minimized or constrained. The processes are described by a stage model. As objective functions for optimization extract purity and productivity are chosen while the raffinate concentration is set to >90%. The maximal pressure drop along the entire process is 70 bar. The optimization variables are feed flow rate, flow rate in section I, flow ratios m_I, m_{II}, and m_{IV}, total feed concentration of the racemic mixture, and unit configuration. For the multiobjective optimization a genetic algorithm is applied.

Figure 7.29 depicts the optimization results for the four- and five-column processes. Varicol, PowerFeed, and Modicon clearly outperform SMB while the performance of PowerFeed and Modicon is slightly better than that of Varicol for five-column processes.

The influence of the number of columns on the extract purity for a given productivity is shown in Table 7.12. These data are extracted from results published by Zhang, Mazzotti, and Morbidelli (2004a). The data should be generalized cautiously, but they show definite tendencies: SMB performance is clearly lower than for the nonstandard SMB processes that change their operating mode during the switching period. For three-column processes PowerFeed and Modicon achieve higher extract purities than SMB and Varicol. But for all processes the profitability decreases as the extract purity is much lower than for higher number of columns. For three-column processes it has also to be kept in mind that during a switching period only Varicol is able to operate with a fraction of a column in all four zones while the column configuration for all other multicolumn processes is 1/1/1/0.

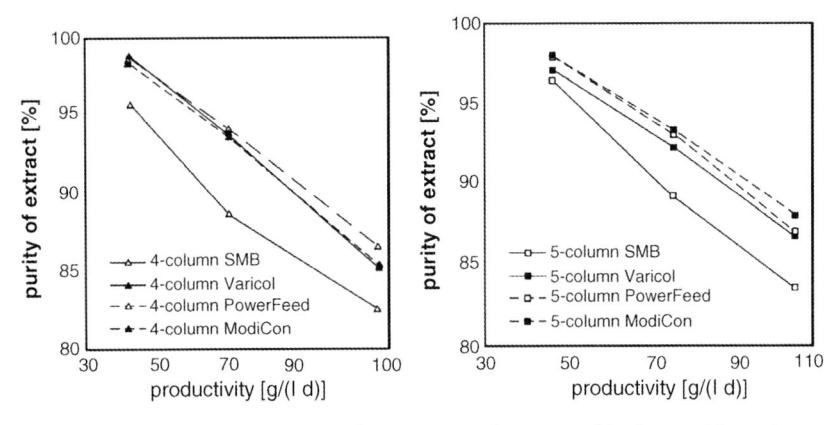

Figure 7.29 Comparison of the optimal separation performances of the four- and five-column SMB, Varicol, PowerFeed, and Modicon processes (reproduced from Zhang, Mazzotti, and Morbidelli, 2004a).

Based on theoretical simulations, Modicon is advantageous, but apart from additional hardware the practical realization is limited as the feed concentration is upper bounded by the solubility and in general the feed concentration of optimized SMB processes will already be close to this limit.

For the same nonlinear chiral separation as reported by Ludemann-Hombourger, Nicoud, and Bailly (2000b), Zhang, Mazzotti, and Morbidelli (2003a, 2003b) compare in another paper SMB, PowerFeed, and Varicol that are simultaneously optimized concerning extract and raffinate purity for all purities >90%. The authors also describe different alternatives for the operation of PowerFeed. In one case only the feed flow rate is changing in the subintervals and in a second case it is investigated to what extent the process can be improved if the flow rates of feed and eluent as well as the flow rate in section II are changed in the subintervals. These data are compared to equivalent Varicol and SMB processes. In all cases the the average feed and eluent flow-rates are the same and constant maximal pressure drop is

Table 7.12 Influence of the number of columns on extract purity for a given productivity.

Columns	Productivity	SMB	Varicol	PowerFeed	Modicon
				Extract purity (%)	
3	50	90.4	91.4	94.9	92.7
4	50	94.0	97.4	97.3	97.0
5	50	95.8	96.8	97.7	97.8
3	90	81.3	82.7	86.7	85.8
4	90	86.2	90.1	90.9	90.5
5	90	86.6	89.7	90.3	91.2

Data extracted from Zhang, Mazzotti, and Morbidelli (2004a).

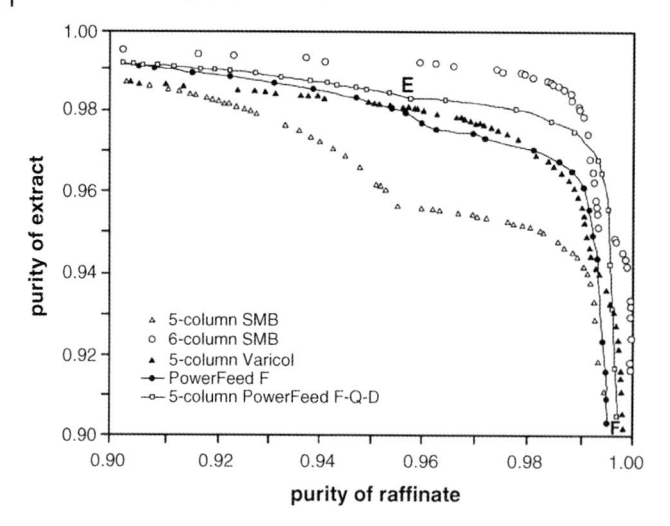

Figure 7.30 Pareto optimal curves PowerFeed and Varicol and SMB processes (reproduced from Zhang, Mazzotti, and Morbidelli, 2003a,2003b).

reached by fixing the flow rate in section I. All processes are optimized for five columns; SMB is additionally designed for six columns.

The multiobjective optimization is based on a stage model and a NSGA. The Pareto optimal results are shown in Figure 7.30. The Pareto curve illustrates how the two conflicting objective functions influence each other; that is, when one concentration is increased, the other will decrease. All points in this curve are equally good (nondominating) optimal solutions. Points below the Pareto curve represent nonoptimal sets of purities while all points above the Pareto curve do not fulfill the constraints for the optimization, that is, constant productivity and eluent consumption. The discontinuities of the Pareto curves are due to changes of the optimal column configuration. In case of the five-column SMB the configuration changes, for instance, from 1/2/1/1 to 1/1/2/1, with increasing raffinate concentration as the fifth column must be located to the port where the highest purity is required. The corresponding configuration of the five-column Varicol is optimized in three steps: from C-C-B-B to C-C-C-B to C-C-C-A with $A = 2/1/1/1$, $B = 1/2/1/1$, and $C = 1/1/2/1$.

The lower continuous curve represents optimal PowerFeed-F separation with feed rate as the only variable, while the upper curve (PowerFeed-F-Q-D) stands for an optimization where the flow rates of feed, section II, and eluent are variable in each subinterval. For PowerFeed-F and Varicol the results are nearly the same while the more sophisticated optimization with three variables is slightly better for the sake of a more difficult operation because three flow rates have to be controlled in each subinterval instead of only one. All in all, the investigations of Zhang, Mazzotti, and Morbidelli (2003a, 2003b) show that for the same productivity, eluent consumption and stationary phase standard SMB processes can be substantially improved by Varicol or PowerFeed operation. This is particularly significant for small numbers of columns. For a better performance it is necessary to run a six-column SMB. In another paper Zhang, Morbidelli, and Mazzotti (2004b) prove experimentally that PowerFeed outperforms a corresponding SMB process in terms of purity.

Another well-known group headed by Ray investigated *inter alia* the enantio-separation of pindolol with *S*-pindolol being the desired component (Zhang, Hidajat, and Ray, 2009). Five-column SMB and Varicol processes are compared; the latter is divided into four subintervals. The rigorous multiobjective optimizations are based on the equilibrium dispersion model and bi-Langmuir isotherms that have been proven by experiments (Zhang, Hidajat, and Ray, 2007).

The authors present Pareto optimal solutions for different cases:

- **Simultaneous maximization of raffinate and extract purity for a given capacity (product of feed concentration and feed flow rate) and two different feed concentrations:** Purities for Varicol are better than for SMB with the same number of columns, but the differences are smaller than those reported by Zhang, Mazzotti, and Morbidelli (2003a, 2003b) who did comparable investigations for a chiral separation of 1,2,3,4-tetrahydro-1-naphtol. A comparison of low and high feed concentrations confirms that "higher feed concentration and lower feed flow rate is superior to that of lower feed concentration and higher feed flow rate in achieving higher productivity without consuming more desorbent." This confirms the general rule to choose high feed concentrations.
- **Maximization of recovery and minimization of desorbent flow for a 97.1% purity of the raffinate:** As shown in Figure 7.31a the recovery increases nearly linearly with the eluent consumption and for a given eluent flow rate the recovery of a Varicol process is approximately 0.5% higher as for SMB with the same number of columns and the same maximum pressure drop. Hence, Varicol can save operating cost as it achieves higher recovery for the same eluent consumption or, vice versa, saves eluent for the same recovery.

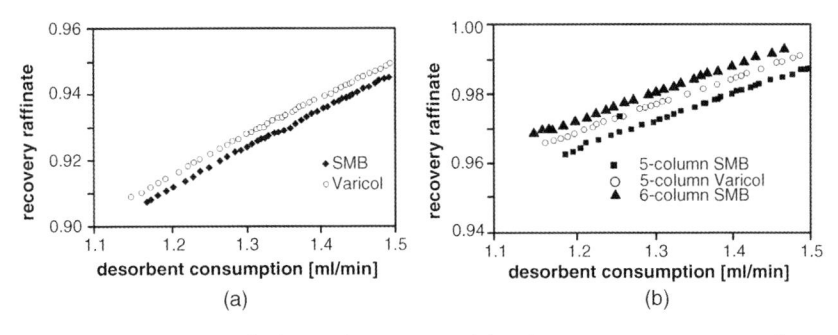

Figure 7.31 Pareto optimal solutions for recovery and desorbent consumption: (a) given design and (b) optimal design of the chromatographic plant (reproduced from Zhang, Hidajat, and Ray, 2009).

- **Maximization of recovery and minimization of desorbent flow at design stage:**
  The optimization task is similar to the second case, but here additionally the column design was optimized with the constraints of raffinate and extract purities higher than 99.0% and 95.0%, respectively. Figure 7.31b depicts that a five-column ($L = 12.9$ cm) Varicol outperforms a five-column ($L = 14.4$ cm) SMB but performs inferior than a six-column ($L = 10.6$ cm) SMB. In this context it is worth noting that the column length is lowest for a six-column SMB. Compared to the five-column SMB and Varicol the recovery increases for a six-column SMB. Compared to about 1% and 0.4%, respectively. Therefore, Zhang, Hidajat, and Ray (2009) conclude that "better performance can be achieved by increasing the number of columns while having a reduced length."

Zhang, Y., et al. (2009) also used a laboratory SMB set-up to carried out three experimental runs to verify the optimization results corresponding to the points in the Pareto sets of case 2. It was found that for SMB as well as Varicol operation the recoveries and the extract purities are much lower then the optimization calculations. The authors explain that the major cause for such deviations stems from limitations of the pump performance and hence severe fluctuations of the recycling flow rate and point out that their "experimental results indicate the great importance of flow rate control and pump performance in SMB and Varicol operation". Therefore, they also recommend that "operating conditions for SMB and Varicol processes should be selected as a compromise between separation performance and robustness".

In preparative chromatography standard SMB processes usually comprise up to eight columns in total. In open literature the application of nonstandard SMB processes is mainly discussed for four or five columns as these configurations offer interesting opportunities for saving investment cost along with production cost savings. In this context it is worth noting an investigation by Kurup et al. (2005a) about an industrial Parex process that typically has a relative large number of columns ($N = 24$). The authors identified several opportunities to optimize the standard SMB process but also point out that Varicol performs much better concerning eluent consumption and product recovery than an equivalent SMB.

Based on experimental experience, different authors indicate that besides an optimized design robustness has to be kept in mind before starting a commercial production. Mota, Araújo, and Rodrigues (2007) investigated this question in a paper on optimal design of multicolumn processes under flow uncertainties. For four-column SMB, Varicol, and PowerFeed processes they optimized, respectively, the feed rate, productivity, and the eluent consumption under uncertainties of the flow rates of ± 2.5%. Their example system is a linear separation of two nucleotides. The two-objective optimization problem, in which productivity is to be maximized and eluent consumption is to be minimized, was solved as nonlinear programming problem as described by Kawajiri and Biegler (2006a, 2006b). Figure 7.32 shows the nominal and robust Pareto curves for four-column SMB, Varicol, and PowerFeed processes.

7.5 Isocratic SMB Chromatography under Variable Operating Conditions

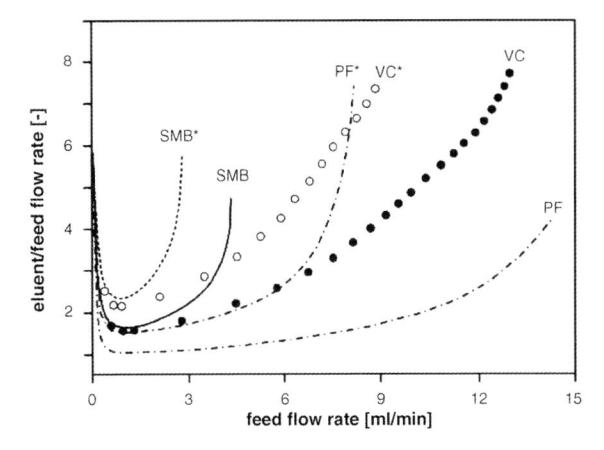

Figure 7.32 Nominal and robust Pareto curves for four-column SMB, Varicol, and PowerFeed processes; (∗) the robust Pareto curves (reproduced from Mota, Araújo, and Rodrigues, 2007).

Each point in the Pareto curve represents an optimal combination of the objective functions. The selection of the operating conditions for a production is a matter of additional economic considerations. "Any point above the Pareto curve has worse performance than the points constituting the Pareto curve, because there is always one element of the Pareto set which has a lower solvent consumption for the same productivity. The points located below the Pareto curve are not feasible because the purity requirements are not fulfilled."

"The overall results show that PowerFeed is the best operating scheme because it not only exhibits the lowest Pareto curve but also the one with the widest extent, i.e. its specific eluent consumption is consistently lower than those of the two other schemes, and the system withstands higher feed throughput without violating the purity constraints. Varicol also clearly outperforms the standard SMB process, both in productivity and eluent consumption, but it is not as efficient as PowerFeed for the separation studied by Mota, Araújo, and Rodrigues (2007). When both schemes are operated under robust conditions, however, Varicol outperforms PowerFeed in the upper end of the feed throughput region." In Figure 7.32 the performance of all processes degrades significantly for small feed rates. Mota, Araújo, and Rodrigues (2007) point out that this has to be traced back to the process model that includes perfectly mixing cells between the columns. All in all, Figure 7.32 exemplifies that robust performance is gained at the expense of higher eluent consumption and/or lower feed throughput.

For the sake of completeness it has to be mentioned that further improvements of nonstandard simulated moving bed processes by combining Varicol, PowerFeed, and/or Modicon to hybrid processes have also been investigated (e.g., Rodrigues, Araújo, and Mota, 2007b; Zhang, Mazzotti, and Morbidelli, 2004a). The presented results indicate further improvements, but these hybrids also increase the requirements on process control and robustness.

The above-mentioned investigations do not cover the whole range of SMB processes with variable operating conditions, so it has to be kept in mind that conclusions drawn on the basis of these results and other publications are of limited validity. But the following trends become apparent:

- For the same number of columns Varicol, PowerFeed, and Modicon can outperform standard SMB processes concerning product purities, productivity, and eluent consumption, respectively.
- Although theoretical optimizations indicate slight advantages of PowerFeed, at the stage of process design it should be considered equivalent to Varicol. The same is to say about Modicon and FF-SMB.
- Modicon is only an alternative if there is sufficient solubility.
- There are indications that combinations of nonstandard SMB processes (e.g., Varicol and PowerFeed) are superior. But before taking into account such a design process operability and robustness have to be proven.
- The number of columns has important influence on process performance, for example, it can be expected that SMB with an additional column can outperform the nonstandard processes.
- Before Varicol, PowerFeed, or Modicon is taken into account for process design make sure that appropriate optimization tools are at your disposal. In contrast to SMB, no shortcut methods such as "triangle theory" are available for these processes.
- The final decision on processes design has to take into account process economics. These are ruled primarily by savings in production costs, compared to SMB, and additional hardware and software for process control and additional pump systems in case of PowerFeed and Modicon. For Varicol only additional hardware and software for process control is necessary.
- The process conditions for production-scale applications have to guaranty robustness, in particular against flow rate fluctuations.

7.6
Gradient SMB Chromatography

If the selectivity of the components is very large, process efficiency can be increased by changing the elution strength of the solvent during the separation (Section 2.6.3). Such gradients of the solvent composition are realized linearly, stepwise, or in other modes. In case of SMB processes a step gradient as described in Section 5.2.6.1 is easily accomplished as no changes of the SMB plant itself are required. The process scheme for gradient SMB chromatography is depicted in Figure 5.23. Ideally, there is a step gradient between sections II and III while the solvent strength in sections I and II, respectively, III and IV is constant. In practice, certain disturbances occur because of the column switching as shown in Figure 6.45.

Ziomek *et al.* (2005a) and Ziomek and Antos (2005b) developed a procedure based on a random search routine to optimize the productivity and eluent consumption for purity constraints of $Pu_{ext,raff} > 90\%$. They applied a stage model and

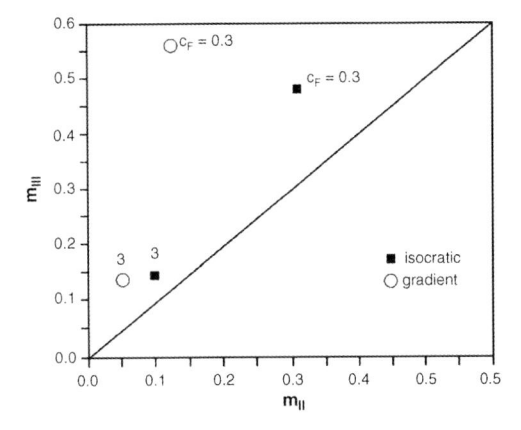

Figure 7.33 Operating points for optimal productivity in the m_{II}–m_{III} plane (reproduced from Ziomek et al., 2005a).

competitive Langmuir isotherms. The modifier concentration was optimized in a stepwise search. Figure 7.33 compares the operating points for optimal productivity in the m_{II}–m_{III} plane. Parameters are the feed concentration and isocratic as well as gradient operation. The corresponding operating points for minimum eluent consumption are rather close to these operating conditions. The data emphasize again the advantage of gradient SMB by increasing productivity and reducing eluent consumption. The authors also point out that based on their model calculations the modifier concentration is crucial for successful separation as its influence on the sensitivity of extract and raffinate purities is approximately one order of magnitude higher compared to the flow rate (Ziomek and Antos, 2005b).

As shown in Section 6.6.2.2.3 validated models for rigorous simulations of gradient SMB processes are available. But again the estimation of favorable process conditions is going to be a problem as the range of process parameters is large and rigorous simulation is too time consuming. In the following a shortcut method based on the TMB model is applied for parameter estimation for solvent gradient SMB processes. Knowing these parameters it is much easier to evaluate the final operation conditions for practical productions by targeted rigorous simulation.

The mass balance for the TMB process under the assumption of ideal conditions is given by Equation 7.90. For isotherms with constant selectivity such as multi-Langmuir and modified competitive multi-Langmuir isotherms analytical solutions are available (Section 7.4.1.1). For isotherms with nonconstant selectivity such as multi-bi-Langmuir and the IAS theory model Migliorini, Mazzotti, and Morbidelli (2000) developed numerical methods to calculate the triangle plane of complete separation for two-component systems. They also simplified this procedure to a shortcut method that makes the applicability for complex isotherms easier without significantly affecting the accuracy of the triangle plane.

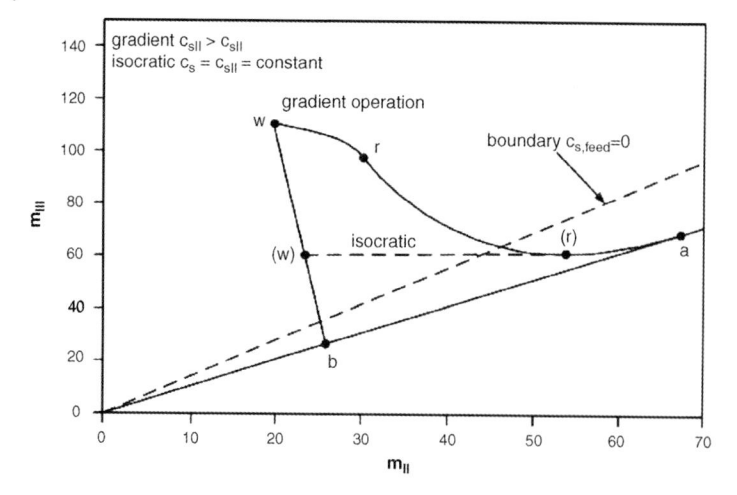

Figure 7.34 Operating map for a solvent gradient TMB process (reproduced from Wekenborg, 2009).

Wekenborg, Susanto, and Schmidt-Traub (2004,2005) and Wekenborg (2009) applied this method to estimate the operating plane for the nonisocratic SMB separation of β-lactoglobulin A and B proteins. The isotherm is described by the steric mass action (SMA) model developed for ion exchange chromatography by Brooks and Cramer (1992) (Section 2.5.2.4).

The complete operating plane is shown in Figure 7.34. As for isocratic operation shown in Figure 7.22 the operating plane reduces with increasing feed concentration. For comparison Figure 7.34 also includes the wr line for an isocratic separation. In this case the constant salt concentration in sections III and IV is equal to the value in sections I and II. Comparing both areas illustrates the advantage of the solvent gradient operation.

For a constant feed concentration of $0.1\,\mathrm{g\,l^{-1}}$ and increasing step gradients Figure 7.35a and b shows how the operating area enlarges and process

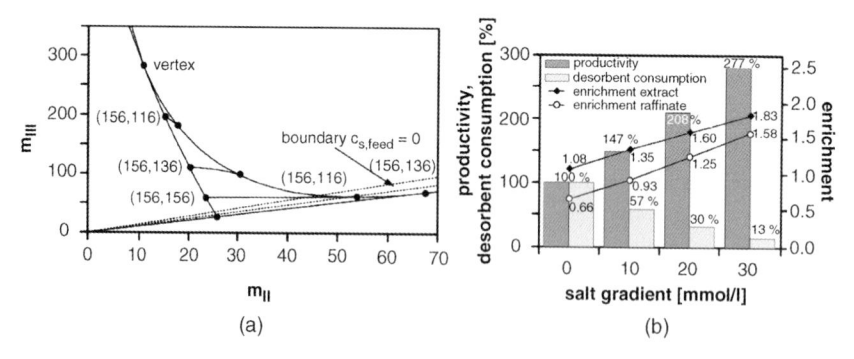

Figure 7.35 Influence of salt concentration on (a) the operating map and (b) productivity and eluent consumption of a solvent gradient process (reproduced from Wekenborg, 2009).

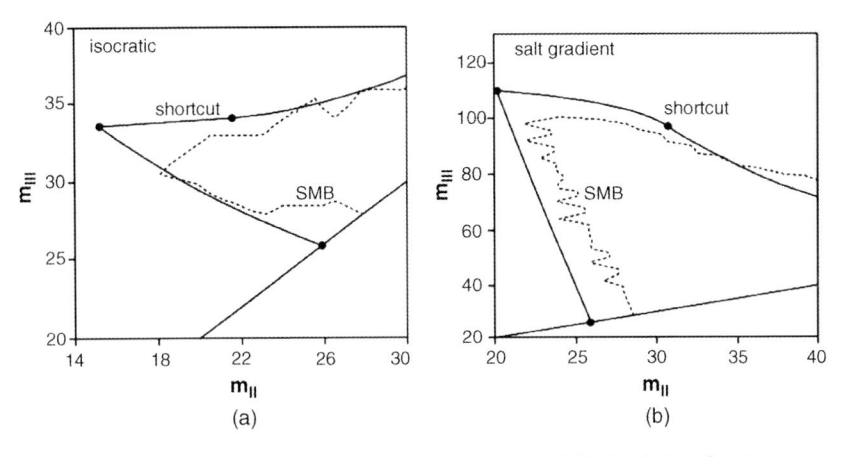

Figure 7.36 Comparison of shortcut calculations and rigorous SMB simulations for (a) an isocratic process and (b) a nonisocratic process (reproduced from Wekenborg, 2009).

efficiency increases in comparison to an isocratic separation at 156 mM salt concentration. Comparing the theoretically optimal values of the different triangle planes represented by the vertex illustrates the significant advantage of nonisocratic chromatography in comparison to isocratic chromatography. For an increasing step gradient the optimal values of m_{II} are reducing while the optimal values of m_{III} are increased, which leads to higher productivity and reduced eluent consumption.

Figure 7.36a and b proves the applicability of shortcut calculations based on the ideal equilibrium model for the estimation of process conditions. The results of rigorous process simulation based on the transport-dispersive model are in very good agreement with the shortcut calculation for isocratic (a) as well as nonisocratic (b) SMB processes. Expectedly safety margins have to be taken into account when the process conditions of an SMB process are estimated by shortcut calculation. The scattering of the numerical data results from an increased grid size for the numerical calculations that has been chosen in order to reduce computer time. The model parameters coincide with the data for the protein separation presented in Section 6.6.2.2.3; the separation quality of the SMB process was set to 99.9% purity.

The above describes a method how to estimate possible operating points and has proven that a 100% purity chromatographic separation is reached within the triangle plane if a certain safety margin is taken into consideration. But in practice such purities are often not economic and not necessary for both components. In such cases it is the aim to identify operating points beyond the triangle plane. This can be done by case studies based on rigorous simulation; a more direct procedure would be an optimization procedure based on a less time-consuming model.

First steps to investigate protein separations by solvent gradient SMB trace back to the research group of van der Wielen (Jensen et al., 2000) based on an extension of the "triangle method"; it was found that compared to isocratic SMB an up to 50% decrease of eluent consumption and a twofold enrichment of the product can be achieved. Later, the separation of BSA and myoglobin on Q-Sepharose FF was investigated experimentally (Houwing et al., 2002). This system, which is described by linear isotherms and an azeotrope with increasing salt concentration, was adopted by the research group of Rodrigues for further studies on gradient SMB configurations with open loop, closed loop, as well as closed loop with holding vessel (Li, Xiu, and Rodrigues, 2007, 2008). They recommend the following process configurations for binary separation of proteins: if closed loop gradient SMB is favored in order to reduce the solvent consumption, a hold-up vessel should be installed between sections IV and I. In this vessel the desorbent and the solvent recycle are mixed during the switching period in order to prevent solvent strength fluctuations in section I. Open loop gradient SMB is advantageous if weak adsorbing impurities are present. A sufficiently high salt concentration is necessary in sections III and IV to prevent adsorption of the impurities and to dispose them with the solvent.

Recently a multidiscipline research group in Magdeburg investigated a three-section SMB process for the purification of recombinant proteins by hydrophobic interaction chromatography (Keßler et al., 2007; Palani et al., 2011; Gueorguieva et al., 2011). They developed an equivalent TMB scheme for the three-section SMB process as shown in Figure 7.37a. The aim of the open loop gradient process is to purify the target product recombinant streptokinase from impurities. The product is drawn off at the extract port while the impurities go with the raffinate stream. As cheap aqueous buffers are used, the fourth section for regeneration of the mobile

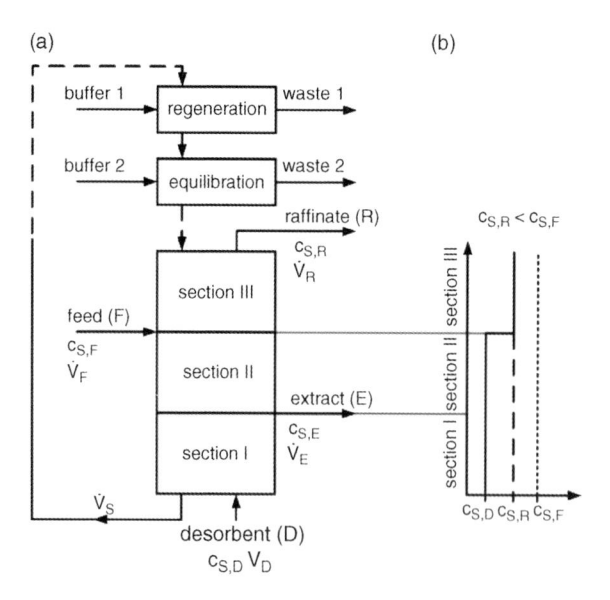

Figure 7.37 (a) Scheme of a three-section open loop gradient TMB process and (b) two-step salt gradient for HIC chromatography (Gueorguieva et al., 2011).

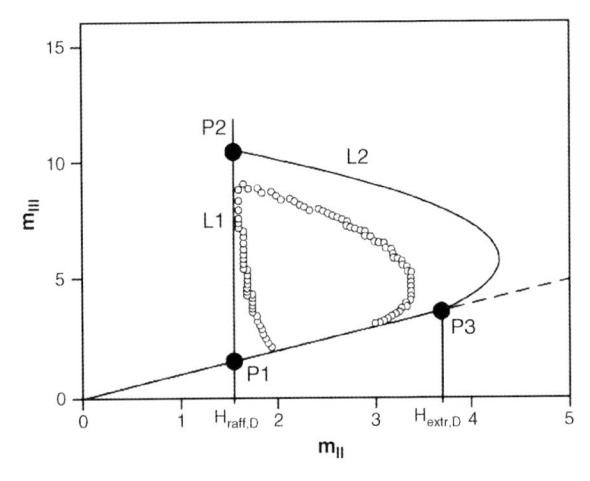

Figure 7.38 Separation region based on equilibrium theory and equilibrium stage model (Gueorguieva et al., 2011).

phase is not required. The step gradient occurs between sections II and III. Due to the HIC mode the salt concentration in sections I and II is lower than in section III (Figure 7.37b). The process design includes two extra sections for regeneration and equilibration of the solid phase. These sections are operated independently and use different mobile phases. For the estimation of the operating parameters of the corresponding SMB process Gueorguieva et al. (2011) used the ideal model based on the equilibrium theory to calculate the m_{II}–m_{III} operating plane as shown in Figure 7.38. This operating plane (P1–P2–P3) corresponds to a number of stages $N \rightarrow \infty$. The smaller operating plane in Figure 7.38 that results from the equilibrium stage model for $N = 20$ in each section is source for the selection of operating parameters of experiments. It was possible to separate streptokinase with relatively high purities, but due to insufficiencies of the plant the predicted enrichment was not reached. Depending on purity and yield constraints the described continuous purification process can be applied for initial capture or intermediate purification of recombinant proteins.

7.7
Multicolumn Systems for Bioseparations

The separation of proteins and other biopolymers has some distinctly different features in comparison with the separation of low molecular weight molecules. Biopolymers have a molecular weight (MW) ranging from several thousand to several million. They are charged and characterized by their isoelectric point. More importantly, they have a dynamic tertiary structure that can undergo changes known as conformation. These changes can influence or even destroy the bioactivity if a protein denaturates permanently. Biopolymers are separated in aqueous buffered eluents under conditions maintaining their bioactivity or under denaturing conditions. Moreover, one has to deal with large molecules that exhibit approximately 100 times lower diffusion coefficients and consequently slower mass transfer than small molecules. In order to resolve these species by selective solute–surface interactions in a chromatographic column, the linear velocity of the eluent at optimum conditions also has to be 100 times lower in the setup. However, access of biopolymeric solutes to the functional groups at the pore surface (stationary phase) is mandatory. While LC of small molecules is performed using adsorbents with 10 nm average pore diameter, peptides require packings with pores of approximately 30 nm and proteins need materials with pore diameters in excess of 50 nm to enable a suitable mass transfer (Unger et al., 2010).

Monoclonal antibodies such as immunoglobulin (IgG) play an important role among new pharmaceutical drugs. An established method to purify mAbs is protein A affinity batch chromatography, but these stationary phases are rather expensive and are insensitive to mAb variants. An alternative to purify charged mAb variants is ion exchange batch chromatography, which has the drawback to achieve high purities at low yield or, vice versa, lower purities at higher yield. A general principle to increase productivity and yield of separation processes is a countercurrent mass transport that, for instance, characterizes SMB processes. The purification of proteins from clarified cell culture supernatants (sCCS) normally requires the separation of impurities, especially host cell impurities (HCP) and mAb variants that are, respectively, weaker and stronger adsorbing than the product. Therefore, a single-step process has to deal with three product streams like the MCSGP process as described in Section 5.2.8.2. Compared to a gradient SMB process the MCSGP process also has the advantage that instead of a steep step gradient a shallow gradient elution is possible.

Müller-Späth et al. 2010a, 2010b, 2011 compared the MCSGP process and batch chromatography for different ion exchange separations. As one example Figure 7.39 depicts the yield–purity relation for batch and MCSGP ion exchange chromatography of an industrial titer from Merck Serono. A fourth column for continuous loading and CIP was added to the three-column MCSGP process. The captured mAb was an IgG_2.

The results of the batch process show the typical trade-off between yield and purity. High purity can only be achieved at the expense of yield and vice versa (dashed line) because in biochromatography baseline separation of product and impurities is not achieved even under shallow gradient or isocratic conditions with large retention times. Müller-Späth et al. (2010b) also compared the state-of-the-art protein A batch process including two polishing steps (CIEX and AIEX) with the MCSGP process followed by an AIEX polishing step (filled square). Based on experimental results and a simplified economic comparison, the authors come to the conclusion that the downstream costs of the two-step MCSGP process are 25% lower than for the three-step affinity chromatography. The main factor for these differences was the difference in material costs of 50%. The authors also expect that this difference will increase when the mAb titers are increased.

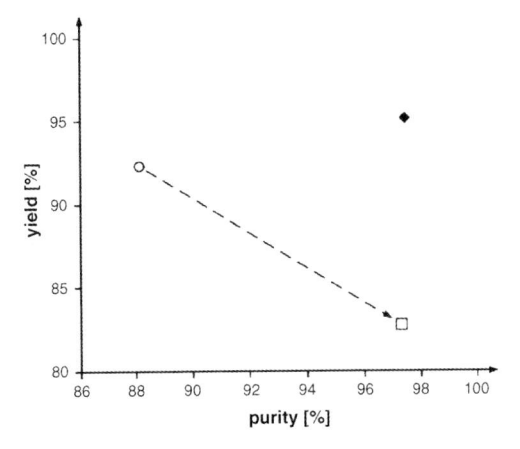

Figure 7.39 Yield–purity relation for batch and MCSGP capture of an IgG$_2$ from an industrial titer using cation exchanger Fractogel SO$_3$ (M) (reproduced from Müller-Späth *et al.*, 2010b).

For the design of an MCSGP process Aumann and Morbidelli (2007) describe an empirical method that is based on a single solvent gradient batch chromatogram. The basic idea is to slice the chromatogram into intervals that represent certain steps of the MCSGP process and to transfer characteristic data of these intervals such as flow rates and solvent gradients into operating data for the MCSGP steps. This empirical method has the advantage that knowledge of isotherms and mass transfer data is not necessary.

To enhance the empirically designed process theoretical simulations are necessary. Müller-Späth *et al.* (2011) explain the simulation of an MCSGP process based on the lumped kinetic, transport-dispersive model, and the determination of required model parameters. Initial process parameters for the simulation were taken from the empirical procedure mentioned above. Figure 7.40 presents a comparison of simulated and measured data based on the same chromatographic system as in Figure 7.39. The model predicts product yield and concentration within error of <6% and is therefore a very good tool to fine-tune the MCSGP process including process start-up and shutdown.

7.8
Advanced Process Control

Chromatographic separations usually are operated manually, that is, the main operating parameters are determined or optimized as discussed in Section 6.5 and implemented at the separation unit, and from the observation of the resulting performance, small iterative changes are made in order to meet the specifications and to gradually improve the process efficiency, for example, by increasing the throughput or by decreasing the solvent consumption. Advanced process control aims at performing this adaptation automatically, hence more reliably, faster, and with a

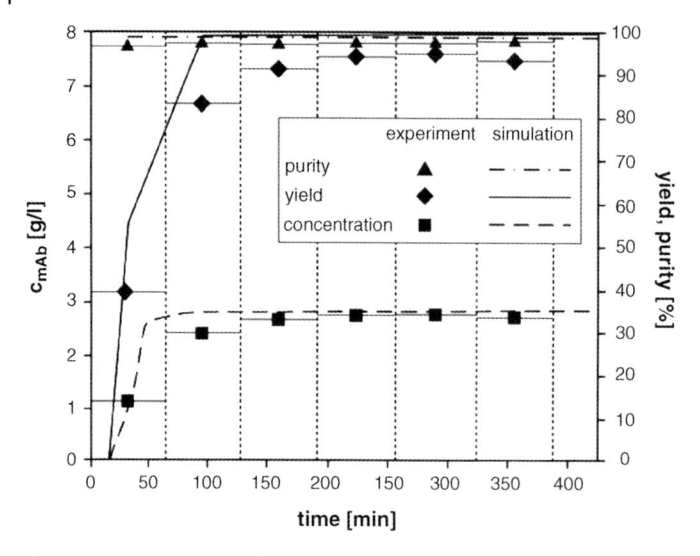

Figure 7.40 Comparison of experimental and simulated data for start-up and steady state of the MCSGP process (reproduced from Müller-Späth et al., 2011).

better reproducibility. Advanced control schemes always rely on the use of a process model, but the required precision of the models differs among the different proposed schemes. Due to the feedback nature of advanced control – the controller reacts, for example, to the observed purities – model inaccuracies can be compensated to some extent. Despite intense academic research and successful demonstrations at real plants, advanced control however has not yet been implemented in industry on a large scale.

7.8.1
Online Optimization of Batch Chromatography

Online optimization of batch chromatography is based on the same basic considerations as described for process design in Section 7.2 for the choice of the operating parameters, but this optimization is performed during the operation of the process. The operational degrees of freedom, for example, injection time and cutting times, are adapted automatically to increase the productivity while maintaining the desired product purities and recoveries. The process models used are dynamic in nature, but as the operating parameters can only be varied for the next injection, the optimization problem that results is static in nature. The model predicts the resulting purities, recoveries, and productivity as a function of the operational degrees of freedom and a nonlinear optimization problem is solved to compute the optimal operating parameters.

The crucial issue here is the handling of the inevitable discrepancies between the behavior of the plant and the predictions by the model. For example, if a mismatch of the predicted purities is observed, this can be used to modify the target values such that the resulting true purity is as specified (bias correction). While this approach assures the meeting of the specifications, it is not sufficient to obtain optimality of the operating parameters in the case of model errors. For optimality, either the process model has to be adapted by online parameter estimation or a suitable modified scheme has to be used (see below).

It is possible to estimate key separation parameters, in particular the parameters of the isotherms, from batch chromatograms, using nonlinear optimization techniques. However, usually only a small number of parameters can be estimated simultaneously and the estimation has to be initialized carefully, for example, by first performing an estimation based on shortcut models. Gao and Engell (2005b) demonstrated that it is even possible to estimate black box models of the isotherms from batch experiments so that no assumptions on the mathematical form of the isotherms have to be made. Parameter adaptation however leads to a small mismatch between the plant model and the real process only if the model is structurally correct, that is, able to predict the observations for suitably chosen parameters, which is not always the case. Especially for high purity requirements, small deviations of the chromatograms, for example, due to the presence of components that are not considered in the model, can lead to significant errors in the predicted purities and therefore to the need of measurement-based adaptation of the operating policy.

In the iterative online optimization scheme proposed by Gao and Engell (2005a), the gradients of the cost function and of the constraints (e.g., purity requirements) are modified based on the available measurements in a fashion that ensures convergence to the true optimum for the real plant despite the error between the behavior of the real plant and the prediction of the model that is used in the optimization. This also reduces the necessary effort for modeling and model parameter estimation because coarse models are sufficient to perform the iterative online optimization. For the estimation of the true gradients of the constraints and of the performance criterion with respect to the operating parameters, the observed outcomes of the recent iterations are used, but in some cases additional experiments have to be performed to improve the estimation that temporarily reduce the performance but are necessary to reach true optimum. The number of iterations (injections) needed to approach the optimum is typically around 10. The online optimization algorithm is depicted graphically in Figure 7.41.

Simulation results of the algorithm for an enantiomer separation with bi-Langmuir isotherms are shown in Figure 7.42. The discrepancies between the chromatograms that are predicted by the optimization model and that are assumed for the real plant are shown in Figure 7.43.

The first iterations $u^{(-2)}$–$u^{(0)}$ are needed to collect the information that is required to estimate the empirical gradients. Thereafter, the iterative improvement starts. It can be seen that despite a significant error in the chromatograms, the iterative optimization converges to the true optimum and establishes the desired purities and recoveries of the components. In recent work, this approach has been applied to continuous annular electrochromatography (Behrens and Engell, 2011).

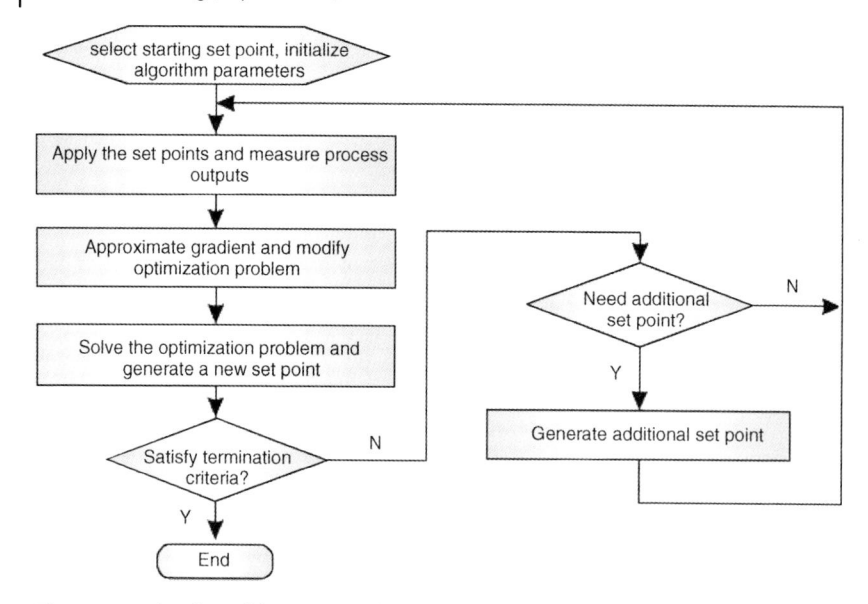

Figure 7.41 Flowchart of the iterative online optimization algorithm for batch chromatography proposed in Gao and Engell (2005a).

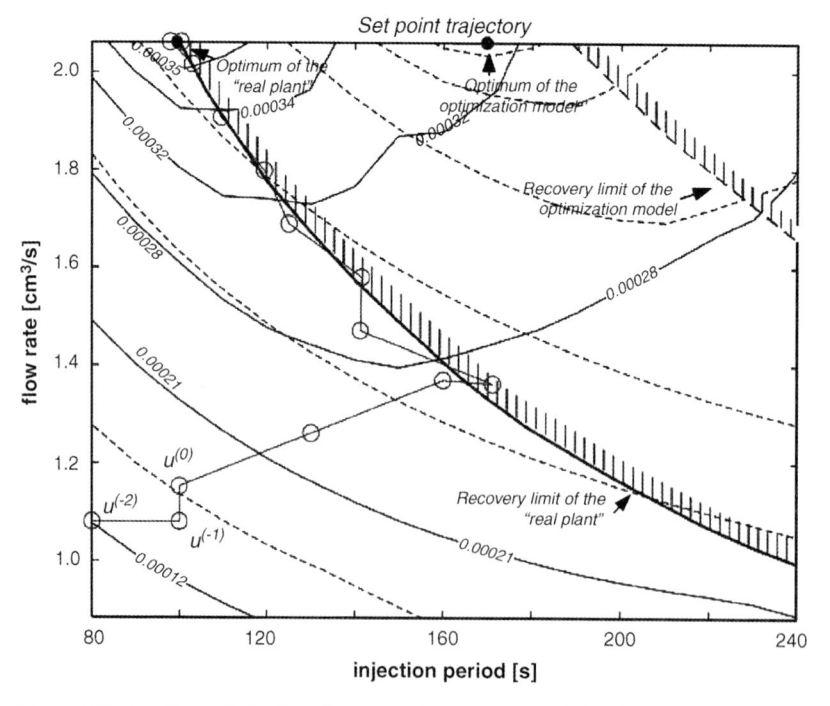

Figure 7.42 Iterative optimization of the operating parameters of a batch separation with plant–model mismatch.

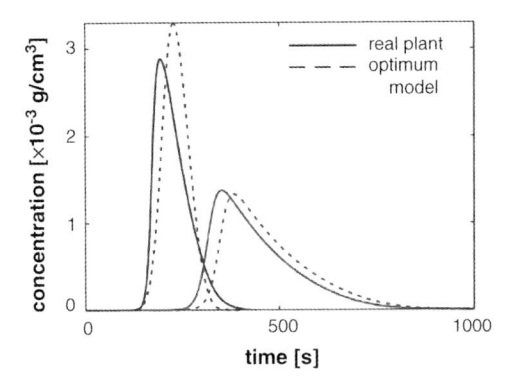

Figure 7.43 Discrepancy between real and predicted chromatograms for the iterative optimization shown in Figure 7.42.

7.8.2
Advanced Control of SMB Chromatography

The basic controllers that are necessary to operate an SMB plant have already been discussed in Section 4.5. In this section we are concerned with the higher control level of guaranteeing the desired product purities and recoveries and optimizing the performance of the plant.

As discussed in several chapters of this volume, the use of detailed process models in the design and in the choice of the operating parameters of chromatographic separation processes may lead to considerable improvements. However, optimal settings of the operating parameters for simulation models do not guarantee an optimal operation of the real plant. This is due, for example, to nonidealities of the flows and of the column packings, unmodeled peripherals, the presence of additional substances in the mixture, inaccuracies in the actuators (pumps and valves), and changes of the plant behavior over time. As an optimal operating point always is at the constraint of at least one performance criterion, the specifications are most likely not met at the real plant if the operation parameters that were obtained from off-line optimization are applied.

In principle, there are three possible remedies for this problem. First, the plant can be operated with a safety margin – the most common but not the most economic approach. Second, some critical specifications can be controlled online, by feeding back the measured values to a controller, especially the product purities as the most critical parameters. This feedback can also be realized manually, which requires the continuous presence of skilled operators. The most advanced approach is to establish the optimality of the operation continuously and automatically, based on more or less sophisticated plant models and the available measurements.

If the variation of the flow rates with the aim to reach the desired product purity is done manually, the modification of the operating parameters is based on either heuristic rules or relying on the expertise of the operators. Antia (2003) proposed the following practical scheme:

- Start with low feed concentrations in order to achieve linear separation conditions.
- Increase $\dot{V}_I$ to a large value and decrease $\dot{V}_{IV}$ to a low value so that the design criteria for the sections I and IV are satisfied by a large margin. Then, the attention is focused on the appropriate choice of the flow rates in the central sections II and III.
- Increase the concentration of the feed in steps. Determine which outlet is polluted and correct the flow rates according to predefined rules. This can be repeated until the feed concentrations reach their upper limits.
- Once the flow rates in the central sections are chosen appropriately, increase $\dot{V}_{IV}$ and decrease $\dot{V}_I$. This ensures that a minimal flow of eluent is used and thus near-optimal process performance is reached.

For more details on the background of these rules we refer to Section 7.4.

In practice, SMB processes are controlled using similar manual schemes (Küsters, Gerber, and Antia, 1995; Juza, 1999; Miller et al., 2003). Antia (2003) suggested to include these heuristic rules into a fuzzy controller in order to achieve full automatic control of SMB processes.

7.8.2.1 Purity Control for SMB Processes

For automatic purity control of SMB processes it is assumed that the plant is operated near a fixed, well-designed or optimized, operating point. The degrees of freedom available for online control are the (external or internal) flow rates and the switching time; these can be adapted to meet the purity specifications.

Automatic control of purities is difficult due to the long time delays and the complex dynamics that are described by nonlinear distributed parameter models and switching of the inputs, leading to mixed discrete and continuous dynamics, small operating windows, and a pronouncedly nonlinear response of the purities to input variations. Because of the complex nonlinear dynamics of SMB processes, their automatic control has attracted the interest of many academic research groups and many different control schemes have been proposed; however, few of them have been tested in experimental work for real plants with limited sensor information.

Automatic purity control was reported for the separation of aromatic hydrocarbons where online Raman spectroscopy can be used to measure the concentrations of the compounds at the outlet of the chromatographic columns (Marteau et al., 1994). This approach, as well as the geometrical nonlinear control concept described by Kloppenburg and Gilles (1999), is based on a model that describes the corresponding TMB process, so the cyclic port switching is neglected.

Natarajan and Lee (2000) investigated the application of a repetitive model predictive control (RMPC) technique to SMB processes. RMPC is a model-based control technique that results from incorporating the basic concept of repetitive control into the model predictive control framework (Qin and Badgwell, 2003). The switching period of the process is assumed to be constant. This is limiting, since the switching time can also be manipulated to control the process. In this approach, a rigorous model discretized process model is linearized along the optimal trajectory.

Schramm *et al.* (2001) and Schramm, Grüner, and Kienle (2003) presented a model-based control approach for the direct control of the product purities of SMB processes. Based on wave theory, relationships between the front movements and the flow rates of the equivalent TMB process were derived. Using these relationships, a simple control concept with two PI controllers was proposed.

Cox, Khattabi, and Dapremont (2003) reported a successful control and monitoring system for the separation of an enantiomer mixture based on the concentration profiles in the recycle loop.

Klatt *et al.* (2000) and Klatt, Hanisch, and Dünnebier (2002) proposed a two-layer control architecture where on the upper level the optimal operating regime is calculated at a low sampling rate by dynamic optimization based on a rigorous nonlinear process model. The low-level control task is to keep the process at the optimal operating point, which is realized by stabilizing the concentration fronts. This leads to a multivariable control problem, which for linear isotherms can be approximately decoupled by using the m-parameters or the β-factors as manipulated variables instead of flow rates and the switching time. For the stabilization of the concentration fronts, in the case of linear isotherms linear SISO IMC controllers were used. The dynamic models for the stabilization of the concentration fronts are obtained by identifying black box models using simulation data of the rigorous process model for small changes of the inputs around the optimal operating point. The scheme was validated experimentally. It turned out that the localization of the concentration fronts based on measured chromatograms is difficult even when two independent sensors are available per outlet stream and in the recycle (UV detectors and polarimeters) due to backmixing effects, and that stabilizing the front positions does not suffice to guarantee the product purities due to plant–model mismatch such that an additional control layer for purity control is needed (Hanisch, 2002). In the case of nonlinear isotherms, dynamic neural network models were used in a nonlinear MPC scheme (Wang *et al.*, 2003).

The same idea of bilevel optimization is pursued in Kim, Lee, and Lee (2010a) and demonstrated experimentally using an RMPC scheme on the lower (purity control) level.

Song *et al.* (2006) proposed a multivariable purity control scheme using the m-parameters as manipulated variables and a model predictive control scheme based on linear models that are identified from nonlinear simulations. The approach proposed by Schramm, Grüner, and Kienle (2003) for purity control has been modified by several authors (Kleinert and Lunze, 2008; Fütterer, 2008). It gives rise to relatively simple, decentralized controllers for the front positions, but an additional purity control layer is needed to cope with plant–model mismatch and sensor errors. Vilas and Van de Wouwer (2011) augmented it by an MPC controller based on a POD (proper orthogonal collocation) model of the plant for parameter tuning of the local PI controllers to cope with the process nonlinearity.

7.8.2.2 Direct Optimizing Control of SMB Processes

Due to the nonlinear behavior of SMB processes in particular if the adsorption isotherms are nonlinear, simple purity control schemes only work in a narrow range around a fixed operating point and can only react slowly to disturbances or changes of the set points. More sophisticated schemes make use of nonlinear or adapted models and are relatively complex. As only two purities are to be controlled, purity control defines only 2 of the at least 4 degrees of freedom that can be used online. A new idea that was proposed independently by the research groups at TU Dortmund and ETH Zürich in 2003/2004 (Erdem *et al.*, 2004; Toumi and Engell, 2004) is to perform a direct model-based optimization of the operating regime online, using the purity requirements not as set points for the controller but as constraints in the optimization of the operating point. This principle can be generalized to other processing units (Engell, 2007).

Both approaches are based on the same principle: a sequence of future operating parameters is computed employing a model of the SMB process such that a performance criterion is optimized over the so-called prediction or look-ahead horizon, respecting constraints, for example, on the product purities and the maximum column pressure. The two approaches differ in the type of models used and the degrees of freedom that are considered in the optimization. In the work of the ETH group, linear prediction models are used and only the flow-rates but not the switching time are used as manipulated variables. Changes of the manipulated variables are realized at shorter intervals than the switching time. In the Dortmund approach, a discretized nonlinear general rate model (Gu, 1995) is employed and a rigorous nonlinear online optimization of all degrees of freedom is performed, including the switching time. Changes of the controlled variables are only implemented after a switching of the ports has occurred. In the implementation, a sequential approach was employed where the process is simulated over the prediction horizon and the degrees of freedom are optimized based on the simulation results using an SQP solver.

Both groups have validated their approaches successfully at laboratory-scale SMB plants. At the ETH the research teams of Morari, Morbidelli, and Mazzotti have recently modified their control scheme to a cycle-to-cycle control scheme where the control is based on averaged online HPLC-purity measurements that are available once per cycle (Grossmann *et al.*, 2008, 2010a) and combined it with the previous scheme to a multirate controller that also uses UV measurements that are available at a much faster sampling rate (Grossmann *et al.*, 2010b). Here, a simple model of the SMB process is used, so the only needed model parameters are the Henry coefficients and the bed porosity. Despite the use of a quite simple model, processes with nonlinear isotherms can be controlled well over a significant range of operating conditions. Recently, the control scheme has also been applied to the multicolumn solvent gradient purification process (MCSGP process) (Grossmann *et al.*, 2010c). The Dortmund approach, due to its general nature, can be applied without conceptual changes also to reactive separations. Below we briefly report on the application to the isomerization of glucose to fructose. For further details see Toumi and Engell (2004).

7.8.2.2.1 **Process Description** The reactive simulated moving bed process considered here is the isomerization of glucose to fructose. The plant consists of six reactive chromatographic fixed beds that are interconnected to form a closed loop arrangement. As shown in Figure 7.44, a pure glucose solution is injected to the

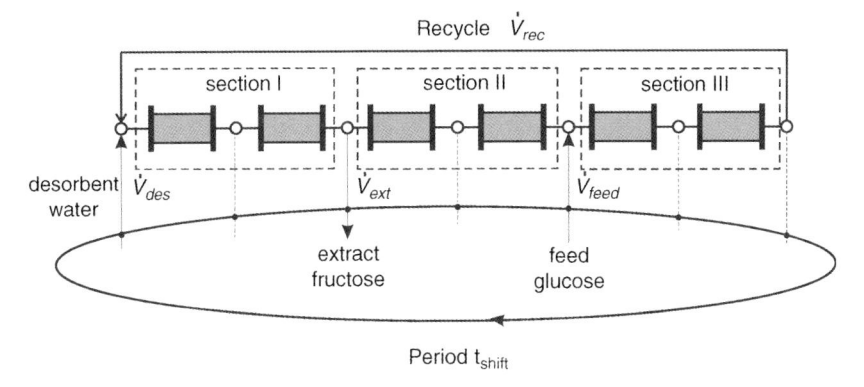

Figure 7.44 Three-section reactive SMB process for glucose isomerization.

system at the feed line. At the extract line, a mixture of glucose and fructose, called high-fructose corn syrup (HFCS), is withdrawn. Water is used as solvent and is fed continuously to the system at the desorbent line. In this special SMB process, no attempt is made to achieve a complete separation of glucose and fructose since the most common type of fructose syrup, usually called high-fructose syrup, is mainly produced either as HFCS42 (42% fructose) or as HFCS55 (55% fructose). For some purposes, syrup with more than 55% fructose, called a higher-fructose syrup, is desirable. In any case, the objective is to transform a feed containing pure glucose to a stream where the glucose is partially converted to fructose.

7.8.2.2.2 Formulation of the Online Optimizing Controller
A schematic representation of the optimizing controller is shown in Figure 7.45. As discussed above, the task of the controller is to optimize the performance of the process over a certain horizon in the future, the prediction horizon. The specifications of the product purities, the limitations of the equipment, and the dynamic process model (a full hybrid model of the process including the switching of the ports and a general rate model of all columns) appear as constraints. The control algorithm solves the following nonlinear optimization problem online (Equation 7.113):

$$\min_{[\beta_k, \beta_{k+1},\ldots,\beta_{k+H_r}]} \Gamma = \sum_{i=k}^{k+H_p} (\text{Cost}(i) + \Delta\beta_i^T \mathbf{R}_i \Delta\beta_i)$$

$$\text{subject to} \quad 0 = \mathbf{f}(\dot{\mathbf{x}}_i, \mathbf{x}_i, \boldsymbol{\beta}_i, \mathbf{p}_i)$$

$$\frac{1}{H_r} \sum_{j=k}^{k+H_r} \text{Pu}_{\text{ext},j} + \Delta\text{Pu}_{\text{ext},i} \geq \text{Pu}_{\text{ext,min},i}$$

$$\frac{1}{H_p} \sum_{j=k}^{k+H_p} \text{Pu}_{\text{ext},j} + \Delta\text{Pu}_{\text{ext},i} \geq \text{Pu}_{\text{ext,min},i} \quad (7.113)$$

$$\dot{V}_1^i \leq \dot{V}_{\max}$$

$$g(\boldsymbol{\beta}_i) \geq 0$$

$$i = k, (k+1), \ldots, (k+H_p)$$

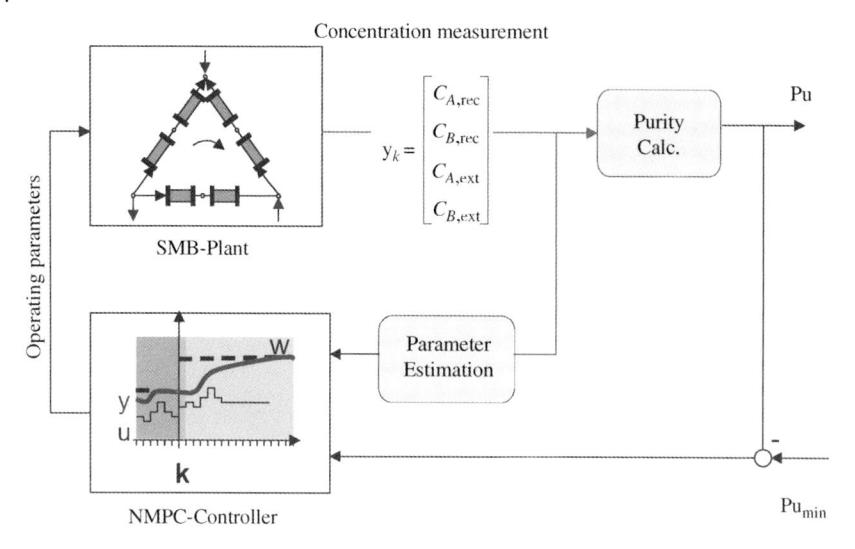

Figure 7.45 Online optimizing control structure.

The natural degrees of freedom of the process are the flow rates of desorbent $\dot{V}_{\text{des}}$, $\dot{V}_{\text{feed}}$, and recycle $\dot{V}_{\text{IV}}$, and the switching period t_{shift}. However, this results in an ill-conditioned optimization problem. The numerical tractability is improved by introducing the so-called β-factors via a nonlinear transformation of the natural degrees of freedom (Equation 7.114):

$$\dot{V}_{\text{solid}} = \frac{1-\varepsilon}{V_s t_{\text{shift}}}, \quad \beta_1 = \frac{1}{H_A}\left(\frac{\dot{V}_I}{\dot{V}_{\text{solid}}} - \frac{1-\varepsilon}{\varepsilon}\right)$$
$$\beta_2 = \frac{1}{H_B}\left(\frac{\dot{V}_{II}}{\dot{V}_{\text{solid}}} - \frac{1-\varepsilon}{\varepsilon}\right), \quad \beta_3 = \frac{1}{H_A}\left(\frac{\dot{V}_{III}}{\dot{V}_{\text{solid}}} - \frac{1-\varepsilon}{\varepsilon}\right) \quad (7.114)$$

Here, $\dot{V}_{\text{solid}}$ is the apparent solid flow rate, and H_A and H_B describe the slopes of the adsorption isotherm, which are calculated in the nonlinear case by linearization of the adsorption isotherm for the feed concentration $c_{\text{feed},i}$. The *prediction horizon* is discretized in cycles, where a cycle is a switching time t_{shift} multiplied by the total number of columns. Equation 7.113 constitutes a dynamic optimization problem with the transient behavior of the process as a constraint. It describes the continuous dynamics of the columns based on the GRM as well as the discrete switching from period to period.

For the solution of the PDE models of the columns, a Galerkin method on finite elements is used for the liquid phase and orthogonal collocation for the solid phase. The switching of the node equations is considered explicitly, that is, a full hybrid plant model is used. The objective function Γ is the sum of the costs incurred for each cycle (e.g., the desorbent consumption) and a regularizing term that is added in order to smooth the input sequence in order to avoid high fluctuations of the inputs from cycle to cycle. The first equality constraint represents the plant model evaluated over the finite prediction horizon H_p. Since the maximal attainable pressure drop by the pumps must not be exceeded, constraints are imposed on the flow rates in section I. Further inequality constraints $g(\beta)$ are added in order to avoid negative flow rates during the optimization.

The objective to meet the product specifications is reflected by the purity constraint over the control horizon H_r that is corrected by a bias term $\delta\,Pu_{ext}$ resulting from the difference between the last simulated and the last measured process output to compensate unmodeled effects (Equation 7.115):

$$\Delta Pu_{ext,i} = Pu_{ext,(i-1),meas} - Pu_{ext,(i-1)} \qquad (7.115)$$

The second purity constraint over the whole prediction horizon acts as a terminal (stability) constraint forcing the process to converge toward the optimal cyclic steady state. The goal of feedback control in a standard control approach (i.e., to fulfill the extract purity) is introduced as a constraint here.

The concentration profiles in the recycle line are measured and collected during a cycle. Since this measurement point is fixed in the closed loop arrangement, the sampled signal includes information from all three sections. During the start-up phase, an online estimation of the actual model parameters is started in every cycle. The quadratic cost functional (Equation 7.116)

$$J_{est}(p) = \sum_{i=1}^{n_{sp}} \left(\int_0^{N_{col}} (c_{i,meas}(t) - c_{i,Re}(t))^2 dt \right) \qquad (7.116)$$

is minimized with respect to the parameter **p**. The model parameters are equal for all columns.

For the prediction of the evolution of the process, the actual state (i.e., the concentration profiles over the columns) is needed. It is computed by simulation of the process model using the measurements in the recycle line as input functions. As the model is adapted online, this is sufficiently accurate.

7.8.2.2.3 Experimental Study For experimental investigations a Licosep 12–50 plant (Novasep, France) was used. An analysis based on the Fisher information matrix showed that the process model is highly sensitive to the values of the Henry coefficients, the mass transfer resistances, and the reaction rate. These parameters are therefore reestimated online at every cycle. Figure 7.46 shows the evaluation of these parameters. In order to change the model parameters during process operation, the water temperature was reduced by 10 °C at the end of the sixth cycle. At the end of the experiment, all system parameters have converged toward stationary values as shown in Figure 7.46. The mathematical model thus describes the behavior of the SMBR process sufficiently well.

The desired purity for the experiment reported below was set to 55.0% and the controller was started at the 60th period. Figure 7.47 shows the evaluation of the product purity as well as of the controlled variables. In the open loop mode where the operating point was calculated based on the initial model, the

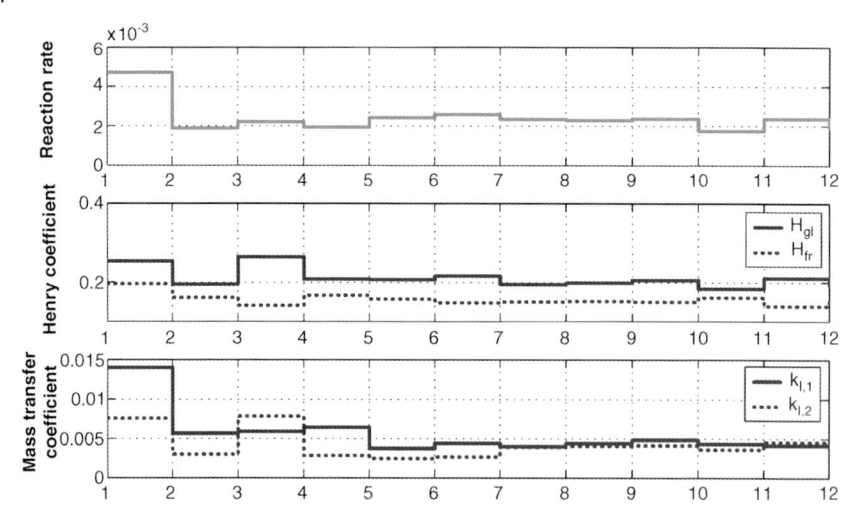

Figure 7.46 Evaluation of the estimated parameters (Toumi and Engell, 2004).

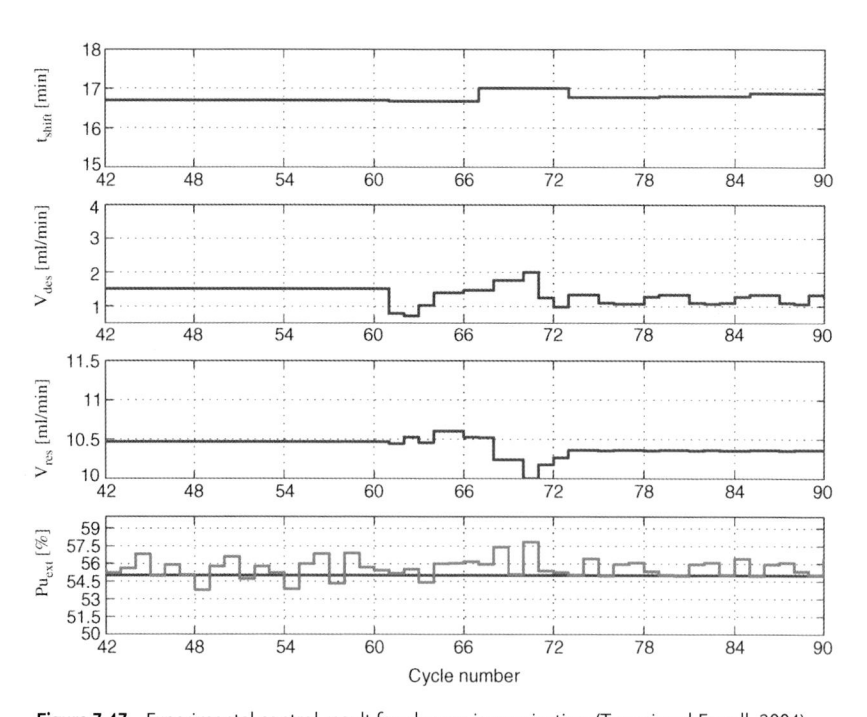

Figure 7.47 Experimental control result for glucose isomerization (Toumi and Engell, 2004).

product purity was violated at periods 48 and 54. After one cycle, the controller drove the purity above 55.0% and kept it there. The controller first reduces the desorbent consumption. This action seems to be in contradiction to the intuitive idea that more desorbent injection should enhance the separation. In the presence of a reaction, this is not true, as shown by this experiment. The controlled variables converge toward a steady state, but they still change from period to period, due to the nonideality of the plant.

In the same spirit, Alamir, Ibrahim, and Corriou (2006) proposed an online optimizing controller that switches between different performance criteria.

The optimizing control scheme based on a rigorous nonlinear process model has been successfully applied also to the racemization of Troeger's base in a Hashimoto SMB process (Küpper and Engell, 2007, 2009). For this process, simulation studies revealed that the optimal operation of the process with respect to the solvent consumption is a nonconventional one where the impurities in the extract stream result from an impure recycle stream. At this operation point it was observed that a parameter error of 5/10% in the Henry parameters leads to large discrepancies in the chromatograms that cannot be compensated by the usual bias correction scheme, whereas if the operation is constrained to the conventional operation mode, it can be controlled also in the presence of plant–model mismatch without problems (Küpper and Engell, 2008), however at the expense of a significantly higher solvent consumption. The robustness of advanced control schemes continues to be a topic of ongoing research.

7.8.3
Advanced Parameter and State Estimation for SMB Processes

For monitoring and advanced control of SMB processes, the estimation of important column parameters and of the concentration profiles along the columns is needed. This problem is characterized by a large number of unknown variables (typically several hundred states, the discretized concentration profiles along the columns, and some other parameters) and very scarce and time-varying measurement information. It has been demonstrated recently (Küpper *et al.*, 2009, 2010) that the concentration profiles and the isotherms can be reconstructed from this measurement information using a moving horizon estimation scheme (Rao and Rawlings, 2000). The estimation scheme works well also in the start-up phase of an SMB plant so that a coarse model can be improved online for use in the optimization of the operating conditions in the cyclic steady state. This work also showed that changes of the porosity and of the isotherms cannot be distinguished in practice. In this approach, uniform parameters of all columns are assumed. For monitoring of an SMB plant, information on the parameters of individual columns is desirable. A switched estimation scheme for state and parameter estimation for individual columns in an SMB plan has been proposed in Küpper and Engell (2006). Figure 7.48 shows a simulation where one Henry parameter of one column was changed and the change was correctly identified by the switched estimation scheme.

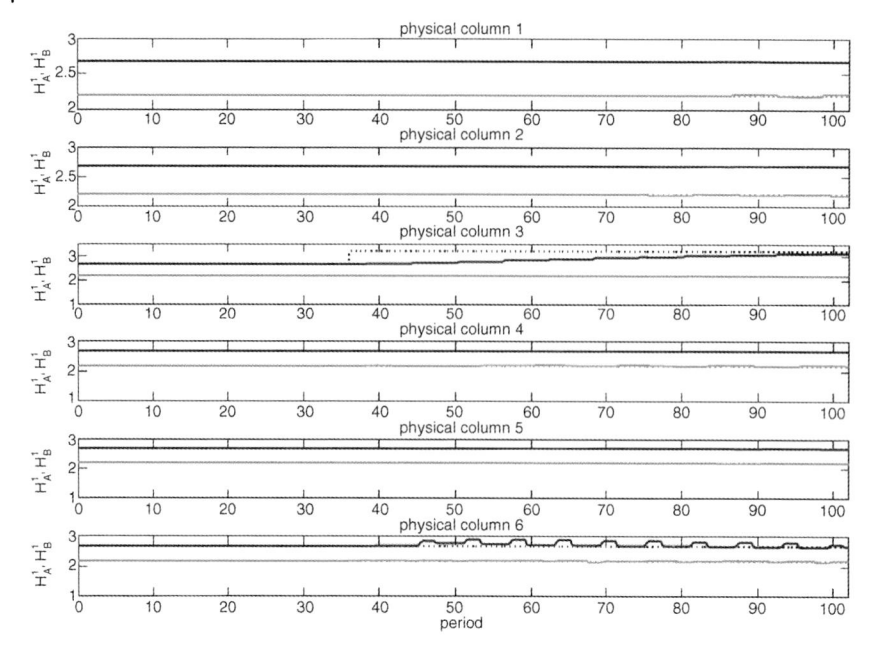

Figure 7.48 Reaction of the switched online parameter and state estimation scheme to a change in one Henry parameter of column 3 (Küpper and Engell, 2006).

References

Alamir, M., Ibrahim, F., and Corriou, J.P. (2006) A flexible nonlinear model predictive control scheme for quality/performance handling in nonlinear SMB chromatography. *J. Process Control.*, **16**, 333–344.

Antia, F. (2003) A simple approach to design and control of simulated moving bed chromatographs. *Chromatogr. Sci. Ser*, **88**, 173–202.

Aumann, L., Morbidelli, M., A continuous multicolumn countercurrent solvent gradient purification (MCSGP) process, *Biotech. and Bioeng.* 2007, **98**, 1043–1055

Bailly, M. and Tondeur, D. (1982) Recycle optimization in non-linear productive chromatography – I. Mixing recycle with fresh feed. *Chem. Eng. Sci.*, **37**, 1199–1212.

Behrens, M. and Engell, S. (2011) *Iterative set-point optimization of continuous annular electro-chromatography.* Proceedings of the 18th IFAC World Congress, Milano, Volume 18, Part 1, pp. 3665–3671.

Beste, Y.A., Lisso, M., Wozny, G., and Arlt, W. (2000) Optimization of simulated moving bed plants with low efficient stationary phases: separation of fructose and glucose. *J. Chromatogr. A*, **868**, 169–188.

Biressi, G., Ludemann-Hombourger, O., Mazzotti, M., Nicoud, R.M., and Morbidelli, M. (2000) Design and optimisation of a simulated moving bed unit: role of deviations from equilibrium theory. *J. Chromatogr. A*, **976**, 3–15.

Brooks, C. and Cramer, S. (1992) Steric-mass-action ion exchange: displacement profiles and induced salt gradients. *AIChE J.*, **38**, 1968–1978.

Chan, S., Titchener-Hooker, N., and S⌀rensen, E. (2008) Optimal economic design and operation of single- and multi-column chromatographic processes. *Biotechnol Prog.*, **24**, 389–401.

Charton, F. and Nicoud, R.-M. (1995) Complete design of a simulated moving bed. *J. Chromatogr. A*, **702**, 97–112.

Cox, G., Khattabi, S., and Dapremont, O. (2003) *Real-time monitoring and control of a small-scale SMB unit from a polarimeter-derived internal profile*. 16th International Symposium on Preparative Chromatography, June 29 to July 2, 2003, San Francisco, USA, pp. 41–42.

Dünnebier, G. (2000) Effektive Simulation und mathematische Optimierung chromatographischer Trennprozesse. Dissertation. Universität Dortmund, Shaker Verlag.

Dünnebier, G., Fricke, J., and Klatt, K.-U. (2000a) Optimal design and operation of simulated moving bed chromatographic reactors. *Ind. Eng. Chem. Res.*, **39**, 2292–2304.

Dünnebier, G., Jupke, A., and Klatt, K.-U. (2000b) Optimaler betrieb von smb-Chromatographieprozessen. *Chem. Ing. Tech.*, **72**(6), 589–593.

Dünnebier, G. and Klatt, K.-U. (1999) Optimal operation of simulated moving bed chromatographic separation processes. *Comp. Chem. Eng.*, **23**, 189–192.

Dünnebier, G., Weirich, I., and Klatt, K.-U. (1998) Computationally efficient dynamic modelling and simulation of simulated moving bed chromatographic processes with linear isotherms. *Chem. Eng. Sci.*, **53**, 2537–2546.

Engell, S. (2007) Feedback control for optimal process operation. *J. Process Contr.*, **17**, 203–219.

Engell, S. and Toumi, A. (2005) Optimization and control of chromatography. *Comp. Chem. Eng.*, **29**, 1243–1252.

Engell, S. and Toumi, A. (2005) Optimisation and control of chromatography, *Comp. Chem. Eng.* 29, 1243–1252.

Epping, A., Modellierung, Auslegung und Optimierung chromatographischer Batch-Trennung, Shaker-Verlag, Aachen, 2005.

Erdem, G., Abel, S., Morari, M., Mazzotti, M., Morbidelli, M., and Lee, J.H. (2004) Automatic control of simulated moving beds. *Ind. Eng. Chem. Res.*, **43**, 405–421.

Felinger, A. and Guiochon, G. (1996) Optimizing preparative separations at high recovery yield. *J. Chromatogr. A*, **752**, 31–40.

Felinger, A. and Guiochon, G. (1998) Comparing the optimum performance of the different modes of preparative liquid chromatography. *J. Chromatogr. A*, **796**, 59–74.

Fricke, J. (2005) Entwicklung einer Auslegungsmethode für chromatographische SMB-Reaktoren. Fortschritts-Berichte VDI: Reihe 3 Nr. 844. VDI Verlag GmbH, Düsseldorf.

Fricke, J. and Schmidt-Traub, H. (2003) A new method supporting the design of simulated moving bed chromatographic reactors. *Chem. Eng. Prog.*, **42**, 237–248.

Fütterer, M. (2008) An adaptive control concept for simulated moving bed plants in case of complete separation. *Chem. Eng. Technol.*, **31**, 1438–1444.

Gao, W. and Engell, S. (2005a) Iterative set-point optimization of batch chromatography. *Comput. Chem. Eng.*, **29**, 1401–1410.

Gao, W. and Engell, S. (2005b) Estimation of general nonlinear adsorption isotherms from chromatograms. *Comput. Chem. Eng.*, **29**, 2242–2255.

Gentilini, A., Migliorini, C., Mazzotti, M., and Morbidelli, M. (1998) Optimal operation of simulated moving-bed units for non-linear chromatographic separations, II. Bi-Langmuir isotherm. *J. Chromatogr. A*, **805**, 37–44.

Grossmann, C., Erdem, G., Morari, M., Amanullah, M., Mazzotti, M., and Morbidelli, M. (2008) 'Cycle to cycle' optimizing control of simulated moving beds. *AIChE J.*, **54**, 194–208.

Grossmann, C., Langel, C., Morbidelli, M., Morari, M., and Mazotti, M. (2010a) Experimental implementation of automatic 'cycle to cycle' control to a nonlinear chiral SMB separation. *J. Chromatogr. A*, **2117**, 2013–2021.

Grossmann, C., Langel, C., Mazzotti, M., Morbidelli, M., and Morari, M. (2010b) Multi-rate optimizing control of simulated moving beds. *J. Process Contr.*, **20**, 490–505.

Grossmann, C., Strohlein, G., Morari, M., and Morbidelli, M. (2010c) Optimizing model predictive control of the chromatographic multi-column solvent gradient purification (MCSGP) process. *J. Process Contr.*, **20**, 618–629.

Gu, T. (1995) *Mathematical Modelling and Scale Up of Liquid Chromatography*, Springer, New York.

Gueorguieva, L., Palani, S., Rinas, U., Jayaraman, G., and Seidel-Morgenstern, A. (2011) Recombinant protein purification using gradient-assisted simulated moving bed hydrophobic interaction chromatography, part II: process design and experimental validation. *J. Chromatogr. A*, **1218**, 6402–6411.

Guiochon, G., Felinger, A., Shirazi, D.G., and Katti, A.M. (2006) *Fundamentals of Preparative and Nonlinear Chromatography*, Elsevier, Amsterdam.

Guiochon, G. and Golshan-Shirazi, S. (1989) Theory of optimizing of the experimental conditions of preparative elution using the ideal model of liquid chromatography. *Anal. Chem.*, **61**, 1276–1287.

Hanisch, F. (2002) Prozessführung präparativer Chromatographieverfahren. Dr.-Ing. Dissertation. Universität Dortmund, Department of Biochemical and Chemical Engineering, Shaker-Verlag, Aachen (in Germany).

Hashimoto, K., Adachi, S., Noujima, H., and Maruyama, H. (1983) Models for separation of glucose–fructose mixtures using a simulated moving bed adsorber. *J. Chem. Eng. Jpn.*, **16** (4), 400–406.

Hashimoto, K., Adachi, S., and Shirai, Y. (1993) Operation and design of simulated moving-bed adsorbers, in *Preparative and Production Scale Chromatography* (eds G. Ganetsos and P.E. Barker), Marcel Dekker Inc, New York.

Helfferich, F. G., Klein, G. *Multicomponent Chromatography – Theory of Interference*, Marcel Dekker Inc., New York, 1970.

Heuer, C., Küsters, E., Plattner, T., and Seidel-Morgenstern, A. (1998) Design of the simulated moving bed process based on adsorption isotherm measurements using a perturbation method. *J. Chromatogr. A*, **827**, 175–191.

Hotier, G. (1998) Process for simulated moving bed separation with a constant recycle rate. US Patent 5.762.806.

Houwing, J., van Hateren, S., Billiet, H., and van der Wielen, L. (2002) Effect of salt gradients on the separation of dilute mixtures of proteins by ion-exchange in simulated moving beds. *J. Chromatogr. A*, **952**, 85–98.

Jensen, T., Reijns, G., Billiet, H., and van der Wielen, L. (2000) Novel simulated moving-bed method for reduced solvent consumption. *J. Chromatogr. A*, **873**, 149–162.

Jupke, A. (2004) Experimentelle modellvalidierung und modellbasierte auslegung von simulated moving bed (SMB) chromatographieverfahren. Fortschrittbericht VDI: Reihe 3 Nr. 807. VDI Verlag GmbH, Düsseldorf.

Jupke, A., Epping, A., and Schmidt-Traub, H. (2002) Optimal design of batch and simulated moving bed chromatographic separation processes. *J. Chromatogr. A*, **944**, 93–117.

Juza, M. (1999) Development of a high-performance liquid chromatographic simulated moving bed separation form an industrial perspective. *J. Chromatogr. A*, **865**, 35–49.

Kaspereit, M. and Sainio, T. (2011) Simplified design of steady-state recycling chromatography under ideal and nonideal conditions. *Chem. Eng. Sci.*, **66**, 5428–5438.

Kaspereit, M., Swernath, S., and Kienle, A. (2012) Evaluation of competing process concepts for the production of pure enantiomers. *Org. Process Res. Dev.*, **16(2)**, 353–363

Katti, A.M. and Jagland, P. (1998) Development and optimisation of industrial scale chromatography for use in manufacturing. *Anal. Mag.*, **26** (7), 38–46.

Kawajiri, Y. and Biegler, L. (2006a) Optimization strategies for simulated moving bed and PowerFeed processes. *AIChE J.*, **52**, 1343–1350.

Kawajiri, Y. and Biegler, L. (2006b) Nonlinear programming superstructure for optimal operation of simulated moving bed processes. *Ind. Eng. Chem. Res.*, **45**, 8503–8513.

Kawajiri, Y. and Biegler, L. (2006c) Large scale nonlinear optimization for asymmetric operation and design of simulated moving beds. *J. Chromatogr. A*, **1133**, 226–240.

Keßler, L.C., Gueorguieva, L., Rinas, U., and Seidel-Morgenstern, A. (2007) Step gradients in 3-zone simulated moving bed chromatography, application to the purification of antibodies and bone morphogenetic protein-2. *J. Chromatogr. A*, **1176**, 69–78.

Keßler, L.C. and Seidel-Morgenstern, A. (2008) Improving performance of simulated moving bed chromatography by fractionation and feed-back of outlet streams. *J. Chromatogr. A*, **1207**, 55–71.

Kim, K., Kim, J.I., Kim, H., Yang, J., Lee, K.S., and Koo, Y.M. (2010b) Experimental verification of bilevel optimizing control for SMB technology. *Eng. Chem. Res.*, **49**, 8593–8600.

Kim, K., Lee, K.S., and Lee, J.H. (2010a) Bilevel optimizing control structure for a simulated moving bed process based on a reduced-order model using the cubic spline collocation method. *Ind. Eng. Chem. Res.*, **49**, 3689–3699.

Klatt, K.-U., Hanisch, F., and Dünnebier, G. (2002) Model-based control of a simulated moving bed chromatographic process for the separation of fructose and glucose. *J. Process Contr.*, **12**, 203–219.

Klatt, K.-U., Hanisch, F., Dünnebier, G., and Engell, S. (2000) Model-based optimization and control of chromatographic processes. *Comput. Chem. Eng.*, **24**, 1119–1126.

Kleinert, T. and Lunze, J. (2008) Decentralised control of chromatographic simulated moving bed processes based on wave front reconstruction. *J. Process Contr.*, **18**, 780–796.

Kloppenburg, E. and Gilles, E.D. (1999) Automatic control of the simulated moving bed process for C8 aromatics separation using asymptotically exact input/output linearization. *J. Process Contr.*, **9**, 41–50.

Küpper, A., Diehl, M., Schlöder, J., Bock, H.-G., and Engell, S. (2009) Efficient moving horizon state and parameter estimation for SMB processes. *J. Process Contr.*, **19**, 785–802.

Küpper, A. and Engell, S. (2006) *Parameter and State Estimation in Chromatographic SMB Processes with Individual Columns and Nonlinear Adsorption Isotherms*. Proceedings of the IFAC Symposium ADCHEM, Gramado, pp. 611–616.

Küpper, A. and Engell, S. (2007) *Optimizing Control of the Hashimoto SMB Process: Experimental Application*. Proceedings of the 8th International Symposium on Dynamics and Control of Process Systems (DYCOPS), Cancun, pp. 151–157.

Küpper, A. and Engell, S. (2008) *Engineering of Online Optimizing Control – A Case Study: Reactive SMB Chromatography*. Proceedings of the 17th IFAC World Congress, Seoul, pp. 964–969.

Küpper, A. and Engell, S. (2009) Optimierungsbasierte regelung des hashimoto-SMB-prozesses. *at-Automatisierungstechnik*, **57**, 360–370.

Küpper, A. and Engell, S. (2011) Optimization of simulated moving bed processes, in *Constrained Optimization and Optimal Control of Partial Differential Equations* (eds G. Leugering*et al.*), Birkhäuser, Basel, pp. 559–582.

Küpper, A., Wirsching, L., Diehl, M., Schlöder, J.P., Bock, H.G., and Engell, S. (2010) Online identification of adsorption isotherms in SMB processes via efficient moving horizon state estimation. *Comput. Chem. Eng.*, **34**, 1969–1983.

Kurup, A., Hidajat, K., and Ray, A. (2005a) Optimal operation of an industrial-scale Parex process for the recovery of *p*-xylene from a mixture of C_8 aromatics. *Ind. Eng. Res.*, **44**, 5703–5714.

Kurup, A., Hidajat, K., and Ray, A. (2006) Comparative study of modified simulated moving bed systems at optimal conditions for the separation of ternary mixtures of xylene isomers. *Ind. Eng. Chem. Res.*, **45**, 6251–6265.

Kurup, A., Subramani, H., Hidajat, K., and Ray, A. (2005b) Optimal design and operation of SMB bioreactor for sucrose. *Chem. Eng. J.*, **108**, 19–33.

Küsters, E., Gerber, G., and Antia, F. (1995) Enantioseparation of a chiral epoxide by simulated moving bed processes. *AIChE J.*, **42**, 154–160.

Li, P., Xiu, G., and Rodrigues, A. (2007) Protein separation and purification by salt gradient ion-exchange. *AIChE J.*, **53**, 2419–2431.

Li, P., Xiu, G., and Rodrigues, A. (2008) Separation region and strategies for protein separation by salt gradient ion-exchange. *Sep. Sci. Technol.*, **43**, 11–28.

Li, S., Kawajiri, Y., Raisch, J., and Seidel-Morgenstern, A. (2010a) Optimization of simulated moving bed chromatography with fractionation and feedback: part I. Fractionation of one outlet. *J. Chromatogr. A*, **1217**, 5337–5348.

Li, S., Kawajiri, Y., Raisch, J., and Seidel-Morgenstern, A. (2010b) Optimization of simulated moving bed chromatography with fractionation and feedback: part II. Fractionation of both outlets. *J. Chromatogr. A*, **1217**, 5349–5357.

Li, S., Kawajiri, Y., Raisch, J., and Seidel-Morgenstern, A. (2011) Optimization of startup and shutdown operation of simulated moving bed chromatographic processes. *J. Chromatogr. A*, **1218**, 3876–3889.

Lode, F. (2002) A simulated moving bed reactor (SMBR) for esterifications. Dissertation. ETH Zürich.

Lode, F., Francesconi, G., Mazzotti, M., and Morbidelli, M. (2003b) Synthesis of methylacetate in a simulated moving bed reactor: experiments and modelling. *AIChE J.*, **49** (6), 1516–1524.

Lode, F., Houmard, M., Migliorini, C., Mazzotti, M., and Morbidelli, M. (2001) Continuous reactive chromatography. *Chem. Eng. Sci.*, **56**, 269–291.

Lode, F., Mazzotti, M., and Morbidelli, M. (2003a) Comparing true countercurrent and simulated moving-bed chromatographic reactors. *AIChE J.*, **49** (4), 977–990.

Ludemann-Hombourger, O., Bailly, M., and Nicoud, R.-M. (2000a) Design of a simulated moving bed: optimal size of the stationary phase. *Sep. Sci. Technol.*, **35** (9), 1285–1305.

Ludemann-Hombourger, O., Nicoud, R. M., and Bailly, M. (2000b) The "VARICOL" process: a new multicolumn continuous chromatographic process. *Sep. Sci. Technol.*, **35**, 1829–1862.

Mallmann, T., Burris, B., Ma, Z., and Wang, N.-H. (1998) Standing wave design for nonlinear SMB systems for fructose purification. *AIChE J.*, **44**, 2628–2646.

Marteau, P., Hotier, G., Zanier-Szydlowski, N., Aoufi, A., and Cansell, F. (1994) Advanced control of C_8 aromatics separation process with real-time multiport on-line Raman spectroscopy. *Process Qual.*, **6**, 133–140.

Mazzotti, M., Storti, G., and Morbidelli, M. (1994) Robust design of countercurrent adsorption separation processes: 2. Multicomponent systems. *AIChE J.*, **40**, 1825–1842.

Mazzotti, M., Storti, G., and Morbidelli, M. (1996) Robust design of countercurrent adsorption separation processes: 3. Nonstoichiometric systems. *AIChE J.*, **42**, 2784–2796.

Mazzotti, M., Storti, G., and Morbidelli, M. (1997) Robust design of countercurrent adsorption separation processes: 4. Desorbent in the feed. *AIChE J.*, **43**, 64–72.

Migliorini, C., Mazzotti, M., and Morbidelli, M. (2000) Robust design of countercurrent adsorption separation processes: 5. Nonconstant selectivity. *AIChE J.*, **46**, 1384–1399.

Miller, L., Grill, C., Yan, T., Dapremont, O., Huthmann, E., and Juza, M. (2003) Batch and simulated moving bed chromatographic resolution of a pharmaceutical racemate. *J. Chromatogr. A*, **1006**, 267–280.

Mota, J., Araújo, J., and Rodrigues, R. (2007) Optimal design of simulated moving-bed processes under flow rate uncertainty. *AICHE J.*, **53**, 2630–2642.

Müller-Späth, T., Aumann, L., Ströhlein, G., Kornmann, H., Valax, P., Delegrange, L., Charbaut, E., Baer, G., Lamproye, A., Jöhnck, M., Schulte, M., and Morbidelli, M. (2010b) Two step capture and purification of IgG2 using multicolumn countercurrent solvent gradient purification (MCSGP). *Biotechnol. Bioeng.*, **107**, 974–984.

Müller-Späth, T., Krättli, M., Aumann, L., Ströhlein, G., and Morbidelli, M. (2010a) Increasing the activity of monoclonal antibody therapeutics by continuous chromatography (MCSGP). *Biotechnol. Bioeng.*, **107**, 652–662.

Müller-Späth, T., Ströhlein, G., Aumann, L., Kornmann, H., Valax, P., Delegrange, L., Charbaut, E., Baer, G., Lamproye, A., Jöhnck, M., Schulte, M., and Morbidelli, M. (2011) Model simulation and experimental verification of a cation-exchanger IgG capture step in batch and continuous chromatography. *J. Chromatogr. A*, **1218**, 5195–5204.

Natarajan, S. and Lee, J.H. (2000) Repetitive model predictive control applied to a simulated moving bed chromatography system. *Comput. Chem. Eng.*, **24**, 1127–1133.

Nicoud, R.M. (1992) The simulated moving bed: a powerful chromatographic process. *Mag. Liquid Gas Chromatogr.*, **5**, 43–47.

Nicoud, R.M. (1998) Simulated moving bed (SMB): some possible applications for biotechnology, in *Bioseparation and Bioprocessing: Volume I* (ed. G. Subramanian), Wiley-VCH Verlag GmbH, Weinheim.

Palani, S., Gueorguieva, L., Rinas, U., Seidel-Morgenstern, A., and Jayaraman, G. (2011) Recombinant protein purification using gradient-assisted simulated moving bed hydrophobic interaction chromatography, part I: selection of chromatographic system and estimation of adsorption isotherm. *J. Chromatogr. A*, **1218**, 6396–6401.

Pröll, T. and Küsters, E. (1998) Optimization strategy for simulated moving bed systems. *J. Chromatogr. A*, **800**, 135–150.

Qin, S. and Badgwell, T. (2003) A survey of industrial model predictive control technology. *Control Eng. Pract.*, **11**, 733–764.

Rao, C. and Rawlings, J. (2000) Nonlinear moving horizon state estimation, in *Progress in Systems and Control Theory*, vol. **26** (eds F. Allgöwer and A. Zheng), Birkhäuser-Verlag, Basel/Switzerland, pp. 45–69.

Rhee, H.-K., Aris, R., and Amundson, N.R. (1970) On the theory of multicomponent chromatography. *Philos. Trans. R. Soc. Lond. A*, **267**, 419–455.

Rodrigues, R., Araújo, J., Eusébio, M., and Mota, J. (2007a) Experimental assessment of simulated moving bed and Varicol processes using a single-column setup. *J. Chromatogr. A*, **1142**, 69–80.

Rodrigues, R., Araújo, J., and Mota, J. (2007b) Optimal design and experimental validation of synchronous, asynchronous and flow-modulated, simulated moving-bed processes using a single-column setup. *J. Chromatogr. A*, **1162**, 14–23.

Ruthven, D.M. and Ching, C.B. (1989) Review article no. 31: counter-current and simulated counter-current adsorption separation processes. *Chem. Eng. Sci.*, **44** (5), 1011–1038.

Sainio, T. and Kaspereit, M. (2009) Analysis of steady state recycling chromatography using equilibrium theory. *Sep. Purif. Technol.*, **66**, 9–18.

Scherpian, P. (2009) Zur Auswahl und Auslegung chromatographischer Prozesskonzepte mit Recycling. Verlag Dr. Hut, München.

Scherpian, P. and Schembecker, G. (2009) Scaling-up recycling chromatography. *Chem. Eng. Sci.*, **64**, 4068–4080.

Schramm, H., Grüner, S., and Kienle, A. (2003) Optimal operation of simulated moving bed processes by means of simple feedback control. *J. Chromatogr. A*, **1006**, 3–13.

Schramm, H., Grüner, S., Kienle, A., and Gilles, E.D. (2001) Control of moving bed chromatographic processes. Proceedings of the European Control Conference, pp. 2528–2533.

Schramm, H., Kaspereit, M., Kienle, A., and Seidel-Morgenstern, A. (2003a) Simulated moving bed process with cyclic modulation of the feed concentration. *J. Chromatogr. A*, **1006**, 77–86.

Schramm, H., Kienle, A., Kaspereit, M., and Seidel-Morgenstern, A. (2003b) Improved operation of simulated moving bed processes through cyclic modulation of feed flow and feed concentration. *Chem. Eng. Sci.*, **58**, 5217–5227.

Seidel-Morgenstern, A., Keßler, L., and Kaspereit, M. (2008a) Neue entwicklungen auf dem gebiet der simulierten gegenstromchromatographie. *Chem. Ing. Techn.*, **80**, 725–740.

Seidel-Morgenstern, A., Keßler, L., and Kaspereit, M. (2008b) New developments in simulated moving bed chromatography. *Chem. Eng. Technol.*, **31**, 826–837.

Shan, Y. and Seidel-Morgenstern, A. (2004) Analysis of the isolation of a target component using multicomponent isocratic preparative elution chromatography. *J. Chromatogr. A*, **1041**, 53–62.

Siitonen, J. and Sainio, T. (2011) Explicit equations for the height and position of the first component shock for binary mixtures with competitive Langmuir isotherms under ideal conditions, *J. Chromatography*, A 1218, 6379-6387

Siitonen, J., Sainio, T., and Kaspereit, M. (2011) Theoretical analysis of steady state recycling chromatography with solvent removal. *Sep. Purif. Technol.*, **78**, 21–32.

Song, I.H., Lee, S.B., Rhee, H.K., and Mazzotti, M. (2006) Identification and predictive control of a simulated moving bed process: purity control. *Chem. Eng. Sci.*, **61**, 1973–1986.

Storti, G., Baciocchi, R., Mazzotti, M., and Morbidelli, M. (1995) Design of optimal conditions of simulated moving bed adsorptive separation units. *Ind. Eng. Res.*, **34**, 288–301.

Storti, G., Mazzotti, M., Morbidelli, M., and Carrà, S. (1993) Robust design of binary countercurrent adsorption separation processes. *AIChE J.*, **39**, 471–492.

Strube, J. (2000) Technische Chromatography: Auslegung, Optimierung, Betrieb und Wirtschaftlichkeit. Dissertation. Universität Dortmund, Shaker-Verlag.

Strube, J., Altenhöner, U., Meurer, M., and Schmidt-Traub, H. (1997a) Optimierung kontinuierlicher Simulated-Moving-Bed-Chromatographie-Prozesse durch dynamische simulation. *Chem. Ing. Techn.*, **69** (3), 328–331.

Strube, J., Altenhöner, U., Meurer, M., Schmidt-Traub, H., and Schulte, M. (1997b) Dynamic simulation of simulated moving-bed chromatographic processes for the optimisation of chiral separations. *J. Chromatogr. A*, **769**, 328–331.

Strube, J., Haumreisser, S., Schmidt-Traub, H., Schulte, M., and Ditz, R. (1998) Comparison of batch elution and continuous simulated moving bed chromatography. *Org. Process Res. Dev.*, **2** (5), 305–319.

Strube, J., Jupke, A., Epping, A., Schmidt-Traub, H., Schulte, M., and Devant, R. (1999) Design, optimization and operation of smb chromatography in the production of enantiomerically pure pharmaceuticals. *Chirality*, **11**, 440–450.

Subramani, H., Hidajat, K., and Ray, A. (2003) Optimization of reactive SMB and Varicol systems. *Comp. Chem. Eng.*, **27**, 1883–1901.

Subramani, H., Zhang, Z., Hidajat, K., and Ray, A. (2004) Multiobjective optimization of simulated moving bed reactor and its modification – Varicol. *Can. J. Chem. Eng.*, **82**, 590–598.

Toumi, A. and Engell, S. (2004) Optimization-based control of a reactive simulated moving bed process for glucose isomerisation. *Chem. Eng. Sci.*, **59**, 3777–3792.

Toumi, A., Engell, S., Diehl, M., Bock, H., and Schlöder, J. (2007) Efficient optimization of simulated moving beg processes. *Chem. Eng. Proc.*, **46**, 1067–1084.

Toumi, A., Engell, S., and Hanisch, F. (2002) Asynchron getaktete Gegenstromchromatographie – Prinzip und optimaler Betrieb. *Chem. Ing. Techn.*, **74**, 1483–1490.

Toumi, A., Engell, S., Ludemann-Hombourger, O., Nicoud, R.M., and Bailly, M. (2003) Optimization of simulated moving bed and VARICOL processes. *J. Chromatogr. A*, **1006**, 15–31.

Unger, K., Ditz, R., Machtejevas, E., and Skudas, R. (2010) Liquid chromatography – its development and key role in life sciences applications. *Angew. Chem.*, **49**, 2300–2312.

Vilas, C. and Van de Wouwer, A. (2011) Combination of multi-model predictive control and the wave theory for the control of simulated moving bed plants. *Chem. Eng. Sci.*, **66**, 632–641.

von Langermann, J., Kaspereit, M., Shakeri, M., Lorenz, H., Hedberg, M., Jones, M.J., Larson, K., Herrschend, B., Arnell, R., Temmel, E., Bäckvall, J.-E., Kienle, A., and Seidel-Morgenstern, A. (2012) Design of an integrated process of chromatography, crystallisation and racemisation for the resolution of 2′,6′-pipecoloxylidide (PPX). *Org. Process Res. Dev.*, **16**(2), 343–352.

Wang, C., Klatt, K., Dünnebier, G., Engell, S., and Hanisch, F. (2003) Neural network based identification of SMB chromatographic processes. *Control Eng. Pract.*, **11**, 949–959.

Wekenborg, K. (2009) *Kontinuierliche Trennung von Proteinen durch nicht-isokratische SMB-Chromatographieprozesse*, Shaker-Verlag, Aachen.

Wekenborg, K., Susanto, A., and Schmidt-Traub, H. (2004) Nicht-isokratische SMB-Trennung von Proteinen mittels Ionenaustauschchromatographie. *Chem. Ing. Techn.*, **76** (6), 815–819.

Wekenborg, K., Susanto, A., and Schmidt-Traub, H. (2005) Modelling and validated simulation of solvent-gradient simulated moving bed (SG-SMB) processes for protein separation, in *Computer-Aided Chemical Engineering*, vol. **20A**, 313–318.

Wongso, F., Hidajat, K., and Ray, A. (2004) Optimal operating mode for enantioseparation of SB-553261 racemate based on simulated moving bed technology. *Biotechnol. Bioeng.*, **87**, 704–722.

Yu, W., Hidajat, K., and Ray, A. (2005) Optimization of reactive simulated moving bed and Varicol systems for hydrolysis of methyl acetate. *Chem. Eng. J.*, **112**, 55–72.

Zhang, Y., Hidajat, K., and Ray, A.K. (2007) Enantio-separation of racemic pindolol on α1-acid glycoprotein chiral stationary phase by SMB and Varicol. *Chem. Eng. Sci.*, **62**, 1364–1375.

Zhang, Y., Hidajat, K., and Ray, A.K. (2009) Multi-objective optimization of simulated moving bed and Varicol processes. *Sep. Purif. Technol.*, **65**, 311–321.

Zhang, Z., Hidajat, K., Ray, A.K., and Morbidelli, M. (2002) Multiobjective optimization of SMB and Varicol process for chiral separation. *AIChE J.*, **48**, 2800–2816.

Zhang, Z., Mazzotti, M., and Morbidelli, M. (2003a) PowerFeed operation of simulated moving bed units: changing flow-rates during the switching interval. *J. Chromatogr. A*, **1006**, 87–99.

Zhang, Z., Mazzotti, M., and Morbidelli, M. (2003b) Multiobjective optimization of simulated moving bed and Varicol processes using a genetic algorithm. *J. Chromatogr. A*, **989**, 95–108.

Zhang, Z., Mazzotti, M., and Morbidelli, M. (2004a) Continuous chromatographic processes with a small number of columns: comparison of simulated moving bed with Varicol, PowerFeed and ModiCon. *Korean J. Chem. Eng.*, **21**, 454–464.

Zhang, Z., Morbidelli, M., and Mazzotti, M. (2004b) Experimental assessment of PowerFeed chromatography. *AIChE J.*, **50**, 625–632.

Zhong, G. and Guiochon, G. (1996) Analytical solution for the linear ideal model of simulated moving bed chromatography. *Chem. Eng. Sci.*, **51** (18), 4307–4319.

Ziomek, G. and Antos, D. (2005b) Stochastic optimization of simulated moving bed process; sensitivity analysis for isocratic and gradient operation. *Comp. Chem. Eng.*, **29**, 1577–1589.

Ziomek, G., Kaspereit, M., Jezowski, J., Seidel-Morgenstern, A., and Antos, D. (2005a) Effect of mobile phase composition on the SMB processes efficiency; stochastic optimization of isocratic and gradient operation. *J. Chromatogr. A*, **1070**, 111–124.

Appendix A: Data of Test Systems

A.1
EMD53986

EMD53986 [5-(1,2,3,4-*tetra*-hydroquinolin-6-yl)-6-methyl-3,6-dihydro-1,3,4-thiadiazin-2-one] (Figure A.1) is a chiral precursor for a pharmaceutical reagent. After chemical synthesis, it is present as a racemic mixture of the *R*- and *S*-enantiomers. The target component for the separation is the *R*-enantiomer.

Given below are the separation conditions (Table A.1) and typical model parameters (Tables A.2 and A.3) used in this book (Jupke, 2004; Epping, 2005).

Eluent: ethanol (gradient grade), LiChrosolv®, 99–100%, Merck (Darmstadt)
Adsorbent: Chiralpak AD, Daicel (Japan)
(Amylose-tris[3,5-dimethylphenylcarbamet]-phase)
Temperature: 25 °C

HETP correlation:

R-enantiomer: $\quad HETP_R = 5.03 \cdot 10^{-3} + 1.9 \cdot u_{int}$ \hfill (A1)

S-enantiomer: $\quad HETP_S = 1.26 \cdot 10^{-2} + 1.63 \cdot u_{int}$ \hfill (A2)

Isotherms for 25 °C (Section 6.5.7.5)

R-enantiomer: $\quad q_R = 2.054 \cdot c_R + \dfrac{5.847 \cdot c_R}{1 + 0.129 \cdot c_R + 0.472 \cdot c_S}$

S-enantiomer: $\quad q_S = 2.054 \cdot c_S + \dfrac{19.902 \cdot c_S}{1 + 0.129 \cdot c_R + 0.472 \cdot c_S}$ \hfill (A3)

Alternatively, the isotherm can also be expressed by a multi-Langmuir equation:

R-enantiomer: $\quad q_R = \dfrac{7.859 \cdot c_R}{1 + 0.29543 \cdot c_R + 0.08293 \cdot c_S}$

S-enantiomer: $\quad q_S = \dfrac{19.75 \cdot c_S}{1 + 0.29543 \cdot c_R + 0.08293 \cdot c_S}$ \hfill (A4)

Pressure drop correlations:

$$\Delta p = 1233.4 \cdot \frac{u_{int} \cdot \eta_l \cdot L_c}{d_p^2}. \tag{A5}$$

Preparative Chromatography, Second Edition. Edited by H. Schmidt-Traub, M. Schulte, and A. Seidel-Morgenstern.
© 2012 Wiley-VCH Verlag GmbH & Co. KGaA. Published 2012 by Wiley-VCH Verlag GmbH & Co. KGaA.

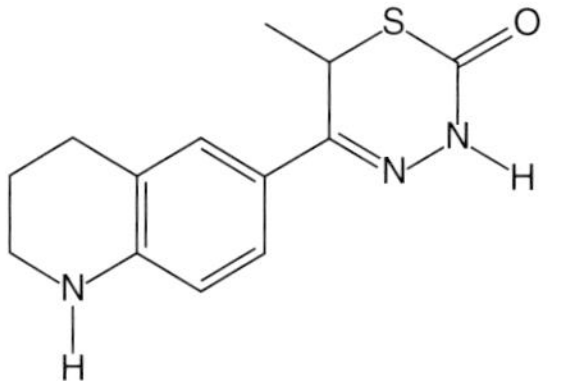

Figure A.1 Chemical structure of EMD53986.

Table A.1 Typical design and model parameters for EMD53986.

Design parameter	Value
Particle diameter	20 μm
Column diameter (d_c)	2.5 cm
Column length (L_c)	11.5 cm
Division (for SMB)	2/2/2/2

Model parameter	Value
Void fraction (ε)	0.355
Total porosity (ε_t)	0.72
Dispersion coefficient (D_{ax})	Equation 6.174 (Section 6.5.6.2)
Viscosity (η_l)	$0.0119 \text{ g cm}^{-1}\text{ s}^{-1}$
Density (ϱ_l)	0.799 g cm^{-3}
Effective transfer coefficient (k_{eff}) R-enantiomer	$1.50 \times 10^{-4} \text{ cm s}^{-1}$
Effective transfer coefficient (k_{eff}) S-enantiomer	$2.00 \times 10^{-5} \text{ cm s}^{-1}$

Table A.2 Additional parameters for model validation batch column (EMD53986).

Typical plant parameters (batch experiments)	Value
Pipe volume (V_{pipe})	12.3 cm^3
Dispersion coefficient pipe ($D_{ax,pipe}$)	$\sim 3000 \text{ cm}^2\text{ s}^{-1}$
Pipe diameter (d_{pipe})	0.05 cm
Tank volume (V_{tank})	3.1 cm^3

A.2
Tröger's Base

Tröger's base (Figure A.2) was the first example of a chromatographic separation of an enantiomeric mixture. Since then it has been the subject of several publications.

Given below are the separation conditions and typical model parameters used in this book (Tables A.4 and A.5) (Jupke, 2004; Mihlbachler et al., 2001, 2002).

A.2 Tröger's Base

Table A.3 Parameters for model validation SMB (EMD53986).

SMB experiments (Figure 6.42)	Unit	Value
Shifting time (t_{shift})	s	431
Asynchronous shift time	s	40.4
Feed flow rate	ml min^{-1}	5.54
Desorbent flow rate	ml min^{-1}	57.5
Extract flow rate	ml min^{-1}	47.97
Raffinate flow rate	ml min^{-1}	15.06
Recycle flow rate	ml min^{-1}	72.0
m_{I}	—	30.14
m_{II}	—	8.34
m_{III}	—	10.86
m_{IV}	—	4.02

Eluent: isopropanol (HPLC grade), Fisher Scientific (Pittsburgh, USA)
Adsorbent: Chiralpak AD, Daicel (Japan)
amylose-tris[3,5-dimethylphenylcarbamet]-phase
Solute: Tröger's base, Aldrich (Milwaukee, USA)
Temperature: 25 °C

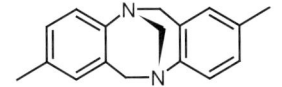

Figure A.2 Chemical structure of Tröger's base.

Table A.4 Typical design and model parameters for Tröger's base.

Design parameter	Value
Particle diameter	20 μm
Column diameter (d_c)	1 cm
Column length (L_c)	10 cm
Division (for SMB)	2/2/2/2

Model parameter	Value
Void fraction (ε)	0.355
Total porosity (ε_t)	0.648
Dispersion coefficient (D_{ax})	Equation 6.174 (Section 6.5.6.2)
Viscosity (η_l)	0.012 g cm^{-1} s^{-1}
Density (ϱ_l)	0.785 g cm^{-3}
Effective transfer coefficient (k_{eff}) R-enantiomer	5×10^{-4} cm s^{-1}
Effective transfer coefficient (k_{eff}) S-enantiomer	4.5×10^{-5} cm s^{-1}

Table A.5 Parameters for model validation SMB (Tröger's base).

SMB experiment (Figure 6.44)	Unit	Value
Shifting time (t_{shift})	s	962
Feed flow rate	ml min^{-1}	0.30
Desorbent flow rate	ml min^{-1}	0.46
Extract flow rate	ml min^{-1}	0.33
Raffinate flow rate	ml min^{-1}	0.43
Recycle flow rate	ml min^{-1}	1.05
m_I	—	4.24
m_{II}	—	2.50
m_{III}	—	3.43
m_{IV}	—	1.70

Isotherms from breakthrough experiments (Section 6.5.7.8):

$$R\text{-enantiomer:} \quad q_R = \frac{54 \cdot c_R \cdot (0.035 + 0.0046 \cdot c_S)}{1 + 0.035 \cdot c_R + 0.062 \cdot c_S + 0.0046 \cdot c_R \cdot c_S + 0.0052 \cdot c_S^2}$$

$$S\text{-enantiomer:} \quad q_S = \frac{54 \cdot c_S \cdot (0.062 + 0.0046 \cdot c_R + 2 \cdot 0.0052 \cdot c_S)}{1 + 0.035 \cdot c_R + 0.062 \cdot c_S + 0.0046 \cdot c_R \cdot c_S + 0.0052 \cdot c_S^2}$$

(A6)

Isotherms from perturbation experiments (Section 6.5.7.8) are

$$R\text{-enantiomer:} \quad q_R = \frac{0.0311 \cdot c_R \cdot (54 + 0.732 \cdot c_S)}{1 + 0.0311 \cdot c_R} + \frac{0.732 \cdot 0.0365 \cdot c_R \cdot c_S}{1 - 0.0365 \cdot c_R}$$

$$S\text{-enantiomer:} \quad q_S = \frac{27 \cdot c_S \cdot (0.1269 + 2 \cdot 0.0153 \cdot c_S)}{1 + 0.1269 \cdot c_S + 0.0153 \cdot c_S^2}$$

(A7)

A.3
Glucose and Fructose

Glucose (Figure A.3) and fructose (Figure A.4) are monosaccharides. Separation of their isomeric mixture is of industrial importance to produce sugar syrups or for further synthesis of sorbitol, gluconic acid, and vitamin C. Due to higher sweetness, fructose is often used as an alternative sweetener for sucrose.

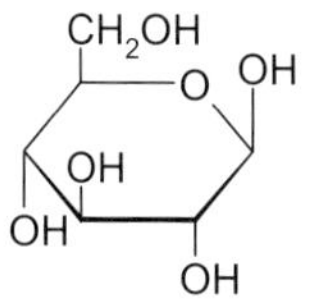

Figure A.3 Chemical structure of glucose.

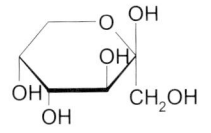

Figure A.4 Chemical structure of fructose.

Table A.6 Typical design and model parameters for fructose–glucose.

Design parameter	Value
Particle diameter	325 μm
Column diameter (d_c)	2.6 cm
Column length (L_c)	~56.5 cm[a]
Division (for SMB)	2/2/2/2
Model parameter	**Value**
Void fraction (ε)	~0.36
Total porosity (ε_t)	~0.36
Dispersion coefficient (D_{ax})	Equation 6.174 (Section 6.5.6.2)
Viscosity (η_l)	4.7×10^{-3} g cm^{-1} s^{-1}
Density (ϱ_l)	0.983 g cm^{-3}
Effective transfer coefficient (k_{eff}) glucose	7×10^{-4} cm s^{-1}, 5.7×10^{-5} cm s^{-1}
Effective transfer coefficient (k_{eff}) fructose	6.5×10^{-5} cm s^{-1}, 8.9×10^{-5} cm s^{-1}
Equilibrium constant (K_{eq})	1.079
Reaction rate constant (k_{reac})	1.427×10^{-3} s^{-1}

a) Due to slightly different adsorbent amounts and packing compressions for different experiments.

Table A.7 Parameters for model validation SMB (fructose–glucose).

SMB experiment	Unit	Value
Shifting time (t_{shift})	s	1590
Asynchronous shift time	s	171.8
Feed flow rate	ml min^{-1}	0.56
Desorbent flow rate	ml min^{-1}	4.24
Extract flow rate	ml min^{-1}	3.35
Raffinate flow rate	ml min^{-1}	1.45
Recycle flow rate	ml min^{-1}	10.0
m_I	—	0.82
m_{II}	—	0.36
m_{III}	—	0.43
m_{IV}	—	0.23

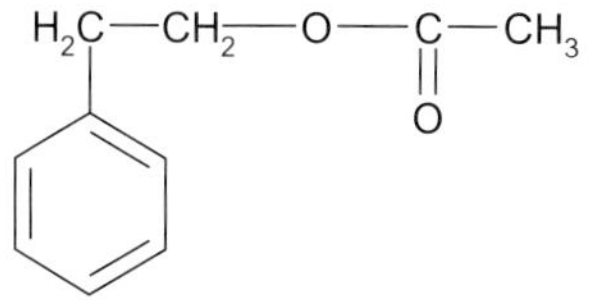

H₂C—CH₂—O—C(=O)—CH₃ (attached to phenyl ring)

Figure A.5 Chemical structure of β-phenethyl acetate.

Separation conditions and typical model parameter used in this book are given below (Tables A.6 and A.7) (Jupke, 2004):

Eluent: deionized and microfiltered water
Adsorbent: ion exchange resin Amberlite CR 1320 Ca (Rohm u. Haas, Frankfurt)
Solute: glucose, fructose (pure, Merck, Darmstadt)
Temperature: 60 °C

For chromatographic reactors additionally:

Catalyst: immobilized glucose invertase Sweetzyme IT (Novozymes, Mainz)

HETP correlations for 60 °C

glucose: $\quad HETP_{glu} = 0.1188 + 8.997 \cdot u_{int}.$ \hfill (A8)

fructose: $\quad HETP_{fru} = 0.1163 + 25.155 \cdot u_{int}.$ \hfill (A9)

Isotherms for 60 °C: (Section 6.5.7.5):

$$q_{glu} = 0.27 \cdot c_{glu} + 0.000122 \cdot c_{glu}^2 + 0.103 \cdot c_{glu} \cdot c_{fru}$$
$$q_{fru} = 0.47 \cdot c_{fru} + 0.000119 \cdot c_{fru}^2 + 0.248 \cdot c_{glu} \cdot c_{fru}$$
\hfill (A10)

A.4
β-Phenethyl Acetate

β-Phenethyl acetate (Figure A.5) is used for the production of scents and perfumes.
Separation conditions and model parameter used in this book are given below (Table A.8):

Eluent: 1,4-dioxan
Adsorbent and catalyst: ion exchange resin Amberlite 15 (Rohm u. Haas)
Solute: acetic acid and β-phenethyl alcohol
Temperature: 85 °C

Isotherms for 85 °C:

$$q_{alcohol} = \frac{0.884 \cdot c_{alcohol}}{1 + 24.16 \cdot c_{alcohol}}$$
$$q_{acid} = 0.687 c_{acid}$$
$$q_{ester} = 0.479 c_{ester}$$
$$q_{water} = \frac{11.126 \cdot c_{water}}{1 + 568.54 \cdot c_{water}}$$

Table A.8 Typical design and model parameters for β-phenethyl acetate.

Design parameter	Value
Particle diameter	590 μm
Column diameter (d_c)	1 cm
Column length (L_c)	30 cm
Model parameter	**Value**
Total porosity (ε_t)	0.36
Dispersion coefficient (D_{ax})	Equation 6.174 (Section 6.5.6.2)
Viscosity (η_l)	5.8×10^{-3} g cm^{-1} s^{-1}
Density (ϱ_l)	1.03 g cm^{-3}
Effective transfer coefficient (k_{eff}) alcohol	4.39×10^{-4} cm s^{-1}
Effective transfer coefficient (k_{eff}) acid	4.82×10^{-4} cm s^{-1}
Effective transfer coefficient (k_{eff}) ester	4.17×10^{-4} cm s^{-1}
Effective transfer coefficient (k_{eff}) water	1.11×10^{-3} cm s^{-1}
Equilibrium constant (K_{eq})	3.18
Reaction rate constant (k_{reac})	0.165326 cm^3 s^{-1} g^{-1}

A.5
β-Lactoglobulin A and B

Separation conditions and model parameter used in this book are given below (Tables A.9–A.12):

  Adsorbent: Source30Q, Amersham Biosciences Europe GmbH

Table A.9 Design and model parameters.

Design parameter	Value
Particle diameter	28.7–32.8 μm
Column diameter (d_c)	1.0 cm
Column length (L_c)	8.75 cm
Model parameter	**Value**
Total porosity (ε_t)	0.74
Porosity solid phase	0.57
Void fraction	0.4
Dispersion coefficient (D_{ax})	Equation 6.174 (Section 6.5.6.2)
Viscosity (η_l)	5.8×10^{-3} g cm^{-1} s^{-1}
Density (ϱ_l)	1.03 g cm^{-3}

Table A.10 Parameter of the SMA isotherms and mass transfer resistance.

	NaCl	ß-Lact A	ß-Lact B
$K_{m,i}$	1	1.45×10^{-3}	3.5×10^{-3}
N_i	1	6.2	5.14
σ_i	0	40	40
i_{eff} (cm s^{-1})	5×10^{-3}	1×10^{-4}	1×10^{-4}

Table A.11 Operating parameters (Figure 6.45).

	Unit	Value
Shifting time (t_{shift})	min	25
Asynchronous shift time	min	7
Feed flow rate	ml min^{-1}	2.17
Desorbent flow rate	ml min^{-1}	4.04
Extract flow rate	ml min^{-1}	3.68
Raffinate flow rate	ml min^{-1}	2.52
Zone 1	ml min^{-1}	5.45
Zone 2	ml min^{-1}	1.81
Zone 3	ml min^{-1}	3.97
Zone 4	ml min^{-1}	1.46
m_I	—	74.4
m_{II}	—	22.6
m_{III}	—	53.1
m_{IV}	—	17.7

Table A.12 Feed concentrations (Figure 6.45).

	NaCl	ß-Lact A	ß-Lact B
Feed (mmol l^{-1})	116	0.0243	0.0224
Desorbent (mmol l^{-1})	156	—	—
Zones I and II (mmol l^{-1})	149.7		
Zones III and IV (mmol l^{-1})	130.5		

References

Epping, A. (2005) *Modellierung, Auslegung und Optimierung chromatographischer Batch-Trennung*, Shaker, Aachen.

Jupke, A. (2004) Experimentelle Modellvalidierung und modellbasierte Auslegung von Simulated Moving Bed (SMB) Chromatographieverfahren, *Fortschrittbericht VDI: Reihe* 3 Nr. 807, VDI Verlag GmbH, Düsseldorf.

Mihlbachler, K., Fricke, J., Yun, T., Seidel-Morgenstern, A., Schmidt-Traub, H., and Guiochon, G. (2001) Effect of the homogeneity of the column set on the performance of a simulated moving bed unit: I. Theory. *J. Chromatogr. A*, **908**, 49–70.

Mihlbachler, K., Jupke, A., Seidel-Morgenstern, A., Schmidt-Traub, H., and Guiochon, G. (2002) Effect of the homogeneity of the column set on the performance of a simulated moving bed unit: II. Experimental study. *J. Chromatogr. A*, **944**, 3–22.

Index

a

accumulation   325, 340
accuracy, parameter   359, 363, 368, 369, 373, 379, 397
activated carbon   48
adsorbent   8, 9, 326, 359, 379
– activity parameter   137
– classification   47
– customized   48
– designed   48
– generic   48
– low affinity   341, 346
– morphology   95
– storage   108
adsorption   9
– chemical   10
– competitive   136, 388, 402
– enthalpy   10
– equilibrium   15, 136, 325, 329, 331, 335, 337, 341, 359
– kinetic   325, 329, 337, 338, 339, 340, 346, 365
– model   134
– physical   9
– rate   329, 341
adsorption-desorption method   384, 386
advanced process control   237, 497
affinity selector sorbents   57
affinity sorbent   54
– protein A   54
– protein G   57
– protein L   57
agarose   73
alkylsilica   66, 153, 159, 167
analysis, moment   360, 361, 362, 370, 375, 376, 377, 382
analytical chromatography   110, 154
analytical solution   335, 336, 345, 360, 369, 370, 391
anion exchangers   77
annular chromatography   283
approach, model   322, 346
asynchronous shifting   233
at-column-dilution   130
automatized screening   177
axial
– compression   19
– diffusion   27
– dispersion   18, 228, 327, 334, 338, 340, 348, 359, 360, 361, 365, 377, 431, 434

b

back mixing   227
balance   321, 325, 405
– differential   331, 337, 338, 340, 341, 349, 408
– overall   336, 345, 385
band broadening   18, 20, 334, 335, 336, 345, 350, 365, 461
– artificial   354
baseline   363
– separation   274, 443
batch chromatography   442
batch method   382
bioprocess, economics   200
bio-separation   215, 220
boundary
– conditions   329, 335, 342, 343, 349, 350, 351, 405
– layer   328, 339
breakthrough   333, 374, 390
– curves   346, 359, 385
brush type adsorbent   153
bubble formation   226
buffer, systems   123

c

calibration   359, 362, 374, 390, 391, 393, 396
– detector   359, 362, 374
capacity factors   12, 28
capacity scenarios   204
capture step   298
catalyst   344
– distribution of   281
cation exchangers   77
cavitation   226
cGMP   201
chiral selector   89, 92
chiral separation   167
– enantioselectivity   169
– optimization   171
– practical hints   172
– racemization   172
chiral stationary phases   85, 168
chromatogram   15, 361, 370, 375
chromatographic batch reactor   281
– application   282
chromatographic reactor   281
– Hashimoto process   302
– modeling   344
– simulated moving bed reactor   301
chromatographic separation
– development of   113
– goal   110
– working area   183
chromatography systems   8, 109, 358
– analytical   110, 154
– batch   438, 442
– column switching   283
– construction material   219
– FDA Code   219
– flash   115
– membranes   267
– normal phase   100, 132
– optimization   181
– preparative   110
– properties   117
– requirements   218
– selection   109, 125, 282
– supercritical-fluid   8
– thin layer   8, 134, 139, 149
– USP test   219
circulation method   384, 386
cleaning in place   103, 215, 220
– ion exchange resins   105
– protein A-resins   105
clinical phase   202
closed loop recycling chromatography   276, 457
coefficient

– effective mass transfer   366
– film mass transfer   365
coherence condition   334, 393
collocation   356
column   9, 446
– chemical resistance   211
– clean-ability   216
– combined stall (DAC packing   250
– dead time   12, 363, 377
– efficiency   24, 256, 346
– equilibration   254
– flow packing   246
– high pressure   210
– lifting pistons   235
– low pressure   210, 214
– maintenance   103
– model   322, 394, 401, 405
– mover   235
– packing   243
– parameter   350
– permeability   21
– pre-packed   266
– preparation   246
– pressure drop   21
– rating   102
– regeneration   103
– sanitary design   215
– semi-preparative   379
– stainless steel   210
– stall packing   250
– storage   108, 257
– testing   254
– vacuum packing   252
– vibration packing   253
column design   207, 208
column efficiency   28
column porosity   12
competitive, adsorption   388, 402
component   9
compression
– dynamic axial   209
– radial   209
computerized process model   203
concentration
– average   326
– distribution   326, 340
– plateau   374, 385, 392
– pore   326, 329, 331, 339
– profile   302, 362
– shock   387
conceptual design   203
condition, boundary   349, 405
conditioning, silica surfaces   106
continuously stirred tank   345, 352, 375

control
– advanced   237
– robustness   238
– standard   236
convection   325, 327, 330, 337, 340, 348, 408
correlation   360, 378, 398
corrosion requirements   203
costs   358, 379
– annual depreciation   428
– annual operating   428
– fixed   428, 446
– hardware   200
– mobile phase   201
– stationary phase   200
– total separation   428, 446, 452
– variable   428, 444
cross-linked polymer   47, 153
crystallization   115
C18 silica   159
CSP (chiral stationary phase)
– amylose-based   89
– antibiotic   91
– brush-type   93
– cellulose-based   89
– monolithic   95
– preparative   85
– selectivity range   169
– synthetic polymeric   91
curves
– breakthrough   346, 359, 385
– fitting   370, 395, 397
customized adsorbent   62
cut strategy   442
cycle time   288
CycloJet   278

d

DAC packing   249
Darcy equation   21, 378, 446, 472
dead time   331, 360
– column   12, 363, 377
– plant   12, 333, 352, 363, 375, 386
– total   11
dead volume   228, 375, 404
– plant   375, 404
– simulated moving-bed   407, 410
deconvolution   371
degasser   227
designed adsorbent   54
detector   240, 350, 386, 411
– calibration   359, 362, 374, 390, 391, 393, 396
– limit   122, 240
– mass spectrometry   157
– model   352

– noise   240, 363, 365, 369, 393, 396
– polarimetry   387
– refractive index   122
– ultra violet   387
determination, parameter   357, 359, 397
differential
– balance   331, 337, 338, 340, 341, 349, 408
– equation   349, 353
– total   334
– volume element   322, 325
diffusion   207
– coefficients   496
– eddy   20, 377, 433, 436
– Fickian   23, 341
– film   23
– – pore   398
– intraparticle   327, 342, 399
– macro   327
– molecular   377, 433
– pore   23, 325, 327, 340, 346, 365
– surface   23, 325, 327, 340, 346
Dirac pulse   345, 364, 365
direct optimizing control   503
discretization   354
disperse front   333, 390, 393
dispersion   325
– axial   327, 334, 338, 340, 348, 359, 360, 361, 365
– coefficient
– – apparent   336, 366
– – axial   335, 338, 358, 367, 377, 398
– numerical   354
displacement effect   42, 117, 183, 332, 387, 402
distortion, peak   350
divinylbenzene   76
downstream processing   115
– of Mabs   174
driving force   328, 344
Dyax phage display technology   63
dynamic methods   385

e

economic criteria   428, 429
eddy diffusion   20, 27, 377
effect, extra column   350, 359
efficiency, column   346
effusion, molecular   399
eluent   451
– consumption   479
– specific consumption   427, 451
eluotropic series   136
elution   298
– gradient   43, 131, 156, 279

– isocratic 43, 157, 274, 279
– order 117, 154, 156, 171
– profile 336, 345, 363, 397, 402
endcapping 68, 167
equation, differential 343, 353
equilibrium 331, 382, 384
– adsorption 359
– dispersive model 334, 346, 366, 367
– model, ideal 330
– stage model 495
– theory 332, 390, 392
Ergun equation 21
estimation
– of column parameters 509
– parameter 360, 369, 375, 377, 394, 398
explosion requirements 202
exponential-modified-Gauss function 370
extended model 415
extra column 350, 364, 375, 404, 407
– effects 18, 350, 359, 361, 391

f

factor, loading 431, 435, 447
feedback control 236
fermentation processes 199
FF-SMB 294, 482
Fickian diffusion 23
filter 235
fine chemicals 201
finite difference method 355
first, moment 377
fitting, peak 361, 369, 370
flash, chromatography 115
flip-flop chromatography 275
flow
– dispersive 352
– ideal plug 352
– packing 246
– rate ratio 435
– rate ratio m_j 464
flowsheet, simulation 350
fluid distribution 19, 20, 213
forced elution step 184
fractionation-feedback 293
fractionation mode (cut strategy) 442
fraction collection 386
frits design 211
frontal analysis 393

g

Galerkin method 357
Gaussian distribution 25
Gaussian equation 336, 346, 370
general rate model 340, 365, 504

gradient
– chromatography 222
– generic 131, 157
– mixing 224
gradient elution 43, 131, 156, 279, 280
– linear gradient 280
– step gradient 279
gradient SMB
– chromatography 295, 415, 490
– hold-up vessel 494
– with open loop 494
guidelines for process concepts 305

h

Hashimoto process 302
Henry constant 349, 360, 365, 367, 379, 382, 391, 397, 410
HETP 24, 336, 361, 362, 367, 368, 371, 398, 433
high pressure injection 225
high-throughput screening 178
His-tagged protein 59
hold up, piping 221
HPLC chromatography 207
HPLC pumps 226
HPLC systems 218, 224
– feed injection 224
– pipes 227
– valves 227
hydrophilicity 48, 65
hydrophobic interaction chromatography 416
hydrophobicity 48, 64, 66, 70, 153, 167
hygienic requirements 203

i

ideal adsorbed solution theory 395, 402
ideal adsorbed solution (IAS) theory 35
ideal model 330, 346, 365
IgM isolation 64
immobilized metal affinity chromatography 57
improved SMB 293
inflection point 386, 387, 393
initial, condition 335, 343
injection 343, 345, 363
– amount 129, 336
– high pressure 225
– loop 225
– low pressure 225
– solvent 129
– system 350, 351
– time 333, 343, 349, 351, 447
– – dimensionless 432

inorganic oxide 51
integrated process, partial deintegration 302
intraparticle diffusion 83
ion exchange 77, 416
– chromatography 176
ion-exchange
– separation 496
ISMB 288
isocratic elution 43, 274, 280
isomerisation of glucose 504
isotherm 15, 32, 329, 358, 379, 396
– concave 394
– convex 386
– effect on elution profile 333, 390, 397
– Langmuir 33, 388, 402, 431
– – anti-Langmuir 389
– – multicomponent 35, 329, 334, 349, 388, 395, 402
– linear 329, 335, 339, 345, 364, 382, 397
– peak 359
– steric mass action (SMA) 38
– Toth model 34
isotherm determination 360, 379
– accuracy 379, 390, 394, 396
– adsorption-desorption method 382, 386
– analysis of disperse fronts 390, 393
– batch method 382
– circulation method 386
– curve fitting of the chromatogram 394
– dynamic methods 385, 396
– frontal analysis 385, 393
– minor disturbance method 391
– neural networks 399
– peak-maximum method 391
– pertubation method 391
– pulse response 385
– static method 382
– step response 385

k
kinetic, adsorption 339, 365

l
large scale manufacturing 202
linear driving force 338, 344
linear gradient 79
lipophilicity(lipophobicity 48
lipophobicity 133
loading 15, 326, 329, 331, 339, 382
– ability 127, 128
– average 326
– distribution 326, 340
– factor 431, 435, 447
– solid 31

– total 31
low pressure injection 225
LPLC system 218, 220
– chromatography 207
– piping systems 220
– valves 220
lumped rate model 338, 346, 365, 366

m
macropores 73, 100
mass balance, differential 344
mass loadability 101
mass spectrometry 156
mass transfer 22, 27, 325, 328, 336, 338, 340, 342, 347, 349, 361, 365, 398, 400, 431, 434, 436, 447
MCSGP
– advanced control 504
– process 299, 496
membranes
– disposable 267
– single use 267
– technologies 269
mercury intrusion 100
mesopores 100
method development 8
micropores 73, 100
minor disturbance method 392
mixed recycle 454
mobile phase 8, 118
– detection property 121
– selection of 127
– water content 133
model 321
– approach 322, 346
– classification of 322
– column 346, 394, 401, 405
– equilibrium dispersive 334, 346, 366, 367
– general rate 340, 346, 365
– ideal 330, 346, 365
– lumped rate 338, 346, 365, 366
– parameter 321, 347, 348, 358, 359, 365, 375, 395, 398, 410
– plant 350, 375, 394
– pore diffusion 341
– reaction 337, 346
– – dispersive 339
– stage 344, 346
– theoretical plate 344
– transport 336
– – dispersive 338, 347, 358, 359, 366, 367, 407, 410, 413, 438, 440
– validation 401
ModiCon 294, 407, 482, 484

moment
- analysis   360, 361, 362, 370, 375, 376, 377, 382, 398
- first   25, 336, 345, 364, 377
- second   26, 336, 345, 364, 373, 377
monoclonal antibodies   496
- purification   174
monoclonal antibody production   203
monolithic column   96
MR-SSR processes   454
multicolumn bioseparations   298
multicolumn countercurrent solvent gradient purification (MCSGP)   299
multicomponent SMB separations   296
multiport switching valve chromatography   284
multipurpose flexibility   202
multi-rate controller   504

n

neural networks   400
nitrogen sorption   100
noise, detector   363
nonlinear optimization   498
normal phase chromatography, practical hints   106
normal phase silica   106
normal phase system
- gradient   132, 148
- optimization   139, 145
- practical hints   128, 136
- retention   134
- selectivity   138, 145
- solvent strength   146
- water content   136
number of stages   336, 345, 346, 367, 371, 433, 435, 446, 447
numerical simulation   321, 344, 353, 369, 401, 413

o

objective function   369, 397, 429
online optimization   499
online parameter estimation   499
open-loop SMB   295
operating parameter   472
optimization   360, 369, 396, 440, 442
- batch process   444
- FFSQP algorithm   482
- genetic algorithm   482
- genetic algorithms, jumping genes   482
- MINLP   483
- model-based   437
- multiobjective   482, 484

- multiple shooting   482
- simulated moving bed   462
overload   40
- concentration   40
- mass   40
- volume   40

p

packing   9, 19, 47
- characterization factor   377
- column   243
- non-ideality   19
- procedure   19, 410
- properties   95, 335, 338, 347, 350, 376
- technology   243
parameter
- accuracy   359, 363, 368, 369, 373, 379, 397
- column   350
- design   357
- determination   357, 359, 397
- dimensionless   431
- estimation   360, 369, 375, 377, 394, 398
- model   321, 347, 348, 358, 359, 365, 375, 395, 398, 410
- operating   358
- optimization   430
- plant   358
Pareto optimal results   486
partial-feed   292
particle size   96
peak   11
- asymmetry   17, 336, 339, 370
- distortion   164, 350, 391
- fitting   361, 369, 370
- isotherm   359
- maximum   391
- resolution   25, 42
- shape   130
- shaving   276, 457
Péclet number   348, 433
performance criteria   426, 429, 435
permeability   21
pertubation method   392
phase system   125
- selection   127
physisorption   10
PID-controller   236
piston, electrically or hydraulically moved   209
piston, manually moved   209
plant
- dead time   333, 352, 363, 375
- dead volume   375, 404
- model   350, 375
- parameter   358

plateau, concentration 374
plate height 24
plate number 24
polyacrylamide packing 73
polymeric packing, polymerization 75
polymerization 75
polymers
– artificial 65
– hydrophilic 76, 77
– hydrophobic 76
– natural 65
– organic 72
– porous 72
– synthetic 91
polysaccharide-type packing 73
pores 326
– connectivity 100
– diffusion 347
– – model 341
– interparticle 98
– intraparticle 98
– macro 73, 100, 341
– meso 100, 327
– micro 73, 100, 327, 341
– structural parameters 99
– texture 97
porogens 47
porosity 358, 359, 367, 396
– column 12
– particle 15
– solid phase 13
– total 13, 15, 331, 367, 376
porous glass 51
porous oxide 51
porous silica 51, 68
PowerFeed 291, 482, 484
practical hints 164
pre-column 275
prediction models 504
pre-packed columns 266
preparative chromatography 110
preparative column 208
pre-purification
– crystallization 115
– extraction 115
pressure drop 19, 21, 213, 243, 324, 359, 378, 446, 447, 472
– maximum allowable 453, 472
PRISMA model 139
– selectivity optimization 142
procedure, packing 410
process design 201
process development 111
process scheduling 204

production rate 426, 447
productivity 427, 451, 479
– cross-section specific 427
– volume specific 427
profile, concentration 362
protein A affinity sorbent 54
protein binding 215
protein separation 416, 492, 493, 495
protein titers 199
pumps 350
– diaphragm 221
– membrane (PTFE) 226
– peristaltic 222
– piston 226
– positive-displacement 221
– slurry 234
– working range 226
purification costs 200
purity 427

r

rate, adsorption 341
reaction dispersive model 339
reaction model 337, 346
reciprocal design approach 92
recycle pumps 288
recycling chromatography 453
regeneration 103, 295, 298
regulatory exigencies 203
resolution 27, 181
retardation factor 134
retention factor 12, 306, 365
retention time 11, 12, 16, 332, 366, 375, 390, 392
– linear isotherm 333, 345
retention-time method 391
reversed phase system 155, 164
– chromatography 153
– gradient 132, 157, 161
– optimization 157, 163
– practical hints 128, 157
– retention 154
– selectivity 154, 160
– silica 66, 153
robotic system 178
robust performance 489

s

salt gradient 416
sample
– characterization 114
– purity 130
sanitization 215
– in place 108

saturation capacity 32, 329, 341, 379
scale
– laboratory 111, 125, 305
– production 111, 127, 305
– technical 111, 125, 148, 150
scale-up 202, 321, 351, 457
– batch chromatography 461
– CLCR processes 458
scheduling logistic software 202
second, moment 373
section 286
selectivity point 141
selectivity range 169
self-displacement effect 42
separation factor 12
separation scenario 116
sequential multicolumn
  chromatography 298
SFC
– feed injection 230
– mobile phases 230
– piping 231
– pumps 231
SFC-SMB 295
shape selectivity 70
shifting
– asynchronous 407, 413
– interval 288, 411
– time 288, 404, 407, 408, 410, 432
– – dimensionless 432
silanisation 67
silanol activity 70, 165
simulated moving bed 286, 301
– advanced control 501
– configuration 476
– dead volume 407, 410, 413
– direct optimizing control 503
– improved 293
– isocratic 461
– model 336, 405, 409, 413
– – validation 410, 413
– optimal axial profile 462
– optimization 440, 462, 472
– – of column geometry 477
– – of configuration 480
– periodic steady state 463
– process 231
– profile 410, 413
– purity control 502
– reactor 301, 302
– – application 302
– – design 302
– – step gradient 304
– recycling strategy 232

– robustness 483
– short-cut for Langmuir Isotherms 469
– solvent gradient 490
– solvent-gradient
– – TMB short cut 491
– start-up 414
– supercritical fluid 296
– three-section process 289, 494
– valves 233
simulation 413
size distribution 96
size exclusion 179
– chromatography 72
– matrices 181
slurry
– preparation 244
– – tank 234
– stirrer 246
– ultrasonification 246
– volume 245
sol-gel process 51
solubility 127
– limitation 128, 130
solute 117, 363, 364, 367, 368
solvent
– flammability 118
– miscibility 121
– mixture 123, 127
– optimization 118
– removal 456
– RI detection 122
– stability 118
– strength 44, 136, 137
– toxicity 121
– UV detection 121
– viscosity 121, 123
specific surface area 100, 328
stage model 344, 346, 486
stall packing 250
Stanton number 349
static method 382
steady state recycling chromatography 278, 454
steam cleaning 220
step gradient 490
stereoisomer separation 127, 133, 143
steric mass action (SMA)
– isotherms 38
– model 492
stirred tank continuously 345
styrene 76
supercritical fluid 8
– chromatography 228
surface chemistry 100, 133

surface group 133
switching
– interval 288
– time 288
synthetic zeolites 50

t

tag-along effect 42, 117, 183, 332
tailing 23, 336
targeted selector design 92
temperature
– gradients 296
– influence of 123, 156
tentacular structure 84
theory, equilibrium 390, 392
time
– injection 349
– shifting 404
tortuosity 377, 399
touching band 453
tracer 12, 363, 364, 365, 367, 368, 375, 376
transfer
– coefficient 398
– – effective 338, 358, 359, 366, 367, 395
– – film 328, 365, 398
– – lumped film 338
transport model 336
– transport dispersive model 338, 347, 358, 359, 366, 367, 401, 407, 410, 413, 497
– – dimensionless 349
triangle method 494
triangle theory 464, 472
troubleshooting 257
– column stability 265
– loss of column efficiency 262
– loss of performance 259
– loss of purity/yield 264
– pressure increrase 259

– technical failures 258
– variation of elution profile 263
true moving bed 286, 463
– model 407, 409
– profile 410
two-step gradient SMB 295

u

uncertainties 488
UV cut-off 121
UV detection 121

v

vacuum packing 252
van Deemter equation 26, 368
van der Waals forces 9
variable, costs 444
variance 26, 345
Varicol process 233, 288, 290, 407, 481, 483
velocity
– effective 331, 335
– interstitial 13, 327, 331, 335, 348, 408
– superficial 21
vibration packing 253
viscosity 123
voidage 20
void fraction 12, 13, 15, 326, 358, 359, 361, 367, 376, 396
volume element, differential 322

w

washing 298
waste fraction 443, 447
water content 134, 136

y

yield 426, 451